a LANGE medical book

Basic and Clinical
Biostatistics

Beth Dawson-Saunders, PhD
National Board of Medical Examiners
Philadelphia, PA
Formerly
Professor of Biostatistics
Southern Illinois University
School of Medicine

Robert G. Trapp, MD
Medical Director
The Arthritis Center
Springfield, IL
Formerly
Assistant Professor of Medicine
Chief of Rheumatology
Southern Illinois University
School of Medicine

APPLETON & LANGE
Norwalk, Connecticut/San Mateo, California

0-8385-6200-0

Notice: Our knowledge in clinical sciences is constantly changing. As new
information becomes available, changes in treatment and in the use of drugs
become necessary. The authors and the publisher of this volume have taken
care to make certain that the doses of drugs and schedules of treatment are
correct and compatible with the standards generally accepted at the time of
publication. The reader is advised to consult carefully the instruction
and information material included in the package insert of each drug or
therapeutic agent before administration. This advice is especially
important when using new or infrequently used drugs.

90 91 92 93 94 / 10 9 8 7 6 5 4 3 2 1

Prentice Hall International (UK) Limited, *London*
Prentice Hall of Australia Pty. Limited, *Sydney*
Prentice Hall Canada, Inc., *Toronto*
Prentice Hall Hispanoamericana, S.A., *Mexico*
Prentice Hall of India Private Limited, *New Delhi*
Prentice Hall of Japan, Inc., *Tokyo*
Simon & Schuster Asia Pte. Ltd., *Singapore*
Editora Prentice Hall do Brasil Ltda., *Rio de Janeiro*
Prentice Hall, *Englewood Cliffs, New Jersey*

ISBN 0-8385-6200-0
ISSN 1045-5523

Production Editor: Christine Langan
Designer: Steve Byrum

PRINTED IN THE UNITED STATES OF AMERICA

Table of Contents

Preface _____ **vii**

1. Introduction to Medical Research _____ **1**
1.1 The Scope of Biostatistics & Epidemiology 1
1.2 Biostatistics in Medicine 1
1.3 The Design of This Book 3
1.4 The Organization of This Book 4

2. Study Designs in Medical Research _____ **6**
2.1 Classification of Study Designs 6
2.2 Observational Studies 6
2.3 Experimental Studies or Clinical Trials 12
2.4 Advantages & Disadvantages of Different Study Designs 15
2.5 Summary 18

3. Exploring & Presenting Data _____ **20**
3.1 Purpose of the Chapter 20
3.2 Scales of Measurement 20
3.3 Tables & Graphs for Nominal & Ordinal Data 22
3.4 Tables & Graphs for Numerical Data 23
3.5 Graphs for Two Characteristics 29
3.6 Examples of Misleading Charts & Graphs 30
3.7 Computer Programs That Produce Tables & Graphs 33
3.8 Summary 37

4. Summarizing Data in Medical Research _____ **43**
4.1 Purpose of the Chapter 43
4.2 Measures of the Middle (Central Tendency) 44
4.3 Measures of Spread (Dispersion) 46
4.4 Measures to Use With Nominal Data 50
4.5 Measures to Describe Relationships Between Two Characteristics 54
4.6 Variation in Data 57
4.7 Computer Programs That Summarize Data 59
4.8 Summary 60

5. Probability, Sampling, & Probability Distributions _____ **64**
5.1 Purpose of the Chapter 65
5.2 The Meaning of the Term _Probability_ 66
5.3 Basic Definitions & Rules of Probability 66
5.4 Populations & Samples 69
5.5 Random Variables & Probability Distributions 73
5.6 Summary 80

6. Drawing Inferences From Data _____ **82**
6.1 Purpose of the Chapter 82
6.2 Population Distributions & the Sampling Distribution of the Mean 83
6.3 Estimation 90
6.4 Hypothesis Testing 92
6.5 Summary 97

7. Estimating & Comparing Means _____ **99**
7.1 Purpose of the Chapter 100
7.2 Decisions About Single Means 101
7.3 Decisions About Paired Groups 107
7.4 Decisions About Two Independent Groups 111
7.5 Determination of Sample Size 118
7.6 Computer Programs Illustrating Comparison of Means 119
7.7 Summary 121

8. Comparing Three or More Means — 124
8.1 Purpose of the Chapter 124
8.2 Intuitive Overview to ANOVA 125
8.3 Traditional Approach to ANOVA 128
8.4 Multiple-Comparison Procedures 132
8.5 Additional Illustrations of the Use of ANOVA 135
8.6 Computer Programs Illustrating ANOVA 139
8.7 Summary 140

9. Estimating & Comparing Proportions — 142
9.1 Purpose of the Chapter 143
9.2 Proportions in Single Groups 143
9.3 Comparing Two Independent Proportions 146
9.4 Comparing Proportions in More Than Two Groups 153
9.5 Comparing Proportions in Paired Groups 154
9.6 Other Applications of Chi-Square 155
9.7 Sample Sizes for Proportions 156
9.8 Computer Programs Illustrating Comparison of Proportions 157
9.9 Summary 159

10. Correlation & Regression — 161
10.1 Purpose of the Chapter 162
10.2 An Overview of Correlation & Regression 162
10.3 Correlation 163
10.4 Other Measures of Correlation 166
10.5 Linear Regression 170
10.6 Use of Correlation & Regression 177
10.7 Computer Programs Illustrating Correlation & Regression 180
10.8 Summary 182

11. Methods for Analyzing Survival Data — 186
11.1 Purpose of the Chapter 186
11.2 Why Special Methods Are Needed to Analyze Survival Data 188
11.3 Actuarial, or Life Table, Analysis 190
11.4 Kaplan-Meier Product Limit Estimates of Survival 192
11.5 The Hazard Function in Survival Analysis 193
11.6 Comparing Two Survival Curves 194
11.7 Interpreting Survival Curves From the Literature 200
11.8 Computer Programs Illustrating Survival Analysis 202
11.9 Summary 202

12. Statistical Methods for Multiple Variables — 207
12.1 Purpose of the Chapter 209
12.2 Predicting With More Than One Variable: Multiple Regression 210
12.3 Confounding Variables: Analysis of Covariance 215
12.4 Predicting a Censored Outcome: Proportional Hazards Model 218
12.5 Predicting Nominal or Categorical Outcomes 219
12.6 Combining the Results From Several Studies: Meta-Analysis 222
12.7 Other Methods for Multiple Variables 224
12.8 Summary of Advanced Methods 226

13. Evaluating Diagnostic Procedures — 229
13.1 Purpose of the Chapter 229
13.2 Evaluating Diagnostic Procedures With the Threshold Model 230
13.3 Measuring the Accuracy of Diagnostic Procedures 231
13.4 Using Sensitivity & Specificity to Revise Probabilities 232
13.5 ROC Curves 240
13.6 Illustration of Physicians' Abilities to Revise Probabilities 241
13.7 Summary 242

14. Clinical Decision Making — 245
14.1 The Decision Process 245
14.2 Making a Decision for an Individual Patient 247
14.3 Making a Decision on Health Policy 250
14.4 Using Decision Analysis to Evaluate Several Protocols 253
14.5 Extensions of Decision Theory 257
14.6 Summary 259

15. Reading the Medical Literature _____ **264**

15.1 Purpose of the Chapter 264

15.2 Review of Major Study Designs 264

15.3 The Abstract & the Introduction Section of a Research Report 265

15.4 The Method Section of a Research Report 266

15.5 The Results Section of a Research Report 272

15.6 The Discussion & Conclusion Sections of a Research Report 274

15.7 A Checklist for Reading the Literature 274

Glossary _____ **283**

References _____ **291**

Appendix A: Tables _____ **297**

Appendix B: Answers to Exercises _____ **304**

Appendix C: Flowcharts _____ **320**

Index _____ **325**

Preface

Basic & Clinical Biostatistics introduces the medical student, researcher, or practitioner to the study of statistics applied to medicine. The authors, a statistician and a physician, have incorporated their experiences in medicine and statistics to develop a comprehensive text covering the traditional topics of biostatistics and the quantitative methods in epidemiology and decision making used in clinical practice and in research. Particular emphasis is given to study design and the interpretation of results of medical research.

OBJECTIVE

This book's objective is to help the reader become an informed user and consumer of statistics, and we have endeavored to make our presentation lively and interesting. You can expect to achieve the following goals:

- Develop sound judgment about data applicable to clinical care.
- Read the medical literature critically, understanding potential errors and fallacies contained therein, and apply confidently the results of medical studies to patient care.
- Interpret commonly used vital statistics and understand the ramifications of epidemiologic information for patient care.
- Reach correct conclusions about diagnostic procedures and laboratory test results.
- Interpret manufacturers' information about drugs, instruments, and equipment.
- Evaluate study protocols and articles submitted for publication, and participate in medical research.

APPROACH & DISTINGUISHING FEATURES

We have attempted to keep the medical professional's interests, needs, and perspective in mind. Thus, our approach incorporates the following features:

- Presenting statistical calculations only when they illustrate the logic behind certain statistics and tests.
- Offering a genuine medical context for the subject matter. Most chapters begin with several Presenting Problems—discussions of studies that have been published in the medical literature—which illustrate the statistics discussed in the chapter and sometimes carry a line of reasoning throughout several chapters or to the exercises.
- Defining terms as we go, whenever practical, because statistics may be a new language to you. In addition, a glossary of statistical and epidemiologic terms is provided at the end of the book.
- Providing, on the inside back cover, a table of all symbols used in the book.
- Discussing a simple classification scheme of study designs used in medicine (Chapter 2). We employ this scheme throughout the book as we discuss the Presenting Problems.
- Using flowcharts to relate research questions to appropriate statistical methods (inside front cover and Appendix C).
- Explaining, step by step, how to read the medical literature critically (Chapter 15)—a necessity for the modern health professional.
- Addressing decision making in a clinical context (Chapters 13 and 14). Clinicians will be called upon increasingly to make decisions based on statistical information.
- Dividing the reference section at the end of the book into five categories to facilitate your search for the source of a Presenting Problem or for a text or an article on a specific topic in statistics.
- Providing numerous end-of-chapter exercises (Chapters 2-14) and complete solutions to these exercises (Appendix B).
- Including a posttest of multiple-choice questions (Chapter 15) similar to those used in course final examinations or licensure examinations.
- Presenting, where appropriate, computer-generated analyses to illustrate results of statistical procedures and tests.

Beth Dawson-Saunders, PhD
Robert G. Trapp, MD
October, 1989

Acknowledgments

As we worked on this book, we became increasingly grateful for the enthusiastic support of Alex Kugushev, president of Lange Medical Publications. His insights into both medicine and statistics give him a unique perspective from which to provide guidance and suggestions. We also owe a special debt of thanks to the editors (Carol Beal, the primary editor, who made important substantive and stylistic suggestions; Kristen Coston, who supervised the editing process; and Alan Venable, who reviewed edited chapters) and to Hal Keith of K & S Graphics, who prepared the artwork.

In order to use actual published studies to illustrate statistical concepts and calculations, we asked many authors and journal editors for permission to use their data or cite their articles and often to reproduce tables or graphs from the articles as well. We wish to express our appreciation to all these authors and journal editors for their generosity in sharing with us the results of their scholarly endeavors. We hope readers will consult many of the full articles we use as the basis for Presenting Problems and illustrations to enhance their understanding of medicine as well as the application of statistics in medicine. We especially want to thank the investigators who shared complete sets of data with us: Dr. Holly Howe and Beverly Evans from the Illinois Department of Public Health, Division of Epidemiologic Studies, for providing data from the Illinois tumor registry; Dr. Michael Irwin from the University of California at San Diego for observations on depression and immune responses in women; and Dr. Alan Birtch from Southern Illinois University School of Medicine for data on patients receiving kidney transplants. We particularly want to mention the kindness of the editors of *The New England Journal of Medicine,* who permitted us to use a large number of studies published in their journal, and the *Biometrika* trustees for permission to reproduce statistical tables.

Today, all researchers use computers to analyze data, and graphics from computer programs are often published in journal articles. We believe physicians should have some experience in reading and interpreting the results of computer analysis. Several producers of statistical software were kind enough to provide us with programs so that we could present descriptions and printouts of various programs. Our sincere thanks go to BMDP Statistical Software, Inc.; Minitab, Inc.; SPSS, Inc.; Statistix; and SYSTAT, Inc.

This book has profited from the critical insight of colleagues who made suggestions for examples or read parts of the text in draft form and made constructive comments. We especially appreciate the assistance of Dr. Gabriella D'Elia, Linda Distlehorst, Dr. Kevin Dorsey, Dr. Earl Loschen, Dr. John Murphy, Dr. Terry Travis, and Dr. Steven Verhulst, all of Southern Illinois University School of Medicine; Dr. Richard Reznick of Toronto Western Hospital; Dr. David Swanson of the National Board of Medical Examiners; and Dr. Judy Shea of the American Board of Internal Medicine. Many students also helped by using the book at various stages of its development; Dr. Mary Bourland and Dr. Susan Rausch, both currently in residency programs at Southern Illinois, and Dr. Thomas Sebo, in residency at Mayo Hospital in Rochester, were particularly generous with their comments. Our efforts were assisted tremendously by the careful and comprehensive external reviews from Dr. Raymond Greenberg of Emory University School of Medicine; Dr. Ronald Marks of the University of Florida; and Dr. Donna Katzman McClish of Richmond, Va., all of whom continually challenged us to concentrate on important concepts and to state them clearly and concisely. We appreciate their insights and encouragement and trust that they can see the impact of their suggestions.

We thank our respective institutions and supervisors for their support and encouragement during our writing of this book, especially Dr. Glen Davidson, Professor and Chair of the Department of Medical Humanities, and Dr. Sergio Rabinovich, Professor and previous Chair of the Department of Internal Medicine, both at Southern Illinois University School of Medicine; and Dr. Ronald Nungester, Vice President of Pyschometrics at the National Board of Medical Examiners. In addition, we thank the administrative and clerical staff. First and foremost, we thank Ronda Cox, who organized the production of the manuscript, preparing most of the drafts (making the word-processing software perform great feats in the process) and helping us obtain permissions from authors and journal editors. We also thank Marilyn Clarke, Lynne Cleverdon, Melinda Weinrich, Carol Faingold, and Judy Riech of Southern Illinois, and Janis McBreen and Bernadette Brennan of the National Board, for their help.

Finally, we express our appreciation of the encouragement and good natures of our spouses, Dr. Jeffrey Saunders and Kathleen Trapp, who persevered with us; and to Gregory and Curtis Dawson and Matthew, Caitlin, Leanne and Claire Trapp, who had to contend with working parents on many evenings and weekends.

Introduction to Medical Research

1

The goal of this text is to provide you with the tools and skills you need to be a smart user and consumer of medical statistics. This goal has guided us in the selection of material and in the presentation of information. This chapter outlines the reasons physicians and medical students should know biostatistics. It also describes how the book is organized, what you can expect to find in each chapter, and how you can use it most profitably.

1.1 THE SCOPE OF BIOSTATISTICS & EPIDEMIOLOGY

The word *statistics* has several meanings: data or numbers, the process of analyzing the data, and the description of a field of study. It derives from the Latin word *status*, meaning "manner of standing" or "position." Statistics were first used by tax assessors to collect information for determining assets and assessing taxes—an unfortunate beginning and one the term has not entirely lived down.

Everyone is familiar with the statistics used in baseball and other sports, such as a baseball player's batting average, a bowler's game point average, and a basketball player's free throw percentage. In medicine, some of the statistics most often encountered are called means, standard deviations, proportions, and rates. Working with statistics involves using statistical methods to summarize the data (to obtain means, standard deviations, etc.) and using statistical procedures to reach certain conclusions that can be applied to patient care or public health planning. The subject area of statistics is the set of all the statistical methods and procedures used by those who work with statistics. The application of statistics is broad indeed and includes business, marketing, economics, agriculture, education, psychology, sociology, anthropology, and biology, in addition to our special interest, medicine. In medicine, the terms **biostatistics** and **biometrics** are also used to refer to the application of statistics.

Although the focus of this text is on biostatistics, topics related to epidemiology are included as well. The term *epidemiology* refers to the study of health and illness in human populations, or,

more precisely, to the patterns of health or disease and the factors that influence these patterns; it is based on the Greek words for "upon" *(epi)* and "people" *(demos)*. Once knowledge of the epidemiology of a disease is available, it is used to understand the cause of the disease, determine public health policy, and plan treatment. The application of population-based information to decision making about individual patients is often referred to as **clinical epidemiology.** The tools and methods of biostatistics are an integral part of epidemiology.

1.2 BIOSTATISTICS IN MEDICINE

Physicians must evaluate and use new information throughout their lives. The skills you learn in this text will assist in this process, because they concern modern knowledge acquisition methods. In the following subsections, we list the most important reasons for learning medical biostatistics. (The most widely applicable reasons are mentioned first.)

1.2.1 Evaluating the Medical Literature

Reading the medical literature begins early in medical school and continues throughout physicians' medical careers, so physicians must understand biostatistics to decide whether they can believe the results presented in the medical literature. Journal editors simply cannot perform this task for physicians. All editors try to screen out articles that are improperly designed or analyzed, but few of them have formal statistical training and they naturally focus on the content of the research rather than the method. Sometimes, investigators for large, expensive studies consult statisticians in project design and data analysis. But even then physicians must be aware of possible shortcomings in the way a study was designed or carried out. In smaller research projects, which make up 90–95% of published studies, investigators rarely consult with statisticians, either because the investigator is not aware of the need for statistical assistance or because the biostatistical resources are not available or affordable.

The problems with studies in the medical literature have been amply documented, and we give only a few illustrations here. Williamson, Goldschmidt, and Colton (1986) reviewed 28 articles (published in the medical literature) that had assessed the scientific adequacy of study designs, data collection, and statistical methods in more than 4200 published medical reports. The reports assessed drug trials and surgical, psychotherapeutic, and diagnostic procedures published in more than 30 journals, many of them well-known and prestigious. For example, the *British Medical Journal, The Journal of the American Medical Association, The New England Journal of Medicine, The Canadian Medical Association Journal, The Lancet, The American Journal of Psychiatry, Annals of Internal Medicine, Archives of Neurology and Psychiatry, The Journal of Nervous and Mental Disease,* and *Psychiatric Quarterly* were all included in three or more assessment articles. Williamson and his colleagues determined that, on the average, only about 20% of 4235 research reports met the assessors' criteria for validity. Eight of the assessment articles had gone a step further and looked at the relationship between the frequency of positive findings and the adequacy of the methods used in research reports. In the research reports evaluated in these eight articles, approximately 80% of those that were inadequately designed and analyzed had reported positive findings; however, only 25% of those adequately designed and analyzed had positive findings. Thus, evidence indicates that positive findings are reported more often in poorly conducted studies than in well-conducted studies.

Other articles indicate that the problems cited by Williamson, Goldschmidt, and Colton have not improved substantially. For example, Avram et al (1985) found that only 15% of more than 200 articles in two anesthesia journals were without major errors in analysis or design. Reviewers of oncology journals found that authors did not even identify the statistical technique they had used in 3–8% of published articles (Hokanson, Luttman, and Weiss, 1986). Continuing problems prompted DerSimonian et al (1982) to take a slightly different approach. They reviewed articles in four leading medical journals: *British Medical Journal, The Lancet, The Journal of the American Medical Association,* and *The New England Journal of Medicine.* They defined ten elementary aspects that authors of medical journal articles should include in their report of clinical trials, such as eligibility criteria to enter the study, how patients were randomized, what statistical tests were used, whether any patients were lost to follow-up, and whether assessments were made in a blinded fashion. The authors did not attempt to determine whether these actions were properly done in the study but merely whether they were reported for the reader's use. The percentage of articles that reported these elementary facts ranged from 12 to 93%, depending on the particular aspect.

The above findings are both shocking and frightening, because physicians depend heavily on the medical literature to stay up to date. Medical journal editors have generally responded positively to these problems and have been willing to publish articles that discuss shortcomings in study design and analysis (eg, see Garfunkel, 1986). But medical journals have been slow to invoke formal statistical review procedures (Altman, 1986), although some evidence suggests that the situation is changing. For example, *The New England Journal of Medicine* lists the position of statistical consultant among its editorial staff, the *British Medical Journal* increasingly uses statisticians as reviewers, and the *Journal of Rheumatology* requests that a statistical work sheet be included with a submitted manuscript to aid in its review.

Williamson, Goldschmidt, and Colton (1986) made suggestions to practitioners for dealing with the issue: Recognize that the problem exists, avoid relying on abstracts as a primary source of information, seek reviews based on validated research results, and make at least a moderate effort to develop the skills necessary for evaluating the medical literature. We agree with these authors that readers must assume the responsibility for determining whether the results of a published study are valid. Our development of this book has been guided by the study designs and statistical methods found in the medical literature, and we have selected topics to provide physicians with the skills they need to determine whether a study is valid and should be believed. Chapter 15 focuses specifically on how to read the medical literature and provides checklists for flaws in studies and problems in analysis.

1.2.2 Applying Study Results to Patient Care

Applying the results of medical studies to patient care is the major reason practicing physicians read the medical literature. They want to know which diagnostic procedures are best, which methods of treatment are optimal, and how the treatment regimen should be designed and implemented. Of course, sometimes physicians read journals simply to stay aware of what is going on in medicine in general as well as in their specific area of interest. Chapters 13 and 14 discuss techniques for applying the concepts of earlier chapters to decisions about the care of individual patients.

1.2.2.a Interpreting Vital Statistics: Physicians must be able to interpret vital statistics in order to diagnose and manage patients effectively. Vital statistics are based on data collected from the on-going recording of vital events, such as births and deaths. A basic understanding of how vital statistics are determined, what they mean, and how they are used facilitates their use. Chapter 4 provides information on these statistics.

1.2.2.b Understanding Epidemiologic Problems: Practitioners must understand epidemiologic problems because this information helps them make diagnoses and develop management plans for patients. Epidemiologic data reveal the prevalence of a disease, how it varies by season of the year and by geographic location, and how it is influenced by certain risk factors. In addition, epidemiologic information helps society make informed decisions about the deployment of health resources—eg, whether a community should begin a surveillance program, whether a screening program is warranted and can be designed to be efficient and effective, and whether community resources should be used for specific health problems such as immunization programs or prenatal care. Describing and using data in making decisions is highlighted in Chapters 3, 4, 5, 13, and 14.

1.2.2.c Interpreting Information About Drugs and Equipment: Physicians continually evaluate information about drugs and medical instruments and equipment. This material may be provided by company representatives, sent through the mail, or published in journals. Because of the high cost of developing drugs and medical instruments, companies do all they can to recoup their investments. In order to sell their products, they must convince physicians that their products are better than those of their competitors. To make their points, they use graphs, charts, and the results of studies comparing their products with others on the market. Every chapter in this text is related to the skills needed to evaluate these materials, but Chapters 3, 4, and 15 are especially relevant.

1.2.2.d Using Diagnostic Procedures: Knowing the correct diagnostic procedure to use is a necessity in making decisions about patient care. In addition to knowing the prevalence of a given disease, physicians must know how sensitive a diagnostic test is in detecting the disease when it is present and how often the test correctly indicates no disease in a well person. These characteristics are called the sensitivity and specificity of a diagnostic test. Information in Chapters 5 and 13 relates particularly to skills for interpreting diagnostic tests.

1.2.2.e Being Informed: Keeping abreast of current trends and being critical about data are more general skills and ones that are difficult to mea-

sure. They are also not easy tasks for physicians, because there are many competing responsibilities that vie for their time. One of the by-products of working through this text is a heightened awareness of the many threats to the validity of information—ie, physicians should be on the alert for statements that do not seem quite right.

1.2.2.f Evaluating Study Protocols and Articles: Physicians associated with universities, medical schools, or major clinics are often called upon to evaluate material submitted for publication in medical journals and to decide whether it should be published. Physicians, of course, have the expertise to evaluate the medical content of a protocol or article, but they often feel uncomfortable about critiquing the design and statistical methods of a study. No study, however important, will provide valid information about the practice of medicine and future research unless it is properly designed and analyzed. Careful attention to the concepts covered in this text will provide physicians with the skills to evaluate the design of studies should they wish to serve as journal referees.

1.2.2.g Participating in or Directing Research Projects: Physicians participating in research will find knowledge about biostatistics and research methods indispensable. The comprehensive coverage of topics in this text should provide most physicians with the information they need to be active participants in all aspects of research.

Directing research projects also calls for some expertise in biostatistics. This text may not be detailed enough in some areas for project directors. In our discussions of more advanced topics, however, we frequently provide references to other resources for those interested in pursuing a subject in greater depth.

1.3 THE DESIGN OF THIS BOOK

We have used the terms *basic* and *clinical* in the title of this text because we emphasize both the basic concepts of biostatistics and the use of these concepts in clinical decision making. We have designed a comprehensive text covering the traditional topics in biostatistics plus the quantitative methods of epidemiology used in research. For example, we have included commonly used ways to analyze survival data in Chapter 11; illustrations of computer printouts in chapters in which they are appropriate, because researchers today use computers to calculate statistics; and applications of the results of studies to the diagnosis and care of individual patients, sometimes referred to as medical decision making.

Our approach deemphasizes calculations and

uses computer programs to illustrate the results of statistical tests. In most chapters, we include the calculations of some statistical procedures, mainly because we wish to illustrate the logic behind the tests, not because we believe you need to be able to calculate them. Some exercises involve calculations because we have found that some students wish to work through a few problems in detail in order to understand better the procedures. The major focus of the text, however, is on the interpretation and use of research methods.

A word of caution regarding the accuracy of the calculations is in order. Many examples and exercises require several steps. The accuracy of the final answer depends on the number of significant decimal places to which figures are extended at each step of the calculation; we generally extend them to two or three places. However, a calculator or computer that uses a greater number of significant decimal places at each step yields an answer different from that obtained using only two or three places. The difference usually will be small, but readers should not be concerned if their calculations vary slightly from ours.

With very few exceptions, the examples used are taken from studies published in the medical literature. Sometimes we use simple hypothetical data to illustrate a more complex procedure. In addition, we sometimes focus on only one aspect of the data analyzed in a published study in order to illustrate a concept or statistical test. To illustrate certain concepts, we occasionally reproduce tables and graphs as they appear in a published study. These reproductions may contain symbols that are not discussed until a later chapter in this book. These symbols should simply be ignored for the time being. We chose to work with published studies for two reasons: First, they convince medical students of the relevance of statistical methods in medical research; and second, they give students an opportunity to learn about some interesting studies along with the statistics.

We have also made an effort to provide insights into the coherency of statistical methods. We often refer to both previous and upcoming chapters to help tie concepts together and point out connections. This technique requires us to use definitions somewhat differently than many other statistical texts do; ie, terms are often used within the context of a discussion without a precise definition, which is given later. Several examples appear in the foregoing discussions (eg, vital statistics, means, standard deviations, proportions, rates, validity). Using terms properly within several contexts helps people to learn complex ideas, and many ideas in statistics become clearer when viewed from different perspectives. Some terms are defined as we go along, but providing definitions for every term would inhibit our ability to point out the connections between ideas. To assist the reader, beginning in Chapter 2 we boldface terms (the first few times they are used) that appear in the Glossary of statistical and epidemiologic terms provided at the end of the book.

1.4 THE ORGANIZATION OF THIS BOOK

Physicians deal with patients who have health problems requiring their attention. In describing their patients, they commonly say, "The patient presents with . . ." or "The patient's presenting problem is. . . ." We use this terminology in the text to emphasize the similarity between medical practice and the research problems discussed in the medical literature. Almost all chapters begin with Presenting Problems, which discuss studies taken from the medical literature; these research problems serve to motivate and illustrate the concepts and methods presented in the chapter. In chapters in which statistics are calculated (such as the mean in Chapter 4) or statistical procedures are illustrated (such as the t test in Chapter 7), data from the Presenting Problems are used in the calculations.

Each chapter also has an introduction that contains an **overview** of the ideas discussed in the chapter. This overview is designed to be an advanced organizer to help the reader visualize the ideas to be discussed. At the conclusion of each chapter is a **summary** that integrates the statistical concepts with the presenting problems used to illustrate them. When flowcharts or diagrams are useful, we include them to help illustrate how different procedures are related and when they are relevant. The flowcharts are grouped in Appendix C for easy reference.

Where appropriate, illustrations of **computer-generated analyses** are included at the conclusion of a chapter. In most cases, the actual computer printout is reproduced to give practitioners and students some familiarity with printouts. Generally, the same data used to illustrate calculations in the chapter are used for the computer examples. This practice reinforces the use of a particular statistic or procedure within the context of an application and extends the practitioner's experience with the types of statistics appropriate for different research questions.

Exercises are provided with all chapters; answers are given in Appendix B, most with complete solutions. We include different kinds of exercises to meet the different needs of students. Some exercises call for calculating a statistic or a statistical test. Some focus on the Presenting Problems or other published studies and ask about the design (as in Chapter 2) or about the use of charts, graphs, tables, statistical methods, etc. Occasionally, exercises extend a concept discussed in the chapter. This additional development is not critical for all readers to understand, but it provides further insights for those who are interested. Some exercises refer to topics discussed in previous chapters to provide reminders and reinforcements. Finally, in response to a plea from our own students, a collection of multiple-choice questions is given in Chapter 15; these questions provide a useful posttest for students who want to be sure they have mastered the material presented in the text.

The **symbols** used in statistics are sometimes a source of confusion. These symbols are listed on the inside back cover for ready access. When more than one symbol for the same item is encountered in the medical literature, we use the most common one and point out the others. Also, as we mentioned in the previous section, a **glossary** of biostatistical and epidemiologic terms is provided at the end of the book.

2   Study Designs in Medical Research

This chapter introduces the different kinds of studies commonly used in medical research. Many of the illustrations in this chapter are Presenting Problems in subsequent chapters, where they are discussed in greater detail. We believe that knowing how a study is designed is important for an understanding of the conclusions that can be drawn and have therefore chosen to devote considerable attention to the topic of study designs. Also, if you are acquainted with this topic, we can describe the type of design used in each Presenting Problem in this book.

This chapter first describes and illustrates study designs found in the medical literature and then presents advantages and disadvantages of each design. If you are familiar with the medical literature, you will recognize many of the terms used to describe different study designs. If you are just beginning to read the literature, you should not be dismayed by all the new terminology; the terms are used repeatedly in the remaining chapters, so there will be ample opportunity to review and become familiar with them. Note also the glossary of terms at the end of the book.

In the final chapter of the text, study designs will be discussed once more, within the context of reading journal articles. At that time, pointers will be given on how to look for possible biases that can occur in medical studies, such as in patient selection or in types of data collected, and possible errors or biases in measurement, in methods used to analyze data, and in conclusions drawn.

2.1 CLASSIFICATION OF STUDY DESIGNS

There are several different schemes for classifying study designs. We have adopted one that divides studies into whether the subjects were merely observed, sometimes called observational studies, or whether some intervention was performed, generally called experiments. This approach is simple and reflects the sequence an investigation sometimes takes. With a little practice, you should be able to read medical articles and classify studies according to this outline with no difficulty.

Table 2–1 contains the scheme for classifying published research reports. Each design is illus-

Table 2-1. Classification of Study Designs.

I. Observational studies
  A. Descriptive or case-series
  B. Case-control studies (retrospective)
    1. Causes and incidence of disease
    2. Identification of risk factors
  C. Cross-sectional studies (prevalence)
    1. Disease description
    2. Diagnosis and staging
    3. Disease processes, mechanisms
  D. Cohort studies (prospective)
    1. Causes and incidence of disease
    2. Natural history, prognosis
    3. Identification of risk factors
  E. Historical cohort studies
II. Experimental studies
  A. Controlled trials
    1. Parallel or concurrent controls
      a. Randomized
      b. Not randomized
    2. Sequential controls
      a. Self-controlled
      b. Crossover
    3. External controls (including historical)
  B. Studies with no controls

trated in this chapter, using some of the studies that are Presenting Problems in upcoming chapters. In **observational studies,** one or more groups of patients are observed and characteristics about the patients are recorded for analysis. **Experimental studies** involve an **intervention**—an investigator-controlled maneuver, such as a drug, a procedure, or a treatment—and interest lies in the effect the intervention has on study subjects. Of course, both observational and experimental studies may involve animals or objects, but the vast majority of studies in medicine (and the ones discussed most frequently in this text) involve people.

2.2 OBSERVATIONAL STUDIES

There are four main types of observational studies: **case-series, case-control, cross-sectional,** and **cohort studies.** When certain characteristics of a group (or series) of patients (or cases) are described in a published report, the result is called a case-series study. The simplest design is a set of case reports in which the author describes some interesting or intriguing observations that occurred in a small number of patients.

Case-series studies frequently lead to generation of hypotheses that are subsequently investigated in a case-control, cross-sectional, or cohort study. The latter three studies are defined by the length of time the study covers and by the direction or focus of the research question. Cohort and case-control studies involve an extended period of time defined by the point when the study begins and the point when it ends; some process occurs, and a certain amount of time is required to assess it. For this reason, both cohort and case-control studies are sometimes also called **longitudinal studies.** The major difference between them is the direction of the inquiry or the focus of the research question: Cohort studies are forward-looking, from a risk factor to an outcome, but case-control studies are backward-looking, from an outcome to risk factors. The third type of observational study, the cross-sectional study, analyzes data collected on a group of subjects at one time. Kleinbaum, Kupper, and Morgenstern (1982, Chapter 5) describe a number of hybrids or combinations of these designs if you are interested in greater detail than we give in this chapter.

2.2.1 Case-Series Studies

As we stated above, a **case-series** report is a simple descriptive account of interesting characteristics observed in a group of patients. For example, Kalman and Laskin (1986) presented information on a series of immunocompetent patients who had been referred to a general hospital with a diagnosis of herpes zoster. The authors were interested in virologic, demographic, and clinical features of zosteriform rashes in immunocompetent adults. They also wanted to determine the percentage of zosteriform rashes clinically diagnosed as herpes but actually caused by herpes simplex virus. The authors concluded that physicians should distinguish between infections caused by herpes zoster and herpes simplex virus because of the advent of antiviral drugs and the proper use of epidemiologic isolation procedures. A case-series report like this one can lead to a more extensive study of ways to diagnose and treat these infections.

Case-series reports generally involve patients seen over a relatively short period of time. By definition, a case-series study does not include **control subjects,** persons who do not have the disease or condition being described. Thus, some investigators would not include case-series in a list of types of studies because they are generally not planned studies and do not involve any research hypotheses. We mention case-series studies because of their important descriptive role as a precursor to other studies.

2.2.2 Case-Control Studies

Case-control studies begin with the absence or presence of an outcome and then look backward in time to try to detect possible causes or risk factors that may have been suggested in a case-series report. The *cases* in case-control studies are individuals selected on the basis of some disease or outcome; the *controls* are individuals without the disease or outcome. The history or previous events of both cases and controls are analyzed in an attempt to identify a characteristic or risk factor present in the cases' histories but not in the controls' histories.

The diagram in Fig 2–1 illustrates that subjects in the study are chosen at the onset of the study after they are known to be either cases with the disease (squares) or controls without the disease (diamonds). The histories of cases and controls are examined over a previous period of time in order to detect the presence (shaded areas) or ab-

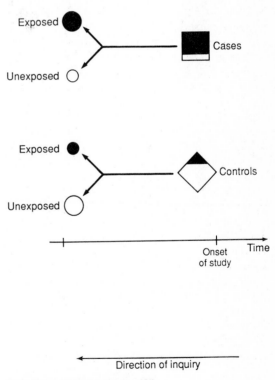

Question: "What happened?"

Figure 2-1. Schematic diagram of case-control study design. Shaded areas represent subjects exposed to the antecedent factor; unshaded areas correspond to unexposed subjects. Squares represent subjects with the outcome of interest; diamonds represent subjects without the outcome of interest. (Adapted and reproduced, with permission, from Greenberg R S: Retrospective studies. In: *Encyclopedia of Statistical Sciences.* Vol 8. Kotz S, Johnson NL [editor]. Wiley, 1988.)

sence (unshaded areas) of predisposing character-istics or risk factors, or, if the disease is infectious, whether the subject has been exposed to the presumed infectious agent. In case-control designs, the nature of the inquiry is backward in time, as indicated by the arrows pointing backward in Fig 2–1 to illustrate the backward or retrospective nature of the research process. We can characterize case-control studies as studies that ask "What happened?" In fact, they are sometimes called **retrospective studies** because of their direction of inquiry. Case-control studies are longitudinal as well, because the inquiry covers a period of time.

Presenting Problem 4 in Chapter 12 is a case-control study aimed at identifying the risk factors for childhood enuresis (Foxman, Valdez, and Brook, 1986). Both cases and controls were identified among children from approximately 2700 families enrolled in a national health insurance experiment. The cases were children who wet their beds at least once during the preceding 3 months, and the controls were children who had not. From information on a questionnaire, the investigators found that children with enuresis were more likely to be younger, were more often boys than girls, and had higher levels of psychologic distress than children who did not wet their beds.

Presenting Problem 3 in Chapter 7 is another example of a case-control study. Ross and Roberts (1985) used autopsy records of patients with carcinoid syndrome to determine which patients had carcinoid heart disease (cases) and which did not (controls). The medical records of these patients were then consulted to learn whether there had been clinical symptoms that distinguish between those who have the disease and those who do not.

In some case-control studies, investigators use **matching** to associate controls with cases on relevant characteristics. For example, although the study of carcinoid syndrome (Ross and Roberts, 1985) did not use matching, the authors stated that patient cases with carcinoid heart disease were similar in age, duration of illness, and systemic blood pressure to control patients who did not have carcinoid heart disease. If an investigator feels that one or more of these characteristics are so important that an imbalance between the two groups of patients would affect any conclusions, the investigator should match controls to cases on the characteristics. For example, each control patient (without carcinoid heart disease) can be matched to a case of the same age. The process of matching ensures that both groups will be similar with respect to important characteristics that may otherwise cloud or confound the conclusions.

Deciding whether a published study is a case-control study or a case-series report is not always easy. Confusion arises because both types of studies are generally conceived and written after the fact rather than having been planned. The easiest

way to differentiate between them is to ask yourself whether the author's purpose was to describe a phenomenon or to attempt to explain it by evaluating previous events. If the purpose is simple description, chances are the study is a case-series report.

2.2.3 Cross-Sectional Studies

The third type of observational study is cross-sectional. **Cross-sectional studies** analyze data collected on a group of subjects at one time rather than over a period of time. Cross-sectional studies are designed to learn "What is happening?" right now. Subjects are selected and information is obtained in a short period of time (Fig 2–2; note the short time line). Because they focus on a point in time, they are sometimes also called prevalence studies. Surveys and polls are generally cross-sectional studies, although surveys can be part of a cohort study. Like case-control studies, cross-sectional studies may be designed to address research questions raised by a case-series, or they may be done without a previous descriptive study.

Cross-sectional studies are most often used in medicine to describe a disease or to provide information regarding diagnosis or staging of a disease. For example, as discussed in Presenting Problem 3 of Chapter 3, Einarsson and colleagues (1985) were interested in learning more about the

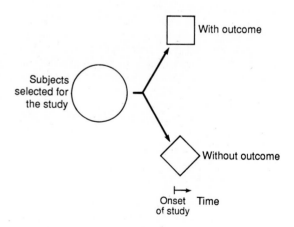

No direction of inquiry

Question: "What is happening?"

Figure 2–2. Schematic diagram of cross-sectional study design. Squares represent subjects with the outcome of interest; diamonds represent subjects without the outcome of interest.

relationship between bile supersaturation with cholesterol and age. That patients with cholesterol gallstones have higher saturation levels of cholesterol is known, but what is not known is whether saturation levels increase as part of the aging process. This question can be addressed by a long-term study that follows a group of subjects over a period of time to see whether bile supersaturation with cholesterol increases as they grow older; however, this study design would take several years to complete. To avoid the extended time period, Einarsson et al collected data on cholesterol saturation levels and ages of a group of healthy subjects at one point in time and examined the relationship between these two factors.

Presenting Problem 4 in Chapter 9 is also an example of a cross-sectional study. Its purpose was to evaluate two methods of diagnosing the disease deep venous thrombosis. Watz, Ek, and Bygdeman (1979) compared two methods of diagnosing the disease: venography, which had been used in the past but was invasive and painful, and thermography, a less invasive and less painful imaging procedure. The investigators used both procedures on the same day on a group of patients with clinical signs of acute deep venous thrombo-

sis. They compared the diagnosis based on thermography with that based on venography and concluded that the newer, less invasive thermography identified 95% of the cases found by venography. Cross-sectional studies are used in all fields of medicine, but they are especially common in examinations of the usefulness of a new diagnostic procedure.

2.2.4 Cohort Studies

A cohort is a group of people who have something in common and who remain part of a group over an extended period of time. In medicine, the subjects in **cohort studies** are selected by some defining characteristic (or characteristics) suspected of being a precursor to or risk factor for a disease or health effect. Cohort studies ask the question "What will happen?" and thus, the direction in cohort studies is forward in time. Fig 2–3 illustrates the study design. Researchers select subjects at the onset of the study and then determine whether they have the risk factor or have been exposed. The subjects, both exposed and nonexposed, are followed over a certain period of time in order to observe the effect of these defining

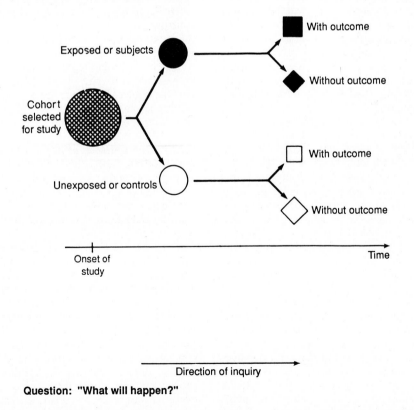

Figure 2–3. Schematic diagram of cohort study design. Shaded areas represent subjects exposed to the antecedent factor; unshaded areas correspond to unexposed subjects. Squares represent subjects with the outcome of interest; diamonds represent subjects without the outcome of interest. (Adapted and reproduced, with permission, from Greenberg RS: Prospective studies. In: *Encyclopedia of Statistical Sciences.* Vol 7. Kotz S, Johnson NL [editors]. Wiley, 1986.)

characteristics. Because the events of interest transpire after the study is begun, these studies are sometimes called **prospective studies.**

A cohort study that most readers are probably familiar with is the Framingham study of cardiovascular disease. This study was begun in 1948 to investigate factors associated with the development of atherosclerotic and hypertensive cardiovascular disease (Gordon and Kannel, 1970). More than 6000 citizens in Framingham, Massachusetts, agreed to participate in this long-term study that involved follow-up interviews and physical examinations every 2 years. Many journal articles have been written about this cohort, and some of the children of the original subjects are now being followed as well.

Cohort studies may examine the causes of a disease, as in the Framingham study, or what happens to the disease over time. For example, Presenting Problem 1 in Chapter 5 describes a longitudinal study to determine the long-term effects of seropositivity for human immunodeficiency virus (HIV) in 250 initially healthy homosexual men (Melbye et al, 1986). The men were evaluated for seropositive status in December 1981, April 1982, February 1983, and at the time of the report in September 1984. Therefore, the risk factor in this cohort study was seropositive status as opposed to seronegativity. The investigators found that the risk of having an inverted T lymphocyte helper-to-suppressor ratio (T4/T8), the outcome of interest, was much greater in men who had been seropositive for 29 months or longer than in men who had been seropositive for 19 months or less.

Cohort studies are also undertaken to identify possible risk factors and thus are a natural follow-up to a case-control study. For example, Presenting Problem 2 in Chapter 11 describes the cohort study by Han et al (1984). They followed 53 patients with chronic lymphocytic leukemia for a year or more after the patients had undergone chromosomal studies. Some patients had abnormal chromosomes (the risk factor in this study), and others did not. The investigators were interested in the relationship between this potential risk factor and survival.

2.2.4.a Historical Cohort Studies: Many cohort studies are prospective; ie, they begin at a specific time, the presence or absence of the risk factor is determined, and then information about the outcome of interest is collected at some future time, as in the two studies described above. However, one can also undertake a cohort study by using information collected in the past and kept in records or files. The direction is still forward, however, beginning with data on the risk factors and looking forward in time to determine their effect on an outcome. For example, suppose Melbye and

his coinvestigators (Presenting Problem 1 of Chapter 5, a study of the long-term consequences of seropositivity for HIV) had decided in 1984 to do this type of study. They would have consulted the records of the men first evaluated in December 1981 and used existing records to follow their course up to the time a report was written. This approach to the study is possible if the records on follow-up are complete and adequately detailed and if the investigators can ascertain the current (1984) status of the men in the 1981 sample.

Some investigators call the type of study just described a **historical cohort study** or **retrospective cohort study** because historical information is used; ie, the events being evaluated actually occurred before the onset of the study (Fig. 2–4). Note that the direction of the inquiry is still forward in time, from a possible cause or risk factor to an outcome. Studies that merely describe an investigator's experience with a group of patients and attempt to identify features associated with a good or bad outcome fall into this category, and many such studies are published in the medical literature.

Sometimes, differentiating between case-series and historical cohort studies is difficult. The easiest way to distinguish them is to determine whether there is a group of unexposed subjects or controls; if so, the study is a historical cohort. If the study describes only cases instead, it is likely to be a case-series. Although historical cohort studies may be planned in advance, often the author decides to write the article after the procedure has been used on a series of patients. The validity of historical cohort studies depends on how complete and clear the historical records are.

2.2.4.b Comparison of Case-Control and Cohort Studies: Both case-control and cohort studies can investigate risks and causes of disease, and which design an investigator selects depends in part upon the research question. To illustrate, consider once more the study by Melbye and colleagues (1986). These investigators undertook a cohort study to look at the ramifications of the length of time a man is seropositive; they learned that long-term seropositivity of 29 months or longer is associated with an increased risk of having an inverted T4/T8 ratio. The investigators could also have designed a case-control study had they asked the research question differently, eg, "Given cases with an inverted T4/T8 ratio and controls with a normal ratio, what are likely precursors or risk factors?" As this illustration shows, a cohort study starts with a risk factor or exposure and looks at consequences; a case-control study takes the outcome as the starting point of the inquiry and looks for precursors or risk factors.

Generally speaking, results from a well-designed

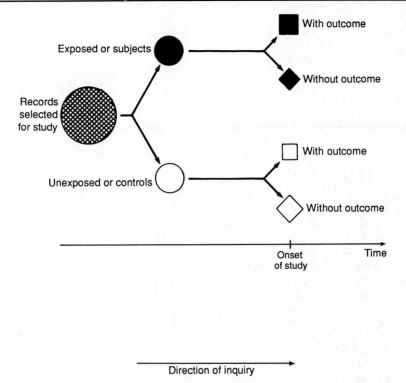

Exposed or subjects

With outcome

Without outcome

Records selected for study

Unexposed or controls

With outcome

Without outcome

Onset of study

Time

Direction of inquiry

Figure 2-4. Schematic diagram of historical cohort study design. Shaded areas represent subjects exposed to the antecedent factor; unshaded areas correspond to unexposed subjects. Squares represent subjects with the outcome of interest; diamonds represent subjects without the outcome of interest. (Adapted and reproduced, with permission, from Greenberg R S: Retrospective studies. In: *Encyclopedia of Statistical Sciences.* Vol 8. Kotz S, Johnson NL [editors]. Wiley, 1988.)

cohort study carry more weight in establishing a cause than do results from a case-control study. A large number of possible biasing factors can play a role in case-control studies, and several of them are discussed at greater length in Chapter 15.

The distinction between the two study designs is clearly illustrated in the controversies that surrounded the studies of tobacco as a risk factor for lung cancer. By the 1950s, many published case-control studies linked lung cancer with prior use of tobacco. The general design of these studies was to begin with a group of patients with lung cancer and a group of control patients without lung cancer and then to compare the smoking histories of both sets of patients. Typically, a greater percentage of those with lung cancer than those without lung cancer had a history of smoking. However, some researchers pointed out a possible bias in these studies, ie, the possibility that it is not smoking itself that is responsible for the cancer but some other factor common both in smokers and in people who develop lung cancer. Although this point can be made for any case-control study, there was considerable financial motivation on the part of the tobacco industry for

raising this issue. To avoid this study bias, investigators in England undertook a cohort study of British physicians. They found that physicians who smoked cigarettes had a subsequent mortality rate from lung cancer about ten times greater than physicians who did not smoke (Doll and Hill, 1950). Also striking was the report 26 years later that there had been a 50% decline in cigarette smoking among British physicians since 1950, accompanied by a 40% reduction in mortality from lung cancer; in the general public, however, there was no decrease in smoking or in the number of lung cancer deaths over this period (Doll and Peto, 1976).

In spite of their shortcomings with respect to establishing causality, case-control studies are frequently used in medicine and can provide useful insights if well designed. They can be completed in a much shorter time than cohort studies and are correspondingly less expensive to undertake. Case-control studies are especially useful for studying rare conditions or diseases that may not manifest themselves for many years. In addition, they are useful for testing an original premise; if the results of the case-control study are promising, the investigator can design and undertake a more involved cohort study.

2.3 EXPERIMENTAL STUDIES OR CLINICAL TRIALS

Experimental studies are generally easier to identify than observational studies in the medical literature. Authors of medical journal articles reporting experimental studies tend to state explicitly the type of study design used more often than do authors reporting observational studies. Experimental studies in medicine that involve humans are called **clinical trials** because their purpose is to draw conclusions about a particular procedure or treatment. Table 2–1 indicates that clinical trials fall into two categories: those with controls and those without controls.

Controlled trials are studies in which the experimental drug or procedure is compared with another drug or procedure, sometimes a placebo and sometimes the previously accepted treatment. **Uncontrolled trials** are studies in which the investigators' experience with the experimental drug or procedure is described, but the treatment is not compared with another treatment, at least not formally. Because the purpose of an experiment is to determine whether the intervention (treatment) makes a difference, studies with controls are much more likely than those without controls to detect whether the difference is due to the experimental treatment or to some other factor. Thus, controlled studies are viewed as having far greater validity in medicine than uncontrolled studies.

2.3.1 Trials With Independent Concurrent Controls

One way a trial can be controlled is to have two groups of subjects, one group receiving the experimental procedure—the experimental group—and the other receiving the placebo or standard procedure—the control group (Fig. 2–5). The experimental and control groups should be treated alike in all ways except for the procedure itself so that if there are differences between the groups, they will be due to the procedure and not to other factors. The best way to ensure that the groups are treated similarly is to plan interventions for both groups for the same time period in the same study. In this way, the study achieves **concurrent control.** To reduce the chances that subjects or investigators see what they expect to see, researchers designed **double-blind trials** in which neither subjects nor investigators know whether the subject is in the treatment or the control group. When only the subject is unaware, the study is called a **blind trial.** In some unusual situations, the study design may call for the investigator to be blinded even when the subject cannot be blinded. Blindedness is discussed in detail Chapter 15, Section 15.4.5.b. Another issue is how to assign some patients to the experimental condition and others to the control condition; the best method of assignment is random assignment. Methods for randomization are discussed in Chapter 5.

2.3.1.a Randomized Clinical Trials: The randomized clinical trial is the epitome of all research designs because it provides the strongest evidence

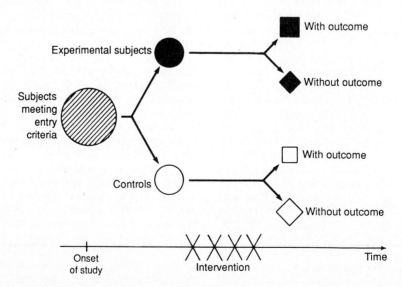

Figure 2–5. Schematic diagram of randomized clinical trial design. Shaded areas represent subjects assigned to the treatment condition; unshaded areas correspond to subjects assigned to the control condition. Squares represent subjects with the outcome of interest; diamonds represent subjects without the outcome of interest.

for concluding causation; it provides the best insurance that the result was due to the intervention.

One of the more noteworthy randomized trials is the Coronary Artery Surgery Study (CASS), discussed in Chapter 5, Presenting Problem 4 (CASS, 1983). This trial was undertaken to evaluate medical and surgical therapy in patients with stable ischemic heart disease. Over a 4-year period, 700 patients with 70% or greater stenosis of one or more operable vessels were randomly assigned to medical or surgical treatment. Patients who had 70% or greater luminal diameter reduction of the left main coronary artery, with an ejection fraction less than 0.35, or who were likely to require additional surgical procedures were excluded from the study. So that an adequate number of patients would be included in the study, several medical centers were involved, resulting in a multicenter trial organized by the National Heart, Lung, and Blood Institute. Outcome measures in this study included quality-of-life indicators and survival. The study found no differences in survival among the patients assigned to the two treatments, but the surgical group enjoyed improved quality of life, as manifested by relief of chest pain, improvement in functional status, and reduced need for drug therapy.

2.3.1.b Nonrandomized Trials: Subjects are not always randomized to treatment options. Studies that do not use randomized assignment are generally referred to as **nonrandomized trials** or simply as clinical trials or comparative studies, with no mention of randomization. Many investigators believe that studies with non-randomized controls are open to so many sources of bias that their conclusions are highly questionable. Studies using nonrandomized controls are considered to be much weaker because they do nothing to prevent bias in patient assignment. For instance, perhaps it is the stronger patients who receive the surgical treatment and the higher-risk patients who are treated conservatively. An example is a nonrandomized study of percutaneous nephrostolithotomy (removing renal calculi under local anesthesia, using dilatation and extraction instruments) versus open surgery for removing renal calculi (Preminger et al, 1985). Both procedures were performed during the same time period, but the study did not describe the criteria for selecting one procedure over another. The investigators concluded that the percutaneous procedure was better with calculus diameters 2.5 cm or smaller, but possibly the nephrostolithotomy and surgical groups differed on important risk factors.

2.3.2 Trials with Self-Controls

An acceptable method of providing control in an experimental study is by using the same group of subjects for both experimental and control options. For example, Arntzenius et al (1985) designed a study to investigate the relationship between diet and progression on coronary atherosclerosis, as described in Chapter 4, Presenting Problem 1. Patients were evaluated at the beginning of the study, and they were then placed on a 2-year vegetarian diet. At the end of this time, patients were reevaluated to determine any changes in coronary artery diameter and cholesterol levels. This study uses patients as their own controls and is called a **self-controlled study.**

2.3.3 Crossover Studies

The self-controlled study design can be modified slightly to provide a combination of concurrent and self-controls. This design uses two groups of patients; one group is assigned to the experimental treatment, and the second group is assigned to the placebo or control treatment (Fig 2–6). After a period of time, the experimental treatment and placebo are withdrawn from both groups for a washout period. During the washout period, the patients generally receive no treatment. The groups are then given the alternative treatment; ie, the first group now receives the placebo, and the second group receives the experimental treatment. This design is called a **cross-over study,** and it provides a powerful design when it is appropriate.

2.3.4 Trials with External Controls

The third method for controlling experiments is to use controls external to the study. Sometimes, the result of another investigator's research is used as a comparison. On other occasions, the controls are patients the investigator has previously treated in another manner, called **historical controls.** The study design is illustrated in Fig 2–7.

An example of external control is a study evaluating the treatment of cirrhotic patients with bleeding esophageal varices (Orloff et al, 1975). The investigators performed a surgical procedure called a portacaval shunt to treat portal hypertension and prevent further bleeding in a group of patients who came to the emergency room with alcoholic cirrhosis and esophageal varices. They compared the survival of these patients with the survival of patients who had been treated with other approaches before their study on the shunt surgery began, and they concluded that the portacaval shunt procedure increases long term survival.

Historical controls are used frequently in oncology research. In studies involving historical controls, researchers should evaluate whether other factors may have changed since the time the historical controls were treated; if so, any differences may be due to these other factors and not the treatment.

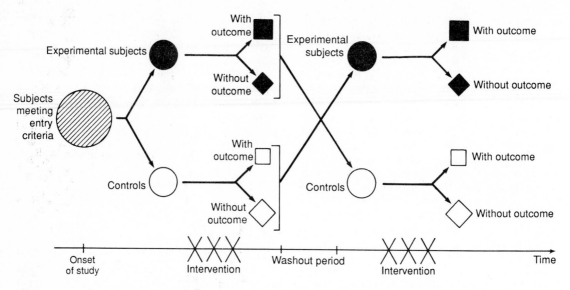

Figure 2-6. Schematic diagram of trial with crossover. Shaded areas represent subjects assigned to the treatment condition; unshaded areas correspond to subjects assigned to the control condition. Squares represent subjects with the outcome of interest; diamonds represent subjects without the outcome of interest.

2.3.5 Uncontrolled Studies

Not all studies involving interventions have controls, and by strict definition they are not really experiments or trials. For example, Anasetti et al (1986), as discussed in Presenting Problem 3, Chapter 5, reported the results of a trial of bone marrow transplants in patients with severe aplastic anemia. The investigators did not give the patients a transfusion of blood until just before the marrow transplantation because of their belief that transfusion-induced sensitization can increase marrow graft rejection. They presented information on the survival of the patients treated in this manner and concluded that their approach results in an excellent probability of long-term survival in patients with aplastic anemia. This study is **uncontrolled study** because there are no comparisons with patients treated in another manner.

Uncontrolled studies are more likely to be used when the comparison involves a procedure than when it involves a drug. The major shortcoming of such studies is that investigators assume that the procedure used and described is the best one. The history of medicine is filled with examples in which one particular treatment is recommended and then discontinued after a controlled clinical trial is undertaken. The most significant problem with uncontrolled trials is that unproved procedures and therapies can become established, making it very difficult for researchers to undertake subsequent controlled studies.

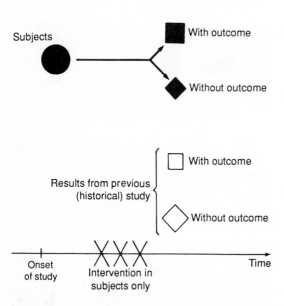

Figure 2-7. Schematic diagram of trial with external controls. Shaded areas represent subjects assigned to the treatment condition; unshaded areas correspond to patients cared for under the control condition. Squares represent subjects with the outcome of interest; diamonds represent subjects without the outcome of interest.

2.4 ADVANTAGES & DISADVANTAGES OF DIFFERENT STUDY DESIGNS

The previous sections introduced the major types of study designs used in medical research, broadly divided into experimental studies, or clinical trials, and observational studies (cohort, case-control, cross-sectional, and case-series designs). Each study design has certain advantages over the others, as well as some specific disadvantages. These advantages and disadvantages are discussed below, beginning with the most powerful study designs, clinical trials, and progressing to the weakest, case-series.

2.4.1 Advantages & Disadvantages of Clinical Trials

The clinical trial is the gold standard, or reference, in medicine—ie, the basic design against which other designs are judged—because it provides the greatest justification for concluding causality and is subject to the least number of problems or biases. (A **bias** in a study is an error that leads to an incorrect conclusion.) Clinical trials are the best type of study to use when the objective is to establish the efficacy of a treatment or a procedure. Clinical trials in which patients are randomly assigned to different treatments, or arms, are the strongest designs of all. One of the treatments is the experimental condition; another is the control condition. The control may be a placebo or a sham procedure; more often, it is the treatment or procedure commonly used, called the standard of care or reference standard. For example, patients with coronary artery disease in the Coronary Artery Surgery Study (Presenting Problem 4 in Chapter 5) were randomized to receive either surgical or medical care; no patient was left untreated or given a placebo.

A good argument for using concurrent randomized controls was made by Sacks, Chalmers, and Smith (1982) in an investigation of the differences in findings in clinical trials utilizing controls. They investigated 6 therapies (cirrhosis with varices, coronary artery surgery, anticoagulants for acute myocardial infarction, 5-FU adjuvant for colon cancer, BCG adjuvant for melanoma, and DES for habitual abortion) involving 50 randomized clinical trials and 56 trials with historical controls. They found that therapy was better than the control regimen in 79% of the historical trials but in only 20% of the trials with concurrent randomized controls. The treated patients had similar outcomes regardless of whether randomized or historical controls were used. The difference between the studies with randomized and historical controls for the same therapy was primarily the result of patients in the historical control groups generally being reported to have worse outcomes than patients in the randomized control groups. Sacks and his colleagues state that "biases in patient selection may irretrievably weight the outcome of historical controlled trials in favor of new therapies." (Biases in patient selection are discussed in detail in Chapter 15.)

The inappropriate use of historical controls has led to serious errors in medicine. For example, Chalmers (1969) describes two erroneous treatments based on studies using historical controls. In one case, patients with peptic ulcers were treated by freezing the gastric area through intubating the patient with a balloon catheter and circulating a coolant in the balloon. In the second case, patients with hepatic failure were treated by using low-protein diets. These treatment methods were abandoned only after studies using concurrent controls showed how ineffective they were.

In some situations, however, historical controls can and should be used. Bailar et al (1984) indicate that historical controls may be useful when preliminary studies are needed or when researchers are dealing with late treatment for an intractable disease, such as advanced cancer. Investigators gain some protection against bias when they clearly state a prior hypothesis and when the same investigators treat the historical controls, thereby providing some assurance of similarity between historical controls and current patients. Nevertheless, the biases demonstrated by Sacks, Chalmers, and Smith (1982) should make you wary of trials with historical controls that report in favor of the experimental treatment.

Although clinical trials provide the greatest justification for concluding causation, obstacles to using them include their great expense and long time period. For instance, a randomized trial comparing total mastectomy, segmental mastectomy, and segmental mastectomy plus radiation treatment for breast cancer (Fisher et al, 1985; Presenting Problem 3 in Chapter 11) was conducted between April 1976 and January 1984, but adequate data for analysis were not available until mid 1984.

Another potential obstacle to using clinical trials is that certain practices become established and accepted by the medical community, even though they have not been properly justified. For example, Swan-Ganz catheters are used to measure pulmonary artery pressure, but their benefit has never been established by clinical trials (Robin, 1985). The tremendous increases in medical technology have not been accompanied by changes in the way new approaches are evaluated, according to Robin. As a result, procedures become established that may be harmful to many patients. Researchers subsequently have difficulty obtaining approval to perform properly designed clinical trials from the human subjects committees that oversee the ethics of research.

2.4.2 Advantages & Disadvantages of Cohort Studies

Cohort studies are the design of choice for studying the causes of a condition, the course of a disease, or the risk factors, because they are longitudinal and follow a group of subjects over a period of time. Causation generally cannot be proved with cohort studies because they are observational and do not involve interventions. However, because they follow a cohort of patients forward through time, they possess the correct time sequence to provide strong evidence for possible causes and effects, as in the smoking and lung cancer controversy. In addition, in prospectively designed cohort studies, as opposed to historical cohort studies, investigators can control many sources of bias related to patient selection and recorded measurements.

Perhaps the most famous large cohort study is the Framingham study of factors associated with development of atherosclerotic and hypertensive cardiovascular disease (discussed in Section 2.2.4). This study has been under way since 1948. Of course, the length of time required in a cohort study depends upon the problem studied. With diseases that develop over a long period of time or with conditions that occur as a result of long-term exposure to some causative agent, many years are needed. Extended time periods make studies costly. They also make it difficult for investigators to argue causation, because other events occurring in the intervening period may have affected the outcome. For example, the long period of time between exposure and effect is one of the reasons it is difficult to study the possible relationship between environmental agents and various carcinomas. Cohort studies that require a long period of time to complete are especially vulnerable to problems associated with patient follow-up, particularly patient attrition (patients discontinue participation in the study) and patient migration (patients move to other communities).

Time for completion is one difficulty with cohort studies. Another difficulty arises if the disease or outcome is rare. For example, carcinoid tumors are very slow growing neoplasms and are relatively uncommon. Investigators who wish to examine the relationship between possible risk factors or clinical features associated with carcinoid heart disease (Presenting Problem 3 in Chapter 7; Ross and Roberts, 1985) in a cohort study would need to examine a very large number of patients and follow them over a very long period of time. The sample size required in a study like this

often makes it prohibitively expensive. In these situations, investigators often turn to case-control studies.

2.4.3 Advantages & Disadvantages of Case-Control Studies

Case-control studies are especially appropriate for studying rare diseases, for examining conditions that develop over a long time, and for investigating a preliminary hypothesis. They are generally the quickest and least expensive studies to undertake. For example, the investigators in the study of clinical factors associated with carcinoid heart disease referred to above reviewed the autopsy findings and medical records of patients who had carcinoid heart disease and those who did not. With this design, they were able to complete the study in a relatively short period of time and with relatively small numbers of subjects. Case-control studies are thus ideal for investigators who need to obtain some preliminary data prior to writing a proposal for a more complete, expensive, and time-consuming study; they are also ideal for house staff required to undertake a clinical research project.

The advantages of case-control studies lead to their disadvantages. Of all study methods, they have the largest number of possible biases or errors, and they depend completely on high-quality existing records. Data availability for case-control studies sometimes requires compromises between what researchers wish to study and what they are able to study. For example, Berry et al (1984) wanted to determine possible risk factors for developing hepatitis B virus infection among health workers (Presenting Problem 3, Chapter 10). They opted to use the number of years since a person graduated from medical training as a measure of risk. A more indicative measure might have been the number of times a health worker was exposed to blood or blood products. Berry and colleagues could not obtain this information from existing records, however, so they had to use a surrogate measure in its place.

One of the greatest problems in a case-control study is selection of an appropriate control group. The cases in a case-control study are relatively easy to identify, but deciding upon a group of persons who provide a relevant comparison is more difficult. To illustrate the importance—and difficulty—of choosing a proper control group, consider a case-control study designed to examine the relationship between use of tobacco, alcohol, tea, and coffee and incidence of pancreatic cancer (MacMahon, Yen, and Trichopoulos, 1981). The cases were hospitalized patients who had a histologic diagnosis of pancreatic cancer. For controls, the investigators selected patients who were under the care of the same physician in the same hospital at the same time. However, many patients were excluded from consideration as controls: patients with diseases of the pancreas, hepatobiliary tract or cardiovascular system and patients with, diabetes mellitus, respiratory cancer, bladder cancer, or peptic ulcer were all excluded. The results of the study indicated that the cases, patients with pancreatic cancer, had consumed greater amounts of coffee beverages than had the controls.

A major shortcoming of this study is the group chosen as controls. The control group included patients with cancer of the stomach and small intestine and patients with nonmalignant gastroenterologic conditions, such as gastritis, enteritis, and colitis; it is quite likely that many of these patients may have discontinued the use of coffee because of their problems. Including these patients in the control group makes it less likely that patients in the control group consumed large amounts of coffee, thereby increasing the chances of an observed relationship between coffee and pancreatic cancer. A better control group would have been patients hospitalized for completely unrelated problems, such as lung cancer. Because of the problems inherent in choosing a control group in a case-control study, some statisticians have recommended the use of two control groups: one control group similar in some ways to the cases (such as having been hospitalized during the same period of time) and another control group of healthy subjects.

2.4.4 Advantages & Disadvantages of Cross-Sectional Studies

Cross-sectional studies are best for determining the status quo of a disease or condition, such as the prevalence of HIV in given populations, and for evaluating diagnostic procedures. Cross-sectional studies are similar to case-control studies in being relatively quick to complete, and they may be relatively inexpensive as well. Their primary disadvantage is that they provide only a "snapshot in time" of the disease or process, which may result in misleading information if the research question is really one of disease process. For example, clinicians used to believe that diastolic blood pressure, unlike systolic pressure, does not increase as patients grow older. This belief was based on cross-sectional studies that had shown mean diastolic blood pressure to be approximately 80 mm Hg in all age groups. However, in the Framingham cohort study mentioned above, the patients who were followed over a period of several years were observed to have increased dia-

stolic blood pressure as they grew older (Gordon et al, 1959).

This apparent contradiction is easier to understand if we consider what happens in an aging cohort. For example, suppose that the mean diastolic pressure in men aged 40 years is 80 mm Hg, although there is individual variation, with some men having a blood pressure as low as 60 and others having a pressure as high as 100. Ten years later there is an increase in diastolic pressure, although it is not an even increase; some men experience a greater increase than others. However, those men who were at the upper end of the **distribution** 10 years earlier and who had experienced a larger increase have died in the intervening period, so they are no longer represented in a cross-sectional study. As a result, the mean diastolic pressure of the men still in the cohort at age 50 is about 80 mm Hg, even though individually their pressures are higher than they were 10 years earlier. Thus, a cohort study, not a cross-sectional study, provides the information leading to a correct understanding of the relationship between normal aging and physiologic processes such as diastolic blood pressure.

Surveys are generally cross-sectional studies. For example, most of the voter preference surveys done prior to an election are one-time samplings of a group of citizens, and different results from week to week are based on different groups of people; ie, the same group of citizens is not followed to determine voting preferences through time. Similarly, consumer-oriented studies on customer satisfaction with automobiles, appliances, health care, etc, are cross-sectional.

A common problem with survey research is obtaining sufficiently large response rates; many people asked to participate in a survey decline because they are busy, not interested, etc. The conclusions are therefore based on a subset of people who agree to participate, and these people may not be **representative** of or similar to the entire population. Note that the problem of representative participants is not confined to cross-sectional studies, however; it can be an issue in other studies whenever subjects are selected or asked to participate and decline or drop out.

2.4.5 Advantages & Disadvantages of Case-Series Studies

Case-series reports have two advantages: They are easy to write, and the observations may be extremely useful to investigators designing a study to evaluate causes or explanations of the observations. But as we noted previously, case-series studies are susceptible to many possible biases related to subject selection and characteristics observed.

In general, you should view them as hypothesis-generating and not as conclusive.

2.5 SUMMARY

This chapter illustrates the study designs most frequently encountered in the medical literature. In medical research, either subjects are observed or experiments are undertaken. Experiments involving humans are called trials. Experimental studies may also use animals and tissue, although we did not discuss them as a separate category; the comments pertaining to clinical trials are relevant to animal and tissue studies as well.

Each type of study discussed has it advantages and disadvantages. Randomized, controlled clinical trials are the most powerful designs possible in medical research, but they are often expensive and time-consuming. Well-designed observational studies can provide useful insights on disease causation, even though they do not constitute proof of causes. Cohort studies are best for studying the natural progression of disease or risk factors for disease; case-control studies are much quicker and less expensive. Cross-sectional studies provide a snapshot of a disease or condition at one time, and we must be cautious in inferring disease progression from them. Case-series studies should be used only to raise questions for further research.

As much as possible, we have used Presenting Problems from later chapters to illustrate different study designs. We will point out salient features in the design of the Presenting Problems as we go along, and we will return to the topic of study design again after all the prerequisites for evaluating the quality of journal articles have been presented.

EXERCISES

Read the descriptions of the following studies and determine the study design used.

1. Kremer et al (1987) designed a study to determine the efficacy of fish oil dietary supplements in patients with rheumatoid arthritis. They were particularly interested in the effect of the fish oil on the inhibition of neutrophil leukotriene levels. The study involved a group of 40 patients with class I, II, or III rheumatoid arthritis; each patient was given either a dietary supplement or a placebo for 14 weeks, but the treatment assignment was not randomized. From weeks 14 to 18, all patients took a placebo for this four-week period; then they were given the opposite treatment (dietary supplement or placebo) from weeks 1 to 14 for the next 14 weeks.

2. A study by O'Malley and Fletcher (1987) looked at the efficacy of the breast self-examination (BSE) as a screening test for breast cancer by reviewing studies published on this topic. The authors found the sensitivity of BSE (the percentage of women with breast cancer who have a positive BSE) to be much lower than the sensitivity of a clinical breast examination or mammography. Although training increases the use of BSE and its sensitivity, the number of false-positives (women without breast cancer who have a positive BSE) also increases. The authors suggest the need for a controlled trial on BSE before advocating its use as a screening device.

3. Kilbourne et al (1983) investigated an epidemic in Spain involving multiple organ systems. Patients presented with cough, dyspnea, pleuritic chest pain, headache, fever, and bilateral pulmonary infiltrates. Although an infectious agent was first suspected, a strong association with food oil sold as olive oil but containing a high proportion of rapeseed oil was detected. Epidemiologic studies found that virtually all patients had ingested such oil but that unaffected persons had rarely done so.

4. Knutson et al (1981) treated wound, burn, and ulcer patients using granulated sugar combined with povidone-iodine. The study was undertaken from January 1976 to August 1980; during that time, 759 patients were treated. Of these, 154 were treated with the standard therapy and the remaining 605 were treated with sugar. Uniformity in treatment and judgment regarding the healing process were enhanced by using three physician-investigators to oversee the process and by documenting wound healing with 35-mm transparencies. The investigators reported that a much lower percentage of patients treated with the sugar and povidone-iodine mixture required skin grafts than those given the standard treatment; the therapy was painless, and changing the burn dressings was facilitated.

5. Colditz et al (1987) reported on the relationship between menopause and risk of coronary heart disease in women. Subjects in the study were selected from the Nurses' Health Study originally completed in 1976; the study included 120,000 married female registered nurses, aged 30–55. Colditz and his colleagues identified 116,000 of these women who were premenopausal or had a known type of menopause and did not have a diagnosis of coronary heart disease at the beginning of the study. The investigators were interested in determining whether the occurrence of menopause alters

the risk of coronary heart disease—specifically, whether the influence of menopausal status is altered by the use of postmenopausal estrogen. The original survey provided information on the subjects' age, parental history of myocardial infarction, smoking status, height, weight, use of oral contraceptives or postmenopausal hormones, and history of myocardial infarction or angina pectoris, diabetes, hypertension, or high serum calcium levels. Follow-up surveys were done in 1978, 1980, and 1982, and the data were 95.4% complete.

6. Bartle, Gupta, and Lazor (1986) designed a study to examine the association between nonsteroidal anti-inflammatory drug use and acute nonvariceal upper gastrointestinal tract bleed-

ing. The association between consumption of acetylsalicylic acid and upper gastrointestinal bleeding is well established; however, no information was available on non-acetylsalicylic acid, nonsteroidal anti-inflammatory drugs. The medical records were reviewed to obtain medication histories of 57 consecutive patients with nonvariceal acute upper gastrointestinal tract hemorrhage presenting at a medical center, and 123 sex-matched and age-matched controls were in the study. (The process of sex and age matching ensures a control group that is similar to the cases with respect to gender and age.) The investigators found that a larger proportion of patients than of controls had taken nonsteroidal anti-inflammatory drugs.

3

Exploring & Presenting Data

PRESENTING PROBLEMS

Presenting Problem 1. A case-control study was undertaken to assess the risk of primary cardiac arrest during vigorous exercise (Siscovick et al, 1984). The wives of 133 men who had experienced a primary cardiac arrest were interviewed. Each woman was asked whether her husband's cardiac arrest occurred during a period of activity (the cases) or not (the controls) and whether or not he had been accustomed to habitual exercise. The investigators need to summarize the findings and evaluate the risk of cardiac arrest during periods of vigorous exercise for men unaccustomed to vigorous activity versus men who had exercised habitually.

Presenting Problem 2. Approximately 145,000 new cases of colorectal cancer develop in the USA each year. For advanced or metastatic disease, the only options are palliative surgery or chemotherapy, because radiation therapy is not generally used on the abdomen. Five-year survival rates range from 90% in patients in whom the disease is detected at an early stage to only about 5% in patients with advanced disease. The magnitude of this problem has led the American Cancer Society to recommend that screening of asymptomatic persons begin at age 50 years. To evaluate progress in the fight against cancer and help determine the need for screening programs, several states and regional cancer centers have developed computerized data banks of information about patients who are diagnosed as having cancer. Data from one year on all the patients diagnosed as having colorectal cancer who live in a midwestern state are given in Table 3–5. How can the information on the age of the patients at the time of diagnosis be displayed in tables and graphs?

Presenting Problem 3. Previous studies have established that supersaturation of bile with cholesterol is a necessary condition for development of cholesterol gallstones. Two major risk factors for the formation of gallstones are female gender and increasing age. Not known, however, is whether these two conditions are also associated with increasing cholesterol saturation of bile. Researchers in Sweden interested in this question studied 31 men and 29 women who were healthy, were not obese, and had no gallbladder disease (Einarsson et al, 1985). They collected concurrent data on each subject's age, gender, and percentage of cholesterol saturation of bile in a cross-sectional study, and they want to display the age and cholesterol saturation information graphically.

3.1 PURPOSE OF THE CHAPTER

The purpose of this chapter is to introduce the different kinds of data collected in medical research and to demonstrate how to organize and display the information collected in a study. Whatever the particular research being done, investigators collect observations and generally want to transform them into tables, graphs, or summary numbers, such as percentages or means. It does not matter whether the observations are made on people, animals, inanimate objects, or events. What does matter is the kind of observation made and how the characteristic observed is measured, because these features determine the types of tables, graphs, and summary tables that best display and communicate the observations to someone else. This chapter discusses how measurements are made and how they are displayed in tables and graphs; it concludes with illustrations from some of the commonly used statistical programs available for computers. The next chapter covers summary numbers used with different kinds of measurements.

3.2 SCALES OF MEASUREMENT

As we indicated above, the scale on which a characteristic is measured has implications for the way information is displayed and summarized. As we will see in later chapters, the **scale of measurement**—the degree of precision with which a characteristic is measured—also determines which statistical methods may be used to analyze the data. Therefore, how characteristics are measured is an important consideration. The three scales of measurement that occur most often in medicine are nominal, ordinal, and numerical.

3.2.1 Nominal Scales

Nominal scales are used for the simplest level of measurement when data values fit into categories. For example, Presenting Problem 1 uses two nominal characteristics to describe men who had a primary cardiac arrest: whether they are engaged in physical activity at the time of their cardiac arrest and whether they were accustomed to vigorous activity. In these examples, the data can take on only one of two values: yes or no—ie, they are **dichotomous variables.**

Many classifications in medical research are evaluated on a nominal scale. Outcomes of a medical treatment or surgical procedure are often given as either occurring or not occurring, as is the presence of possible risk or exposure factors, such as participating in vigorous exercise at the time of cardiac arrest in the above study. Data that can take on more than two values occur frequently as well. Anemias, for example, may be classified as (1) microcytic anemias, including iron deficiency; (2) macrocytic or megaloblastic anemia, including vitamin B_{12} deficiency; and (3) normocytic anemias, often associated with chronic disease. As another example, a study examining the prognosis for patients with lung cancer might sort the type of cancer into several categories, such as small cell, large cell, oat cell, and squamous cell. The easiest way to determine whether observations are measured on a nominal scale is to ask whether the observations are classified or placed into categories.

Data evaluated on a nominal scale are also called **qualitative observations,** because they describe a quality of the person or thing studied, or **categorical observations,** because the values fit into categories. Nominal or qualitative data are generally described in terms of **percentages** or **proportions.** (Percentages and proportions are defined and illustrated in the next chapter.) For example, in the cardiac arrest study, 7% (9 out of 133) of the men were engaged in activity at the time of their cardiac arrest. **Contingency tables** and **bar charts** are most often used to display this type of information. Ways to display nominal scale data are presented later in this chapter.

3.2.2 Ordinal Scales

If there is an inherent order among the categories, the observations are said to be measured on an **ordinal scale.** Observations are still classified, as with nominal scales, but some observations have "more" or are "greater than" other observations. Tumors, for example, are staged according to their degree of development. The international classification for staging of carcinoma of the cervix is an ordinal scale from 0 to IV, where stage 0 represents carcinoma in situ and stage IV represents carcinoma extending beyond the pelvis or involving the mucosa of the bladder and rectum. The inherent order in this ordinal scale is, of course, that stage IV is worse than stage 0 with respect to prognosis. Similarly, the colorectal tumors in Presenting Problem 2 are classified as stage 1, 2, 3, or 4 (corresponding to Dukes stages A, B, C, and D), where 1 is an in situ tumor and 4 is distant or systemic disease.

Classifications based on extent of disease arise in diseases other than carcinoma. For example, patients with arthritis are classified according to the severity of disease into four classes ranging from class 1, normal activity, to class 4, wheelchair-bound. In Presenting Problem 1, the amount of habitual vigorous activity the men had been accustomed to prior to cardiac arrest was classified into four amounts to produce an ordinal scale: 0 minutes per week, 1–19 minutes per week, 20–139 minutes per week, and 140 minutes or more per week.

An important characteristic of ordinal scales is that although order exists among categories, the difference between two adjacent categories is not the same throughout the scale. To illustrate, consider Apgar scores, which describe the maturity of newborn infants on a scale of 0 to 10, with lower scores indicating depression of cardiorespiratory and neurologic functioning and higher scores indicating good cardiorespiratory and neurologic functioning. The difference between a score of 8 and a score of 10 is probably not of the same magnitude, however, as the difference between a score of 0 and a score of 2.

Brief mention should be made of a special type of ordered scale called a **rank-order scale,** in which observations are ranked from the highest to the lowest (or vice versa). The rank of students in a high school senior class, for instance, is an ordinal measure of interest to college admissions officers. Also, the duration of surgical procedures might be converted to a rank scale to obtain one measure of difficulty of the procedure—eg, the most difficult, the second most difficult, and so forth. In this example, the difference in times for the first and second procedures is not necessarily the same as the difference in times for any two other procedures.

As with nominal scales, percentages and proportions are often used with ordinal scales. For example, in the colorectal data we will see that 27% of the patients had a tumor at stage 1 or 2. The entire set of data measured on an ordinal scale is sometimes summarized by the median value. (How and when to use medians is covered in Chapter 4.) The same types of tables and graphs used to display nominal data may also be used with ordinal data.

3.2.3 Numerical Scales

Observations in which the differences between numbers have meaning on a numerical scale are sometimes called **quantitative observations**, because they measure the quantity of something. There are two types of numerical scales—interval,* or continuous, scales and discrete scales. A **continuous scale** has values on a continuum (eg, age); a **discrete scale** has values equal to integers (eg, number of fractures).

If only a certain level of precision is required, continuous data may be reported to the closest integer. The important point, however, is that more precise measurement is possible, at least theoretically. For example, Presenting Problem 2 examines the age at which patients were diagnosed as having colorectal surgery. Age occurs on a continuum and can be any value between zero and the age of the oldest patient, ie, age can be specified as precisely as necessary. In studies of adults, such as the data from the tumor registry, age to the nearest year will generally suffice; for young children, age to the nearest month is better. In infants, age to the nearest hour or even minute may be appropriate, depending on the purpose of the study. Other examples of continuous data include height, weight, length of time of survival, range of joint motion, and many laboratory values, such as serum glucose, sodium potassium, or uric acid.

When an observation can take on only integer

values, the scale of measurement is discrete. For example, counts of things—number of pregnancies, number of previous operations, number of fractures, number of transient ischemic attacks (TIAs) prior to a stroke—are discrete measures.

In the study described in Presenting Problem 3, three patient characteristics were evaluated: percentage of cholesterol saturation of bile, patient age, and patient gender. The first two characteristics are measured on a continuous numerical scale because they can take on any individual value in the possible range of values. Gender of the patient has a nominal scale with only two values. In a cohort study in which patients are followed for a given period of time to see whether they develop gallstones, the number of gallstones might be the data collected; these data are examples of discrete numerical measures.

Characteristics measured on a numerical scale are frequently displayed in tables or graphs. Means and standard deviations are generally used to summarize the values of numerical measures (these summary measures are presented in Chapter 4).

3.3 TABLES & GRAPHS FOR NOMINAL & ORDINAL DATA

To illustrate how to construct tables for nominal data, consider the observations on activity level and cardiac arrest given in Table 3–1 for 20 hypothetical men in Presenting Problem 1. (The original observations were not given in the article, and we have assigned representative numbers to a group of 20 men to illustrate how tables and graphs are constructed.)

The simplest way to present nominal data (or ordinal data, if there are not too many points on the scale) is to list the categories in one column of the table and the **frequency** (counts) or percentage of observations in another column. Table 3–2 shows a simple way of presenting data for the

---

*Some statisticians differentiate interval scales (with an arbitrary zero point) from ratio scales (with an absolute zero point); examples are temperature on a Celsius scale (interval) and temperature on a Kelvin scale (ratio). There are no differences, however, in how measures on these two scales are treated statistically, so we call them both simply numerical.

Table 3-1. Activity level of men having cardiac arrest.[1]

Subject	Active at Time of Arrest	Activity per Week (min)	Subject	Active at Time of Arrest	Activity per Week (min)
1	Yes	1–19	11	No	0
2	No	0	12	No	≥ 140
3	No	1–19	13	Yes	20–139
4	No	20–139	14	No	1–19
5	No	≥ 140	15	No	0
6	No	0	16	No	0
7	No	20–139	17	No	20–139
8	No	1–19	18	No	1–19
9	Yes	20–139	19	No	1–19
10	No	1–19	20	No	0

[1]Hypothetical observations from Siscovick DS et al: The incidence of primary cardiac arrest during vigorous exercise. *N Engl J Med* 1984; **311**:874.

Table 3-2. Activity of men having cardiac arrest.

Active at Time	Number of Men
Yes	3
No	17

number of men who were or were not active at the time of their cardiac arrest.

When two nominal scale characteristics are examined, a common way to display the data is in a **contingency table,** a cross-classification of counts. Contingency tables are easy to construct and interpret. The first step in developing a contingency table for the observations in Table 3–1 is to list the eight cells, or categories of counts, that will appear in the table. The eight cells for our example are shown in Table 3–3. Tallies are then placed in the appropriate cell for each subject. Subject 1 has a tally in the cell ''Active at arrest, 1–19 min/ wk''; Subject 2 has a tally in the cell ''Not active at arrest, 0 min/wk.'' Table 3–3 contains tallies for the subjects listed in Table 3–1.

The sum of the tallies in each cell is then used to construct a contingency table such as Table 3–4, which contains cell counts for all 133 subjects in the study. Percentages may be given along with the cell counts.

For a graphic display of nominal or ordinal data, bar charts are most commonly used. In a **bar chart,** counts or percentages of the characteristics of interest are shown as bars. A bar chart illustrating the number of men who were not active at the time of cardiac arrest is shown in Fig. 3–1. The categories of activity are placed along the horizontal, *X*-axis, the number of subjects along the vertical, *Y*-axis. Bar charts may also be constructed with the categories along the *Y*-axis and the numbers along the *X*-axis. Other graphic devices such as pie charts and pictographs are often used in newspapers, magazines, and advertising brochures; they are used in the health field primarily to display resource information—eg,

Table 3-3. Step 1 in constructing contingency table for cardiac arrest data.

Cell	Tally
Active at arrest, 0 min/wk	
Active at arrest, 1–19 min/wk	/
Active at arrest, 20–139 min/wk	//
Active at arrest, ≥ 140 min/wk	
Not active at arrest, 0 min/wk	//////
Not active at arrest, 1–19 min/wk	//////
Not active at arrest, 20–139 min/wk	///
Not active at arrest, ≥ 140 min/wk	//

Table 3-4. Contingency table for cardiac arrest data.[1]

Habitual High-Intensity Activity (min/wk)	Incidence of Primary Cardiac Arrest	
	During Activity	Not During Activity
0	0	30
1–19[2]	2	44
20–139	3	32
≥ 140[3]	4	18

[1]Adapted and reproduced, with permission, from Table 1 in Siscovick DS et al: The incidence of primary cardiac arrest during vigorous exercise. *N Eng J Med* 1984; **311:**874.
[2]Corresponds to less than 112 kcal of high-intensity activity per week.
[3]Corresponds to more than 868 kcal of high-intensity activity per week.

the portion of the gross national product devoted to health expenditures or the geographic distribution of primary-care physicians. However, graphics for nominal data in the clinical literature are generally restricted to bar charts. As you can see in Fig 3–1, bar charts are effective communication devices; furthermore, they are easy to construct and interpret.

3.4 TABLES & GRAPHS FOR NUMERICAL DATA

Numerical data may be presented in a variety of ways, and we will use the data from Presenting Problem 2 to illustrate some of the more common

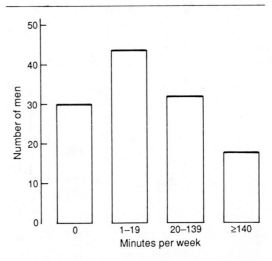

Figure 3-1. Habitual high-intensity activity levels in men with primary cardiac arrest. (Adapted, with permission, from Siscovick DS, et al: The incidence of primary cardiac arrest during vigorous exercise. *N Engl J Med* 1984; **311:** 874.)

methods. The age and stage of disease in 118 patients diagnosed with colorectal cancer are given in Table 3–5.

3.4.1 Stem & Leaf Plots

Stem and leaf plots are graphs devised in 1977 by Tukey, a statistician interested in meaningful ways to communicate by visual display. They provide a convenient means of tallying the observations and can be used as a direct display of data or as a preliminary step in constructing a frequency table.

A glance at the observations in Table 3–5 indicates that most of the patient ages are in the 50-, 60-, 70-, and 80-year categories. Closer inspection reveals that the youngest patient is 28 years old and the oldest 93. The first step in organizing data is to decide on the number of subdivisions, called

classes, that you want to use (it should generally be between 6 and 14; more details on this decision are given in Section 3.4.2); for a first attempt, we will categorize observations by 10s, from 20 to 29, 30 to 39, 40 to 49, etc, up to 90 to 99.

To form a **stem and leaf plot,** draw a vertical line, and place the first digit of each category—called the *stem*—on the left side of the line, as in Fig 3–2. The numbers on the right side of the vertical line represent the second digit of each observation; they are the *leaves*. The steps in building a stem and leaf plot are as follows: (1) take the age of the first patient, 51, and write the second digit, or leaf—1—on the *right* side of the vertical line, opposite the first digit, or stem, corresponding to 5; (2) for the second patient, 56 years of age, write the 6 (leaf) on the right side of the vertical line opposite 5 (stem); (3) for the third

Table 3–5. Age and stage of disease of colorectal cancer patients.[1]

Patient	Age	Stage	Patient	Age	Stage	Patient	Age	Stage
1	51	3	41	60	3	81	70	2
2	56	4	42	60	2	82	73	2
3	81	3	43	61	2	83	82	3
4	64	3	44	63	4	84	87	3
5	82	—	45	67	2	85	86	3
6	88	2	46	67	3	86	51	3
7	58	1	47	70	3	87	58	3
8	56	3	48	71	3	88	75	4
9	61	1	49	76	3	89	47	4
10	64	—	50	91	2	90	81	2
11	68	3	51	73	3	91	73	3
12	45	2	52	77	2	92	63	3
13	70	2	53	84	2	93	63	3
14	58	2	54	73	3	94	80	3
15	54	3	55	80	3	95	84	4
16	52	2	56	85	3	96	73	3
17	60	3	57	76	4	97	80	4
18	63	3	58	75	4	98	58	4
19	74	3	59	77	3	99	59	3
20	73	2	60	65	2	100	59	3
21	71	2	61	83	2	101	69	3
22	83	3	62	63	2	102	70	4
23	58	4	63	65	3	103	71	2
24	70	2	64	75	3	104	73	4
25	83	2	65	89	—	105	61	3
26	88	3	66	71	2	106	65	3
27	73	3	67	43	3	107	74	3
28	79	2	68	73	4	108	65	3
29	62	3	69	28	3	109	70	2
30	73	2	70	47	3	110	63	3
31	75	2	71	50	4	111	77	3
32	90	3	72	71	3	112	86	3
33	76	4	73	71	3	113	90	3
34	85	4	74	72	3	114	58	3
35	53	3	75	79	4	115	80	3
36	72	3	76	93	3	116	75	3
37	72	4	77	76	1	117	61	3
38	77	4	78	82	3	118	82	4
39	80	3	79	83	3			
40	48	2	80	64	4			

[1]Data used, with permission, from the Illinois State Cancer Registry, Illinois Department of Public Health, Division of Epidemiologic Studies.

2	
3	
4	
5	16
6	4
7	
8	12
9	

Figure 3-2. Constructing stem and leaf plot of colorectal cancer patient ages (by 10s).

patient, 81 years old, write the 1 (leaf) opposite 8 (stem); and so on. When more than one observation begins with the same initial digit, such as ages 51 and 56 for patients 1 and 2, write the second digit (leaf) to the right of any previous leaves. The leaves for the first five patients are given in Fig 3-2.

The complete stem and leaf plot for the ages of all 118 patients in the tumor registry is given in Fig 3-3. Not only does the plot provide a tally of observations, but it also shows how the ages are distributed.

With a large number of patients, another reasonable technique is to display their ages by using classes that are only 5 years wide, rather than 10 years wide as in the previous graphs. In addition, the leaves are generally reordered from lowest to highest within each class (stem). A stem and leaf

plot for these data, using class widths of 5 years and reordering within each class, is given in Fig 3-4. After the reordering, you can easily locate the middle of the distribution by simply counting in from either end. With 118 observations, the middle is halfway between the 59th and 60th observations. A caret (^) has been placed at this point in Fig 3-4. (The + signs in the figure are discussed later.)

3.4.2 Frequency Tables

Scientific journals often present information in frequency distributions or frequency tables. Tables are more difficult to construct for numerical data than for nominal data because the scale of the observations must first be divided into classes, as in stem and leaf plots. The observations in each class are then counted. The steps for constructing a frequency table (or a stem and leaf plot) are as follows:

1. Identify the largest and smallest observations.

2. Subtract the smallest observation from the largest to obtain the **range** of the data.

3. Determine the number of classes. Several rules are available for calculating the number of classes, but common sense is usually adequate for making this decision. The following guidelines are useful in determining the number of classes.

a. For most applications, experience has shown that between 6-14 classes are adequate to provide enough information without being overly detailed.

b. There should be enough classes to demonstrate the shape of the distribution but not so many that minor fluctuations are noticeable.

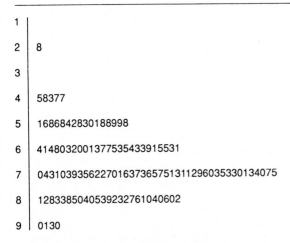

Figure 3-3. Stem and leaf plot of colorectal cancer patient ages (by 10-year classes). (Adapted, with permission, from data supplied by the Illinois State Cancer Registry, Illinois Department of Public Health, Division of Epidemiologic Studies.)

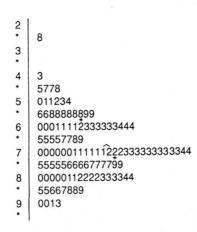

Figure 3-4. Stem and leaf plot of colorectal cancer patient ages (by 5-year classes, arranged from lowest to highest). (Adapted, with permission, from data supplied by the Illinois State Cancer Registry, Illinois Department of Public Health, Division of Epidemiologic Studies.)

4. Divide the range of observations by the number of observations to obtain the width of the classes. For some applications, deciding on the class width first may make more sense; then use the class width to determine the number of classes. Following are some guidelines for determining class width.

a. **Class limits** (beginning and ending numbers) must not overlap. For example, they must be stated as "40–49" or "40 up to 50," not as "40–50," "50–60," etc. Otherwise, you cannot tell which class an observation like 50 belongs to.

b. If possible, the class widths should be equal; unequal class widths present graphing problems. Unequal widths should be used only when there are large gaps in the data.

c. If possible, open-ended classes should be avoided because they do not accurately communicate the range of the observations. Examples of open-ended classes are "49 years or less" and "90 years or more."

d. If possible, class limits should be chosen so that most of the observations in the class are closer to the midpoint of the class than to either end of the class.

5. Tally the number of observations in each class. If you are constructing a stem and leaf plot, the actual value of the observation is noted. If you are constructing a frequency table, you need use only the number of observations that fall within the class.

Table 3–6 is a frequency table for age of colorectal cancer patients. Some tables present only frequencies (number of patients); others present percentages as well. **Percentages** are found by dividing the number of observations in a given class, n_i, by the total number of observations, n, and

then multiplying by 100. For example, for the class from 60 to 64 years of age, the percentage is

$$\frac{n_i}{n} \times 100 = \frac{17}{118} \times 100 = 0.144 \times 100 = 14.4\%$$

(rounded to one decimal place).

For some applications, cumulative frequencies or percentages may be desirable. The **cumulative frequency** (or percentage) is the frequency (or percentage) of observations for a given value plus all lower values. The cumulative value in the last column of Table 3–6, for instance, shows that the percentage of patients diagnosed with colorectal cancer who are under age 70 years is 39.8%.

Frequency tables may also be constructed for data measured on an ordinal scale. For example, if we want to examine the number of patients diagnosed at each stage of disease, we can use the stages as classes and tally the number of patients in each stage.

3.4.3 Histograms, Box & Whisker Plots, & Frequency Polygons

Graphs are used extensively in medicine—in journals in the health field, in slides for presentations at professional meetings, and in advertising literature directed toward health professionals. Three graphic devices especially useful in medicine are histograms, box and whisker plots, and frequency polygons or line graphs. The first step in constructing any of these graphs is to create a stem and leaf plot or a frequency table, as illustrated in the previous two sections.

3.4.3.a Histograms: A histogram of the ages of patients at the time of diagnosis of colorectal cancer is shown in Fig 3–5. **Histograms** usually present the measurement of interest along the X-axis and the number or percentage of observations along the Y-axis (although some computer programs do just the opposite, as illustrated later, in Fig 3–17). Whether numbers or percentages are used depends on the purpose of the histogram. For example, percentages are needed when two histograms based on different numbers of subjects are compared.

Note that the area of each bar is proportional to the percentage of observations in that interval; eg, the 17 observations in the interval from 60 to 64 years of age account for 17/118, or 14.4%, of the area covered by this histogram. Therefore, the information a histogram communicates is one of *area*. The area concept is one reason the width of classes should be equal; otherwise the heights of columns in the histogram must be appropriately modified to maintain the correct area. For example, in Fig 3–5, if the lowest class were 20

Table 3–6. Frequencies of ages of colorectal cancer patients[1]

Years of Age	Number of Patients	Percentage	Cumulative Percentage
25 through 29	1	0.8	0.8
30 through 34	0	0	0.8
35 through 39	0	0	0.8
40 through 44	1	0.8	1.7
45 through 49	4	3.4	5.1
50 through 54	6	5.1	10.2
55 through 59	10	8.5	18.6
60 through 64	17	14.4	33.1
65 through 69	8	6.8	39.8
70 through 74	27	22.9	62.7
75 through 79	15	12.7	75.4
80 through 84	17	14.4	89.8
85 through 89	8	6.8	96.6
90 through 94	4	3.4	100.0
Total	118	100.0	

[1]Data used, with permission, from the Illinois State Cancer Registry, Illinois Department of Public Health, Division of Epidemiologic Studies.

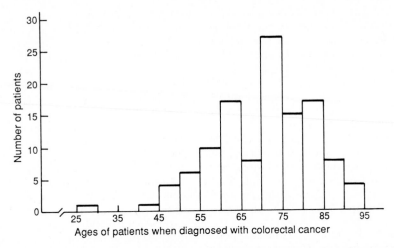

Figure 3-5. Histogram of colorectal cancer patient ages. (Adapted, with permission, from data supplied by the Illinois State Cancer Registry, Illinois Department of Public Health, Division of Epidemiologic Studies.)

years wide (from 25 to 44) while all other classes remained 5 years wide, there would be two observations in the interval and the height of the column for that interval would be only 0.5 unit (instead of 2 units) to compensate for its quadrupled width. Note also the broken line on the X-axis, indicating a gap in the scale between 0 and 25, where the histogram begins. A break in the line is not required on the X-axis, but it is required on the Y-axis in certain cases in which the frequencies are large numbers. Failure to indicate a gap in the scale is discussed in Section 3.6 along with other common errors in graphs.

3.4.3.b Box and Whisker Plots:

A **box and whisker plot,** sometimes called simply a **box plot,** is another way to display information when the objective is to illustrate certain locations in the distribution (Tukey, 1977). It is constructed from the information in a stem and leaf plot. Refer again to the stem and leaf plot for ages in Fig 3–4. The caret (**^**) denotes the midpoint of the distribution and, as illustrated earlier, is found by counting up (or down) 59 observations from the bottom (or top) of the distribution. The *midpoint* of a distribution is the number that divides the distribution into halves; in this example, 71.5 years is the midpoint of the distribution. The plus signs (+) further divide each of the upper and lower halves of the distribution into two equal parts. Therefore, 62 years is the approximate value that divides the lowest 25% of the observations from the highest 75%, and 79 years is the value that divides the highest 25% of the observations from the lowest 75%. These numbers are also called the **first** and **third quartiles,** respectively, of the distribution, and they are used in constructing box and whisker plots. (Quartiles are discussed in more detail in Sections 4.3.1.d and 4.3.1.e of Chapter 4.)

A box and whisker plot for the cancer data (Fig 3–6) is constructed by first drawing a vertical scale representing ages of patients with colorectal cancer. (This placement is different from placement for the histogram, in which age was represented on the horizontal axis.) A box is then drawn, with the top of the box at the third quartile and the bottom at the first quartile; quartiles are sometimes referred to as *hinges* in box plots. The width of the box is not important. The location of the midpoint of the distribution is indicated with a horizontal line in the box. Finally, straight lines, or *whiskers,* are drawn from the center of the top

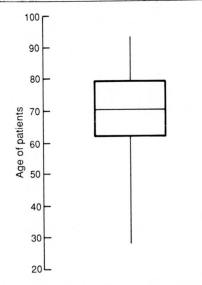

Figure 3-6. Box and whisker plot of colorectal cancer patient ages. (Adapted, with permission, from data supplied by the Illinois State Cancer Registry, Illinois Department of Public Health, Division of Epidemiologic Studies.)

of the box to the largest observation and from the center of the bottom of the box to the smallest observation. With a box and whisker plot, we can easily see that the range of ages of patients diagnosed with colorectal cancer is from about 28 to about 93 years, that 50% of the cancers occur in patients between 62 and 79 years of age, and that half of those cancers occur in patients who are 71 or 72 years of age or less.

Box and whisker plots are also effective when there is more than one set of observations and the objective is to compare them. For example, box plots of the distribution of ages of patients diagnosed at the four different stages can be displayed to permit a visual comparison of the distributions.

3.4.3.c Frequency Polygons: Frequency polygons are line graphs similar to histograms that are used to compare two distributions on the same graph. Table 3–7 gives the data on age and percentage saturation of bile for patients in the study in Presenting Problem 3. There were 31 men and 29 women in this study. As a first step, stem and leaf plots for the percentage saturation of bile in men and women are shown in Fig 3–7.

The data on percentage saturation of bile in men from the stem and leaf plot in Fig 3–7 has been used to construct the frequency polygon in Fig 3–8. A histogram of the data is represented by the dashed lines in Fig 3–8 and shows that frequency polygons are constructed by connecting the midpoints of the columns of a histogram. Therefore, the same guidelines for constructing frequency tables and histograms hold for constructing frequency polygons. Note that the line extends from the midpoint of the first and last columns to the X-axis in order to close up both ends of the distribution and indicate zero frequency of any values beyond the extremes. Because a frequency polygon is based on a histogram, approximately the same area is under the curved line as is under the histogram. Thus, frequency polygons also portray area.

Percentage polygons are especially useful in comparing two frequency distributions. Percentage polygons for saturation of bile in both men

Table 3–7. Age and percentage supersaturation of bile.[1]

	Men			Women	
Subject	Age	% Supersaturation	Subject	Age	% Supersaturation
1	23	40	1	40	65
2	31	86	2	33	86
3	58	111	3	49	76
4	25	86	4	44	89
5	63	106	5	63	142
6	43	66	6	27	58
7	67	123	7	23	98
8	48	90	8	56	146
9	29	112	9	41	80
10	26	52	10	30	66
11	64	88	11	38	52
12	55	137	12	23	35
13	31	88	13	35	55
14	20	80	14	50	127
15	23	65	15	47	77
16	43	79	16	36	91
17	27	87	17	74	128
18	63	56	18	53	75
19	59	110	19	41	82
20	53	106	20	25	69
21	66	110	21	57	84
22	48	78	22	42	116
23	27	80	23	49	73
24	32	47	24	60	87
25	62	74	25	23	76
26	36	58	26	48	107
27	29	88	27	44	84
28	27	73	28	37	120
29	65	118	29	57	123
30	42	67			
31	60	57			

[1]Observations reproduced, with permission, from Figure 1 in Einarsson K et al: Influence of age on secretion of cholesterol and synthesis of bile acids by the liver. *N Eng J Med* 1985; **313**:277.

Men			Women	
3			3	5
4	07		4	
5	2678		5	258
6	567		6	569
7	3489		7	35667
8	00667888		8	0244679
9	0		9	18
10	66		10	7
11	00128		11	6
12	3		12	0378
13	7		13	
14			14	26

Figure 3–7. Stem and leaf plots of percentage saturation of bile. (Adapted, with permission, from Einarsson K et al: Influence of age on secretion of cholesterol and synthesis of bile acids by the liver. *N Engl J Med* 1985; **313**:277.

3.5 GRAPHS FOR TWO CHARACTERISTICS

Most studies in medicine involve measuring more than one characteristic, and graphs displaying the relationship between two characteristics are common in the literature. There are no graphs for displaying a relationship between two characteristics when both are measured on a nominal scale; the numbers are simply presented in contingency tables. When one of the characteristics is nominal and the other is numerical, the data can be displayed in a box and whisker plot like the one for percentage saturation of bile (a numerical variable) for men and women (a nominal variable). Alternatively, both frequency distributions can be displayed on the same graph in a **dot plot,** as in Fig 3–11, in which each dot (•) designates one observation. The center of each frequency distribution is often denoted in some manner; in Fig 3–11, a short, thin line has been drawn at the center but an X or some other symbol could be used instead.

Also common in medicine is the use of **bivariate plots** (also called **scatterplots** or scatter diagrams), which illustrate the relationship between two characteristics measured on a numerical scale. In the same study on percentage saturation of bile, information was collected on the age of each patient to see whether a relationship existed between the two measures. Fig 3–12 is a scatterplot of age and percentage saturation of bile for the women in the study. A scatterplot is constructed by drawing X- and Y-axes; the characteristic hypothesized to explain or predict or the one that occurs first—sometimes called the risk factor—is placed on the X-axis, and the characteristic to be explained or predicted or that occurs second is placed on the Y-axis. In applications in which a relationship is hypothesized but is not one of causation, placement for the X- and Y-axes does not matter. Each

and women are shown in Fig 3–9. Frequencies must be converted to percentages when the two distributions being compared are unequal in size, and this conversion has been made for Fig 3–9. Examination of the polygons indicates that the distribution of percentage saturation of bile does not appear to be very different for men and women; although some observations are higher among women than among men, most of the area in one polygon is overlapped by that in the other.

Box and whisker plots are also appropriate for displaying the frequency distributions of percentage saturation of bile in men and women and are illustrated in Fig 3–10. The graphs indicate the similarity of the distributions, just as do the frequency polygons. Again, we see that percentage saturation of bile is a bit more spread out among women, but the box and whisker plots show clearly that the midpoints of the distributions are almost the same and that most of the spread in values in women occurs in the upper half of the distribution.

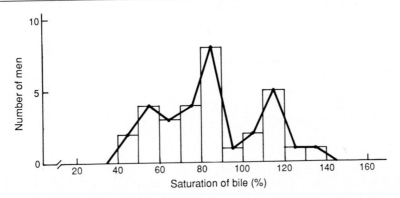

Figure 3–8. Frequency polygon for percentage saturation of bile in men. (Adapted, with permission, from Einarsson K et al: Influence of age on secretion of cholesterol and synthesis of bile acids by the liver. *N Engl J Med* 1985; **313**:277.)

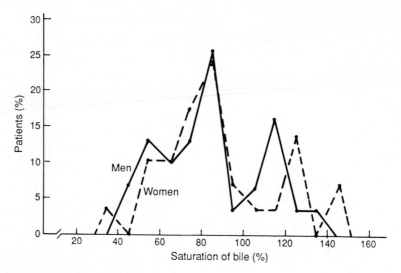

Figure 3-9. Percentage polygons for percentage saturation of bile in men and women. (Adapted, with permission, from Einarsson K et al: Influence of age on secretion of cholesterol and synthesis of bile acids by the liver. *N Engl J Med* 1985; **313**:277.)

observation is represented by a point or dot (•); eg, the point located at the X in Fig 3–12 represents female patient 5, who is 63 years of age and has a percentage saturation of bile equal to 142. More information on interpreting scatterplots is presented in Chapter 10, but we see here that the data in Fig 3–12 indicate the possibility of a positive relationship between age and percentage saturation of bile in women. At this point, we cannot say whether the relationship is a significant rela-

tionship or one that simply occurs by chance; this topic is covered in Chapter 10.

3.6 EXAMPLES OF MISLEADING CHARTS & GRAPHS

The quality of charts and graphs published in the medical literature is higher than similar displays in the popular press. The most significant prob-

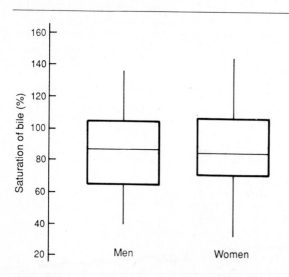

Figure 3-10. Box and whisker plots for percentage saturation of bile. (Adapted, with permission, from Einarsson K et al: Influence of age on secretion of cholesterol and synthesis of bile acids by the liver. *N Engl J Med.* 1985; **313**:277.)

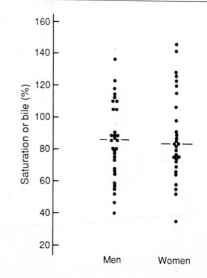

Figure 3-11. Two dot plots for percentage saturation of bile. (Adapted, with permission, from Einarsson K et al: Influence of age on secretion of cholesterol and synthesis of bile acids by the liver. *N Engl J Med* 1985: **313**:277.)

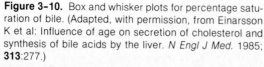

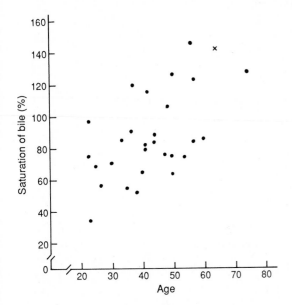

Figure 3-12. Scatterplot of age and percentage saturation of bile for women. (Adapted, with permission, from Einarsson K et al: Influence of age on secretion of cholesterol and synthesis of bile acids by the liver. *N Engl J Med* 1985; **313**:277.)

lem with graphs (and tables as well) in medical journal articles is complexity; many authors attempt to present too much information, and it may take the reader a long time to figure out what the graph is really saying. In these situations, many of us give up and consequently miss or misinterpret important information the authors are trying to communicate.

Before illustrating misleading displays, we wish to emphasize that the purpose of tables and graphs is to present information (often based on large numbers of observations) in a concise way so that the reader can comprehend and remember the information. Therefore, the two cardinal rules are that (1) charts, tables, and graphs should be simple and easily understood by the reader, and (2) concise but complete labels and legends must accompany them.

Knowing about common errors helps you correctly interpret information in articles and presentations, provides a hint about the research skills of the investigators, and helps you design your own charts and graphs. We illustrate four errors we have seen with sufficient frequency to warrant their discussion. We use hypothetical examples to illustrate these errors and do not imply that they necessarily occurred in the Presenting Problems used in this text. An interesting and entertaining report by Wainer (1982) draws on published tables and graphs from various nonmedical sources and is recommended if you would like more information on this topic.

A researcher can easily make a change appear more or less dramatic by selecting a starting time for a graph either before or after the change begins. Fig 3–13A shows the decrease in annual mortality from a disease, beginning in 1940. The major decrease in mortality from this disease occurred in the 1950s. Although not incorrect, a graph that begins in 1960 (Fig 3–13B) deemphasizes the decrease and sends a message to the reader that the change has been small.

If the values on the *Y*-axis are large, the entire scale cannot be drawn. For example, suppose an investigator wants to illustrate the number of deaths from cancer each decade from 1950, when there were 200,000 deaths, to 1990, when there are 400,000 projected deaths. Even if the vertical scale is in thousands of deaths, it must range from 200 to 400. If the *Y*-axis is not broken, the message given to the person looking at the graph is not accurate; the change appears larger than it really is. This error, called **suppression of zero,** is common in histograms and line graphs. Fig 3–14A illustrates the effect of suppression of zero on number of deaths from cancer per year; Fig 3–14B illustrates the correct construction.

The magnitude of change can also be enhanced or minimized by the choice of scale on the vertical axis. For example, suppose a researcher wishes to compare the ages at death of a group of men and

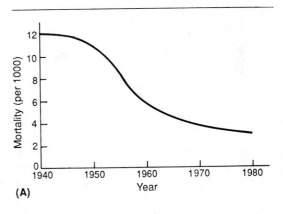

(A)

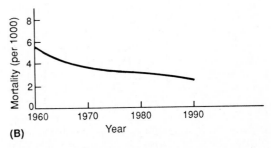

(B)

Figure 3-13. Illustration of effect of portraying change at two different times. (*A*) Mortality from a disease since 1940. (*B*) Mortality from a disease since 1960.

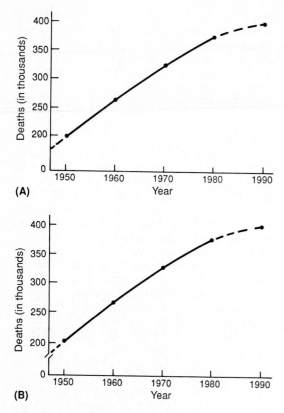

Figure 3-14. Illustration of effect of supression of zero on *Y*-axis in graphs showing deaths from cancer. (*A*) No break in line on *Y*-axis (*B*) Break in line correctly placed on *Y*-axis.

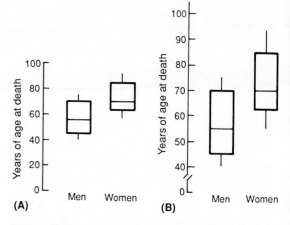

Figure 3-15. Illustration of effect of suppressing or stretching the scale in plots showing age at death. (*A*) Suppressing the scale. (*B*) Stretching the scale.

tus, as in Table 3–8B, in which percentages in each row total 100%. From Table 3–8B, one can conclude that 55% of patients with no insurance coverage have a low level of compliance. In other words, the format of the table should reflect the questions asked in the study. If one measure is examined to see whether it explains another measure, such as insurance status explaining compliance, investigators should look at percentages within the explanatory measure (insurance status, in our example).

a group of women. Fig 3–15A, by suppressing the scale, indicates that the ages of men and women at death are similar; Fig 3–15B, by "stretching" the scale, makes the differences in age at death between men and women seem large.

Our final example is a table that gives irrelevant percentages, a somewhat common error in the medical literature. Suppose that the investigators are interested in the relationship between levels of patient compliance and the type of insurance coverage they have. When there are two or more measures of interest, the purpose of the study generally determines which measure is viewed within the context of the other. Table 3–8A shows the percentage of patients with different types of insurance coverage within three levels of patient compliance, so the percentages in each column total 100%. The percentages in Table 3–8A make sense if the investigator wishes to compare the type of insurance coverage of patients who have specific levels of compliance; it is possible to conclude, for example, that 35% of patients with low levels of compliance have no insurance.

Contrast this interpretation with that obtained if percentages are calculated within insurance sta-

Table 3-8. Effect of calculating column percentages versus row percentages for study of compliance with medication versus insurance coverage.

A. Percentages Based on Level of Compliance (Column %)

Insurance Coverage	Level of Compliance With Medication		
	Low	Medium	High
Medicaid	30	20	15
Medicare	20	25	30
Medicaid and Medicare	5	5	5
Other insurance	10	30	40
No insurance	35	20	10

B. Percentages Based on Insurance Coverage (Row %)

Insurance Coverage	Level of Compliance With Medication		
	Low	Medium	High
Medicaid	45	30	25
Medicare	25	35	40
Medicaid and Medicare	33	33	33
Other insurance	15	35	50
No insurance	55	30	15

3.7 COMPUTER PROGRAMS THAT PRODUCE TABLES & GRAPHS

Beginning in this chapter, we give examples of output from **computer packages**—ie, computer programs especially designed to analyze statistical data. We briefly describe each computer package the first time it is used as an example. We reproduce the actual output obtained in analyzing observations from the Presenting Problems, even though it frequently contains statistics not yet discussed, so that you have the opportunity to see what it really looks like. We will discuss here the important aspects of the output. For the time being, you can simply ignore unfamiliar information on the printout; in subsequent chapters, we will explain many of the statistics. Note, however, that statistical computer programs are designed to meet the needs of researchers in many different fields, so some of the statistics in the printouts may rarely be used in medical studies and hence are not included in this book.

The computer output from a MINITAB program using the observations of Presenting Problem 1 is given in Fig 3–16. MINITAB is a set of data analysis programs commonly used in introductory statistics courses in many colleges and universities, so some readers may be familiar with it. MINITAB stores the data in rows and columns (rather like a spreadsheet, if you are familiar with this type of computer program), and different analyses are requested by specifying the type of analysis followed by the columns to which the analysis is to be applied. MTB > and SUBC > followed by information refers to the different commands.

When we analyzed the data, we put the observations from Table 3–1 in columns 16 and 17 in the MINITAB program; these observations include the amount of time (in minutes) each subject spent in habitual exercise each week (called "time") and whether the subject was engaged in activity at the time of cardiac arrest (called "active"). Two commands produced the two tables in Fig 3–16. The first table contains the frequencies for all 20 men; the total in column 1 (3 men) is the number who were active at the time of their cardiac arrest, and the total in column 2 (17) is the number who were not. The second table contains the percentage of the total represented by each frequency; eg, 15% of the men were active at the time of arrest.

Two outputs for the data on age of colorectal cancer patients are given: one from SPSS (Statistical Package for Social Sciences) are one from SAS (Statistical Analysis System). In the SPSS program called FREQUENCIES (Fig 3–17), ages are divided into classes of 5-year width (25 though 29, 30 through 34, etc) to produce a frequency table and a histogram. Unlike MINITAB, SPSS does not print commands as part of the output. The frequency table provides a great deal of informa-

```
MTB > table c17 c16

  ROWS: time        COLUMNS: active

                 1          2        ALL

    0            0          6          6
    1            1          6          7
   20            2          3          5
  140            0          2          2
  ALL            3         17         20

   CELL CONTENTS --
                      COUNT

MTB > table c17 c16;
SUBC> totpercent.

  ROWS: time        COLUMNS: active

                 1          2        ALL

    0           --      30.00      30.00
    1         5.00      30.00      35.00
   20        10.00      15.00      25.00
  140          --       10.00      10.00
  ALL        15.00      85.00     100.00

   CELL CONTENTS --
                      % OF TBL
```

Figure 3-16 MINITAB program to produce tables for nominal data using observations from Presenting Problem 1 on level of activity and cardiac arrest. (Adapted, with permission, from Siscovick DS et al: The incidence of primary cardiac arrest during vigorous exercise. *N Engl J Med* 1984;**311**:874. MINITAB is a registered trademark of Minitab, Inc. Used with permission.)

tion: the frequency of patients with ages in each 5-year interval, the percentage of patients, and the cumulative percentage. The column labeled "Valid Percent" corresponds to observations for which values are known. If the age of a patient is not available, a number that is impossible, such as 999, is used and specified as a "Missing Value"; SPSS then ignores all the patients with "Missing Values" in the calculation of percentages or statistics. SPSS prints histograms horizontally (which is easier for printers) with the frequency on the horizontal axis; compare the histogram in Fig 3–17 to the histogram we made in Fig 3–5.

The stem and leaf plot and the box plot in Fig 3–18 are from an SAS computer program called UNIVARIATE. As part of the UNIVARIATE program, SAS provides several statistics called "Moments," "Quantiles," and "Extremes" above the plot; these statistics are discussed in Chapter 4. Also, disregard the "Normal Probabil-

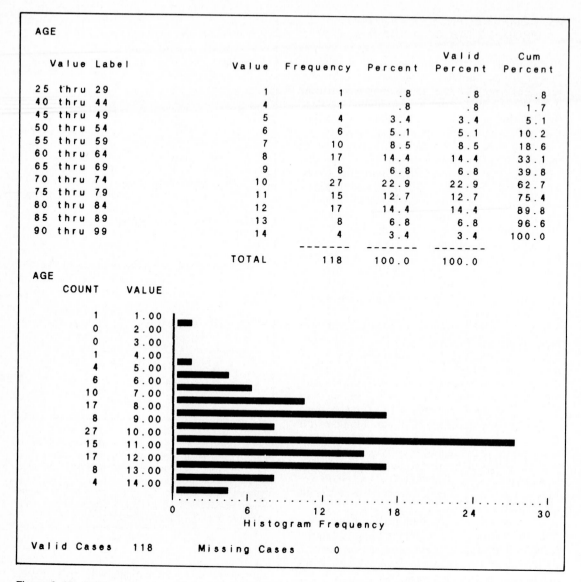

Figure 3-17. SPSS program to produce frequency tables and histograms for numerical data, using observations on ages of colorectal cancer patients from Presenting Problem 2. (Adapted, with permission, from data supplied by the Illinois State Cancer Registry, Illinois Department of Public Health, Division of Epidemiologic Studies. SPSS is a registered trademark of SPSS, Inc. Used with permission.)

ity Plot'' for now; we refer to it in Section 7.4.2 of Chapter 7. Turning to the stem and leaf plot and the box plot, note that SAS makes the stem and leaf with the older patients at the top and the younger patients at the bottom of the plot. Otherwise, the plot is identical to the one in Fig 3–4. SAS also provides the number of patients in each age group (printed to the right of the stem and leaf plot under the symbol ''#''). The box plot is similar to the one in Fig 3–6.

We used SYSTAT and SPSS to analyze the observations on subjects' age and percentage cholesterol saturation of bile from Presenting Problem

3. The SYSTAT STEM AND LEAF PLOT and BOX PLOT programs give the output shown in Fig 3–19. SYSTAT produces the same stem and leaf plots illustrated in Fig 3–7 (ie, in order of ascending numbers). The minimum and maximum values are also given, along with lower hinge (H), which is the first-quartile value; the median or middle value (M); and the upper hinge (H), which is the third-quartile value. The box plots are produced on a horizontal scale (in contrast, SAS uses a vertical scale for box plots). Otherwise, with the exception of the ''( )'' marks inside the box, the plots illustrating the percentage saturation of bile

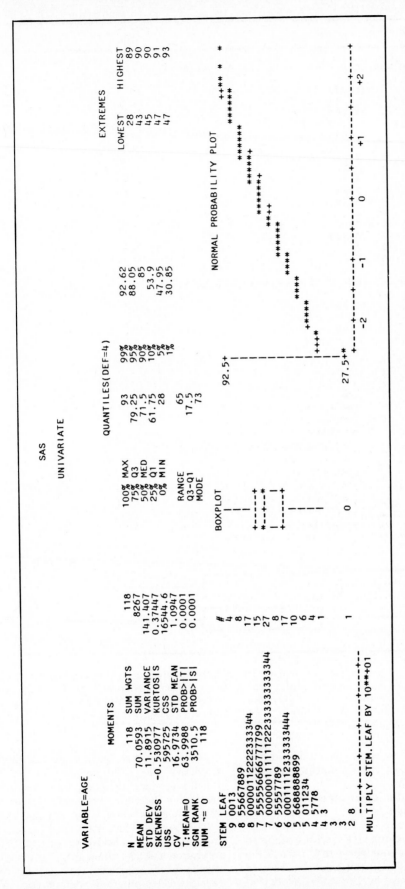

Figure 3-18. SAS program to generate stem and leaf plots and box plots for numerical data, using observations on age of colorectal cancer patients from Presenting Problem 2. (Adapted, with permission, from data supplied by the Illinois State Cancer Registry, Illinois Department of Public Health, Division of Epidemiologic Studies. SAS is a registered trademark of SAS Institute, Inc. Used with permission.)

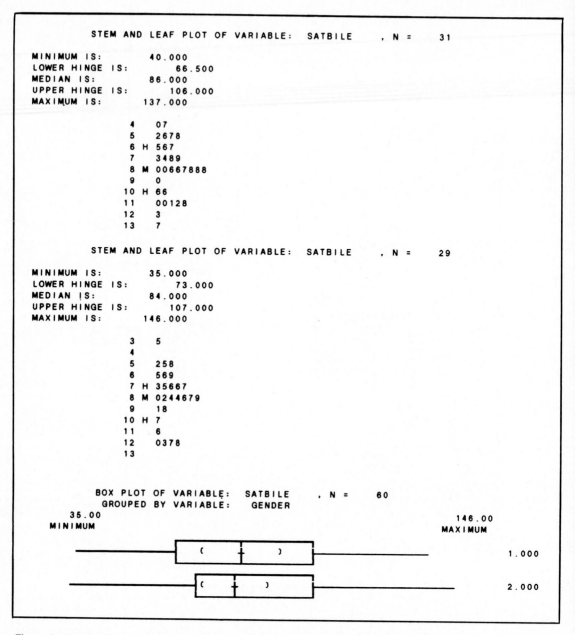

Figure 3-19. SYSTAT program to produce stem and leaf plots for numerical data, using observations on percentage saturation of bile in men and women from Presenting Problem 3. (Adapted, with permission, from Einarsson K et al: Influence of age on secretion of cholesterol and synthesis of bile acids by the liver. *N Engl J Med* 1985;**313**:277. SYSTAT is a registered trademark of SYSTAT, Inc. Used with permission.)

in men and women are equivalent to those in Fig 3-10. The "( )" marks, called "notches," are discussed in Chapter 7.

A scatterplot of the observations on age and percentage saturation of bile generated by the PLOT program in SPSS is reproduced in Fig 3-20. This program lets the user specify symbols to identify the observations when they come from more than one group. In this example, we used "f" for females and "m" for males. Note that

the data points for females resemble those in the graph in Fig 3-12. For comparison, see Fig 3-21, which reproduces the graph published in the article cited in Presenting Problem 3 (Einarsson et al, 1985). (Ignore Ps and rs for now.)

SSPS automatically determines the scale needed for the vertical and horizontal axes for percentage saturation and age, although an option allows the user to specify a different scale if so desired. Note that the SSPS program's default for scaling the

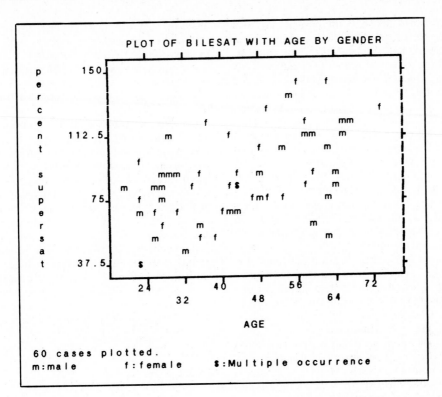

Figure 3-20. SPSS program to generate scatterplots for numerical data, using observations on percentage saturation of bile in men and women from Presenting Problem 3. (Adapted, with permission, from Einarsson K, et al: Influence of age on secretion of cholesterol and synthesis of bile acids by the liver. *N Engl J Med* 1985;**313**:277. SPSS is a registered trademark of SPSS, Inc. Used with permission.)

Relation between age and cholesterol saturation of bile. Open circles denote women (n = 29) and closed circles men (n = 31). There was no significant difference between the regression lines for men and women (P of slope = .15, P of level = .49).

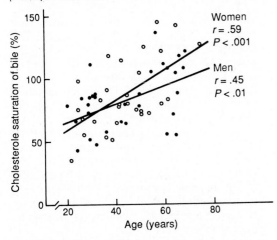

Figure 3-21. Relationship between age and cholesterol saturation of bile as published in original article. (Modified and reproduced, with permission, from Einarsson K, et al: Influence of age on secretion of cholesterol and synthesis of bile acids by the liver. *N Engl J Med* 1985; **313**:277.)

plot in Fig 3–20 makes the error of suppression of zero, an especially bad practice on the vertical axis; this error is one disadvantage in using default scales in most computer programs for graphs. We should also note that these programs provide graphs for analysis and are not intended to produce publication quality output. In addition, some programs are more flexible than others in allowing the user to provide labels for the graphs.

3.8 SUMMARY

The Presenting Problems in this chapter were used to demonstrate different methods of displaying data. The authors of the study on cardiac arrest (Siscovick et al, 1984) reported that the risk of cardiac arrest for men with low levels of habitual activity was 56 times greater than at other times. They concluded that "although the risk of primary cardiac arrest is transiently increased during vigorous exercise, habitual vigorous exercise is associated with an overall decreased risk of primary cardiac arrest."

Our analysis of age of patients with colorectal

cancer determined that half the patients are 71.5 years of age or older when their cancer is diagnosed; in addition, 25% are 62 years or younger, and 25% are 79 years or older. Further analysis of the data is included in the exercises.

In Presenting Problem 3, Einarsson and colleagues (1985) determined cholesterol saturation in a group of patients without gallstones. They concluded that cholesterol saturation of bile does increase with age as a consequence of increased secretion of cholesterol combined with decreased bile acid synthesis. Their findings may help explain the role of age in the development of cholesterol gallstones; however, as discussed in Chapter 2, Section 2.4.4, observations relating changes to age may be misleading when they are based solely on cross-sectional studies.

This chapter presented two important biostatistical concepts: the different scales of measurement used in medicine and appropriate methods for displaying information, depending on the measurement scale. The simplest level of measurement is a nominal scale, also called a categorical or qualitative scale. Nominal scales measure characteristics that can be classified into categories; the number of observations in each category is counted. Examples include gender, race, and type of disease. Dichotomous characteristics have only two values: a genetic marker is either present or absent. Nominal characteristics are displayed in contingency tables and in bar charts.

The next level of measurement is an ordinal scale, which is used for characteristics that have an underlying order. The differences between values on the scale are not equal throughout the scale, however; the numbers are arbitrary and have no inherent meaning. Examples are Apgar scores, which range from 0 to 10, and many cancer staging schemes, which have four or five categories corresponding to the severity and invasiveness of the tumor. There is no reason that the Apgar scale must range from 0 to 10; it could just as logically go from 90 to 100. Ordinal characteristics, like nominal characteristics, are displayed in contingency tables and bar charts.

Numerical scales are the highest level of measurement; they may also be called interval or qualitative scales. Characteristics measured on a numerical scale can be continuous (taking on any value on the number line) or discrete (taking on only integer values). Frequency tables summarize numerical observations; the scale is divided into classes, and the number of observations in each class is counted. Both frequencies and percentages are commonly used in frequency tables.

Several graphic devices can be used to present numerical observations, including histograms, frequency polygons or line graphs, and box and whisker plots (or, simply, box plots). Although each method provides the reader with information on the distribution of the observations, box plots are especially useful as concise displays because they show at a glance the values that define the middle, the upper 25%, and the lower 25% of a distribution. Stem and leaf plots combine features of frequency tables and histograms; they show the frequencies as well as the shape of the distribution.

Contingency tables display two nominal characteristics measured on the same set of subjects. Two numerical measures taken on the same subjects are illustrated with scatterplots. When measurements consist of one nominal and one numerical characteristic, frequency polygons, box plots, or dot plots may illustrate the distribution of numerical observations for each value of the nominal characteristic.

Remember that the purpose of tables and graphs is to communicate information. To be effective, they must be concise and intelligible. The most important features of good tables and graphs are clarity, complete labeling, and accuracy of the message they give to readers.

EXERCISES

The first three questions describe studies summarized in the exercises for Chapter 2; you may wish to refer to the fuller descriptions given there.

1. Kremer et al (1987) examined the relationship between decreases in neutrophil leukotriene B_4 production and decreases in the number of tender joints in patients with rheumatoid arthritis who were given fish oil dietary supplements. What graphic device best illlustrates this information?

2. During their investigation of an unusual new illness involving multiple-organ systems, Kilbourne et al (1983) reviewed the medical records of 121 patients in a severely affected town near Madrid. The epidemiologic investigation linked the occurrence of illness with ingestion of an unlabeled, illegally marketed cooking oil. The illness was self-limited in many patients and could not be determined in some, but severe neuromuscular manifestations occurred late in the disease in 23% of the patients. The investigators suspect that onset of illness early in the epidemic is associated with progression to muscular illness and wish to display this relationship graphically. What type(s) of graphic display should they use?

3. In the study of nonsteroidal anti-inflammatory drug use and acute nonvariceal upper gastrointestinal bleeding, Bartle, Gupta, and Lazor (1986) had the following information to present to readers:

Of the 57 patients with acute nonvariceal upper GI bleeding, 14 had taken acetylsalicylic acid and 10 had taken nonacetylsalicylic acid. Among the 123 control subjects, the numbers were 16 and 7, respectively.

What is the best way to display this information in a table?

4. Nathan et al (1984) evaluated the clinical information provided in the glycosylated hemoglobin assay by comparing it with practitioners' estimates of glucose control. Their ten-week study included 216 patients with diabetes.

   a. What graphic method is most effective for comparing the estimated and actual blood glucose levels in the 216 patients?

   b. The percentage of patients for which the practitioners overestimated and underestimated actual blood glucose levels is given in Table 3-9. Draw an appropriate graph to display this information.

5. Jacobson and Weström (1969) performed a landmark study on the diagnostic and prognostic values of routine laparoscopy in acute pelvic inflammatory disease. The number of symptoms and signs of acute salpingitis in 905 women was determined and compared according to whether the women were subsequently diagnosed as normal, as having gonococcal salpingitis, or as having nongonococcal salpingitis. Frequency polygons illustrating the number of signs and symptoms in each group were generated and are reproduced in Fig 3-22. (Ignore symbols in heading for now.)

Comparison of visually normal cases, gonococcal salpingitis cases, and nongonococcal salpingitis cases as to percentage distribution of number of symptoms and signs out of nine registered. Differences: Normal – Nongonococcal salpingitis, $\chi^2 = 45.14$; 7 d.f.: $P < .001$. Normal – Gonococcal salpingitis, $\chi^2 = 70.37$; 7 d.f.; $P < .001$. Nongonococcal salpingitis – gonococcal salpingitis, $\chi^2 = 12.52$; 7 d.f.; $P < .05$.

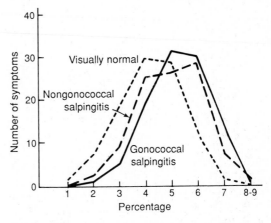

Figure 3-22. Frequency polygons of number of signs and symptoms in three groups of women. (Modified and reproduced, with permission, from Jacobson L, Weström L: Objectivized diagnosis of acute pelvic inflammatory disease. *Am J Obstet Gynecol* 1969; **105**:1088.)

   a. Which group of patients, in general, tends to have the largest number of signs and symptoms?

   b. Is the number of signs and symptoms useful diagnostically? Explain why or why not.

6. Use the data on colorectal cancer (Table 3-5) to form four stem and leaf plots of the ages of patients whose disease was diagnosed at the four stages.

7. Draw frequency polygons for the stem and leaf plots of stages 2-4 in Exercise 6.

8. Hornig, Dorndorf, and Agnoli (1986) studied 65 patients with ischemic cerebral infarction. Data on 28 patients who experienced a hemorrhagic transformation of the infarction are given in Table 3-10. The clinical score, ranging from one to 29, was determined by assigning points for consciousness, aphasia, orientation, hemianopia, facial power, motor strength, and sensory disturbances.

   a. Develop a graph to compare the before-hemorrhage scores for patients who experienced a hemorrhage during the first seven days and for those who experienced a hemorrhage after seven days.

   b. Form a contingency table to illustrate the maximum systolic blood pressure for different localization areas of the middle cerebral artery (MCA).

Table 3-9. Overestimates and underestimates of blood glucose levels in 216 patients.[1]

Glucose Level[2] (mg/dl)	Frequency of Estimate ($n = 216$) (%)
Overestimates	
≥75	10.6
74 to 50	7.9
49 to 25	15.7
24 to 0	11.6
Underestimates	
1 to 25	17.6
26 to 50	14.4
51 to 75	8.3
> 75	13.8
Absolute values	
Mean 53	
Median 40	
Range 0-265	

[1]Reproduced, with permission, from Table 1 in Nathan DM et al: The clinical information value of the glycosylated hemoglobin assay. *N Engl J Med* 1984; **310**:341.
[2]To convert values for blood glucose to millimoles per liter, multiply by 0.05551.

Table 3-10. Data on 28 patients with hemorrhagic cerebral infarction.[1]

Patient No.	Age	Sex	Etiology of Infarction	Localization in Area of MCA	Hemorrhagic Transformation Before Day	Pattern of Hemorrhage	First Contrast Enhancement	Clinical Score		Maximum Systolic Blood Pressure	Therapy (LD = Low-Dose Heparin)[2]
								Before Hemorrhage	After Hemorrhage		
19	54	F	Atrial fibrill. nonrheumatic	Superficial	3	Heterogenous	—	20	18	Normal	LD
54	71	M	Unknown	Deep	3	Hematoma	n.d.[3]	18	21	180	LD
58	66	F	Atrial fibrill. nonrheumatic	Whole	3	Hematoma	3	14	21	180	LD
65	25	M	Atrial carditis viral genesis	Superficial	3	Hematoma	14	15	13	Normal	LD
14	56	M	Ipsilateral carotid occlusion	Superficial	7	Cortical	3	10	9	180	LD
16	59	M	Ipsilateral carotid occulsion	Whole	7	Hematoma	7	14	11	Normal	LD
27	43	M	Endocarditis	Whole	7	Heterogenous	—	18	21	Normal	LD
32	79	M	Atrial fibrill. nonrheumatic	Superficial	7	Hematoma	—	19	18	180	LD
44	54	M	Rheumatic heart disease	Superficial	7	Cortical	3	9	6	Normal	LD
46	41	F	Ipsilateral carotid stenosis	Whole	7	Cortical	7	21	18	Normal	LD
53	62	M	Atrial fibrill. nonrheumatic	Whole	7	Cortical	7	20	17	180	LD
15	45	M	Ipsilateral carotid occlusion	Superficial	14	Hematoma	7	13	12	Normal	ASS

18	74	F	Unknown	Superficial	14	Hematoma	7	7	4	220	ASS
20	55	F	Ipsilateral carotid occlusion	Superficial	14	Cortical	14	7	5	Normal	ASS
23	56	F	Atrial fibrill. nonrheumatic	Superficial	14	Heterogenous	14	16	16	Normal	LD
30	61	F	Unknown	Superficial	14	Hematoma	7	19	19	Normal	LD
31	61	F	Unknown	Superficial	14	Heterogenous	14	16	15	180	LD
33	70	M	Unknown	Whole	14	Hematoma	14	17	16	220	LD
34	45	M	Ipsilateral carotid stenosis	Deep	14	Heterogenous	3	18	17	180	LD
35	74	M	Unknown	Superficial	14	Cortical	14	9	9	180	LD
43	50	M	Ipsilateral carotid occlusion	Superficial	14	Heterogenous	3	8	8	180	ASS
47	51	M	Unknown	Superficial	14	Heterogenous	14	6	5	Normal	ASS
50	53	M	Unknown	Superficial	14	Heterogenous	14	12	12	200	ASS
60	49	M	Unknown	Superficial	14	Heterogenous	7	8	8	Normal	ASS
63	65	F	Ipsilateral carotid occlusion	Superficial	14	Cortical	7	12	2	180	LD/ASS
64	49	M	Ipsilateral carotid occlusion	Superficial	14	Cortical	14	15	14	200	LD/ASS
11	51	M	Unknown	Whole	21	Heterogenous	—	15	14	180	LD
13	53	M	Ipsilateral carotid occlusion	Whole	21	Heterogenous	14	14	13	200	LD

[1] Adapted and reproduced, with permission, from Table 1 in Hornig CR, Dorndorf W, Angoli AL: Hemorrhagic cerebral infarction: A prospective study. *Stroke* 1986; **17**:180.
[2] ASS = acetylsalcyilic acid.
[3] n.d. = not done.

Rates of total coronary heart disease (CHD) per 100,000 person-years vertical scale) among women, according to cigarette use and age.

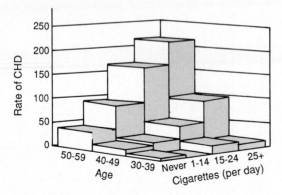

Figure 3-23. Three dimensional histogram of rates of coronary heart disease (CHD). (Modified and reproduced, with permission, from Willett WC et al: Relative and absolute excess risks of coronary heart disease among women who smoke cigarettes. *N Engl J Med* 1987;**317**:1303.)

9. Graphs can also illustrate relationships when more than two characteristics are of interest. In a study of the risk of coronary heart disease in woman, Willett et al (1987) used three-dimensional histograms; Fig 3–23 shows the rate of coronary heart disease (CHD) by age and number of cigarettes smoked per day. What conclusions are possible from this graph?

10. Mantzouranis, Rosen, and Colten (1988) studied patients with cystic fibrosis and other inflammatory lung diseases to learn more about the Fc receptor-mediated clearance of autologous erythrocytes. They presented their results in a dot plot, reproduced here in Fig 3–24. (Ignore the bars in Fig 3–24 for now.)
    a. Describe the impressions given by Fig 3–24.

Fc-mediated reticuloendothelial clearance of autologous erythrocytes in normal adults, in patients with cystic fibrosis (CF), chronic obstructive pulmonary disease (COPD), immune deficiencies, or systemic lupus erythematosus (SLE), and in those who had undergone splenectomy. The large open corcles denote data obtained in patients receiving corticosteroid therapy. The bars indicate mean values ± SEM.

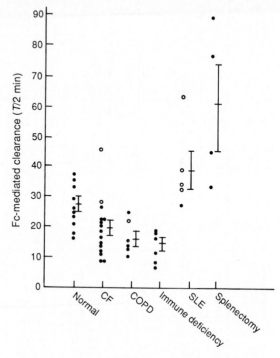

Figure 3-24. Fc receptor–mediated clearance of autologous erythrocytes. (Modified and reproduced, with permission, from Mantzouranis EC, Rosen FS, Colten HR: Reticuloendothelial clearance in cystic fibrosis and other inflammatory lung diseases. *N Engl J Med* 1988;**319**:338.)

    b. Which group included the largest number of patients?
    c. What two groups of patients have the highest levels of Fc-mediated clearance?

Summarizing Data

4

PRESENTING PROBLEMS

Presenting Problem 1. A study investigated the relationship between diet, serum lipoprotein concentration, and progression of coronary atherosclerotic lesions (Arntzenius et al, 1985). Thirty-nine patients with stable angina pectoris and at least one vessel with 50% obstruction as shown by coronary angiography were placed on a two-year vegetarian diet, with a ratio of polyunsaturated fatty acid to saturated fatty acid of at least 2 and less than 100 mg of cholesterol per day. These dietary changes were associated with decreases in body weight, systolic blood pressure, total serum cholesterol, and total/HDL (high-density lipoprotein) cholesterol. After 24 months of dietary intervention, coronary angiography was performed to reassess coronary artery diameter. Coronary lesion growth was associated with total/HDL cholesterol. Typical observations on 18 patients who had no lesion growth are given in Table 4–1. The investigators wish to summarize the measurements for cholesterol levels and for mean change in vessel diameter over the two-year period of the study, and they want to illustrate graphically the relationship between these two variables.

Presenting Problem 2. Does long-term treatment with a beta-adrenergic drug in patients surviving acute myocardial infarction reduce mortality and reinfarction rates? Earlier studies that had not shown a significant reduction have been criticized because the numbers of patients enrolled were not large enough to exclude the possibility that a beneficial effect was being overlooked. Perhaps differences in the pharmacologic properties of various beta-adrenergic blocking agents or differences in the severity of illness in postinfarction patients studied could account for different study outcomes. To examine this question further, the Norwegian Multicenter Study Group (1981) performed a multicenter double-blind randomized trial comparing the effect of placebo to timolol maleate (beta-adrenergic drug). Outcome measures included mortality and reinfarction rates among 1885 patients who had recently survived an acute myocardial infarction. After a minimum follow-up of 12 months, the investigators found a 44.6% reduction of death and a 28.4% reduction of reinfarction in patients receiving timolol. Some

of the results are shown in Table 4–4. How can the investigators summarize these numbers and describe the relationship?

Presenting Problem 3. A high prevalence of hypertension has been reported in obese subjects. The authors of a study reported in *The Lancet* suggest that this high prevalence is likely to be overestimated because of error in the blood pressure recording introduced by using an inappropriate cuff size for a given arm circumference (Maxwell et al, 1982). They measured blood pressure in 1240 obese subjects by using the three cuff sizes available in clinical practice and found that regardless of arm circumference, recorded blood pressure rose with decreasing cuff size. When the cuff is too wide, measured blood pressure is often underestimated; when the cuff is too narrow, blood pressure is overestimated. They compared blood pressure measurements in a large number of moderately obese patients and found that when an appropriately large cuff was used, rather than the standard-sized cuff, 37% of patients originally thought to be hypertensive were actually normotensive. This study highlights the practitioner's responsibility to consider carefully sources of error that may be introduced in measurement.

4.1 PURPOSE OF THE CHAPTER

This chapter illustrates the common statistics used to summarize data or describe attributes of a set of data—often called **descriptive statistics.** Descriptive statistics are often used in medical journals; reviews indicate that their use ranges from 42% in psychiatric articles (Hokanson et al, 1986) to more than 60% in otolaryngology articles (Hokanson, Luttman, and Weiss, 1986).

We first consider **measures of central tendency**—statistics that describe the location of the center of a distribution of numerical and ordinal measurements. A **distribution** consists of values of a characteristic and the frequency of their occurrence. **Measures of dispersion,** statistics that describe the spread of numerical data, are examined next. **Rates** and **proportions,** the statistics used to summarize nominal data, are then discussed, and some common **vital statistics** rates are defined. Next, statistics that describe a relationship between two measurements on the same

Table 4-1. Typical total/HDL cholesterol and mean change in vessel diameter for 18 patients with no lesion growth.[1]

Patient	Total/HDL Cholesterol (mmol/L)	Change in Vessel Diameter (mm)	Patient	Total/HDL Cholesterol (mmol/L)	Change in Vessel Diameter (mm)
1	6.8	0.13	10	6.0	0.06
2	5.3	0	11	7.2	−0.19
3	6.1	−0.18	12	6.4	0.39
4	4.3	−0.15	13	6.0	0.30
5	5.0	0.11	14	5.5	0.18
6	7.1	0.43	15	5.8	0.11
7	5.5	0.41	16	8.8	0.94
8	3.8	−0.12	17	4.5	−0.07
9	4.6	0.06	18	5.9	−0.23

[1]Hypothetical data based on the study by Arntzenius AC et al: Diet, lipoproteins, and the progression of coronary atherosclerosis. *N Engl J Med* 1985; **312**:805.

group of subjects are illustrated. We conclude with a brief discussion of variation in observations, steps that can be taken to reduce variation, and ways to assess the **reliability** of measurements. The examples include the steps involved in calculating the statistics, because seeing the steps will help you understand them; most people analyzing large amounts of data will use a computer, however.

4.2 MEASURES OF THE MIDDLE (CENTRAL TENDENCY)

When an investigator collects many observations, such as the serum cholesterol levels of patients in Presenting Problem 1, a sensible technique is to use indexes or summary numbers to communicate information about the data. One of the most useful summary numbers is an index of the center of a distribution of observations that tells us what the middle or average value in the data is. The three measures of central tendency used frequently in medicine and epidemiology are the mean, the median, and the mode. All three are used for numerical data, and the median and the mode can be used for ordinal data as well.

4.2.1 Calculating Measures of Central Tendency

4.2.1.a The Mean: Although several means may be mathematically calculated, the arithmetic, or simple, mean is used most frequently in statistics and is the one generally referred to by the term *mean*. The **mean** is the arithmetic average of the observations. It is symbolized by $\overline{X}$ *(called X-bar)* and is calculated as follows:

1. Add the observations to obtain their sum.
2. Divide the sum by the number of observations.

The formula for the mean in written $\Sigma X/n$, where Σ (Greek letter sigma) means to add, X represents

the individual observations, and n is the total number of observations.

Table 4-1 gives data for 18 patients in the diet and lipoprotein study. These values are typical ones that might be observed in patients who experienced no lesion growth (in the study, values for patients were not given). The mean total/HDL cholesterol value (millimoles per liter) for these 18 patients is

$$\overline{X} = \frac{\Sigma X}{n} = \frac{6.8 + 5.3 + 6.1 + \cdots + 4.5 + 5.9}{18}$$

$$= \frac{104.6}{18} = 5.81$$

The mean is used when the numbers can be added—ie, when the characteristics are measured on a numerical scale; it should not be used with ordinal data because of the arbitrary nature of an ordinal scale. Note, also, that the mean is sensitive to extreme values in a set of observations. For example, the value of 8.8 for Patient 16 is high relative to the values for other patients in this group. If this value were not present, the mean would be 5.64 instead of 5.81.

If the original observations are not available, the mean can be estimated from a frequency table. A **weighted average** is formed by multiplying the midpoint of each interval by the number of observations that fall in the interval. Table 4-2 is a frequency table for the data in Table 4-1. The weighted-average estimate of the mean, using the number of patients and the midpoints in each interval, is

$$\frac{(1 \times 3.45) + (3 \times 4.45) + \cdots + (1 \times 8.45)}{n}$$

$$= \frac{105.1}{18} = 5.84$$

Table 4-2. Frequency table for typical total/HDL cholesterol in 18 patients with no lesion growth.

Total/HDL Cholesterol	No. of Patients	Midpoint of Interval
3.0 to 3.9	1	3.45
4.0 to 4.9	3	4.45
5.0 to 5.9	6	5.45
6.0 to 6.9	5	6.45
7.0 to 7.9	2	7.45
8.0 to 8.9	1	8.45
Total	**18**	

The value of $\overline{X}$ calculated from a frequency table is not always the same as the value obtained with raw numbers, although the two values are close in this example. Investigators who must calculate the mean for presentation in a paper have the original observations, of course, and should use the exact formula. The formula for use with a frequency table is helpful when you do not have access to the raw data but want an estimate of the mean.

4.2.1.b The Median: The **median** is the middle observation; ie, half the observations are smaller and half are larger. The median is sometimes symbolized by M or Md, but it has no conventional symbol. The procedure for calculating the median is as follows:

1. Arrange the observations from smallest to largest (or vice versa).

2. Count in to find the middle value. The median is the middle value for an *odd* number of observations; it is defined as the mean of the two middle values for an *even* number of observations.

For example, in rank-order (from lowest to highest), the total/HDL cholesterol observations in Table 4-1 are as follows:

3.8, 4.3, 4.5, 4.6, 5.0, 5.3, 5.5, 5.5, 5.8, 5.9, 6.0, 6.0, 6.1, 6.4, 6.8, 7.1, 7.2, 8.8

For 18 observations, the median is the mean of the ninth and tenth values (5.8 and 5.9), or 5.85. The median tells us that half the total/HDL cholesterol values in this group of patients are less than 5.85 and half are greater than 5.85. Recall from Chapter 3 that the median is easy to determine from a stem and leaf plot of the observations.

The median is less sensitive to extreme values than is the mean. For example, if the largest observation, 8.8, were excluded from the sample, the median would be the ninth value, or 5.8, which is not very different from 5.85. Another useful feature of the median is that it can be used with ordinal observations, because its calculation does not use actual values of the observations.

The median, like the mean, may also be estimated from a frequency table. This procedure is rarely needed, however; and because the formula is complicated, we will not illustrate that computation.

4.2.1.c The Mode: The **mode** of a distribution is the value of the observations that occurs most frequently. It is commonly used for a large number of observations when the researcher wants to designate the "most popular" value. No single observation occurs most frequently among the total/HDL data in Table 4-1. Both 5.5 and 6.0 occur twice, and all other observations occur only once; so technically, the total/HDL cholesterol level in this group of patients has two modes and is called *bimodal.*

For frequency tables or a small number of observations, the mode is estimated by the **modal class**, which is the interval having the largest number of observations. For example, the modal class for the data in Table 4-2 is 5.0-5.9.

4.2.1.d The Geometric Mean: Another measure of central tendency is the geometric mean; however, it is not used as often as the arithmetic mean or the median. The **geometric mean,** sometimes symbolized as GM or G, is defined as the nth root of the product of the observations. In symbolic form, for n observations $X_1, X_2, X_3, \ldots, X_n$, the geometric mean is

$$GM = \sqrt[n]{(X_1)(X_2)(X_3)\cdots(X_n)}$$

The geometric mean is generally used with data measured on a logarithmic scale. Note that if we take the logarithm of both sides of the above equation, we obtain

$$\log GM = \Sigma \frac{\log X}{n}$$

That is, the logarithm of the geometric mean is equal to the mean of the logarithms of the observations.

4.2.2 Using Measures of Central Tendency

Given a set of observations, an investigator may naturally ask which measure of central tendency is best to use with the data. Two factors are important in making this decision: the scale of measurement (ordinal or numerical) and the shape of the distribution of observations. Although distributions are discussed in more detail in Chapter 5, we can consider here the notion of whether a distribution is symmetric about the mean or is skewed to the left or the right of the mean. This information helps us decide which measure of central tendency is best.

If there are outlying observations in one direction only—either a few small values or a few large ones—the distribution is said to be a **skewed distribution.** If the outlying values are small, the distribution is skewed to the left, or negatively skewed; if the outlying values are large, the distribution is skewed to the right, or positively skewed. A **symmetric distribution** is one in which the distribution has the same shape on both sides of the mean. Fig 4–1 gives examples of negatively skewed, positively skewed, and symmetric distributions.

The following guidelines help an investigator to decide which measure of central tendency is best with a given set of data.

1. The mean is used for numerical data and for symmetric (not skewed) distributions.

2. The median is used for ordinal data or for numerical data whose distribution is skewed.

3. The mode is used primarily for bimodal distributions.

4. The geometric mean is used primarily for observations measured on a logarithmic scale.

The following points help a reader know the shape of a distribution without actually seeing it.

1. If the mean and the median are equal, the distribution of observations is symmetric, as in Figs 4–1C and 4–1D.

2. If the mean is larger than the median, the distribution is skewed to the right, as in Fig 4–1B.

3. If the mean is smaller than the median, the distribution is skewed to the left, as in Fig 4–1A.

4.3 MEASURES OF SPREAD (DISPERSION)

Suppose all you knew about the 18 patients with no lesion growth in Presenting Problem 1 was that

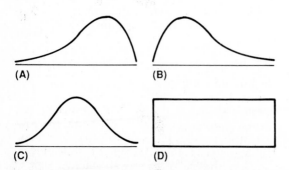

Figure 4–1. Shapes of common distributions of observations. (*A*) Negatively skewed. (*B*) Positively skewed. (*C*) and (*D*) Symmetric.

the mean total/HDL cholesterol was 5.81. The mean is useful information, but in order to have a better idea of the distribution of total/HDL cholesterol values in these patients, you also need to know something about the spread or the variation of the observations. Several statistics are frequently used in medicine to describe the dispersion of data: range, standard deviation, coefficient of variation, percentile rank, and interquartile range. All are described in the following subsections.

4.3.1 Calculating Measures of Spread (Dispersion)

4.3.1.a The Range: The **range** is the difference between the largest observation and the smallest observation. It is easy to determine once the data have been organized in a stem and leaf plot or arranged in rank order. For example, the smallest total/HDL cholesterol value among the 18 patients with no lesion growth is 3.8, and the highest is 8.8; thus, the range is 8.8 minus 3.8, or 5. Many authors give minimum and maximum values instead of the range, and in some ways these values are more useful information. The actual range cannot be determined from data presented in a frequency table, but a close estimate can be obtained by using the lower boundary of the lowest class interval and the upper boundary of the highest class interval.

4.3.1.b The Standard Deviation: The standard deviation is the most commonly used measure of dispersion with medical and health data; although its meaning and its computation are somewhat complex, it is definitely worth knowing about. Undoubtedly, most of you will use a computer (or calculator) to determine the standard deviation, but we present the steps involved in its calculation because they lead to a greater understanding of this statistic.

The standard deviation is a measure of the spread of data about their mean. Before we present the formula, however, let us briefly discuss an approach that would seem to be ideal for calculating an index of deviation—but is not. If what we desire is a measure of how observations are dispersed about the mean, an ''average,'' or ''mean,'' deviation seems like a good idea. For example, we can compute the deviation of each observation from the mean, add these deviations, and divide the sum by n to form an analogy to the mean itself. In symbols, the mean deviation is

$$\frac{\Sigma(X - \overline{X})}{n}$$

The problem with this index is that the sum of deviations of observations from their mean is always zero, and the value of the index will be zero in all cases (see Exercise 3). This problem can be solved in two ways: by summing the absolute values of the deviations or by squaring the deviations before they are added. The **absolute value** of a number is its positive value and is denoted by vertical bars on each side of the number. For example, the absolute value of 5, $|5|$, is 5, and the absolute value of -5, $|-5|$, is also 5. For the absolute value approach, the mean deviation is

$$\frac{\Sigma \, |X - \overline{X}|}{n}$$

Although there is nothing wrong with this approach conceptually, it does not have the properties needed to make statistical inferences, and it is not used for this reason. Therefore, the second approach of squaring deviations is used—with two slight modifications: $n - 1$ replaces n in the denominator (see the discussion that follows), and the square root is taken so that the original scale of measurement of the observations results. The standard deviation is symbolized as *s, sd,* or *SD,* and its formula is then

$$s = \sqrt{\frac{\Sigma \, (X - \overline{X})^2}{n - 1}}$$

The name of the statistic before the square root is taken is the **variance,** but the standard deviation is the statistic of primary interest.

The reason for using $n - 1$ instead of n in the formula for the standard deviation is complicated. We can simply tell you that $n - 1$ in the denominator produces a more accurate estimate of the true population standard deviation and has desirable mathematical properties for statistical inferences. A more precise explanation involves restrictions imposed on the data by the definition of standard deviation; ie, the quantities squared and then summed are deviations from the mean of the data. If there are n observations, there are also n deviations from the mean. However, because $\Sigma(X - \overline{X}) = 0$, once $n - 1$ of the deviations are specified, the last deviation is already determined as the value that will cause the sum of the deviations to be zero (see Exercise 4). Hence, the denominator uses the number of independent quantities ($n - 1$ in this case), which is called the **degrees of freedom,** a concept you will encounter in succeeding chapters on statistical tests.

To make matters more complex, the above formula for standard deviation is not the formula usually presented in introductory textbooks as the best one for calculating the value of the standard deviation. The above formula is called the **definitional formula,** and another formula, the **computational formula,** is generally used. However, under the assumption that most of you will not actually need to compute a standard deviation very often (statisticians do not compute them

manually either), the illustrations in this text use the more meaningful but computationally less efficient definitional formula for calculations. The computational formula is presented in Exercise 7, however.

Now let us try a calculation. The observations on change in vessel diameter for 18 patients with no lesion growth (Table 4–1) are repeated in Table 4–3. The table also shows two calculations needed to compute standard deviation. The steps for its calculation follow.

1. Let X be the change in vessel diameter for each patient; find the mean of these changes: $\overline{X} = 2.18/18 = 0.12$.

2. Subtract the mean change, 0.12, from each observation to form the deviations $X - \overline{X}$ (see column 3 in Table 4–3).

3. Square each deviation to form $(X - \overline{X})^2$. (Note that the calculations in column 4 are carried to four places to avoid round-off error when the square root is taken in step 6.)

4. Add the squared deviations to form the sum: $\Sigma(X - \overline{X})^2 = 1.4586$.

5. Divide the result in step 4 by $n - 1$; we get 0.0858. This value is the variance.

6. Take the square root of the answer in step 5; we get 0.2929, or 0.29. This value is the standard deviation.

The standard deviation of the change in vessel diameter is 0.29 mm. But note the large squared deviation, 0.6724, for Patient 16 in Table 4–3. This observation alone contributes almost half the variation in the data, because the sum of the squared deviations is 1.4586. The standard deviation of the remaining 17 patients (after elim-

Table 4-3. Calculations for standard deviation of typical change in vessel diameter (X) of 18 patients with no lesion growth.

Patient	X	$X - \overline{X}$	$(X - \overline{X})^2$
1	0.13	0.01	0.0001
2	0	−0.12	0.0144
3	−0.18	−0.30	0.0900
4	−0.15	−0.27	0.0729
5	0.11	−0.01	0.0001
6	0.43	0.31	0.0961
7	0.41	0.29	0.0841
8	−0.12	−0.24	0.0576
9	0.06	−0.06	0.0036
10	0.06	−0.06	0.0036
11	−0.19	−0.31	0.0961
12	0.39	0.27	0.0729
13	0.30	0.18	0.0324
14	0.18	0.06	0.0036
15	0.11	−0.01	0.0001
16	0.94	0.82	0.6724
17	−0.07	−0.19	0.0361
18	−0.23	−0.35	0.1225
Sums	$\Sigma X = 2.18$		$\Sigma(X - \overline{X})^2 = 1.4586$

inating Patient 16) is substantially smaller, 0.22, demonstrating the overwhelming effect that even one outlying observation can have on the value of the standard deviation.

The standard deviation, like the mean, requires numerical data. Also like the mean, it is very important in statistics. First, it is an essential part of many **statistical tests** (which are discussed in detail in later chapters). Second, the standard deviation is very useful in describing the spread of the observations about the mean value. Two rules of thumb for using the standard deviation follow.

1. Regardless of how the observations are distributed, at least 75% of the values *always* lie between these two numbers: $\overline{X} - 2s$ and $\overline{X} + 2s$. In the vessel diameter example, the mean change $\overline{X}$ is 0.12 mm and the standard deviation s is 0.29 mm; therefore, at least 75% of the 18 observations, or 14 of them, are between $0.12 - 2(0.29)$ and $0.12 + 2(0.29)$, or between -0.46 and $+0.70$ mm. In this example, 17 of the 18 observations (94%) actually are between these limits.

2. If the distribution of observations is a **bell-shaped distribution,** then even more can be said about the percentage of observations that lie between the mean and ± 2 standard deviations. For a bell-shaped distribution, the following rules hold:

~67% of the observations lie between $\overline{X} - 1s$ and $\overline{X} + 1s$.

~95% of the observations lie between $\overline{X} - 2s$ and $\overline{X} + 2s$.

~99.7% of the observations lie between $\overline{X} - 3s$ and $\overline{X} + 3s$.

For further discussion on the use of the mean and standard deviations with a bell-shaped distribution, see Chapter 5.

4.3.1.c The Coefficient of Variation: The coefficient of variation is a useful measure of relative spread in data and is used frequently in the biologic sciences. For example, suppose the authors of the study on diet and lipoproteins want to compare the variability in the ratio of total/HDL cholesterol with the variability in vessel diameter change for the 18 patients who had no lesion growth. The mean and the standard deviation of total/HDL cholesterol (in millimoles per liter) are 5.81 and 1.20, respectively; for the vessel diameter change (in millimeters), they are 0.12 and 0.29, respectively. A comparison of 1.20 and 0.29 makes no sense because cholesterol and vessel diameter are measured on different scales. The coefficient of variation *standardizes* the variation so that a sensible comparison can be made.

The coefficient of variation is defined as the standard deviation divided by the mean times 100%. It produces a measure of relative variation—variation that is relative to the size of the mean. The formula for the **coefficient of variation** is

$$CV = \left(\frac{s}{\overline{X}} \right)(100\%)$$

From this formula, the CV for total/HDL cholesterol is $(1.20/5.81)$ $(100\%) = 20.7\%$, and the CV for vessel diameter change is $(0.29/0.12)$ $(100\%) = 241.7\%$. Therefore, we can conclude that the relative variation in vessel diameter change is much greater than (more than 10 times as great as) that in cholesterol ratio.

Another frequent application of the coefficient of variation is in laboratory testing and quality control procedures. For example, screening for neural tube defects is accomplished by measuring maternal serum alpha-fetoprotein. DiMaio et al (1987) evaluated the use of this test in a prospective study of 34,000 women. The reproducibility of the test procedure was determined by repeating the assay ten times in each of four pools of serum. They calculated the mean and the standard deviation of the ten assays in each pool of serum and then used them to find the coefficient of variation for each pool. The coefficients of variation for the four pools were 7.4%, 5.8%, 2.7%, and 2.4%. These values indicate relatively good reproducibility of the assay because the variation, as measured by the standard deviation, is small relative to the mean. Therefore, readers of their article can be confident that the assay results were consistent.

4.3.1.d Percentiles: A **percentile** is a number that indicates the percentage of a distribution that is equal to or below that number. For example, consider the standard physical growth chart for girls from birth to 36 months of age given in Fig 4–2 (Hamill et al, 1979). For girls 21 months of age, the 95th percentile of weight is 13.4 kg, as noted by the arrow in the lower half of the chart. This percentile means that in the distribution of weights of 21-month-old girls, 95% weigh 13.4 kg or less and only 5% weigh more than 13.4 kg. The 50th percentile is, of course, the same value as the median; for 21-month-old girls, the median or 50th percentile weight is approximately 11.4 kg.

Percentiles are used most often to compare an individual value with a set of norms, as in the above example. They are used extensively to develop and interpret physical growth charts and measurements of ability and intelligence. They also determine normal ranges of laboratory values; the normal limits of many laboratory values are set by the $2\frac{1}{2}$ and $97\frac{1}{2}$ percentiles, so that the normal limits contain the central 95% of the distribution (see Exercise 9).

For exa
one yea
ing tim
0.1037
year.

4.4.2 V

Rates a
are the
which
Some o
defined

4.4.2.a

a stand
curring
in a m
died du
nomina
risk of
denomi
ber of
through
estimate
USA du

A cru
uals in a
annual
Table 4

Tabl

Age G
Under 1
1–4 year
5–14 yea
15–24 ye
25–34 ye
35–44 ye
45–54 ye
55–64 ye
65–74 ye
75–84 ye
85 years older
Total

¹Data fr
*Statistic
A, Table

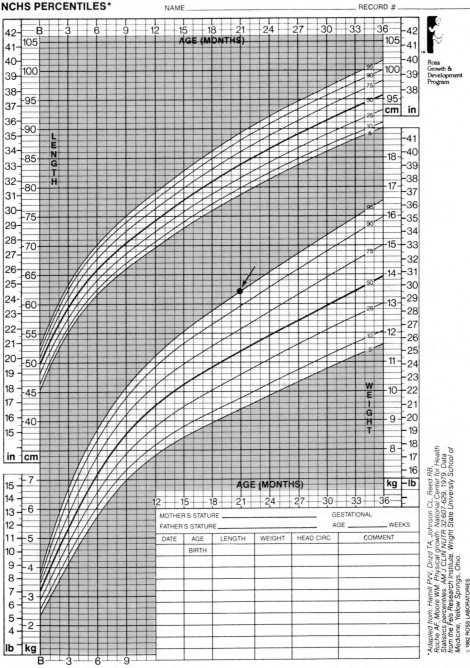

Figure 4–2. Standard physical growth chart. (Reproduced, with permission, from Ross Laboratories.)

4.3.1.e Interquartile Range: A measure of variation that makes use of percentiles is the **interquartile range,** defined as the difference between the 25th and 75th percentiles. The interquartile range contains the central 50% of observations.

For example, the interquartile range of weights of girls 12 months of age (Fig 4–2) is the difference between 10.2 kg (the 75th percentile) and 8.8 kg (the 25th percentile); ie, 50% of infant girls at 12 months weigh between 8.8 and 10.2 kg.

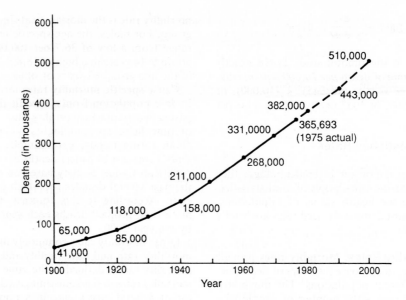

Figure 4–3. Forecast of cancer deaths developed by American Cancer Society. (Modified and reproduced, with permission, from the American Cancer Society.)

vances in care of diabetic patients have led to greater longevity. In contrast, for diseases with a short duration (eg, influenza) or with an early mortality (eg, pancreatic cancer), the incidence rate is generally larger than the prevalence.

4.4.3 Adjusting Rates

We can use crude rates to make comparisons between two different populations only if the populations are similar in all characteristics that might affect the rate. For example, if the populations are different or **confounded** on factors such as age, gender, or race, then either age-, gender-, or race-specific rates must be used, or the crude rates must be adjusted; otherwise, comparisons will not be valid.

A common adjustment to rates in medicine is an age adjustment. Often, two populations of interest have different age distributions; yet many characteristics studied in medicine are affected by age, becoming either more or less frequent as indi-

viduals grow older. If the two populations are to be compared, the rates must be adjusted to reflect what they would be had their age distributions been similar.

4.4.3.a Direct Method of Adjusting Rates: As an illustration, suppose a researcher compares the infant mortality rates from a developed country and a developing country and concludes that the mortality rate in the developing country is almost twice as high as the rate in the developed country. Is this conclusion reasonable, or are there possible confounding factors that affect infant mortality and might be distributed differently in the two countries? Certainly there is a relationship between birth weight and mortality, and in this example, a valid comparison of mortality rates requires that the distribution of birth weights be similar in the two countries. Hypothetical data are given in Table 4–6.

The crude infant mortality rate for the developed country is 12.0 per 1000 infants; for the de-

Table 4–6. Infant mortality rate adjustment: Direct method.

	Developed Country				Developing Country			
	Infants Born		Deaths		Infants Born		Deaths	
Birth Weight	*N* (in 1000s)	%	No.	Rate	*N* (in 1000s)	%	No.	Rate
< 1500 g	20	10	870	43.5	30	21	1860	62.0
1500–2499 g	30	15	480	16.0	45	32	900	20.0
≥ 2500 g	150	75	1050	7.0	65	47	585	9.0
Total	200		2400	12.0	140		3345	23.9

veloping country, it is 23.9 per 1000. The specific rates for the developing country are higher in all birth weight categories. However, the two distributions of birth weights are not the same: The percentage of low-birth-weight infants (< 2500 g) is more than twice as high in the developing country as in the developed country. Because birth weight of infants and infant mortality are related, we cannot determine how much of the difference in crude mortality rates between the countries is due to differences in weight-specific mortality and how much is due to the developing country's higher proportion of low-birth-weight babies. In this case, the mortality rates must be standardized or adjusted so that they are independent of the distribution of birth weight.

Determining an **adjusted rate** is a relatively simple process when information such as that in Table 4–6 is available. For each population, we must know the specific rates. Note that the crude rate in each country is actually a *weighted* average of the specific rates, with the *number of infants* born in each birth-weight category used as the *weights*. For example, the crude mortality rate in the developed country is 2400/200,000 = 0.012, or 12 per 1000, and is equal to

$$\frac{\Sigma(\text{Rate} \times N)}{\text{Total } N}$$

$$= \frac{(43.5 \times 20) + (16.0 \times 30) + (7.0 \times 150)}{20 + 30 + 150}$$

$$= \frac{2400}{200} \quad \text{or} \quad 12 \text{ per } 1000$$

Because the goal of adjusting rates is to have them reflect similar distributions, the numbers in each birth-weight category from one population are applied to the specific rates in the other population. Which population is chosen as the "standard" does not matter; in fact, a set of frequencies corresponding to a totally separate reference population may be used. The point is that the same set of numbers must be applied to both populations.

For example, if the numbers of infants born in each birth-weight category in the developed country are used as the "standard" and applied to the specific rates in the developing country, we obtain

$$\text{Adjusted rate} = \frac{\Sigma(\text{Rate} \times N \text{ in standard}}{\text{Total } N \text{ in standard}}$$

$$= \frac{(62.0 \times 20) + (20.0 \times 30) + (9.0 \times 150)}{20 + 30 + 150}$$

$$= \frac{3190}{200} \quad \text{or} \quad 15.95 \text{ per } 1000$$

Therefore, the crude mortality rate in the developing country would be 15.95 per 1000 (rather than 23.9 per 1000) if the proportion of infant birth weights were distributed as they are in the developed country.

To use this method of adjusting rates, you must know the specific rates for each category in the populations to be adjusted and the frequencies in the reference population for the factor being adjusted. The method is known as the **direct method of rate standardization.**

4.4.3.b Indirect Method of Adjusting Rates: The indirect method may be used when specific rates are not available in one or both of the populations of interest. To use the indirect method, you must know the frequencies of the adjusting factor, such as age or birth weight, for each population and a set of specific rates for whatever you are using as a standard population. The adjusted rate is called the **standardized mortality ratio** and is defined as the number of observed deaths divided by the number of expected deaths.

For example, let us assume we know specific rates for only a standard reference population, but the distribution of birth weights is available for both the developed and the developing countries (Table 4–7). The expected number of deaths is calculated in *each* population by using the specific rates from the standard population. For the developed country, the expected number of deaths is

$$(50.0 \times 20) + (20.0 \times 30) + (10.0 \times 150) = 3100$$

In the developing country, the expected number of deaths is

$$(50.0 \times 30) + (20.0 \times 45) + (10.0 \times 65) = 3050$$

The standard mortality ratio (the observed number of deaths divided by the expected number) for the developed country is 2400/3100 = 0.77. For the developing country, the standard mortality ratio is 3345/3050 = 1.1. If the standard mortality ratio is greater than 1, as it is in the developing country, the population of interest

Table 4–7. Infant mortality rate adjustment: Indirect method.

Birth Weight	No. of Infants Born (in 100s)		Specific Death Rates per 1000 in Standard Population
	Developed Country	Developing Country	
< 1500 g	20	30	50.0
1500–2499 g	30	45	20.0
≥ 2500 g	150	65	10.0
Number of Deaths	2400	3345	

has a mortality rate greater than that of the standard population; if it is less than 1, as it is in the developed country, the mortality rate is less than that of the standard population. Thus, the indirect method allows us to make a relative comparison; in contrast, the direct method allows us to make a direct comparison.

4.5 MEASURES TO DESCRIBE RELATIONSHIPS BETWEEN TWO CHARACTERISTICS

The measures discussed thus far are appropriate for summarizing observations on only one characteristic. Much of the research in medicine, however, concerns the relationship between two or more characteristics. The following discussion focuses on examining the relationship between two variables measured on the same scale: both numerical, both ordinal, or both nominal.

4.5.1 Describing the Relationship Between Two Numerical Characteristics

In Presenting Problem 1, the authors wanted to estimate the relationship between two numerical measures: cholesterol levels and mean change in vessel diameter in 39 patients with stable angina. The **correlation coefficient** (sometimes called the Pearson product moment correlation coefficient, named for the statistician who defined it) is one measure of the relationship between two numerical characteristics, symbolized by X and Y. The

formula for the correlation coefficient, symbolized by r, is

$$r = \frac{\Sigma(X - \overline{X})(Y - \overline{Y})}{\sqrt{\Sigma(X - \overline{X})^2}\ \sqrt{\Sigma(Y - \overline{Y})^2}}$$

Table 4–8 gives the information needed to calculate the correlation between typical values of total/HDL cholesterol and mean vessel diameter change for the 18 angina patients who had no lesion growth. As we did for standard deviation, we give the formula and computation for illustration purposes only. Using the data from Table 4–8, we obtain a correlation of

$$r = \frac{3.6668}{\sqrt{24.6378}\ \sqrt{1.4586}} = \frac{3.6668}{5.9947} = .61$$

4.5.2 Interpreting Correlation Coefficients

The relationship between two numerical variables, such as those in Presenting Problem 1, is often displayed graphically in a scatterplot, as we discussed in Chapter 3. Fig 4–4 is a scatterplot of the cholesterol and vessel diameter data for the 39 patients in the original study by Arntzenius and colleagues. The value of the correlation coefficient was .50 in the original 39 patients (we calculated .61 for our hypothetical observations).

What does the correlation of .50 between cholesterol and change in vessel diameter mean? (In this text, we will follow the generally accepted procedure of reporting correlations to two deci-

Table 4–8. Calculations for correlation coefficient between total/HDL cholesterol (X) and mean change in vessel diameter (Y) of 18 hypothetical patients.

Patient	X	Y	$X - \overline{X}$	$Y - \overline{Y}$	$(X - \overline{X})(Y - \overline{Y})$	$(X - \overline{X})^2$	$(Y - \overline{Y})^2$
1	6.8	0.13	0.99	0.01	0.0099	0.9801	0.0001
2	5.3	0	−0.51	−0.12	0.0612	0.2601	0.0144
3	6.1	−0.18	0.29	−0.30	−0.0870	0.0841	0.0900
4	4.3	−0.15	−1.51	−0.27	0.4077	2.2801	0.0729
5	5.0	0.11	−0.81	−0.01	0.0081	0.6561	0.0001
6	7.1	0.43	1.29	0.31	0.3999	1.6641	0.0961
7	5.5	0.41	−0.31	0.29	−0.0899	0.0961	0.0841
8	3.8	−0.12	−2.01	−0.24	0.4824	4.0401	0.0576
9	4.6	0.06	−1.21	−0.06	0.0726	1.4641	0.0036
10	6.0	0.06	0.19	−0.06	−0.0114	0.0361	0.0036
11	7.2	−0.19	1.39	−0.31	−0.4309	1.9321	0.0961
12	6.4	0.39	0.59	0.27	0.1593	0.3481	0.0729
13	6.0	0.30	0.19	0.18	0.0342	0.0361	0.0324
14	5.5	0.18	−0.31	0.06	−0.0186	0.0961	0.0036
15	5.8	0.11	−0.01	−0.01	0.0001	0.0001	0.0001
16	8.8	0.94	2.99	0.82	2.4518	8.9401	0.6724
17	4.5	−0.07	−1.31	−0.19	0.2489	1.7161	0.0361
18	5.9	−0.23	0.09	−0.35	−0.0315	0.0081	0.1225
Sum	104.6	2.18			3.6668	24.6378	1.4586

Association between change in coronary lesion and ratio of total to high-density lipoprotein (total/HDL) cholesterol, according to individual patient (computer assessment). Values for total/HDL cholesterol are averages of base-line and mean two-year values for 39 patients.

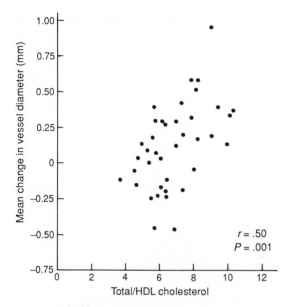

Figure 4–4. Scatterplot of change in coronary lesion and ratio of total to high-density lipoprotein (total /HDL) cholesterol. (Redrawn and reproduced, with permission, from Arntzenius AC et al: Diets, lipoproteins, and the progression of coronary atherosclerosis. 1985; *N Engl J Med* **312**:805)

mal places.) What is the relationship between these two variables? Chapter 10 discusses methods used by investigators to tell whether there is a statistically significant relationship; for now, we will discuss a few characteristics of the correlation coefficient that will help us to interpret its numerical value and describe a relationship.

The correlation coefficient ranges from -1 to $+1$, with -1 describing a perfect negative linear, or straight-line, relationship and $+1$ describing a perfect positive linear, or straight-line, relationship. A correlation of 0 means there is no linear relationship between the two variables. There is a correspondence between the size of the correlation coefficient and a scatterplot of the observations, as will be discussed more fully in Chapter 10. When the correlation is near zero, the "shape" of the pattern of observations is more or less circular. As the value of the correlation gets closer to $+1$ or -1, the shape becomes more elliptical, until, at $+1$ and -1, the observations fall directly on a straight line. With a correlation of .50, we expect a scatterplot of the data to be somewhat oval-shaped, as it is in Fig 4–4.

Sometimes the correlation is squared (r^2) to form a very important statistic called the **coefficient of determination.** For the cholesterol and vessel diameter data, the coefficient of determination is $(.50)^2$, or .25, and means that 25% of the variability in one of the measures, such as vessel diameter change, may be accounted for (or predicted) by knowing the value of the other measure, total/HDL cholesterol. Stated another way, if we knew the value of the patients' total/HDL cholesterol and took that into consideration when examining the change in vessel diameter, the variance (standard deviation squared, from Section 4.3.1.b) of the change in vessel diameter would be reduced by 25%.

There are several characteristics of the correlation coefficient that deserve mention. The value of the correlation coefficient is independent of the particular units used to measure the variables. For example, suppose two medical students measure the heights and weights of a group of preschool children in order to determine the correlation between height and weight. They measure the children's height in centimeters and record their weight in kilograms, and they calculate a correlation coefficient equal to .70. What would the correlation be if they had used inches and pounds instead? It would, of course, still be .70, because the denominator in the formula for the correlation coefficient adjusts for the scale of the units.

The value of the correlation coefficient is markedly influenced by an outlying value, just as is the standard deviation. One patient represented in Fig 4–4 has both a total/HDL cholesterol value and a vessel diameter change greater than those of the other 38 patients, approximately 8.6 and 0.9, respectively. If this one observation is removed and the correlation recomputed for the remaining 38 patients, the correlation is .47 instead of .50. In the hypothetical sample of 18 patients, removing the one patient with high values has an even more striking effect, reducing the correlation from .61, as calculated in the previous section, to .32. We purposely made the hypothetical data more skewed to make the point that even one outlying observation can have a great effect on the correlation. Therefore, the correlation does not provide a good description of the relationship between two variables when the distribution of either variable is skewed or contains outlying values. In this situation, a **transformation** of the data—which changes the scale of measurement and moderates the effect of outlyers (such as a rank or logarithmic transformation, discussed in Chapter 7)— should be done before the correlation is computed.

Students first learning about the correlation coefficient often ask, "How large should a correlation be?" The answer depends on the application.

For example, when physical characteristics are measured and good measuring devices are available, as in many physical sciences, fairly high correlations are possible. However, measurement in the biologic sciences often involves characteristics that are less well defined and measuring devices that are imprecise; in such cases, lower correlations may occur. Colton (1974) gives the following crude rule of thumb for interpreting the size of correlations:

> Correlations from 0 to .25 (or − .25) indicate little or no relationship; those from .25 to .50 (or −.25 to −.50) indicate a fair degree of relationship; those from .50 to .75 (or − .50 to − .75) a moderate to good relationship; and those greater than .75 (or −.75) a very good to excellent relationship.

Colton also cautions against correlations of .95 or higher in the biologic sciences because of the inherent variability in most biologic characteristics. When you encounter a correlation with this value, you should ask whether it is an error or an artifact. An artifact occurs, for instance, when the number of pounds lost by patients in the first week of a diet program is correlated with the number of pounds they lost during the entire two-month program. In this case, the number of pounds lost during the first week is included in the number of pounds lost during the two-month period and results in a spuriously high correlation. The correct comparison is between the number of pounds lost in week 1 and the number of pounds lost in weeks 2–8.

Two final reminders are worth attention. First, the correlation coefficient measures only a straight-line relationship; two factors may, in fact, have a strong curvilinear relationship, even though the correlation is quite small. Therefore, when you analyze relationships between two characteristics, always plot the data (or have a computer do so) and calculate a correlation coefficient. A graph will also help you detect outlyers and skewed distributions. The second reminder is the adage that "correlation does not imply causation." The statement that one characteristic causes another must be justified on the basis of experimental observations or logical argument, not on the basis of the size of a correlation coefficient.

4.5.3 Describing the Relationship Between Two Ordinal Characteristics

The **Spearman rank correlation,** sometimes called Spearman's rho (also named for the statistician who defined it), is frequently used to describe the relationship between two ordinal (or one ordinal and one numerical) characteristics. It is also the appropriate statistic to use with numerical variables when their distributions are skewed and there are outlying observations. The calculation of the Spearman rank correlation, symbolized as r_s, involves rank-ordering the values on each of the characteristics from lowest to highest; then the ranks are treated as though they were the actual values themselves. Although the formula is simple when there are no ties in the values, the computation is quite tedious and is available on many computer programs; thus, we postpone its illustration until Chapter 10, where it is discussed in greater detail.

The Spearman rank correlation may range from −1 to +1, like the Pearson correlation coefficient; but +1 or −1 indicates perfect agreement between the *ranks* of the values rather than between the values themselves. Otherwise, its interpretation is similar to that for the Pearson *r*. An example of its application is a study in which a pediatrician wishes to investigate the relationship between Apgar score and birth weight in a group of premature infants. Because Apgar scores are measured on an ordinal score, the Spearman rank correlation is appropriate for measuring the relationship.

4.5.4 Describing the Relationship Between Two Nominal Characteristics

In many studies involving two characteristics, both measured on a nominal scale, the primary interest is often in determining whether there is a significant relationship between the two, rather than in describing the magnitude of the relationship. These studies are the subject of Chapters 9 and 10. There is a special application in medicine, however, when describing the strength of the relationship between two nominal measures is of interest: studies that measure the risk of a given outcome relative to whether a risk or predisposing factor is present. Two ratios used to estimate risk are the relative risk (sometimes also called the risk ratio) and the odds ratio. For example, in Presenting Problem 2, the investigators may wish to examine the risk for patients taking timolol relative to the risk for patients taking placebo to see whether timolol reduces the risk.

4.5.4.a The Relative Risk (Risk Ratio): The **relative risk** of a disease, symbolized by *RR,* is the ratio of incidence in exposed persons to incidence in nonexposed persons. Table 4–9 gives survival and death data for patients taking timolol and patients taking a placebo. In this example, taking the drug timolol is playing the role of the risk factor. The relative risk can be calculated only from a cohort study in which a group of patients with the risk factor and a group without the risk factor are first identified and then followed

Table 4-9. Data for relative risk for death in patients taking timolol.

	Died	Survived	
Timolol	98	847	945
Placebo	152	787	939

through time to determine which patients develop the disease. In our example, patients given timolol and those given placebo were observed through time to determine mortality. The incidence of death in patients who received timolol is 98/945, or 0.104; the incidence of death in patients who received placebo is 152/939, or 0.162. Therefore, the relative risk of death with timolol, compared with death with placebo, is

$$RR = \frac{98/945}{152/939} = \frac{0.104}{0.162} = 0.641$$

Because fewer patients died on timolol than on placebo, the relative risk is less than one.

4.5.4.b The Odds Ratio: To discuss the odds ratio, we consider another example. In an assessment of the initial electrocardiogram (ECG) as a predictor of acute myocardial infarction (MI), investigators identified 200 men who had had an MI and 200 who had not (Table 4–10). Of the 200 men with an MI, 170 had had a positive ECG; of the 200 who did not have an MI, 90 had also had a positive ECG.

The relative risk cannot be calculated in this study because it is not a cohort study of patients who are followed over a period of time. Rather, it is a case-control study: A set of cases and a set of controls are first identified, and then their histories are examined to determine which patients had the risk factor. In a case-control study, the numbers of cases and controls are predetermined by the investigators who select subjects, and incidence rates cannot be estimated. However, an estimate of the relative risk, called the odds ratio (*OR*) can be determined. The **odds ratio** is the odds that a patient is exposed to the risk factor divided by the odds that a control is exposed.

In this study, the odds that a patient (with an MI) is exposed (eg, has a positive ECG) is

$$\frac{170/200}{30/200} = \frac{170}{30} = 5.667$$

Table 4-10. Data for odds ratio for myocardial infarction with positive electrocardiogram.

	MI	No MI
Positive ECG	170	90
Negative ECG	30	110
	200	200

The odds that a control (without an MI) is exposed (has a positive ECG) is

$$\frac{90/200}{110/200} = \frac{90}{110} = 0.818$$

Therefore, the odds ratio *OR* is

$$OR = \frac{5.667}{0.818} = 6.93$$

Thus, the odds that a patient with an MI has had a positive ECG is 6.93, or almost 7 times greater than for a patient without an MI.

The odds ratio is also called the cross-product ratio because it can be defined as the ratio of the product of the diagonals in a 2 × 2 table:

$$\frac{170 \times 110}{90 \times 30} = 6.93$$

Table 4–11 demonstrates the arrangement of data and the formulas needed for calculation of the relative risk and odds ratio. Readers interested in more detail are referred to the very readable elementary text on epidemiology by Fletcher, Fletcher, and Wagner (1988).

4.6 VARIATION IN DATA

In many clinics, a nurse collects certain information about a patient (eg, height, weight, date of birth, blood pressure, pulse) and records it on the medical record before the patient is seen by a physician. Suppose a patient's blood pressure is recorded as 140/88 on the chart; the physician, taking the patient's blood pressure again as part of the physical examination, observes of reading of 148/96. Which blood pressure reading is correct? What factors might be responsible for the differences in the observation? We use blood pressure and other clinical examples to examine sources of

Table 4-11. Table arrangement and formulas for relative risk (*RR*) and odds ratio (*OR*).

	Disease	No Disease	
Risk factor present	A	B	A + B
Risk factor absent	C	D	C + D
	A + C	B + D	

$$RR = \frac{A/(A + B)}{C/(C + D)}$$

$$OR = \frac{[A/(A + C)]/[C/(A + C)]}{[B/(B + D)]/[D/(B + D)]} = \frac{A/C}{B/D} = \frac{AD}{BC}$$

variation in data and ways to measure the reliability of observations. Two articles in *The Canadian Medical Association Journal* (Sackett, 1980) discuss sources of clinical disagreement and ways disagreement can be minimized; you may wish to consult them for a more detailed discussion.

4.6.1 Factors That Can Cause Variation in Clinical Observation

The causes for **variation**—ie, variability in measurements on the same subject—in clinical observations and measurements can be classified in three categories: (1) variation owing to the person being measured, (2) variation owing to the examiner, and (3) variation owing to the instrument or method used. There may be substantial variability in the measurement of biologic characteristics. For example, a person's blood pressure is not the same from one time to another, and thus, its measurements will vary. A patient's description of symptoms to two different physicians may vary because the patient may forget something. Medications and illness can also affect the way a patient behaves and what information he or she remembers to tell a nurse or physician.

Even when there is no change in the subject, different observers may report different measurements. When examination of a characteristic requires visual acuity, such as the reading on a sphygmomanometer or the features on an x-ray, differences may result from the varying visual abilities of observers. Such differences can also play a role when hearing (detecting heart sounds) or feeling (palpating internal organs) is required. Some individuals are simply more skilled than others in history taking or performing certain examinations. In addition, observers may tend to observe and record what they expect based on other information about the patient. For this reason, physicians and other observers in research studies are often **blinded** (not told about treatments) to minimize the effect of preconceptions.

The instrument used in the examination is another important source of variation. For instance, mercury column sphygmomanometers have less inherent variation than do aneroid models. In addition, the environment in which the examination takes place, including lighting and noise level, presence of other individuals, and room temperature can produce differences that are not true. Finally, variation can result if the correct observation was made but was incorrectly recorded or coded.

Several steps may be taken to reduce variability. Taking a history when the patient is calm and not heavily medicated and checking with family members when the patient is incapacitated are both useful in minimizing errors that result from a patient's illness or the effects of medication. Collecting information and making observations in a proper environment is also a good strategy. Recognizing one's own strengths and weaknesses helps to evaluate the need for other opinions. Blind assessment, especially of subjective characteristics, will guard against errors resulting from preconceptions. Repeating questionable aspects of the examination or asking a colleague to perform a key aspect (blindly, of course) reduces the possibility of error. Ensuring that instruments are properly calibrated and correctly used will eliminate many errors and thus reduce variation. For example, Presenting Problem 3 deals with an attempt to reduce variability in blood pressure readings of obese patients. The investigators found that using the typical blood pressure cuff with these patients caused substantial errors in blood pressure readings; an appropriate-sized cuff reduced errors considerably.

4.6.2 Ways to Measure Reliability of Measurements

A common strategy to ensure **reliability**—ie, reproducibility—of measurements, especially for research purposes, is to replicate the measurements and evaluate the degree of agreement. When one person measures the same item twice and the measurements are compared, an index of intraobserver variability called **intrarater reliability** is obtained. When two or more persons measure the same item and their measurements are compared, an index of interobserver variability called **interrater reliability** is obtained. The index of reliability used depends on the type of measurement made. Let us look at some examples.

Frequently in medicine, a practitioner must interpret a procedure as indicating or not indicating the presence of a disease or abnormality; ie, the observation is a yes or no outcome, which is a nominal measure. For example, suppose two physicians evaluated the results from 100 liver spleen scans and categorized them as either "normal" or "abnormal." The evaluations were made independently; ie, one physician did not know the results of the other physician's determination. The hypothetical data are displayed in a 2 × 2 contingency table (Table 4–12). Physician 1 diagnosed

Table 4-12. Observed agreement on reading liver spleen scans.

Physician 1	Physician 2		
	Abnormal	Normal	
Abnormal	20	15	35
Normal	10	55	65
	30	70	

30 of the scans as abnormal; Physician 2 diagnosed 35 as abnormal; they agreed on 20 of these diagnoses.

How can we describe the degree of agreement between the two physicians? Twenty percent is an underestimate because they also agreed that 55 of the scans were normal. The total observed agreement (20% + 55% = 75%) is an overestimate because it ignores that, with only two categories (normal and abnormal), some agreement will result by chance. The statistic most often used to measure agreement between two observers on a dichotomous variable is kappa (κ), defined as the agreement beyond chance divided by the amount of agreement possible beyond chance:

$$\kappa = \frac{O - C}{1 - C}$$

where O is the observed agreement and C is the chance agreement.

An example should clarify this concept. Chance agreement is found as follows:

1. Calculate how many scans the physicians may agree are positive by chance, determined by multiplying the numbers each found positive and dividing by 100, because there are 100 total scans $(30 \times 35)/100 = 10.5$.

2. Calculate how many scans they may agree are negative by chance, determined by multiplying the numbers each found negative and dividing by the total number: $(70 \times 65)/100 = 45.5$.

3. Add the two numbers found in steps 1 and 2 and divide by 100 to obtain a proportion for chance agreement: $(10.5 + 45.5)/100 = 0.56$.

The observed agreement is 20% + 55% = 75%, or 0.75; therefore, the agreement beyond chance is $0.75 - 0.56 = 0.19$, the numerator of κ.

The potential agreement beyond chance is simply 100% minus the chance agreement of 56%, or, using proportions, $1 - 0.56 = 0.44$. Therefore, κ in this example is $0.19/0.44 = 0.43$. Sackett, Haynes, and Tugwell (1985) point out that the level of agreement varies considerably depending on the clinical task, ranging from 57% agreement with a κ of 0.3 for two cardiologists examining the same elctrocardiograms from different patients, to 97% agreement with a κ of 0.67 for two radiologists examining the same set of mammograms.

As another example, suppose a clinician wants to determine the reliability of measurements of the tracheal diameter in patients having chest x-rays. One approach is to measure a group of x-rays and record the measurements. Then some days or weeks later, measure the same x-rays again without consulting the earlier figures. Tracheal diameter is measured on a numerical scale, and the statistic used to examine the relationship between

two numerical characteristics is the correlation coefficient discussed in Sections 4.5.1 and 4.5.2. Therefore, the correlation between the two sets of tracheal diameter measurements will provide a measure of how reliable the clinician's measurements are.

4.7 COMPUTER PROGRAMS THAT SUMMARIZE DATA

All statistical computer programs have routines to summarize observations; this chapter discusses the output from SPSS, SAS, MINITAB, and STATISTIX to illustrate the kinds of analyses available. Observations from Presenting Problem 1 on total/HDL cholesterol in 18 men placed on a diet for two years are analyzed by the SPSS FREQUENCIES program in Fig 4–5. Frequency tables are discussed in Chapter 3; the printout, however, also gives summary statistics. SPSS permits the user to request a variety of summary statistics; we have chosen those discussed in this chapter. Note that the mean for total/HDL cholesterol agrees with the value we calculated, 5.81 mmol/L, as do the median, maximum, and minimum values. In the information on mean change in vessel diameter, the value 0.293 for standard deviation, abbreviated Std Dev in the printout, agrees with the value 0.2929 calculated in Section 4.3.1.

SAS produces a variety of summary statistics as part of its UNIVARIATE program. Please refer to the printout of ages of patients with colorectal cancer illustrated in Chapter 3 (Fig 3–18). The columns under the heading "Moments" contain some familiar numbers: N, the number of patients in the study; MEAN, the mean age of 70.06 years in the 118 patients; the standard deviation, STD DEV; the coefficient of variation, CV; and in the second column, VARIANCE. In the columns under the heading "Quantiles," SAS produces the values of age that occur at various percentiles. These columns also show that the median age (MED) is 71.5 years; that 95% of the patients are 88.05 years of age or younger; that the range in age is 65 years; and that the interquartile range (Q3 − Q1) is 17.5 years, indicating that 50% of the patients are between 61.75 (Q1) and 79.25 (Q3) years of age. In the far right columns under the heading "Extremes," the five lowest and five highest observations are given.

The relationship between two numerical observations is measured with the correlation coefficient, illustrated here for the data on total/HDL cholesterol and age for the entire set of 39 patients (Presenting Problem 1; Arntzenius et al, 1985). Two MINITAB programs, Plot and Correlation, yield the printout reproduced in Fig 4–6. The plot is similar to the one reproduced from the article

HDL   Total/HDL Cholesterol

Value Label	Value	Frequency	Percent	Valid Percent	Cum Percent
	3.8	1	5.6	5.6	5.6
	4.3	1	5.6	5.6	11.1
	4.5	1	5.6	5.6	16.7
	4.6	1	5.6	5.6	22.2
	5.0	1	5.6	5.6	27.8
	5.3	1	5.6	5.6	33.3
	5.5	2	11.1	11.1	44.4
	5.8	1	5.6	5.6	50.0
	5.9	1	5.6	5.6	55.6
	6.0	2	11.1	11.1	66.7
	6.1	1	5.6	5.6	72.2
	6.4	1	5.6	5.6	77.8
	6.8	1	5.6	5.6	83.3
	7.1	1	5.6	5.6	88.9
	7.2	1	5.6	5.6	94.4
	8.8	1	5.6	5.6	100.0
	TOTAL	18	100.0	100.0	

HDL   Total/HDL Cholesterol

Mean	5.811	Median	5.850	Std Dev	1.204
Variance	1.449	Range	5.000	Minimum	3.800
Maximum	8.800				

Figure 4-5. SPSS program that produces summary statistics for numerical data, using observations on cholesterol and vessel diameter from Presenting Problem 1. (SPSS is a registered trademark of SPSS, Inc. Used with permission.)

in Fig 4–4. The correlation coefficient of 0.505 is within round-off error of the one reported by the authors.

Finally, the 2 × 2 table in Fig 4–7 is produced by STATISTIX. STATISTIX offers fewer labeling options to the user, but this disadvantage is more than compensated for by its ease of use. All other computer packages we illustrate require the user to do some typing to indicate what procedures are to be used. STATISTIX, however, is menu-driven, meaning that the choices of activities are displayed in a menu (a list on the screen), and the user simply presses the carriage return to select the desired procedure. Therefore, STATISTIX is easy to learn and use.

The 2 × 2 table in Fig 4–7 contains the observations from Table 4–10 on myocardial infarction and positive electrocardiograms. As in Table 4–10, the rows contain data for ECG, with row 1 being positive and row 2 negative; column 1 indicates MI and column 2 no MI. Several statistics are presented; we are interested in the CROSS PRODUCT RATIO, another name for the odds ratio. The value printed, 6.926, agrees with the value we calculated.

4.8 SUMMARY

This chapter presented a variety of measures commonly used in medicine to summarize a collection of observations. Some of these measures also form the basis of statistical tests illustrated in subsequent chapters.

We used typical values of total/HDL cholesterol and change in vessel diameter based on Presenting Problem 1 to illustrate measures to use with numerical observations. We recommended that the mean—a measure of the middle of a distribution—be used with observations that have a symmetric distribution. The median, also with a measure of the middle, is used with ordinal observations or numerical observations that have a skewed distribution. When the mean is appropriate for describing the middle, the standard deviation is appropriate for describing the spread, or variation, of the observations. The value of the standard deviation is affected by outlying or skewed values, so percentiles or the interquartile range should be used with observations for which the median is appropriate. The range gives information on the extreme values, but it does not pro-

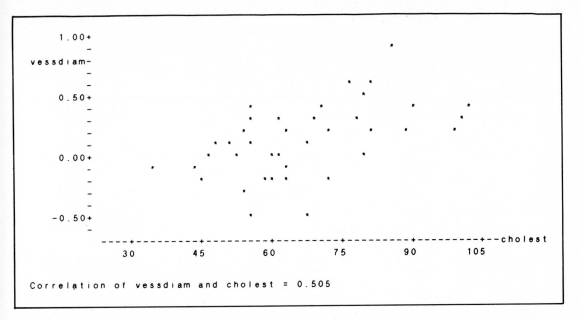

Figure 4-6. MINITAB program that produces scatterplots and correlation for numerical data, using original observations on cholesterol and vessel diameter from Presenting Problem 1. (Adapted, with permission, from Arntzenius AC et al: Diet, lipoproteins, and the progression of coronary atherosclerosis. *N Engl J Med* 1985;**312**:805. MINITAB is a registered trademark of Minitab, Inc. Used with permission.)

vide any insight into how the observations themselves are distributed. The geometric mean is sometimes used with extremely skewed data, such as observations measured on a logarithmic scale. An easy way to determine whether the distribution

of observations is symmetric or skewed is by inspecting a histogram or box plot, as illustrated in Chapter 3.

The results of the study on timolol versus placebo in treating patients who had survived a myo-

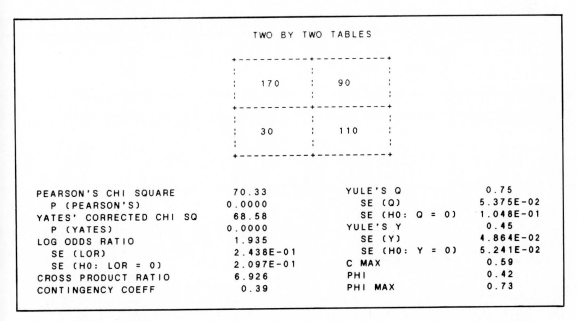

Figure 4-7. STATISTIX program that produces 2 × 2 tables and odds ratio for nominal data, using observations on myocardial infarction and electrocardiograms from Table 4–10. (STATISTIX is a registered trademark of NH Analytical Software. Used with permission.)

cardial infarction (Presenting Problem 2) were used to illustrate that proportions and percentages can be used interchangeably to describe the relationship of a part to the whole; ratios relate the two parts themselves. When a proportion is calculated over a period of time, the result is called a rate. Some of the rates commonly used in medicine were defined and illustrated. For comparison of rates from two different populations, the populations must be similar with respect to characteristics that might affect the rate; adjusted rates are necessary when these characteristics differ between the populations. In medicine, rates are frequently adjusted for disparities in age.

Practitioners often wish to describe the relationship between two measurements made on the same group of individuals. We illustrated the correlation coefficient for the cholesterol and vessel diameter observations from Presenting Problem 1 and gave guidelines for interpreting its value. Spearman's rank correlation is used with skewed or ordinal observations—ie, in the same situations for which medians are appropriate. When the characteristics are measured on a nominal scale and proportions are calculated to describe them, the relative risk or the odds ratio may be used to measure the relationship between two characteristics. The relative risk is appropriate for observations obtained in a cohort study; the odds ratio is appropriate for case-control studies. All these measures of association are discussed again in later chapters.

Finally, we discussed methods for measuring variation agreement. A researcher may wish to demonstrate agreement between two procedures or between two individuals. Various sources of errors causing variation were discussed and illustrated. The statistic κ measures agreement for nominal observations; the correlation coefficients discussed in Section 4.5 measure agreement for numerical or ordinal observations.

EXERCISES

1. In Section 4.2, the mean ratio of total to high-density lipoprotein cholesterol was computed as 5.81 from actual observations and as 5.84 from the frequency distribution in Table 4-12. Generally, the estimate from a frequency distribution is not exact. Explain why.

2. What is the most likely shape of the distribution of observations in the following studies?
   a. The age of subjects in a study of patients with Crohn's disease.
   b. The number of babies delivered by all physicians who deliver babies in a large city during the past year.
   c. The number of patients transferred to a ter-

tiary care hospital by other hospitals in the region.

3. Calculate the sum of $X - \overline{X}$ for the vessel diameter data in Table 4-3, and demonstrate that the sum is zero.

4. To illustrate that only $n - 1$ deviations from the mean $(X - \overline{X})$ are free to vary, consider the ages of 5 patients seen in the clinic on a given afternoon. The mean age has been determined and subtracted from each patient's age in the following table.

Patient	Age in Deviations From Mean Age $(X - \overline{X})$
1	10
2	−7
3	−5
4	3
5	

   a. What is $X - \overline{X}$ for Patient 5?
   b. If the mean age of the patients is 40, what are the ages of the 5 patients?

5. What measures of central tendency and dispersion are the most appropriate to use with the following sets of data?
   a. Salaries of 125 physicians in a clinic.
   b. The test scores of all medical students taking Part I of the National Board Examination in a given year.
   c. Serum sodium levels of healthy patients.
   d. Number of tender joints in 30 joints evaluated on a standard examination for disease activity in rheumatoid arthritis patients.
   e. Presence of diarrhea in a group of infants.
   f. The disease stages for a group of patients with Reye's syndrome (six stages, ranging from 0 = alert wakefulness to 5 = unarousable, flaccid paralysis, areflexia, pupils unresponsive).
   g. The age at onset of breast cancer in females.
   h. The number of pills left in subjects' medicine bottles when investigators in a study counted the pills to evaluate compliance in taking medication.

6. Oren, Kelly, and Shannon (1986) studied a group of 76 infants who had had an unexplained episode of sleep apnea (Table 4-13). Twenty-five infants had repeat sleep apnea episodes, and 7 of these infants died. Among the 51 infants who did not have repeat sleep apnea

Table 4–13. Data on infants with sleep apnea.[1]

	No. of Infants	Mortality Rate (%)
All infants	76	13.2
Infants with repeat sleep apnea episodes Requiring vigorous stimulation	12	25
Requiring resuscitation	13	30
Siblings of SIDS victims	8	25
Infants with seizures and repeat episodes requiring vigorous stimulation and resuscitation	7	57.1

[1]Reproduced, with permission, from Table 2 in Oren J, Kelly D, Shannon DC: Identification of a high-risk group for sudden infant death syndrome among infants who were resuscitated for sleep apnea. *Pediatrics* 1986; **77**:495.

episodes, there were 3 deaths. Determine the risk of death in infants who had seizures versus those who did not.

7. The computational formula for the standard deviation is

$$s = \sqrt{\frac{\Sigma X^2 - (\Sigma X)^2/n}{n - 1}}$$

Illustrate that the value of the standard deviation calculated from this formula is equivalent to that found with the definitional formula using the vessel diameter changes in Table 4–1. From Section 4.3.1.b, the value of the standard deviation of vessel diameters using the definitional formula is 0.2929. (Use Table 4–3 to save some calculations.)

8. Use the observations on percentage supersaturation of bile in men and women from Table 3–7 (based on Presenting Problem 3 in Chapter 3) to determine the 2.5 and 97.5 percentiles, ie, the "normal limits" of supersaturation of bile.

9. Refer to Fig 4–2 to answer the following questions.
   a. What is the mean weight of girls 24 months old?
   b. What is the 90th percentile in length for 12-month-old girls?
   c. What is the 5th percentile in weight for 12-month-old girls?
   d. What conclusions can you draw about a 15-month-old girl who is 29 in long and weighs 26 lb?

10. Find the coefficient of variation of mean change in vessel diameter.

5

Probability, Sampling & Probability Distributions

PRESENTING PROBLEMS

Presenting Problem 1. The acquired immunodeficiency syndrome (AIDS) is caused by the human immunodeficiency virus (HIV). Recent studies suggest that up to 20% of the individuals with antibodies to HIV (HIV-seropositive) in their serum will develop AIDS within 5 years, and researchers are concerned that this percentage underestimates the risk. The long-term effects of the infection on the immune system and general health of asymptomatic persons with HIV infection are not well known.

A recent article in the *Annals of Internal Medicine* reported a 5-year longitudinal study of a cohort of 250 initially healthy homosexual men to determine the natural history of infection with HIV (Melbye et al, 1986). These men were divided into four groups (seronegative, seropositive for less than 19 months, seropositive for 19–29 months, and seropositive for longer than 29 months) and were studied with respect to T lymphocyte subset ratios and clinical symptoms. There was a striking time-related decline in the ratio of T lymphocyte helper cells to suppressor cells (T4/T8). An inverted T4/T8 ratio occurred in 92% of the individuals who were seropositive longer than 29 months but occurred in only 13% of those who remained seronegative. Lymphadenopathy was highly associated with seropositivity among both short-term and long-term seropositive individuals. Diarrhea, oral thrush, and herpes zoster were associated with seropositivity for longer than 29 months but not with seropositivity for less than 19 months. Results for a subset of men followed for 34 months have been collapsed into three groups (seronegative, seropositive for 29 months or less, and seropositive for more than 29 months) in Table 5–1 and are used to illustrate basic concepts of probability and to demonstrate the relationship between time and T4/T8 ratio.

Presenting Problem 2. A local blood bank was asked to provide information on the distribution of blood types among males and females. This information is useful in illustrating some basic prin-

ciples in probability theory. The results are given in Table 5–2.

Presenting Problem 3. The term *aplastic anemia* refers to a severe pancytopenia (anemia, neutropenia, thrombocytopenia) resulting from an acellular or markedly hypocellular bone marrow. It is thought to be due to injury to the common pleuripotential stem cell important in the formation of all subsequent stem cell populations. Although various causes have been identified (drugs, toxins, infections), in more than half the cases in the USA the cause remains unknown. Patients with severe disease have a high risk of dying from bleeding or infections. Allogeneic bone marrow transplantation is probably the treatment of choice for patients under 40 years of age with severe disease who have a human leukocyte antigen (HLA) -matched donor. Experimental data suggest that transfusion of blood products into these patients prior to bone marrow transplants can sensitize bone marrow recipients, resulting in rejection of the marrow graft.

Researchers from the Fred Hutchinson Cancer

Table 5–1. T lymphocyte subset changes and seronegativity or seropositivity in 50 homosexual men followed longitudinally for 34 months.[1]

Seronegative or Seropositive Status	Number of Men		
	Low T4/T8 Ratio	Normal T4/T8 Ratio	Total
Seronegative	4	27	31
Seropositive ≤ 29 months	5	2	7
Seropositive > 29 months	11	1	12
Total	20	30	50

[1]Adapted and reproduced, with permission, from Table 1 in Melbye M et al: Long-term seropositivity for human T-lymphotropic virus type III in homosexual men without the acquired immunodeficiency syndrome: Development of immunologic and clinical abnormalities. A longitudinal study. *Ann Intern Med 1986;* **104**:496.

Table 5-2. Distribution of blood type by gender.

Blood Type	Probabilities		
	Males	Females	Total
O	.21	.21	.42
A	.215	.215	.43
B	.055	.055	.11
AB	.02	.02	.04
Total	.50	.50	1.00

Research Center in Seattle, Washington, reported their results of bone marrow transplantation into 50 patients with severe aplastic anemia who did not receive a transfusion of blood products until just before the marrow transplantation (Anasetti et al, 1986). The probability of 10-year survival in this group of nontransfused patients was 82%; the survival rate was 43–50% for patients studied earlier who had received multiple transfusions. Results of this study are used to illustrate the binomial probability distribution.

Presenting Problem 4. The Coronary Artery Surgery Study (CASS, 1983) was a prospective, randomized, multicenter collaborative trial of medical and surgical therapy in subsets of patients with stable ischemic heart disease. Five-year survival in this group of patients was equally good in the medically treated and surgically (coronary revascularization) treated groups. A second part of the study compared the effects of medical and surgical treatment on the quality of life.

Over a 5-year period, 780 patients with stable ischemic heart disease were subdivided into three clinical subsets. Patients within each subset were randomly assigned to either medical or surgical treatment. All patients enrolled had 50% or greater stenosis of the left main artery or 70% or greater stenosis of the other operable vessels. Group A had mild angina and an ejection fraction of at least 50%; group B had mild angina and an injection fraction less than 50%; group C had no angina after myocardial infarction. History, examination, and treadmill testing were done at 6, 18, and 60 months; a follow-up questionnaire was completed at 6-month intervals. Quality of life was evaluated by assessing chest pain status, heart failure, activity limitation, employment status, recreational status, drug therapy, number of hospitalizations, and risk factor alteration such as smoking status, blood pressure control, and cholesterol. The data on number of hospitalizations will be used to illustrate the Poisson probability distribution.

Presenting Problem 5. An individual's blood pressure has important health implications; hypertension is among the most commonly treated chronic medical problems. What factors influence blood pressure, and what constitutes an abnormal level? The Society of Actuaries (1980) presented data from a study of systolic and diastolic blood pressures in men and women aged 15–69 years. The results are summarized below.

Mean Blood Pressures

Ages	Men		Women	
	Systolic	Diastolic	Systolic	Diastolic
16	115	70	112	69
19	119	71	114	70
24	122	73	115	71
29	122	75	116	72
39	123	76	118	74
49	125	78	123	76
59	128	79	128	79
69	132	79	134	80

For men and women between the ages of 20–39, mean systolic pressure is approximately 120 mm Hg. Standard deviations were approximately 10 for systolic pressure and approximately 8 for diastolic pressure. How can this information be used to calculate probabilities of patients' having blood pressures at different levels?

5.1 PURPOSE OF THE CHAPTER

The previous two chapters presented methods for summarizing information from studies: graphs, plots, and summary statistics. A major reason for performing clinical research, however, is to generalize the findings from the set of observations on one group of subjects to others who are similar to those subjects. For example, in Presenting Problem 3, the authors concluded that "patients with severe aplastic anemia who have transplants before the onset of transfusion-induced sensitization have an excellent probability of long-term survival and a normal life." This conclusion was based on their study and longitudinal follow-up of 50 patients. Studying all patients with severe aplastic anemia is neither possible nor desirable; therefore, the investigators made **inferences** to a larger **population** of patients on the basis of their study of a **sample** of patients. Their use of the term **probability** in their conclusion indicates that they cannot be sure that *all* patients with severe aplastic anemia will respond to the treatment they recommend, but they think the chances are good from their experience with these 50 patients. How did the investigators come to this conclusion, and what methods did they use to determine that the probability of survival is excellent?

This chapter discusses concepts that allow investigators to make statements about the probability of possible outcomes, such as long-term survival. First, probability is defined and discussed, and elementary rules for calculating probabilities are given. Methods for selecting a sample of subjects to be included in a study are then illustrated. Finally, **probability distributions** are introduced, along with examples of how they are used in medicine.

5.2 THE MEANING OF THE TERM *PROBABILITY*

People use the concept and the term *probability* many times each day; yet not many, unless they are scientific investigators, could define what probability means. For our purposes in this text, so that we may use the concept of probability in a meaningful way to interpret results of medical studies, the frequency definition, which makes intuitive sense, will suffice. Assume that an experiment can be repeated many times, with each replication (repetition) called a **trial;** and assume that there are one or more outcomes that can result from each trial. Then, the **probability** of a given outcome is the number of times that outcome occurs divided by the total number of trials. If the outcome is sure to occur, it has a probability of 1; if an outcome cannot occur, its probability is 0.

An estimate of probability may be determined empirically, or it may be based upon a theoretical model. To use a familiar example, we all know that the probability of flipping a fair coin and getting a "tail" is .50, or 50%. If a coin is flipped 10 times, there is no guarantee, of course, that exactly 5 tails will be observed; the proportion of tails can range from 0 to 1, although in most cases we expect it to be closer to .50 than to 0 or 1. If the coin is flipped 100 times, the chances are even better that the proportion of tails will be close to .50, and with 1000 flips, the chances are better still. As the number of flips becomes larger, the proportion of coin flips that result in tails approaches .50; therefore, the probability of a tail on any one flip is .50.

The above definition of probability is sometimes called objective probability, as opposed to **subjective probability,** which is an estimate that reflects a person's opinion, hunch, or best guess about whether an outcome will occur. Subjective probabilities are important in medicine because they form the basis of a physician's opinion about whether a patient has a specific disease. In Chapter 13, we will discuss how this estimate, based on information gained in the history and physical examination, changes as the result of laboratory tests or other diagnostic procedures. Although some people discount subjective probability estimates because they are not based on "hard" data, subjective probabilities are necessary in medicine because they often represent the only information available, and a decision must be made despite the uncertainty in the situation. The probabilities discussed in this chapter are objective and are based on the frequency definition; subjective probabilities are discussed in later chapters.

5.3 BASIC DEFINITIONS & RULES OF PROBABILITY

Probability concepts are helpful for understanding and interpreting data presented in tables and graphs in published articles. In addition, the concept of probability lets us make statements about how much confidence we have in estimates, such as means, proportions, or relative risks (introduced in the previous chapter). Understanding probability is essential for understanding the meaning of *P*-values given in journal articles.

We will use two examples to illustrate some definitions and rules for determining probabilities: Presenting Problem 1 on seropositivity in homosexual men (Table 5–1) and the information given in Table 5–2 on gender and blood type.

5.3.1 Basic Definitions in Probability

In probability, an **experiment** is defined as any planned process of data collection. For Presenting Problem 1, the experiment is the process of determining the seronegative or seropositive status in a group of 50 homosexual men. An experiment consists of a number of independent **trials** (replications) under the same conditions; in this example, a trial consists of determining serum status for an individual man. Each trial can result in one of three outcomes: seronegative, seropositive for 29 months or less, or seropositive for more than 29 months.

The probability of a particular outcome, say outcome *A,* is written *P(A)*. For example, in Table 5–1, if outcome *A* is "seronegative," the probability that a man in the study is seronegative is

$$P(\text{seronegative}) = 31/50 = .62$$

In Presenting Problem 2, the probabilities of different outcomes are already computed. The outcomes of each trial to determine blood type are O, A, B, and AB. From Table 5–2, the probability that a person has type A blood is

$$P(\text{type A}) = .43$$

The blood type data illustrate two important features of probability:

1. The probability of each outcome (blood type) is greater than or equal to 0.

2. The sum of the probabilities of the various outcomes is 1.

Events may be defined either as single outcomes

or as a set of outcomes. For example, the outcomes of serum status are seronegative, seropositive for 29 months or less, and seropositive for more than 29 months; but we may wish to define an event as a man's being seronegative or seropositive. The event "seropositive" contains two outcomes: ≤ 29 months and > 29 months.

Sometimes, we want to know the probability that an event will not happen; an event opposite to the event of interest is called a **complementary event.** For example, the complementary event to "being seronegative" is "not being seronegative." The probability of the complement is

$$P(\text{complement of seronegative}) = P(\text{not being seronegative})$$
$$= P(\text{seropositive})$$
$$= \frac{7 + 12}{50} = \frac{19}{50} = .38$$

Note that the probability of a complementary event may also be found as 1 minus the probability of the event of interest, and this calculation may be easier in some situations. Thus, we obtain

$$P(\text{complement of seronegative} = 1 - P(\text{seronegative})$$
$$= 1 - \frac{31}{50} = 1 - .62 = .38$$

5.3.2 Mutually Exclusive Events & the Addition Rule

Two or more events are **mutually exclusive events** if the occurrence of one event precludes the occurrence of the others. By definition, outcomes are mutually exclusive; ie, a man in Presenting Problem 1 cannot be both seropositive ≤ 29 months and seropositive > 29 months. In addition, all complementary events are also mutually exclusive; however, events can be mutually exclusive without being complementary if there are three or more possible events.

As we indicated earlier, what constitutes an event is a matter of definition. Let us redefine the experiment in Presenting Problem 2 so that each outcome (blood type O, A, B, or AB) is a separate event. The probability of mutually exclusive events occurring is the probability that either one event occurs *or* the other event occurs. This probability is found by adding the probabilities of the two events, which is called the **addition rule** for probabilities. For example, the probability of being either blood type O or blood type A is

$$P(O \text{ or } A) = P(O) + P(A)$$
$$= .42 + .43$$
$$= .85$$

Does the addition rule work for more than two events? The answer is yes, as long as they are all mutually exclusive. We discuss the approach to use with non-mutually exclusive events in Section 5.3.5.

5.3.3 Independent Events & the Multiplication Rule

Two different events are **independent events** if the outcome of one event has no effect on the outcome of the second event. Using the blood type example, let us also define a second event as the gender of the person; this event consists of the outcomes "male" and "female." In this example, gender and blood type are independent events; the sex of a patient does not affect in any way the patient's blood type, and vice versa. The probability of two independent events is the probability that both events occur. This probability is found by multiplying the probabilities of the two events, which is called the **multiplication rule** for probabilities. For example, the probability of being male and of being blood type O is

$$P(\text{male and blood type O})$$
$$= P(\text{male}) \times P(\text{blood type O})$$
$$= .50 \times .42 = .21$$

The probability of being male, .50, and the probability of having blood type O, .42, are both called **marginal probabilities;** ie, they appear on the "margins" of a probability table. The probability of being male and of having blood type O, .21, is called a **joint probability;** it is the probability of both "male" and "type O" occurring jointly.

Is having a low T4/T8 ratio independent of serum status in the subjects in Presenting Problem 1? Table 5–1 shows that these two events are not independent; ie, seropositive patients tend to have lower T4/T8 ratios. If two events are independent, the product of the marginal probabilities will equal the joint probability in all instances. To show that two events are not independent, we need demonstrate only one instance in which the product of the marginal probabilities is not equal to the joint probability. For example, to show that serum status and T4/T8 ratio are not independent, we consider the joint probability of having a low T4/T8 ratio and of being seropositive. Table 5–1 shows that

$$P(\text{low T4/T8 } and \text{ seropositive}) = \frac{5 + 11}{50} = .32$$

However, the product of the marginal probabilities does not yield the same result; ie,

$$P(\text{low T4/T8}) \times P(\text{seropositive}) = \frac{20}{50} \times \frac{7 + 12}{50} = .152$$

We could show that the product of the marginal probabilities is not equal to the joint probability for any of the combinations in this example, but we need show only one instance to prove that two events are not independent.

5.3.4 Nonindependent or Conditional Events & the Modified Multiplication Rule

Finding the joint probability of two events when they are not independent is a bit more complex than simply multiplying the two marginal probabilities. When two events are not independent, the occurrence of one event depends on whether the other event has occurred. Let A stand for the event "low T4/T8 ratio" and B for the event "seropositive." We want to know the probability of event A given event B, written $P(A|B)$ where the vertical line, |, is read as "given." From the data in Table 5–1, the probability of a low T4/T8 ratio, given that a patient is seropositive, is

$$P(\text{low T4/T8}|\text{seropositive}) = \frac{5 + 11}{7 + 12} = .842$$

This probability is called a **conditional probability**; it is the probability of one event given that another event has occurred. Put another way, the probability of a low T4/T8 ratio is conditional on the event of seropositivity; it is substituted for $P(\text{low T4/T8})$ in the multiplication rule. If we put these expressions together, we can find the joint probability of having a low T4/T8 ratio *and* being seropositive:

$$P(\text{low T4/T8 and seropositive})$$
$$= P(\text{low T4/T8}| \text{ seropositive}) \times P(\text{seropositive})$$
$$= \frac{16}{19} \times \frac{19}{50}$$
$$= .842 \times .38 = .32$$

The probability of having a low T4/T8 ratio and being seropositive can also be determined by finding the conditional probability of being seropositive, given a low T4/T8 ratio, and substituting that expression in the multiplication rule for P(seropositive). Thus, we obtain

$$P(\text{low T4/T8 and seropositive})$$
$$= P(\text{seropositive and low T4/T8})$$
$$= P(\text{seropositive}|\text{low T4/T8}) \times P(\text{low T4/T8})$$
$$= \frac{16}{20} \times \frac{20}{50}$$
$$= .80 \times .40 = .32$$

5.3.5 Non-Mutually Exclusive Events & the Modified Addition Rule

Remember that two or more mutually exclusive events cannot occur together, and the addition rule applies for the calculation of the probability that one or another of the events occurs. Let us now examine the situation for finding the probability that either of two events occurs when they are not mutually exclusive. For example, gender and blood type O are **non-mutually exclusive events** because the occurrence of one event does not preclude the occurrence of the other. The addition rule must be modified in this situation; otherwise, the probability that both events occur will be added into the calculation twice.

In Table 5–2, the probability of being male is .50 and the probability of blood type O is .42. However, the probability of being male *or* having blood type O is not .50 + .42, because in this sum, males with type O blood have been counted twice. Therefore, the joint probability of being male *and* having blood type O, .21, must be subtracted. The calculation is

$$P(\text{male or type O})$$
$$= P(\text{male}) + P(\text{type O}) - P(\text{male and type O})$$
$$= .50 + .42 - .21$$
$$= .71$$

Of course, if we do not know that P(male *and* type O) = .21, we must use the multiplication rule (for independent events, in this case) to determine this probability.

5.3.6 Summary of Rules & an Extension

In this section, we give a summary of the rules presented so far and discuss an extension of them. Remember that questions about mutual exclusiveness use the word *or* and the addition rule; questions about independence use the word *and* the multiplication rule. For the summary, we use letters to represent events; ie, A, B, C, and D are four different events with probability $P(A)$, $P(B)$, $P(C)$, and $P(D)$.

5.3.6.a Addition Rule: The addition rule for the occurrence of either of two or more events is as follows: If A, B, and C are mutually exclusive, then

$$P(A \text{ or } B \text{ or } C) = P(A) + P(B) + P(C)$$

If two events such as A and D are not mutually exclusive, then

$$P(A \text{ or } D) = P(A) + P(D) - P(A \text{ and } D)^*$$

5.3.6.b Multiplication Rule: The multiplication rule for the occurrence of both of two or more events is as follows: If *A, B,* and *C* are independent, then

$$P(A \text{ and } B \text{ and } C) = P(A) \times P(B) \times P(C)$$

If two events such as *B* and *D* are not independent, then

$$P(B \text{ and } D) = P(B|D) \times P(D)$$
$$\text{or}$$
$$P(B \text{ and } D) = P(B) \times P(D|B).^*$$

5.3.6.c An Extension—Bayes' Theorem: The multiplication rule for probabilities when events are not independent can be used to derive one form of an important formula called **Bayes' theorem.** Because *P(B and D)* equals both *P(B|D)* × *P(D)* and *P(B)* × *P(D|B)*, these latter two expressions are equal to each other. Assuming *P(B)* and *P(D)* are not equal to zero, we can solve for one in terms of the other. Thus, if

$$P(B|D) \times P(D) = P(B) \times P(D|B)$$

then

$$P(B|D) = \frac{P(D|B) \times P(B)}{P(D)}$$

which is found by dividing both sides of the equation by *P(D)*. Similarly,

$$P(D|B) = \frac{P(B|D) \times P(D)}{P(B)}$$

In the equation for *P(B|D)*, *P(B)* in the right-hand side of the equation is sometimes called the **prior probability,** because its value is known prior to the calculation; *P(B|D)* is called the **posterior probability,** because its value is known only after the calculation.

The two formulas of Bayes' theorem are important because clinicians frequently know only one of the pertinent probabilities and must determine the other. Examples are genetics and diagnostic testing, a topic discussed in detail in Chapter 13.

---

*The probability of three or more events that are not mutually exclusive or not independent involves complex calculations beyond the scope of this book. Interested readers can consult elementary books on probability, such as Miller and Freund (1985).

5.3.7 A Comment on Terminology

Although in everyday use the terms **probability, odds,** and **likelihood** are used synonymously, mathematicians do not use them that way. **Odds** can best be thought of as the probability that an event occurs divided by the probability that the event does not occur—or as the probability of an event divided by the probability of its complement. For example, the odds that a person has blood type O are .42/(1 − .42) = .72 to 1, but "to 1" generally is not stated explicitly. This interpretation is consistent with the meaning of the odds ratio, discussed in Chapter 4, Section 4.5.4.b.

Likelihood may be related to Bayes' theorem for conditional probabilities. Suppose a physician is trying to determine which of three likely diseases a patient has—myocardial infarction, pneumonia, or reflux esophagitis. Chest pain can appear with any one of these three diseases; and the physician needs to know the probability that chest pain occurs with myocardial infarction, the probability that chest pain occurs with pneumonia, and the probability that chest pain occurs with reflux esophagitis. The probabilities of a given outcome (chest pain) when evaluated under different hypotheses (myocardial infarction, pneumonia, and reflux esophagitis) are called **likelihoods** of the hypotheses (or diseases).

5.3 POPULATIONS & SAMPLES

As we stated in the introduction, a major purpose of doing research is to infer, or generalize, from a sample to a larger population. This process of **inference** is accomplished by using statistical methods based on probability. **Population** is the term statisticians use to describe a large set or collection of items that have something in common. In medicine, *population* generally refers to patients or other living organisms, but the term can also be used to denote collections of inanimate objects, such as sets of autopsy reports, hospital charges, or birth certificates. A **sample** is a subset of the population, selected in such a way that it is representative of the larger population.

There are many good reasons for studying a sample instead of an entire population, and four commonly used methods for selecting a sample are discussed in this section. Before turning to those topics, however, we note that the term *population* is frequently misused in medicine to describe what is, in fact, a sample. For example, physicians sometimes refer to the "population of patients in this study." After you have read this book, you will be able to spot such errors when you see them in the medical literature.

5.4.1 Reasons for Sampling

There are at least six reasons for studying samples instead of populations.

1. Samples can be studied more quickly than populations. Speed can be an important factor if a physician needs a quick answer to an important question, as in search of a vaccine or treatment for a new disease.

2. A study of a sample is less expensive than a study of an entire population, because a smaller number of items or subjects are examined. This consideration is especially important in the design of large studies that require a lengthy follow-up.

3. A study of an entire population is impossible in most situations. Sometimes, the process of the study destroys or depletes the item being studied. For example, in a study of cartilage healing in limbs of dogs after 6 weeks of limb immobilization, the animals must be sacrificed in order to perform histologic studies. On other occasions, the desire is to infer to future events, as in the study of patients with severe aplastic anemia summarized in Presenting Problem 3. In both cases, a study of a population is impossible.

4. Sample results are often more accurate than results based on a population. For samples, more time and resources can be spent on training the people who perform observations and collect data. In addition, more expensive procedures that improve accuracy can be used for a sample because fewer procedures are required.

5. If samples are properly selected, probability methods can be used to estimate the error in the resulting statistics. It is this aspect of sampling that permits investigators to make probability statements about observations in a study.

6. Samples can be selected to reduce heterogeneity. For example, systemic lupus erythematosus (SLE) has many clinical manifestations, resulting in a heterogeneous population. A sample of the population with specified characteristics is more appropriate than the entire population for the study of certain aspects of the disease.

To summarize, bigger does not always mean better. For this reason, investigators must plan the sample size appropriate for their study prior to beginning research. This process is called determining the **power** of a study and is discussed in greater detail in later chapters.

5.4.2 Methods of Sampling

Now, let us turn to methods for obtaining samples. The best way to ensure that a sample will lead to reliable and valid inferences is to use **probability samples** in which the probability of being included in the sample is known for each subject in the population. The four commonly used probability sampling methods in medicine are simple random sampling, systematic sampling, stratified sampling, and cluster sampling.

We will use the following example to illustrate each method: Consider a physician applying for a grant for a study that involves measuring the tracheal diameter on x-rays. The physician wants to convince the granting agency that these measurements are reliable. To estimate intrarater reliability, the physician will select a sample of chest x-rays from those performed during the previous year, remeasure the tracheal diameter, and compare the new measurement with the original one on file in the patient's chart. The physician has a population of 3400 x-rays, and let us assume that the physician has decided that a sample of 200 x-rays is sufficient to provide an accurate estimate of intrarater reliability. Now the physician must select the sample for the reliability study.

5.4.2.a Simple Random Sampling: A **simple random sample** is one in which every subject (every film in the x-ray example) has an equal probability of being selected for the study. The recommended way to select a simple random sample is to use a table of random numbers or a computer-generated list of random numbers. For this approach, each x-ray must have an identification (ID) number, and a list of ID numbers, called a **sampling frame,** must be available. For the sake of simplicity, assume that the x-rays are numbered from 1 to 3400. Using a random number table, after first identifying a starting place in the table at random, the physician can select the first 200 digits between 1 and 3400. The x-rays with the ID numbers corresponding to 200 random numbers make up the simple random sample. If a computer program to generate random numbers is available, the physician can request 200 numbers between 1 and 3400. To illustrate the process with a random number table, we reproduce here a portion of Table A–1 in Appendix A as Table 5–3. One way to select a starting point is by tossing a die to select a row and a column at random. Tossing a die twice determines, first, a large block of rows and, second, an individual row within the block. For example, if we throw a 2 and a 3, we begin in the second block, third row, beginning with the number 83. (If we had thrown a 6, we would toss the die again, because there are only five rows.) Now, we must select a beginning column at random, again by tossing the die twice to select a block and a column within the block. For example, if we toss a 3 and a 1, we use the third block of columns and the first column, headed by the number 1. Therefore, the starting point in this example is located where the row beginning with 83 and the column beginning with 1 intersect at the number 6 (underlined in Table 5–3).

Because there are 3400 x-rays, we must read four-digit numbers; so the first ten numbers are

Table 5-3. Random numbers from Table A–1.

927415	956121	168117	169280	326569	266541
926937	515107	014658	159944	821115	317592
867169	388342	832261	993050	639410	698969
867169	542747	032683	131188	926198	371071
512500	843384	085361	398488	774767	383837
062454	423050	670884	840940	845839	979662
806702	881309	772977	367506	729850	457758
837815	163631	622143	938278	231305	219737
926839	453853	767825	284716	916182	467113
854813	731620	978100	589512	147694	389180
851595	452454	262448	688990	461777	647487
449353	556695	806050	123754	722070	935916
169116	586865	756231	469281	258737	989450
139470	358095	528858	660128	342072	681203
433775	761861	107191	515960	759056	150336
221922	232624	398839	495004	881970	792001
740207	078048	854928	875559	246288	000144
525873	755998	866034	444933	785944	018016
734185	499711	254256	616625	243045	251938
773112	463857	781983	078184	380752	492215

6221, 7678, 9781, 2624, 8060, 7562, 5288, 1071, 3988, and 8549. The numbers less than 3401 are the IDs of the x-rays to be used in the sample. In the first ten numbers selected, only two are less than 3401; so we use x-rays with the ID numbers 2624 and 1071. This procedure continues until we have selected 200 x-rays. When the number in the bottom row (7819) is reached, we go to the top of the row and move one digit to the right for numbers 6811, 1465, 3226, etc.

If a number less than 3401 occurs twice, the x-ray with that ID number can be selected for the sample and used in the study a second time. In this case, the final sample of 200 will be 200 measurements rather than 200 x-rays. Frequently, however, when a number occurs twice, it is ignored the second time and the next eligible number is used instead. The differences between these two procedures are negligible when we sample from a large population.

5.4.2.b Systematic Sampling: A **systematic random sample** is one in which every kth item is selected; k is determined by dividing the number of items in the sampling frame by the desired sample size. For example, 3400 x-rays divided by 200 is 17, so every 17th x-ray is sampled. In this approach, however, we must select a number ran-

domly between 1 and 17 first, and we then select every 17th x-ray. Suppose we randomly select the number 12 from a random number table. Then, the systematic sample consists of x-rays with ID numbers 12, 29, 46, 63, 80, etc.; each subsequent number is determined by adding 17 to the last ID number.

Systematic sampling should not be used when a cyclic repetition is inherent in the sampling frame. For example, systemic sampling is not appropriate for selecting months of the year in a study of the frequency of different types of accidents, because some accidents occur more often at certain times of the year. For instance, skiing injuries and automobile accidents most often occur in cold-weather months, whereas swimming injuries and farming accidents most often occur in warm-weather months.

5.4.2.c Stratified Sampling: A **stratified random sample** is one in which the population is first divided into relevant strata (subgroups), and a random sample is then selected from each stratum. In the x-ray example, the physician may wish to stratify on age of patients, because the trachea varies in size with age and measuring the diameter accurately in young patients may be difficult. The population of x-rays may be divided into infants

less than 1 year old, children from 1 year old to less than 6 years old, children from 6 to less than 16 years old, and all other subjects 16 years of age or older; a random sample is then selected from each age stratum. Other commonly used strata in medicine besides age include gender of patient, severity or stage of disease, and duration of disease. Measures selected as strata must be related to the measurement of interest, in which case stratified random sampling is the most efficient—ie, requires the smallest sample size.

5.4.2.d Cluster Sampling: A **cluster random sample** results from a two-stage process in which the population is divided into clusters and a subset of the clusters is randomly selected. Clusters are commonly based on geographic areas or districts, so this approach is used more often in epidemiologic research than in clinical studies. For example, the sample for a household survey taken in a city may be selected by using city blocks as clusters; a random sample of city blocks is selected, and all households (or a random sample of households) within the selected city blocks are surveyed. In multicenter trials, the institutions selected to participate in the study comprise the clusters; patients from each institution can be selected using another random sampling procedure. Cluster sampling is somewhat less efficient than the other sampling methods, requiring a larger sample size, but in some situations it is the method of choice, such as in multicenter trials, in order to obtain adequate numbers of patients.

5.4.2.e Nonprobability Sampling: The sampling methods just discussed are all based on probability, but there are also nonprobability sampling methods, such as convenience samples or quota samples. **Nonprobability samples** are those in which the probability that a subject is selected is unknown. Nonprobability samples often reflect selection biases of the person doing the study and do not fulfill the requirements of randomness needed to estimate sampling errors. When we use the term *sample* in the context of observational studies, we will assume that the sample has been randomly selected in an appropriate way. However, in Chapter 15, we return to the problems involved in using convenience samples and discuss how they affect the interpretation of results.

5.4.2.f Random Assignment: Random sampling methods are used when a sample of subjects is selected from a population of possible subjects in observational studies, such as cohort, case-control, and cross-sectional studies. In experimental studies such as randomized clinical trials, subjects are first selected for inclusion in the study on the basis of appropriate criteria; they are then assigned to different treatment modalities. If the assignment of subjects to treatments is done by using random methods, the process is called **random assignment.** Random assignment may also be defined as randomly assigning treatments to subjects. In either case, random assignment helps ensure that the groups receiving the different treatment modalities are as similar as possible. Thus, any differences in outcome at the conclusion of the study are more likely to be the result of differences in treatments used in the study rather than differences in compositions of the groups.

Random assignment is best carried out by using random numbers. As an example, consider the CASS study (1983), in which patients meeting the entry criteria were divided into clinical subsets and then randomly assigned to either medical or surgical treatment. Random assignment in this study can be accomplished by using a list of random numbers (obtained from a computer or a random number table) and assigning the random numbers to patients as they enter the trial. Perhaps patients assigned an even random number receive medical treatment, and those assigned an odd random number receive surgical treatment. If the study involves several investigators at different sites, such as in a multicenter trial, the investigator preparing to enter an eligible patient in the study may call a central office to learn which treatment assignment is next. As an alternative, separately randomized lists may be generated for each site. Of course, in double-blind studies, someone other than the investigator must keep the list of random assignments.

Suppose investigators in the CASS study want an equal number of patients at each site participating in the study. For this design, the assignment of random numbers may be balanced within blocks of patients of a predetermined size. For example, balancing patients within blocks of 12 will guarantee that every time 12 patients enter the study at a given site, 6 patients receive the medical treatment and 6 receive the surgical treatment. Within the block of 12 patients, however, assignment is random until 6 patients are assigned to one or the other of the treatments.

A study design may also match subjects on important characteristics, such as gender, age group, or severity of disease, and then make the random assignment. This stratified assignment controls for possible confounding effects of the characteristic(s); it is equivalent to stratified random sampling in observational studies.

Many types of biases may result in studies in which patients are not randomly assigned to treatment modalities. For instance, early studies comparing medical and surgical treatment for coronary artery disease did not randomly assign patients to treatments, and they were criticized as a result. Some critics claimed that sicker patients were not candidates for surgery, and thus, the group receiving surgery was biased by having healthier subjects. Other critics stated that the healthier patients were given medical treatment because their disease was not as serious as that of the sicker patients. The problem is that in nonrandomized studies, determining which biases are operating and which conclusions are appropriate is difficult; in fact, the CASS study was designed partly in response to these criticisms. A detailed description of different kinds of biases that threaten the validity of studies is given in Chapter 15.

5.4.2.g Using and Interpreting Random Samples: In actual clinical studies, patients are not always randomly selected from the population from which the investigator wishes to infer. Instead, the clinical researcher often uses all patients at hand who meet the entry criteria for the study. This practical procedure is used especially when studies involve conditions that are not very common. Colton (1974) makes a useful distinction between the target population and the sampled population. The **target population** is the population to which the investigator wishes to generalize; the **sampled population** is the population from which the sample was actually drawn. Figure 5–1 presents a schema of these concepts.

For example, the investigators in Presenting Problem 3 clearly wished to generalize their findings about survival to all patients with severe aplastic anemia, such as patients who live in other locations and perhaps even patients who do not yet have the disease. The population sampled was the set of patients receiving bone marrow transplants in Seattle between September 1972 and September 1984. Thus, the sample in this study may actually be a complete census of all patients who received a bone marrow transplant during that time period or it may be a subset of these patients; the authors do not specify. The role of statistical inference is to permit generalization from the sample to the population sampled. In order to make inferences from the population sampled to the target population, we must ask whether the population sampled is representative of the target population. A population (or sample) is **representative** of the target population if the distribution of important characteristics is the same as in the target population. This judgment is clinical, not statistical. It points to the importance of always reading the "Method" section of journal articles to learn what population was actually sampled so that you can determine the representativeness of that population.

5.4.3 Population Parameters & Sample Statistics

Statisticians use precise language to describe characteristics of populations and samples. Measures of central tendency and variation, such as the mean and the standard deviation, are fixed and invariant characteristics in populations, and they are called **parameters.**[*] In samples, however, the observed mean or standard deviation calculated on the basis of the sample information is actually an estimate of the population mean or standard deviation; these estimates are called **statistics.** Statisticians customarily use Greek letters for population parameters and Latin letters for sample statistics. Some of the frequently encountered symbols used in this text are summarized in Table 5–4.

5.5 RANDOM VARIABLES & PROBABILITY DISTRIBUTIONS

The characteristic of interest in a study is called a **variable.** We have examined two variables for the study described in Presenting Problem 1: serum status of a subject with respect to HIV and T4/T8 ratio. The term *variable* makes sense because

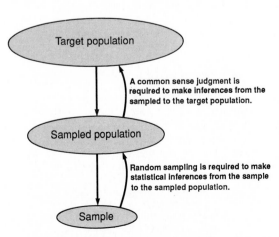

Figure 5-1. Target and sampled populations.

A common sense judgment is required to make inferences from the sampled to the target population.

Random sampling is required to make statistical inferences from the sample to the sampled population.

Table 5-4. Commonly used symbols for parameters and statistics.

Characteristic	Parameter Symbol	Statistical Symbol
Mean	μ	$\bar{X}$
Standard deviation	σ	s
Variance	σ^2	s^2
Correction	ρ	r
Proportion	π	p

[*]The word *parameter* has a much more specific and somewhat different meaning in statistics from that in everyday use. Usually, when people speak of the "parameters involved in the case," they are actually referring to characteristics that vary, not to fixed characteristics in the population.

the value of the characteristic varies from one subject to another. This variation results from inherent biologic variation among individuals and errors made in measuring and recording a subject's value on a characteristic, called measurement error. A **random variable** is a variable in a study in which subjects are randomly selected. If we think of the subjects in Presenting Problem 1 as being a random sample selected from a larger population of homosexual males, then serum status and T4/T8 ratio are examples of random variables.

Just as values of characteristics, such as age of patients with colorectal cancer or percentage supersaturation of bile (see Chapter 3), can be summarized in frequency distributions, values of a random variable can be summarized in a frequency distribution called a **probability distribution.** For example, if X is a random variable defined as the length of time a homosexual man is seropositive for HIV, then from the study described in Presenting Problem 1, X can take on values 0, 1, 2, . . . , 34 months; and we can determine the probability that the random variable X has any given value. For instance, from Table 5–1, the probability $X = 0$ months is $31/50 = .62$. In some applications, a formula or rule will adequately describe a distribution; the formula can then be used to calculate the probability of interest. In other situations, a theoretical probability distribution provides a good fit to the distribution of the variable of interest.

Several theoretical probability distributions are important in statistics, and we shall examine three that are useful in medicine. Both the binomial and the Poisson are *discrete* probability distributions; ie, the associated random variable associated takes only integer values, 0, 1, 2, . . . , n. The normal (gaussian) distribution is a *continuous* probability distribution; ie, the associated random variable has values measured on a continuous scale. We will examine the binomial and Poisson distributions briefly, using examples from the Presenting Problems to illustrate each; then we will discuss the normal distribution in greater detail.

5.5.1 The Binomial Distribution

Suppose an event can have only dichotomous outcomes (eg, yes and no, or positive and negative), denoted A and B. The probability of A is denoted by π, ie, $P(A) = \pi$, and this probability stays the same each time the event occurs. Therefore, the probability of B must be $1 - \pi$, because B occurs if A does not. If an experiment involving this event is repeated n times and the outcome is independent from one trial to another, what is the probability that outcome A occurs exactly X times? Or equivalently, what proportion of the n outcomes will be A? These questions frequently are of interest, especially in laboratory research,

and they can be answered with the binomial distribution.

Basic principles of the binomial distribution were developed by the Swiss mathematician Jacob Bernoulli, who lived in the 17th century and made many contributions to probability theory. He was the author of what is generally acknowledged as the first book devoted to probability, published in 1713. In fact, each trial involving a binomial probability is sometimes called a Bernoulli trial, and a sequence of trials is called a Bernoulli process. The **binomial distribution** gives the probability that a specified outcome occurs in a given number of independent trials. The binomial distribution can be used to model the inheritability of a particular trait in genetics, to estimate the occurrence of a specific reaction such as the single packet (quantal release) of acetylcholine at the neuromuscular junction, or to estimate the death of a cancer cell in an in vitro test of a new chemotherapeutic agent.

We will use Presenting Problem 3 to illustrate the binomial distribution. Assume, for a moment, that the entire population of patients with severe aplastic anemia has been studied, and the probability of year survival is equal to .8 (we use .8 for computational convenience, rather than .82 as observed in the study). Let S represent the event of 10-year survival and D represent death before 10 years; then, $\pi = P(S) = .8$ and $1 - \pi = P(D) = .2$. Consider a group of $n = 2$ patients with severe aplastic anemia. What is the probability that exactly 2 patients live 10 years? That exactly 1 lives 10 years? That none lives 10 years? These probabilities are found by using the multiplication and addition rules outlined earlier in this chapter.

The probability that exactly 2 patients live 10 years is found by using the multiplication rule for independent events. We know that $P(S) = .8$ for patient 1 and $P(S) = .8$ for patient 2. Because the survival of one patient is independent from (has no impact on) the survival of the other patient, the probability of *both* surviving is

$$P(S \text{ for patient 1 } and \text{ } S \text{ for patient 2})$$
$$= P(S)P(S) = (.8)(.8) = .64$$

The event of exactly 1 patient living 10 years can occur in two ways: patient 1 survives 10 years and patient 2 does not, or patient 2 survives 10 years and patient 1 does not. These two events are mutually exclusive; therefore, after using the multiplication rule to obtain the probability of each event, we can use the addition rule for mutually exclusive events to combine the probabilities.

$$P(S \text{ for patient 1 } and \text{ } D \text{ patient 2})$$
$$= P(S)P(D) = (.8)(.2) = .16,$$
$$\text{and } P(D \text{ for patient 1 } and \text{ } S \text{ for patient 2})$$
$$= P(D)P(S) = (.2)(.8) = .16$$
$$\text{Therefore, } P(\text{event 1 } or \text{ event 2}) = .16 + .16 = .32$$

Finally, the probability that neither lives 10 years is

$$P(D \text{ for patient 1 } and \text{ D for patient 2})$$
$$= P(D)P(D) = (.2)(.2) = .04$$

The above computational steps are summarized in Table 5-5. Note that the total probability is

$$(.8)^2 + 2(.8)(.2) + (.2)^2 = 1.0$$

which some of you may recognize as the form of the binomial formula, $a^2 + 2ab + b^2$.

The same process can be applied for a group of patients of any size or for any number of trials, but it becomes quite tedious. An easier technique is to use the formula for the binomial distribution, which follows. The probability of X outcomes in a group of size n, if each outcome has probability π and is independent from all other outcomes, is given by

$$P(X) = \frac{n!}{X!\,(n - X)!}\, \pi^X(1 - \pi)^{n-x}$$

where ! is the symbol for factorial; $n!$ is called n factorial and is equal to the product $n(n - 1)$ $(n - 2) \ldots (3)(2)(1)$. For example, $4! = (4)(3)(2)(1) = 24$. The number $0!$ is defined as 1. The symbol π^X indicates that π is raised to the power X, and $(1 - \pi)^{n-X}$ means that $1 - \pi$ is raised to the power $n - X$. The expression $n!/[X!\,(n - X)!]$ is sometimes referred to as the formula for **combinations** because it gives the number of combinations (or assortments) of X items possible among the n items in the group. Using this formula permits you to calculate such brainteasers as the number of different pairs of socks it is possible to make from among 6 original pairs (or 12 socks).

To verify that the probability that exactly $X = 1$ of $n = 2$ patients survive 10 years is .32, we use the formula:

$$P(1) = \frac{2!}{1!(2 - 1)!}\, (.8)^1(1 - .8)^{2-1}$$

$$= \frac{(2)(1)}{(1)(1)}\, (.8)(.2) = (2)(.8)(.2) = .32$$

Table 5-5. Summary of probabilities for two patients.

Outcome		Number of Ways to Occur	Probability
First Patient	Second Patient		
S	S	1	.8 × .8 = .64
D	S	2	2 × .8 × .2
S	D		= .32
D	D	1	.2 × .2 = .04

To summarize, the binomial distribution is useful for answering questions about the probability of X number of occurrences in n independent trials when there is a constant probability π of success on each trial. For example, suppose a new series of bone marrow transplants is begun with 10 patients. We can use the binomial distribution to calculate the probability that any particular number of them will survive 10 years. For instance, the probability that all 10 will survive 10 years is

$$P(10) = \frac{10!}{10!(10 - 10)!}\, (.8)^{10}(1 - 8)^{10-10}$$

$$= \frac{10!}{(10!)(0!)}\, (.8)^{10}(.2)^0 = (1)(.8)^{10}(1) = .107$$

Similarly, the probability that exactly 8 patients will survive 10 years is

$$P(8) = \frac{10!}{8!(10 - 8)!}\, (.8)^8(1 - .8)^{10-8}$$

$$= \frac{(10)(9)(8!)}{(8!)(2!)}\, (.8)^8(.2)^2$$

$$= \frac{(10)(9)}{(2)(1)}\, (.168)(.04) = (45)(.168)(.04) = .302$$

Table 5-6 lists the probabilities for $X = 0, 1, 2, 3, \ldots, 10$; a plot of the binomial distribution when $n = 10$ and $\pi = .8$ is given in Fig. 5-2. The mean of the binomial distribution is $n\pi$; so $(10)(.8) = 8$ is the mean number of patients surviving 10 years in this example. The standard deviation is $\sqrt{n\pi(1 - \pi)}$, which for this example is $(10)(.8)(.2) = 1.265$. Thus, the only two pieces of information needed to define a binomial distribution are n and π, which are called the parameters of the bionomial distribution. Studies involving dichotomous variables (see Chapter 9) often use proportion rather than number; eg, the propor-

Table 5-6. Probabilities for binomial distribution with $n = 10$ and $\pi = .8$.

Number of Patients Surviving	$\dfrac{n!}{X!(n - X)!}$	π^X	$(1 - \pi)^{n-x}$	$P(X)$[1]
0	1	1	.0000001	0
1	10	.8	.0000005	0
2	45	.64	.0000026	.0001
3	120	.512	.0000128	.0008
4	210	.410	.000064	.0055
5	252	.328	.00032	.0264
6	210	.262	.0016	.0881
7	120	.210	.008	.2013
8	45	.168	.04	.3020
9	10	.134	.2	.2684
10	1	.107	1	.1074

[1]Rounded to four decimal places.

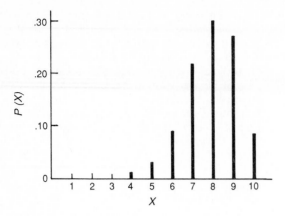

Figure 5-2. Binomial distribution for $n = 10$ and $\pi = .8$.

tion of patients surviving a given length of time rather than the number of patients. When proportion is used instead of number of successes, the same two pieces of information (n and π) are needed. However, because the proportion is found by dividing X by n, the mean of the distribution of the proportion becomes π, and the standard deviation becomes $\sqrt{\pi(1 - \pi)/n}$.

Even using the formula for the binomial distribution becomes time-consuming, especially if the numbers are large. Also, the formula gives the probability of observing exactly X successes, and interest frequently lies in knowing the probability of X or more successes or of X or less successes. For example, to find the probability that 8 or more patients will survive 10 or more years, we must use the formula to find the separate probabilities that 8 will survive, 9 will survive, and 10 will survive, and we must then sum these results; from Table 5-6, we obtain $P(X \geq 8) = P(X = 8) + P(X = 9) + P(X = 10) = .3020 + .2684 + .1074 = .6778$. Tables giving probabilities for the binomial distribution are presented in many elementary texts; if you are interested in analyzing dichotomous outcomes from small samples, you may wish to consult these texts (Daniel, 1974; Mendenhall, 1987; Sokal and Rohlf, 1981). Binomial tables are not included in this text because most research in medicine is conducted with samples sizes large enough to use an approximation to the binomial distribution; this approximation is discussed in Chapter 9.

5.5.2 The Poisson Distribution

The Poisson distribution is named for the French mathematician who derived it, Simeon D. Poisson. Like the binomial, the Poisson distribution is a discrete distribution applicable when the outcome is the number of times an event occurs. The **Poisson distribution** can be used to determine the probability of rare events; ie, it gives the probabil-

ity that an outcome occurs a specified number of times when the number of trials is large and the probability of any one occurrence is small. For instance, the Poisson distribution is used to plan the number of beds a hospital needs in its intensive care unit, the number of ambulances needed on call, or the number of operators needed on a switchboard. It can also be used to model the number of cells in a given volume of fluid, the number of bacterial colonies growing in a certain amount of medium, or the emission of radioactive particles from a specified amount of radioactive material.

Consider a random variable representing the number of times an event occurs in a given time or space interval. Then the probability of exactly X occurrences is given by the following formula:

$$P(X) = \frac{\lambda^X e^{-\lambda}}{X!}$$

in which λ (the Greek letter lambda) is the value of both the mean and the variance of the Poisson distribution, and e is the base of the natural logarithms, equal to 2.718. The term λ is called the parameter of the Poisson distribution, just as n and π are the parameters of the binomial distribution. Therefore, only one piece of information, λ, is needed to characterize any given Poisson distribution.

A random variable having a Poisson distribution was used in the Coronary Artery Surgery Study (CASS, 1983) summarized in Presenting Problem 4. The investigators described the number of hospitalizations for each group of patients (medical and surgical) as following Poisson distributions. This model is appropriate because the chance that a patient goes into the hospital during any one time interval is small and can be assumed to be independent from patient to patient. The results indicated that the 390 patients randomized to the medical group were hospitalized a total of 660 times over the 5-year study period; the 390 patients randomized to the surgical group were hospitalized a total of 950 times. The mean number of hospitalizations, from the Poisson model for medical patients, is $660/390 = 1.69$, and the mean for the surgical patients is $950/390 = 2.44$. We can use this information and the formula for the Poisson model to calculate probabilities of numbers of hospitalizations. For example, the probability that a patient in the medical group has zero hospitalizations is

$$P(X = 0) = \frac{(1.69^0)(e^{-1.69})}{0!} = \frac{(1)(.184)}{1} = .184$$

The probability that a patient has exactly 1 hospitalization is:

$$P(X = 1) = \frac{(1.69^1)(e^{-1.69})}{1!} = \frac{(1.69)(.184)}{1} = .312$$

The calculations for the complete Poisson distribution when $\lambda = 1.69$ are given in Table 5–7.

Fig. 5–3 presents a graph of the Poisson distribution for $\lambda = 1.69$. Note that the mean of the distribution is between 1 and 2 (actually, it is 1.69). Note the positive skew of the Poisson distribution. For readers desiring more information on the Poisson distribution, see the references listed for the binomial distribution.

5.5.3 The Normal (Gaussian) Distribution

We now turn to the most famous probability distribution in statistics, called the normal, or gaussian, distribution (or bell-shaped curve). The normal curve was first discovered by French mathematician Abraham Demoivre and published in 1733. However, two mathematician-astronomers, Pierre-Simon Laplace from France and Carl Friedrich Gauss from Germany, were responsible for establishing the scientific principles of the normal distribution. Many consider Laplace to have made the greatest contributions to probability theory, but Gauss's name was given to the distribution after he applied it to the theory of motions of heavenly bodies. Some statisticians prefer to use the term *gaussian* instead of normal because the latter term has the unfortunate (and incorrect) connotation that the normal curve describes the way characteristics are distributed in populations composed of ''normal''—as opposed to sick—individuals. We use the term *normal* in this text because it is most frequently used in the medical literature.

5.5.3.a Describing the Normal Distribution:
The normal distribution is continuous, so it can take on any value (not just integers, as for the binomial and Poisson distributions). It is a smooth, bell-shaped curve and is symmetric about the mean of the distribution, which is symbolized by μ (Greek letter mu). The curve is shown in Fig 5–4.

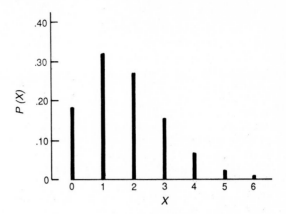

Figure 5–3. Poisson distribution for $\lambda = 1.69$.

The standard deviation of the distribution is symbolized by σ (Greek letter sigma); σ is the horizontal distance between the mean and the point of inflection on the curve—ie, the point where the curve changes from convex to concave. The mean and the standard deviation (or variance) are the two parameters of the normal distribution; ie, μ and σ completely determine the location on the number line and the shape of a normal curve. Thus, there are many different normal curves, one each for every value of μ and σ. Because the normal distribution is a probability distribution, the area under the curve is equal to 1. Because it is a **symmetric distribution,** half the area is on the left of the mean and half is on the right.

Given a random variable X that can take on any value between $-\infty$ and $+\infty$ (∞ represents infinity), the formula for the area under the normal curve is as follows:

$$\int_{-\infty}^{+\infty} \frac{1}{\sqrt{2\pi\sigma^2}} \exp \left[-\frac{1}{2} \left(\frac{x - \mu}{\sigma} \right)^2 \right]$$

where exp stands for the base e of the natural logarithms and $\pi = 3.1416$. Note that the function depends only on μ and σ for a given X, because they are the only components that vary.

Table 5–7. Probabilities for Poisson distribution with $\lambda = 1.69$.

Number of Hospitalizations	1.69^X	$e^{-1.69}$	$X!$	$P(X)$[1]
0	1	.184	0	.184
1	1.69	.184	1	.312
2	2.856	.184	2	.265
3	4.827	.184	6	.148
4	8.157	.184	24	.063
5	13.786	.184	120	.021
6	23.298	.184	720	.006

[1]Rounded to three decimal places.

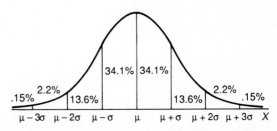

Figure 5–4. Normal distribution and percentage of area under curve.

That the area under the curve is equal to 1 leads to using the curve for calculating probabilities. For example, suppose we want to know the probability that an observation falls between a and b on the curve in Fig 5–5; this probability can be found by integrating the above equation between a and b; ie, in the integration, −∞ is given the value a, and +∞ is given the value b. (Integration is a mathematical technique in calculus used to find area under a curve.)

5.5.3.b The Standard Normal (z) Distribution:

Fortunately, there is no need to integrate this function because tables for it are available. However, so that we do not need a different table for every value of μ and σ, we use the **standard normal curve (distribution),** which has a mean of 0 and a standard deviation of 1, as shown in Fig 5–6. This curve is also called the **z-distribution.** Table A–2 (see Appendix A) gives the area under the curve that is between −z and +z, the sum of the areas to the left of −z and the right of +z, and the area to either the left of −z or the right of +z.

Before we use Table A–2, look at the standard normal distribution in Fig 5–6 and estimate the proportion (or percentage) of these areas:

1. Above 1
2. Below −1
3. Above 2
4. Below −2
5. Between −1 and 1
6. Between −2 and 2

Now turn to Table A–2 and find the designated areas. The answers are given below.

1. .159 of the area is to the right of 1 (from the fourth column in Table A–2).

2. Table A–2 does not list values for z less than 0; however, because the distribution is symmetric about 0, the area below −1 is the same as the area to the right of 1, which is .159.

3. .023 of the area is to the right of 2 (from the fourth column in Table A–2).

4. The same reasoning as in 2 applies; so .023 of the area is to the left of −2.

5. .683 of the area is between −1 and 1 (from the second column in Table A–2).

6. .954 of the area is between −2 and 2 (from the second column in Table A–2).

When the mean of a gaussian distribution is not

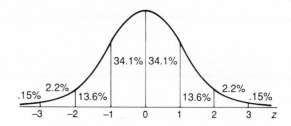

Figure 5–6. Standard normal (z) distribution.

0 and the standard deviation is not 1, a simple transformation, called the z-transformation, must be made so that we can use the standard normal table. The **z-transformation** expresses the deviation from the mean in standard deviation units. That is, any normal distribution can be transformed to the standard normal distribution by using the following two steps:

1. Move the distribution up or down the number line so that the mean is 0. This step is accomplished by subtracting the mean μ from the given variable X.

2. Make the distribution either narrower or wider so that the standard deviation is equal to 1. This step is accomplished by dividing by σ.

To summarize, the transformed value is

$$z = \frac{X - \mu}{\sigma}$$

and is variously called a **z-score,** a normal deviate, a standardized score, or a **critical ratio.**

5.5.3.c Examples Using the Standard Normal Distribution:

To illustrate the standard normal distribution, we consider Presenting Problem 5, in which systolic blood pressure (BP) in normal healthy individuals is normally distributed with μ = 120 and σ = 10 mm Hg. Make the appropriate transformations to answer the following questions. *Hint:* Make sketches of the distribution to make sure you are finding the correct area.)

1. What area of the curve is above 130 mm Hg?
2. What area of the curve is above 140 mm Hg?
3. What area of the curve is between 100–140 mm Hg?
4. What area of the curve is above 150 mm Hg?
5. What area of the curve is either below 90 mm Hg or above 150 mm Hg?
6. What is the value of the systolic blood pressure that divides the area under the curve into the lower 95% and the upper 5%?
7. What is the value of the systolic blood pressure that divides the area under the curve into the lower 97.5% and the upper 2.5%?

The answers, referring to the sketches in Fig 5–7, are given below.

1. z = (130 − 120)/10 = 1.00, and the area above 1.00 is .159. So 15.9% of normal healthy

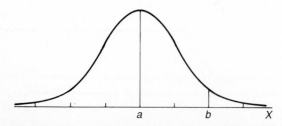

Figure 5–5. Area under normal curve between a and b.

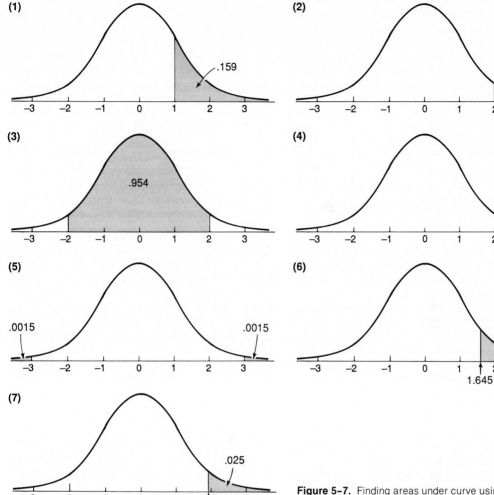

Figure 5-7. Finding areas under curve using normal distribution.

individuals have a systolic blood pressure above one standard deviation (> 130 mm Hg).

2. $z = (140 - 120)/10 = 2.00$, and the area above 2.00 is .023. So 2.3% have a systolic blood pressure above 2 standard deviations (> 140 mm Hg).

3. $z_1 = (100 - 120)/10 = -2.00$, and $z_2 = (140 - 120)/10 = 2.00$; the area between -2 and $+2$ is .954. So 95.4% have a systolic blood pressure between -2 and $+2$ standard deviations (between 100–140 mm Hg).

4. $z = (150 - 120)/10 = 3.00$, and the area above 3.00 is .001. So only 0.1% have a systolic blood pressure above 3 standard deviations (> 150 mm Hg).

5. $z_1 = (20 - 120)/10 = -3.00$, and $z_2 = 3.00$; the area below -3 and above $+3$ is .003. So only 0.3% have a systolic blood pressure either below or above 3 standard deviations (< 90 or > 150 mm Hg).

6. This problem is a bit more difficult and must be worked backward. The value of z, obtained from Table A–2, that divides the lower .95 of the area from the upper .05 is 1.645. Substituting this value for z in the formula and solving for X yields

$$z = \frac{X - \mu}{\sigma}$$

$$1.645 = \frac{X - 120}{10}$$

$$(10)(1.645) = X - 120$$

$$136.45 = X$$

Therefore, a systolic blood pressure of 136.45 mm Hg is at the 95th percentile (granted, this is more specific than the blood pressure measurements generally made). So 95% of normal, healthy people have a systolic blood pressure of 136.45 mm Hg or lower.

7. Working backward again, we obtain the value 1.96 for z. Substituting and solving for X yields

$$1.96 = \frac{X - 120}{10}$$

$$19.6 = X - 120$$

$$139.6 = X$$

Thus, a systolic blood pressure of 139.6 mm Hg divides the distribution of normal, healthy individuals into the lower 97.5% and the upper 2.5%.

From the results of the exercises above, we can state some important guidelines for using the gaussian distribution. As mentioned in Chapter 4, the gaussian distribution has three distinguishing features:

1. The mean ± 1 standard deviation contains approximately 66.7% of the area under the normal curve.

2. The mean ± 2 standard deviations contains approximately 95% of the area under the normal curve.

3. The mean ± 3 standard deviations contains approximately 99.7% of the area under the normal curve.

Although these features indicate that the normal distribution is a valuable tool in statistics, its value goes beyond merely describing distributions. In actuality, few characteristics are normally distributed. The systolic blood pressure data in Presenting Problem 5 surely are not exactly normally distributed in the population at large. In some populations, data are positively skewed: More people are found with systolic pressures above 120 mm Hg than below. And Elveback, Guillier, and Keating (1970) have shown that some very common laboratory values are not normally distributed; consequently, using the guideline about the mean ± 2 standard deviations may cause substantially more or less than 5% of the population to lie outside 2 standard deviations. However, used judiciously, the three guidelines are good rules of thumb about characteristics that have approximately normal distributions.

The major importance of the normal distribution, however, is in the role it plays in statistical inference. In the next chapter, we show that the normal distribution forms the basis for making statistical inferences even when the population is not normally distributed. However, one point is very important and will be made several times: Statistical inference generally involves mean values of a population, not values related to individuals. The applications in this chapter deal with individuals and, if we are to make probability statements about individuals using the mean and standard deviation rules, the distribution of the characteristic of interest must be approximately normally distributed.

5.6 SUMMARY

This chapter focused on three concepts that explain why the results of one study involving a certain set of subjects can be used to draw conclusions about similar subjects. These three concepts are probability, sampling, and probability distributions. After defining probability, we used two examples to illustrate how the rules for calculating probabilities can help us determine the distribution of characteristics in samples of people (eg, the distribution of blood types in men and women; the distribution of the length of time homosexual men are seronegative or seropositive for HIV and its relationship to T4/T8 ratios).

The addition rule, multiplication rule, and modifications of these rules for non-mutually exclusive and nonindependent events were also illustrated. The addition rule is used to add the probabilities of two or more mutually exclusive events. If the events are not mutually exclusive, the probability of their joint occurrence must be subtracted from the sum. The multiplication rule is used to multiply the probabilities of two or more independent events. If the events are not independent, they are said to be conditional; Bayes' theorem should be used to obtain the probability of conditional events. Application of the multiplication rule allowed us to conclude that gender and blood type are independently distributed in people. However, serum status and T4/T8 ratio in healthy homosexual men are not independent characteristics.

The advantages and disadvantages of different methods of random sampling were illustrated for a study involving the measurement of tracheal diameters. A simple random sample was obtained by randomly selecting x-rays corresponding to random numbers taken from a random number table. Systematic sampling was illustrated by selecting each 17th x-ray, and we noted that systematic sampling is easy to use and is appropriate as long as there is no cyclical component to the data. X-rays from different age groups were used to illustrate stratified random sampling. Stratified sampling is the most efficient method and is therefore used in many large studies. In clinical trials, investigators must randomly assign patients to experimental and control conditions (rather than randomly select patients) so that biases threatening the validity of the study conclusions are avoided.

Finally, three probability distributions were presented: binomial, Poisson, and normal (gaussian). The binomial distribution is used to model events that have a dichotomous outcome (ie, either the outcome occurs or it does not) and to determine the probability of outcomes of interest. We used the binomial distribution to obtain the probabilities that a specified number of patients with severe aplastic anemia survive 10 years. We noted that the binomial distribution is often used to determine cumulative probabilities.

The Poisson distribution is used to determine probabilities for rare events. In the CASS study of coronary artery disease, hospitalization of patients during the 5-year follow-up period was relatively rare. We calculated the probability of hospitalization for patients randomly assigned to medical treatment. Exercise 4 asks for calculations for similar probabilities for the surgical group.

The normal distribution is used to determine the probability of characteristics measured on a continuous numerical scale. When the distribution of the characteristics is approximately bell-shaped, the normal distribution can be used to show how representative or extreme an observation is. We used the normal distribution to determine percentages of the population expected to have systolic blood pressures above and below certain levels. We also found the level of systolic blood pressure that divides the population of normal, healthy adults into the lower 95% and the upper 5%. Finally, we noted that the normal distribution also plays a key role in making inferences to other samples, as we shall see in the next chapter.

EXERCISES

1. **a.** Show that gender and blood type are independent; ie, show that the joint probability is the product of the two marginal probabilities for each cell in Table 5–2.
   b. What happens if you use the multiplication rule with conditional probability when two events are independent? Use the gender and blood type data for males, type O, to illustrate this point.

2. Table 5–8 gives the incidence of acute graft-versus-host disease, chronic graft-versus-host disease, and death in subgroups of patients defined according to serum titers of antibodies to cytomegalovirus from Presenting Problem 3. Use the table to answer the following questions.
   a. What is the probability of chronic graft-versus-host disease?

b. What is the probability of acute graft-versus-host disease?
   c. If a patient seroconverts, what is the probability that the patient has acute graft-versus-host disease?
   d. How likely is it that a patient who died was seropositive?
   e. What proportion of patients were seronegative? If this value were the actual proportion in the population, how likely would it be for 4 of 8 new patients to be seronegative?

3. A plastic surgeon wants to compare the number of successful skin grafts in her series of burn patients with the number in other burn patients. A literature survey indicates that approximately 30% of the grafts become infected but that 80% survive. She has had 7 of 8 skin grafts survive in her series of patients and has had one infection.
   a. How likely is only 1 out of 8 infections?
   b. How likely is survival in 7 of 8 grafts?

4. Calculate the Poisson distribution for number of hospitalizations in the surgical group of patients in the CASS study (Presenting Problem 4), and draw a graph of the distribution. (Recall from Section 5.5.2 that there were 950 hospitalizations among the 390 surgical patients.) How does the shape of the Poisson distribution change as λ becomes larger? How likely is it that a surgical patient has exactly 2 hospitalizations in the 5-year follow-up period?

5. The values of serum sodium in healthy adults approximately follow a normal distribution with a mean of 141 meq/L and a standard deviation of 3 meq/L.
   a. What is the probability that a normal healthy adults will have a serum sodium value above 147 meq/L?
   b. What is the probability that a normal healthy adult will have a serum sodium value below 130 meq/L?
   c. What is the probability that a normal healthy adult will have a serum sodium value between 132 and 150 meq/L?
   d. What serum sodium level is necessary to put someone in the top 1% of the distribution?
   e. What serum sodium level is necessary to put someone in the bottom 10% of the distribution?

6. Calculate the binomial distribution for each set of parameters: $n = 6$, $\pi = .1$; $n = 6$, $\pi = .3$; $n = 6$, $\pi = .5$. Draw a graph of each distribution, and state your conclusions about the shapes.

Table 5–8. Incidence of graft-versus-host disease.

Condition	Sero-negative	Sero-converters	Sero-positive
Acute graft-versus-host disease	6	2	2
Chronic graft-versus-host disease	7	8	2
Death	3	3	2
Total patients	17	18	12

6

Drawing Inferences From Data

PRESENTING PROBLEMS

Presenting Problem 1. The arterial blood gases, P_{CO_2}, P_{O_2}, and pH, are useful measures in assessing disturbances in respiratory function. For young healthy adults, the distributions of these measures are approximately normal; the means and standard deviations are 40 and 2.5 mm Hg, 90 and 7 mm Hg, and 7.40 and 0.025, respectively (Scully, McNeely, and Mark, 1986). Information on the distribution of blood gases and pH can be used to make decisions about individual patients and groups of patients. For example, we can answer a question like "What proportion of individuals can be expected to have a P_{O_2} between 83–97 mm Hg?" If a group of six people rather than individuals is of interest, the question becomes "What proportion of the time will the group have a mean P_{O_2} between 83–97 mm Hg?"

Presenting Problem 2. Patients with rheumatoid arthritis are at greater risk of developing osteoporosis, especially women who receive steroid therapy. The reasons are not well understood, partly because of the difficulty in quantitatively assessing bone metabolism and mineral content. Radiologic and nuclear medicine procedures, such as quantitative CT scanning and dual-photon absorptiometry, are expensive to perform repeatedly, can only be applied to certain anatomic areas, and can be technically difficult to perform in a reproducible manner. Biochemical measurements of alkaline phosphatase and urinary hydroxyproline to assess bone metabolism are relatively insensitive and not specific for bone.

Weisman and colleagues (1986) measured plasma levels of bone γ carboxyglutamic acid-containing protein (BGP) in 38 women and 43 men with rheumatoid arthritis. This protein is a specific component of bone, and its concentration in plasma is a sensitive marker for bone formation. In addition, they measured parathyroid hormone, calcitonin and 1,25-dihydroxyvitamin D levels, alkaline phosphatase, and calcium. An unspecified number of age- and sex-matched control subjects were recruited from normal volunteers. The investigators found that mean BGP, human calcitonin, and serum calcium levels were lower in men and women with rheumatoid arthritis than in controls. Women patients who were taking steroids had lower levels of BGP than patients not taking steroids, but men patients did not. Men with rheumatoid arthritis had calcitonin and serum calcium levels lower than controls; women patients did not, although all women had calcitonin levels lower than men. What methods can the researchers use to determine that men patients have lower serum calcium levels than controls?

6.1 PURPOSE OF THE CHAPTER

This chapter completes the process begun in the previous chapter of providing the building blocks needed to understand methods of statistical inference. Although the concepts presented in this chapter may seem difficult, they are absolutely necessary for an understanding of what statistical inference means—or what investigators mean when they make statements like the following: "The difference between treatment and control groups was tested by using a t test and found to be significantly greater than zero." Or "An α value of 0.01 was used for all statistical tests." Or "The sample sizes were determined to give 90% power of detecting a difference of 30% between treatment and control groups."

In the previous chapter we studied the way measurements, such as the number of hospitalizations of patients with coronary artery disease or the blood pressure levels in healthy people, are distributed in the population of individuals of interest. We found that the binomial, Poisson, and normal distributions can be used to determine how likely it is that any specific measurement is in the population. In this chapter, we discuss another type of distribution, called a **sampling distribution,** that is very important in statistics. Understanding sampling distributions is essential for understanding the logic underlying the prototypal statements from the literature given above. Next, we discuss two methods, **estimation** and **hypothesis testing,** that permit investigators working with a sample of observations to generalize study results to the population from which the sample comes.

Our experience indicates that the concepts underlying statistical inference are not easily absorbed in a first reading. We suggest that you read this chapter and become acquainted with the basic ideas. Then, after completing Chapters 7–10, read this chapter again; chances are you will understand the basic ideas of inference much better by using this approach.

6.2 POPULATION DISTRIBUTIONS & THE SAMPLING DISTRIBUTION OF THE MEAN

The distribution of individual measurements is quite different from the distribution of means, which is called a **sampling distribution.** Recall that the major purpose of statistics is to make inferences about a population on the basis of a sample. For example, in Presenting Problem 1 in Chapter 4, the investigators wanted to learn about changes in cholesterol levels and the diameter of coronary arteries in patients with stable angina (Artzenius et al, 1985). From a study of 39 patients, the investigators concluded there is no coronary lesion growth in patients who have low total/HDL cholesterol or who lower their initially high total/HDL cholesterol through diet. The mean decrease in total/HDL cholesterol was 0.6 mmol/L with a standard deviation of 1.2.

In this example, the target population is all patients with stable angina and at least one vessel with 50% obstruction. The sampled population is patients at a hospital in the Netherlands who meet certain criteria, and all patients who met the criteria apparently were included in the study sample. The researchers might like to know the value of the true mean decrease in total/HDL cholesterol if all patients in the target population were placed on a 2-year vegetarian diet, but this study of the entire population is not feasible. Therefore, they use a sample of 39 patients to obtain an estimate of the mean change in the population—ie, $\overline{X}$ equal to 0.6 mmol/L is an estimate of the true, but unknown, mean μ. If another sample of 39 patients with stable angina and at least one 50% obstructed vessel were selected, it is unlikely that exactly the same mean change, 0.6, would be observed. The mean in another group is more likely to be less than 0.6 or more than 0.6, and the researchers wish to know how much more or less can be expected. To find out, they can select many samples from the target population of patients, compute the mean in each sample, and then examine the distribution of means to estimate the amount of variation that can be expected from one sample to another. This distribution of means is called the **sampling distribution of the mean** (or sampling distribution of means), and it has several desirable characteristics that allow us to answer questions about a mean observed in only one sample.

In the next subsection, we use a simple example to illustrate how a sampling distribution can be generated. Then, however, we will show that we need not generate a sampling distribution in practice; instead, we can use theory to answer questions about a single observed mean.

6.2.1 The Sampling Distribution of the Mean

Four features define a sampling distribution. The first feature is the statistic of interest, ie, mean, standard deviation, or proportion. Because the sampling distribution of the mean plays such a key role in statistics, we will use it to illustrate the concept. The second defining feature is random selection of the sample. The third—and very important—feature is the size of the random sample. The fourth feature is specification of the population being sampled.

Let us consider an example to illustrate these concepts. Suppose a physician is trying to decide whether to begin the practice of mailing reminders to patients for their annual examination. The physician examines the files of all patients who have come in for an annual examination during the past month and determines how many months had passed since their previous examination. We will use a very small population size of 5 patients for convenience and simplicity. Table 6–1 lists the number of months since the last examination for the 5 patients in this population. The following subsections present details about generating and using a sampling distribution for this example.

6.2.1.a Generating a Sampling Distribution: To illustrate how a sampling distribution can be generated from this population of 5, we first select random samples of 2 patients per sample and calculate the mean number of months since the last examination for each sample. We replace the observations in the population and repeat the process. For a population of 5, there are 25 different possible samples of 2 that can be selected. That is, patient 1 (12 months since last exam) can be selected as the first observation and returned to the sample; then, patient 1 (12 months), or patient 2 (13 months), or patient 3 (14 months), etc, can be selected as the second observation. The 25 different possible samples and the mean number of months since the patient's last exam are given in Table 6–2.

Table 6–1. Population of months since last examination.

Patient	No. of Months Since Last Exam
1	12
2	13
3	14
4	15
5	16

Table 6–2. 25 samples of size 2 patients each.

Sample	Patients Selected	No. of Months for Each	Mean
1	1, 1	12, 12	12.0
2	1, 2	12, 13	12.5
3	1, 3	12, 14	13.0
4	1, 4	12, 15	13.5
5	1, 5	12, 16	14.0
6	2, 1	13, 12	12.5
7	2, 2	13, 13	13.0
8	2, 3	13, 14	13.5
9	2, 4	13, 15	14.0
10	2, 5	13, 16	14.5
11	3, 1	14, 12	13.0
12	3, 2	14, 13	13.5
13	3, 3	14, 14	14.0
14	3, 4	14, 15	14.5
15	3, 5	14, 16	15.0
16	4, 1	15, 12	13.5
17	4, 2	15, 13	14.0
18	4, 3	15, 14	14.5
19	4, 4	15, 15	15.0
20	4, 5	15, 16	15.5
21	5, 1	16, 12	14.0
22	5, 2	16, 13	14.5
23	5, 3	16, 14	15.0
24	5, 4	16, 15	15.5
25	5, 5	16, 16	16.0

6.2.1.b Comparing the Population Distribution With the Sampling Distribution:

To compare the population distribution and the sampling distribution of means, we display the information in graphs. Fig 6–1 is a graph of the population of patients and the number of months since their last examination. The probability distribution in this population is *uniform,* because every length of time has the same (or uniform) probability of occurrence; because of its shape, this distribution is also referred to as *rectangular.* The mean in this population is 14 months, and the standard deviation is 1.41 months (see Exercise 1).

Fig 6–2 is a graph of the sampling distribution of the mean number of months since the last visit for a sample of 2. The sampling distribution of means is certainly not uniform; it is shaped somewhat like a pyramid. There are several important characteristics of this sampling distribution:

1. The mean of the 25 separate means is 14 months, which is the same as the mean in the population (see Exercise 1).

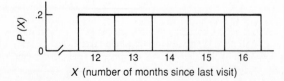

Figure 6–1. Distribution of population values of number of months since last office visit (data from Table 6–1).

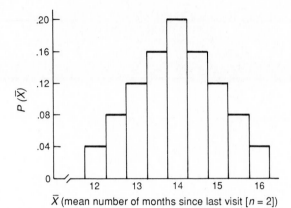

$\bar{X}$ (mean number of months since last visit [$n = 2$])

Figure 6–2. Distribution of mean number of months since last office visit for $N = 2$ (data from Table 6–2).

2. The variability in the sampling distribution of means is less than the variability in the original population. The standard deviation in the population is 1.41; the standard deviation of the means is 1.00 (see Exercise 1).

3. The shape of the sampling distribution of means, even for a sample of 2, is beginning to "approach" the shape of the normal distribution, although the shape of the population distribution is rectangular, not normal.

6.2.1.c Using the Sampling Distribution:

The sampling distribution of means can be extremely useful because it allows us to make probabilistic statements about the likelihood of occurrence of specific observations. For example, using the sampling distribution in Fig 6–2, we can ask questions like "If the mean number of months since a previous exam is really 14, how likely is a random sample of $n = 2$ patients in which the mean is 15 or more months?" From the sampling distribution, we see that a mean of 15 or more can occur 6 times out of 25, or 24% of the time. Therefore, a random sample with a mean of 15 or more is not all that unusual.

In medical studies, the sampling distribution of the mean can answer questions such as "If the proposed hypertensive drug A actually provides no improvement over a placebo, what is the probability of observing a mean difference in blood pressure between treatment A and the control group as great as the one observed in this study?" Or, stated differently, "If there really is no difference, how often would the result (or something more extreme) occur simply by chance?"

6.2.2 The Central Limit Theorem

Generating sampling distributions for the mean each time an investigator wanted to ask a statistical question would be too time-consuming. However, as we noted earlier, this process is not necessary. Instead, statistical theory can be used to determine the sampling distribution of the mean in any particular situation. The properties of the sampling distribution we just observed are the basis for one of the most important theorems in statistics, called the central limit theorem. A mathematical proof of the central limit theorem is not possible in this text, but we will advance some empirical arguments that should convince you of the validity of this theorem. The central limit theorem is as follows:

Given a population with mean μ and standard deviation σ, the sampling distribution of the mean based on repeated random samples of size n has the following properties:

1. The mean of the sampling distribution, or the mean of the means, is equal to the population mean μ based on the individual measurements.

2. The standard deviation in the sampling distribution of means is equal to $\sigma/\sqrt{n}$. This quantity, called the **standard error of the mean**, plays an important role in many of the statistical procedures discussed in later chapters. The standard error of the mean is variously written $\sigma_{\bar{x}}$, $SE(\bar{X})$, or SEM.

3. If the distribution in the population is normal, then the sampling distribution is also normal. More importantly, for sufficiently large sample sizes, the sampling distribution of means is approximately normally distributed, *regardless* of the shape of the original population distribution.

The central limit theorem is illustrated for four different population distributions in Fig 6-3. In row A, the shape of the population distribution is uniform, or rectangular; this distribution can represent, say, the number of months since a previous physical examination. Row B is a bimodal distribution in which extreme values of the random variable are more likely to occur than middle values. Results from opinion polls in which people rate their agreement with political issues sometimes have this distribution, especially if the issue polarizes people. Bimodal distributions also occur in biology when two populations are mixed, as they are for ages of people who have Crohn's disease. Modal ages for these populations are mid 20s and late 40s to early 50s. In row C, the distribution is negatively skewed because of some small outlying values. This distribution can model a random variable over time, such as age of patients diagnosed with colorectal cancer (Presenting Problem 2 of Chapter 3). Finally, row D is similar to the normal (gaussian) distribution.

The second column of distributions in Fig 6-3 illustrates sampling distributions obtained for the means when samples of 2 are randomly selected from the parent populations. In row A, the pyramid shape is quite similar to the sampling distribution of means obtained in the example on months since a patient's last examination. Note that even for the bimodal population distribution in row B, the sampling distribution of means begins to approach the shape of the normal distribution. This bell shape is more evident in the third column of Fig 6-3 in which the sampling distributions are based on sample sizes of 10. Finally, in the fourth row, for sample sizes of 30, all sampling distributions resemble the normal distribution.

A sample of 30 is commonly used as a cutoff; ie, sampling distributions of the mean based on sample sizes of 30 or more are considered to be normally distributed. However, a sample this large is not always needed. If the parent population is normally distributed, the *means of samples of any size* will be normally distributed. In nonnormal parent populations, large sample sizes are required with extremely skewed population distributions; smaller sample sizes can be used with moderately skewed distributions. Fortunately, guidelines about sample size have been developed, and they will be pointed out as they are needed.

In Fig 6-3, also note that in every case the mean of the sampling distributions is the same as the mean of the parent population distribution. The variability of the means decreases as the sample size increases, however, so the standard error of the mean decreases as well. Another feature to note is that the relationship between sample size and standard error of the mean is not linear; it is based on the square root of the sample size, not the sample size itself—ie, SEM = $\sigma/\sqrt{n}$. Therefore, to reduce the standard error by half, one must quadruple, not double, the sample size.

6.2.3 Points to Remember

Several points deserve reemphasis. In actual practice, selecting repeated samples of size n and generating a sampling distribution for the mean is not necessary. Instead, only one sample is selected, the sample mean is calculated as an estimate of the population mean, and, if the sample size is 30 or more, the central limit theorem is invoked to argue that the sampling distribution of the mean is known and does not need to be generated. Then, because the mean has a known distribution, statistical questions can be addressed. Other points to remember are discussed in the following subsections.

6.2.3.a Standard Deviation Versus Standard Error: The value σ measures the standard deviation in the population and is based on measurements of individuals. That is, the standard deviation tells us how much variability can be expected among *individuals*. The standard error of the mean SEM, however, is the standard deviation of the means in a sampling distribution; it tells us how much variability can be expected among *means* in future samples.

For example, in the previous chapter, we used

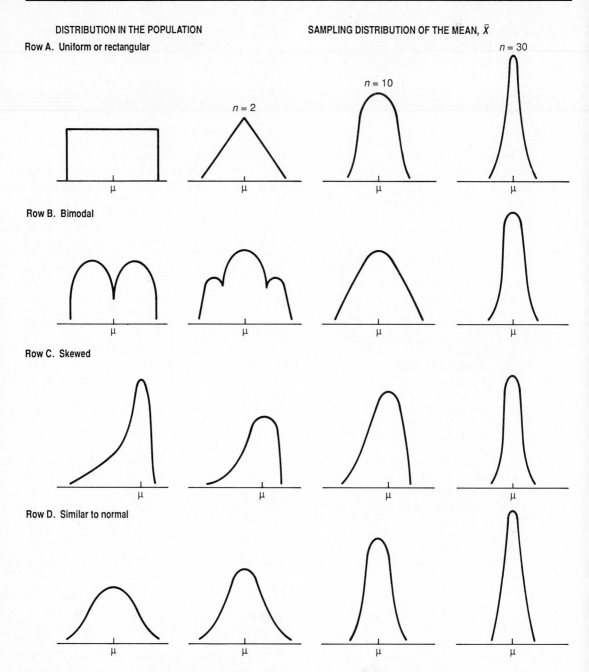

DISTRIBUTION IN THE POPULATION

Row A. Uniform or rectangular

SAMPLING DISTRIBUTION OF THE MEAN, $\bar{X}$

$n = 30$

$n = 10$

$n = 2$

Row B. Bimodal

Row C. Skewed

Row D. Similar to normal

Figure 6-3. Illustration of ramifications of central limit theorem.

the fact that systolic blood pressure is approximately normally distributed in normal healthy populations with mean 120 mm Hg and standard deviation 10 to illustrate how areas under the curve are related to probabilities. We also demonstrated that the interval (mean ± 2 standard deviations) contains approximately 95% of the observations on individuals when the observations have a normal distribution. Because the central limit theorem tells us that a sample mean is normally distributed (when the sample size is 30 or more),

we can use these same properties to relate areas under the normal curve to probabilities when the sample mean instead of an individual value is of interest. Also, we will soon see that the interval (sample mean ± 2 SEM) generally contains about 95% of the *means,* not the individuals, that would be observed if samples of the same size were repeatedly selected.

6.2.3.b The Use of the Standard Deviation in Research Reports: Authors of research reports sometimes present data in terms of the mean and standard deviation. At other times, authors report the mean and standard error of the mean. Some journal editors are beginning to require authors to use the standard deviation and not the standard error of the mean (Bartko, 1985). There are two reasons for this trend. First, the SEM is a function of the sample size, so it can be made smaller simply by increasing n. Second, the interval (mean ± 2 SEM) will contain approximately 95% of the *means* of samples, but it will never contain 95% of the observations on *individuals*; in the latter situation, mean ± 2 standard deviations is needed (as illustrated in Chapter 5). By definition, the SEM pertains to means, not to individuals. Yet when physicians consider applying research results, they generally wish to apply them to individuals in their practice, not to groups of individuals. In most circumstances, the standard deviation is the more appropriate measure to use with the mean.

6.2.3.c Other Sampling Distributions:
Statistics other than the mean also have sampling distributions. For example, there are sampling distributions for standard deviations, medians, proportions, and correlations. In each case, the statistical issue is the same: How can the statistic of interest be expected to vary across different samples of the same size?

Although the sampling distribution of the mean is approximately normally distributed, the sampling distributions of most other statistics are not. In fact, the sampling distribution for the mean assumes that the value of the population standard deviation σ is known. In actuality, it is rarely known; therefore, σ is estimated by the sample standard deviation s, and the SEM, $\sigma/\sqrt{n}$, is estimated by $s/\sqrt{n}$. The sampling distribution of the mean with an estimated standard deviation actually follows a ***t* distribution** instead of the normal distribution. (The t distribution is similar to the normal distribution; it is introduced and discussed in detail in Chapter 7.)

As other examples, the sampling distribution of the ratio of two variances (squared standard deviations) follows an ***F* distribution,** a theoretical distribution presented in Chapters 7 and 8. The proportion, which is based on the binomial distribution, is normally distributed under certain circumstances, as we shall see in Chapter 9. For the correlation to follow the normal distribution, a transformation must be applied, as illustrated in Chapter 10.

One property that all sampling distributions have in common is the concept of standard error: The variation of the statistic in its sampling distribution is called the standard error of the statistic. Thus, the standard error of the mean is just one of many standard errors, albeit the one most commonly referred to in medicine.

6.2.4 Applications Using the Sampling Distribution of the Mean

Let us turn to some applications of the concepts covered so far in this chapter. Recall that the **critical ratio** (or z-score, introduced in the preceding chapter) transforms a normally distributed random variable with mean μ and standard deviation σ to the standard normal *(z)* distribution with mean 0 and standard deviation 1; ie,

$$z = \frac{X - \mu}{\sigma}$$

When we are interested in the sampling distribution of the mean rather than the distribution of individual observations, the mean itself is the entity transformed. According to the central limit theorem, the mean of the sampling distribution is still μ, but the standard deviation (standard error of the mean) is $\sigma/\sqrt{n}$. Therefore, the critical ratio for a mean is

$$z = \frac{\overline{X} - \mu}{\sigma/\sqrt{n}}$$

The use of the critical ratio is illustrated in the following examples.

Example 1. Suppose a physician studies a group of 25 men and women between 20–39 years of age and finds that their mean systolic blood pressure (BP) is 124 mm Hg. How often would a sample of 25 patients have a mean systolic BP this high or higher? Using the data from Presenting Problem 5 in Chapter 5 as a guide, we assume that systolic BP is a normally distributed random variable with a known mean of 120 mm Hg and a standard deviation of 10 mm Hg in the population of normal healthy adults. The physician's question is equivalent to asking: If repeated samples of 25 individuals are selected from the population, what proportion of samples will have mean values greater than 124 mm Hg?

Solution. The sampling distribution of the mean is normal because the population of blood pressures is normally distributed. The mean is 120 mm Hg, and the SEM (based on the known standard deviation) is equal to $10/\sqrt{25} = 10/5 = 2$. Therefore, the critical ratio is

$$z = \frac{124 - 120}{10/\sqrt{25}} = \frac{4}{2} = 2.0$$

From column 4 of Table A–2 (Appendix A) for the normal curve, the proportion of the z distribution area above 2.0 is 0.023; therefore, 2.3% of the samples with $n = 25$ will have a mean systolic BP of 124 mm Hg or higher. Fig 6–4A illustrates how the distribution of means is transformed to the critical ratio.

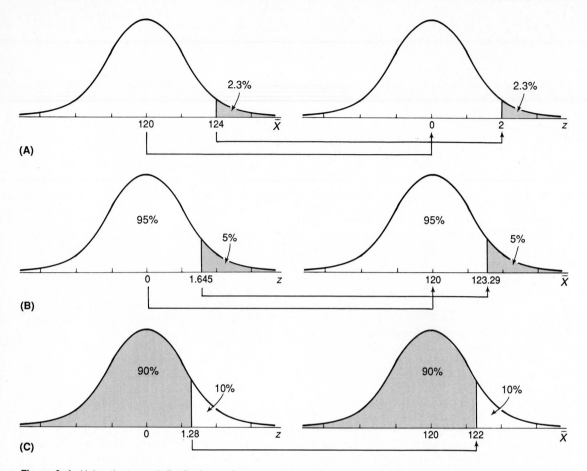

Figure 6-4. Using the normal distribution to draw conclusions about mean systolic blood pressure in healthy adults.

Example 2. Suppose a physician wants to detect adverse effects on systolic BP in a sample of 25 patients using a drug that causes vasoconstriction. The physician decides that a mean systolic BP in the upper 5% is cause for alarm. That is, the physician must determine the value that divides the upper 5% of the sampling distribution from the lower 95%.

Solution. The solution to this example requires working backward from the area under the standard normal curve to find the value of the mean. The value of z that divides the area into the lower 95% and the upper 5% is 1.645 (we find 0.05 in column 4 of Table A-2 and read 1.645 in column 1). Substituting this value for z in the critical ratio and then solving for the mean yields

$$1.645 = \frac{(\overline{X} - 120)}{10/\sqrt{25}} = \frac{\overline{X} - 120}{2}$$

and

$$(1.645)(2) + 120 = \overline{X}$$

or

$$\overline{X} = 123.29$$

Therefore, a mean systolic BP of 123.29 is the value that divides the sampling distribution into the lower 95% and the upper 5%. So there is cause for alarm if the mean in the sample of 25 patients surpasses this value. (See Fig 6-4B.)

Example 3. Continuing with Examples 1 and 2, suppose the physician does not know how many patients should be included in a study of the drug's effect. After some consideration, the physician decides that the mean systolic blood pressure in the sample of patients must nor rise above 122 mm Hg 90% of the time. How large a sample is required so that 90% of the means in samples of this size will be 122 mm Hg or less?

Solution. The answer to this question requires determining n so that only 10% of the sample means exceed $\mu = 120$ by 2 or more, ie, $\overline{X} - \mu = 2$. The value of z in Table A-2 that divides the area into the lower 90% and the upper 10% is 1.28. Using $z = 1.28$ and solving for n yields

$$1.28 = \frac{122 - 120}{10/\sqrt{n}} = \frac{(2)(\sqrt{n})}{10}$$

Therefore,

$$\frac{(1.28)(10)}{2} = \sqrt{n}$$

or $\sqrt{n} = 6.40$ and $n = 6.40^2 = 40.96$

So, a sample of 41 individuals is needed for a sampling distribution of means in which no more than 10% of the mean systolic BPs are above 122 mm Hg. (See Fig 6-4C.)

Example 4. Presenting Problem 1 states that PO$_2$ has a mean of 90 mm Hg with a standard deviation of 7 in normal healthy adults. What proportion of *individuals* can be expected to have a PO$_2$ between 83–97 mm Hg, assuming a gaussian distribution?

Solution. Here, the question involves individuals, and the critical ratio for individual values of X must be used. The mean in the population distribution is 90 mm Hg, and $\sigma = 7$. The trans-

formed values of the z distribution for $X = 83$ and $X = 97$ are

$$X = 97$$

are

$$z = \frac{X - \mu}{\sigma} = \frac{83 - 90}{7} = \frac{-7}{7} = -1.00$$

and

$$z = \frac{X - \mu}{\sigma} = \frac{97 - 90}{7} = \frac{7}{7} = 1.00$$

The proportion of area under the normal curve between -1 and 1, from Table A-2, column 2, is 0.683. Therefore, 68.3% of normal healthy individuals can be expected to have a PO$_2$ between 83–97 mm Hg. (See Fig 6-5A.)

Example 5. If repeated samples of 6 healthy individuals are selected, what proportion will have a *mean* PO$_2$ between 83–97 mm Hg?

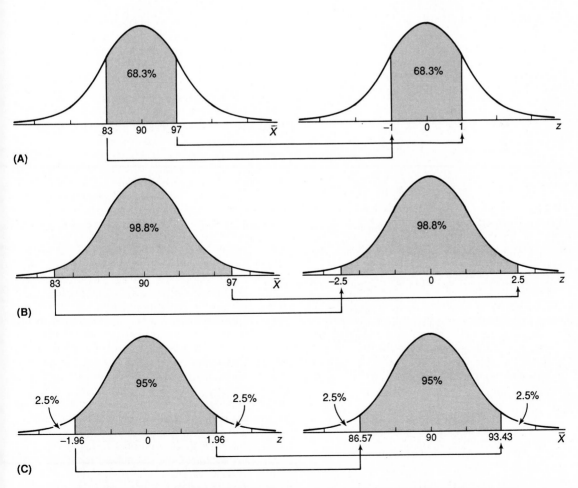

Figure 6-5. Using normal distribution to draw conclusions about levels of PO$_2$ in healthy adults.

Solution. This question concerns means, not individuals, so the critical ratio for means must be used to find appropriate areas under the curve. For $\overline{X} = 83$,

$$z = \frac{\overline{X} - \mu}{\sigma/\sqrt{n}} = \frac{83 - 90}{7/\sqrt{6}} = \frac{-7}{7/2.5} = -2.5$$

Similarly, for $\overline{X} = 97$, $z = +2.5$. Therefore, we must find the area between -2.5 and $+2.5$. From Table A–2, the area is 0.988. Therefore, 98.8% of the area lies between -2.5 and 98.8% of the mean Po_2 values in samples with 6 patients will fall between 83–97 mm Hg. (See Fig 6–5B.)

Examples 4 and 5 illustrate the difference between drawing conclusions about proportions of individuals and drawing conclusions about proportions of means.

Example 6. For 100 healthy individuals in repeated samples, what proportion of the samples will have mean values between 83 and 97 mm Hg?

Solution. We will not do computations for this example; from the previous calculations, we can see that the proportion of means is very large. (The z values are ± 10, which go beyond the scale of Table A–2.)

Example 7. What mean value of Po_2 divides the sampling distribution for 16 individuals into the central 95% and the upper and lower 2.5%?

Solution. The value of z is ± 1.96 from Table A–2. We substitute -1.96 in the critical ratio to get

$$-1.96 = \frac{\overline{X} - 90}{7/\sqrt{16}}$$

and

$$-1.96 \left(\frac{7}{4} \right) + 90 = 86.57 = \overline{X}$$

Similarly, when we use $+1.96$, we get $\overline{X} = 93.43$. Thus, $+93.43$ mm Hg divides the upper 2.5% of the sampling distribution of Po_2 from the remainder of the distribution, and 86.57 mm Hg divides the lower 2.5% from the remainder. (See Fig 6–5C.)

Example 8. What size sample is needed to ensure that 95% of the sample means for Po_2 will be within ± 2 mm Hg of the population mean?

Solution. To obtain the central 95% of any normal distribution, we use $z = 1.96$, as in Example 7. Substituting 1.96 into the formula for z and solving for n yields

$$1.96 = \frac{2}{7/\sqrt{n}}$$

or

$$\sqrt{n} = \frac{(1.96)(7)}{2} = 6.86$$

and

$$n = 6.86^2 = 47.06$$

Thus, a sample of 48 individuals is needed to ensure that 95% of the sample means are within ± 2 mm Hg of the population mean. Note that sample sizes are always rounded up to the next integer.

These examples show how the normal distribution can be used to draw conclusions about distributions of individuals and of means. Although some questions were deliberately contrived in order to illustrate the concepts, the important point is to understand the logic involved in these solutions. The exercises provide additional practice in solving problems of these types.

6.3 ESTIMATION

We have talked about the process of making inferences from data in this chapter and the previous chapter. Now, finally, we have covered all the prerequisite concepts and can begin to illustrate the inference process itself. There are two approaches to statistical inference: estimating parameters and testing hypotheses. In this section we discuss **estimation,** the process of using sample information to draw conclusions about the value of a population parameter; **hypothesis testing** is the subject of the next section.

6.3.1 The Need for Estimates

Suppose we wish to evaluate the relationship between toxic reactions to drugs and falls and resulting fractures among elderly patients. For logistic and economic reasons, we cannot study the entire population of elderly patients to determine the proportion who have toxic drug reactions and fractures. Instead, we conduct a cohort study with a sample of elderly patients followed for a specified period of time. The proportion of patients in the sample who experience drug reactions and fractures can be determined and used as an estimate of the proportion of drug reactions and fractures in the population; ie, the sample proportion p is an estimate of the population proportion π.

In another study, we may be interested in the mean rather than the proportion, so the mean $\overline{X}$ in the sample is used as an estimate of the mean μ in the population. For example, in a study of a low-calorie diet for weight loss, suppose the mean weight loss in a sample of patients is 20 lb; this value is an estimate of the mean weight loss in the population of subjects represented by the sample.

Both p and $\overline{X}$ are called **point estimates** because they involve a specific number rather than an interval or a range. Other estimates include the sample standard deviation s as an estimate of σ and the sample correlation r as an estimate of the population correlation ρ.

6.3.2 Properties of Good Estimates

A good estimate should have certain properties; one is the absence of systematic error, or **unbiasedness.** Recall that when we developed a sampling distribution for the mean, we noted that the mean of the mean values in the sampling distribution is equal to the population mean. Thus, the mean of a sampling distribution of means is an unbiased estimate. Both the mean and the median are unbiased estimates of the population mean μ. However, the sample standard deviation s is not an unbiased estimate of the population standard deviation σ if n is used in the denominator. Recall that the formula for s uses $n - 1$ in the denominator (see Chapter 4, Section 4.3). Using n in the denominator of the sample standard deviation produces an estimate of the population standard deviation that is systematically too small; using $n - 1$ makes the sample standard deviation an unbiased estimate of the population standard deviation.

Another property of a good estimate is small variability from one sample to another; this property is called **minimum variance.** The standard error of the mean is $\sigma/\sqrt{n}$, as demonstrated by the central limit theorem. One reason the mean is used more often than the median as a measure of central tendency is that the standard error of the median is approximately 25% larger than SEM when the distribution of observations resembles the gaussian distribution. Thus, the median has greater variability from one sample to another. Therefore, the chances are greater, in any one sample, of obtaining a median value that is farther away from the population median value than the sample mean is from μ. For this reason, the mean is the recommended statistic when the distribution of observations follows a normal distribution. (If the distribution of observations is quite skewed, however, the median is the better statistic, as we discussed in Chapter 4, because the median has minimum variance in skewed distributions.)

6.3.3 Confidence Intervals and Confidence Limits

Sometimes, instead of giving a simple point estimate, investigators wish to indicate the variability the estimate would have in other samples. To indicate this variability, they use interval estimates. A shortcoming of point estimates, such as a mean weight loss of 20 lb, is that they do not have an associated probability indicating how likely the result is. In contrast, we can associate a probability with an interval estimate, such as the interval from, say, 15 to 25 lb. Interval estimates are called **confidence intervals;** they define an upper limit (25 lb) and a lower limit (15 lb) with an associated probability. The ends of the confidence interval (15 and 25 lb) are called the **confidence limits.**

To illustrate how confidence intervals are determined, we refer to Example 7 of section 6.2.4. We saw that 95% of the area under the standard normal curve is contained between $z = -1.96$ and $z = +1.96$. This result can be used to derive the equation for an interval that has 95% probability of containing the population mean μ. The probability statement, written algebraically, is

$$P(-1.96 \leq z \leq +1.96) =$$

$$P\left(-1.96 \leq \frac{\overline{X} - \mu}{\sigma/\sqrt{n}} \leq +1.96\right) = .95$$

By a series of algebraic manipulations on the terms inside the parentheses (multiplying the expression by $\sigma/\sqrt{n}$, subtracting $\overline{X}$ from all terms, multiplying by -1 to reverse the inequalities, and rearranging terms), the following expression results:

$$P\left(\overline{X} - 1.96 \frac{\sigma}{\sqrt{n}} \leq \mu \leq \overline{X} + 1.96 \frac{\sigma}{\sqrt{n}}\right) = .95$$

$\overline{X} - 1.96(\sigma/\sqrt{n})$ and $\overline{X} + 1.96(\sigma/\sqrt{n})$ are the 95% confidence limits for the population mean μ.

The interpretation of confidence limits is much clearer in the context of an example. In Presenting Problem 2, the investigators wished to draw conclusions about the mean level of serum calcium in a group of 43 male patients with rheumatoid arthritis; the mean in this group was 9.5 mg/dL. They also found a mean of 9.9 mg/dL, with standard deviation 0.66, in a population of healthy male controls.[*] The investigators want to know whether serum calcium measurements in their patients can be viewed as coming from the normal population or whether they represent a different population of patients. The 95% confidence limits for the mean serum calcium in the population from which this group of patients was selected are determined as follows:

$$P\left(\overline{X} - 1.96 \frac{\sigma}{\sqrt{n}} \leq \mu \leq \overline{X} + 1.96 \frac{\sigma}{\sqrt{n}}\right)$$

$$= P\left[9.5 - (1.96)\left(\frac{0.66}{\sqrt{43}}\right)\right.$$

$$\left. \leq \mu \leq 9.5 + (1.96)\left(\frac{0.66}{\sqrt{43}}\right)\right]$$

$$= P[9.5 - (1.96)(0.10) \leq \mu \leq 9.5 + (1.96)(0.10)]$$

$$= P(9.5 - 0.20 \leq \mu \leq 9.5 + 0.20)$$

$$= P(9.3 \leq \mu \leq 9.7)$$

---

[*] The authors actually reported the mean and SEM for control subjects as 9.9 mg/dL and 0.10, respectively. Although the number of control subjects was not specified in the article, we assume there were 43 because the authors stated that controls were sex- and age-matched to the rheumatoid arthritis patients. Then, we worked backward to solve for s, to get $s = \sqrt{43} \times 0.10 = 0.66$.

and 9.3 mg/dL and 9.7 mg/dL are 95% confidence limits.

We interpret the resulting interval from 9.3 to 9.7 mg/dL as having 95% probability of containing the true population mean μ. Because this interval does not contain 9.9, the conclusion is tenable that this group of 43 rheumatoid arthritis patients has a mean serum calcium lower than that in a normal healthy control population. (To be accurate, the confidence interval is interpreted as follows: 95% of such confidence intervals will contain the true population mean μ if repeated samples are taken and 95% confidence intervals are calculated for each sample. However, in actual practice, only one sample is taken, and the resulting confidence interval is interpreted as having 95% probability of containing the true mean.)

There is nothing magical about 95% confidence intervals, although they are used more often than any others. Other confidence limits traditionally used in the medical literature correspond to 90% and 99%.

We also note that we can rewrite the above formula so that it is easier to remember and use. Thus, 95% confidence limits for the true population mean μ are given by

$$\overline{X} \pm 1.96 \frac{\sigma}{\sqrt{n}}$$

where 1.96 is the multiplier for 95% confidence.

To illustrate 99% confidence limits, let us assume that the researchers investigating serum calcium wish to determine 99% confidence limits for the true mean serum calcium in the population from which their study patients came. The standard normal distribution in Table A–2 is used to obtain the number corresponding to 99% confidence, which is then substituted for 1.96 in the above formula. The term *99% confidence* refers to the two values of z that separate the distribution into the central 99% and the upper and lower 1/2%. From Table A–2, the value of z corresponding to an area of 0.005 in the upper tail is 2.575. In the population, the standard deviation is 0.66, and the observed mean in the sample of 43 male patients with rheumatoid arthritis was found to be 9.5 mg/dL. Therefore, 99% confidence limits for the true mean serum calcium are

$$\overline{X} \pm 2.575 \frac{\sigma}{\sqrt{n}} = 9.5 \pm (2.575)\left(\frac{0.66}{\sqrt{43}} \right)$$
$$= 9.5 \pm (2.575)(0.10)$$
$$= 9.5 \pm 0.26$$

or **9.24 and 9.76**

We have 99% confidence that the interval from 9.24 to 9.76 mg/dL includes the true population mean μ. Note that the 99% confidence interval is

wider than the 95% confidence interval found above, reflecting that a greater range of values must be included for greater confidence. Correspondingly, 90% intervals are narrower than 95% intervals; ie, a narrower range of values is possible if we are willing to have less confidence that the interval includes the mean.

Confidence intervals can be established for any population parameter. You may commonly encounter confidence intervals for the mean, proportion, relative risk, odds ratio, and correlation, as well as for the difference between two means, between two proportions, etc. Confidence intervals for these parameters will be introduced in subsequent chapters. We see an increasing use of confidence intervals in the medical literature; so after presenting hypothesis tests, we will return to the subject of confidence intervals and compare them with hypothesis tests.

6.4 HYPOTHESIS TESTING

As with estimation and confidence limits, the purpose of a **hypothesis test** is to permit generalizations from a sample to the population from which it came. Both statistical hypothesis testing and estimation make certain assumptions about the population and then use probabilities to estimate the likelihood of the results obtained in the sample, given the assumptions about the population.

To illustrate the concepts of hypothesis testing, let us refer to Presenting Problem 2 again. From the values calculated for their sample of controls, the investigators assume that serum calcium in normal men has a mean equal to 9.9 mg/dL, with a standard deviation of 0.66. They studied 43 men with rheumatoid arthritis and found a mean serum calcium level of 9.5 mg/dL, and they want to know whether the mean in the study sample is different from the population mean value of 9.9 mg/dL. Another way to state this question is: Can male patients with rheumatoid arthritis be considered to come from the same population as normal, healthy males with respect to serum calcium levels?

Statistical hypothesis testing first assumes that the answer to this question is yes and then determines the probability of a mean serum calcium equal to 9.5 mg/dL in a group of 43 men given this assumption—ie, that the true mean value of their serum calcium is really 9.9 mg/dL. If the probability is large, we conclude that the assumption is justified and that men with rheumatoid arthritis have a mean serum calcium level equal to that in normal men. If the probability is small, however—such as 1 out of 20 (.05) or 1 out of 100 (.01)—then we conclude that the assumption is not justified and that there really is a difference, ie, that men with rheumatoid arthritis have a mean serum calcium level that is different from that in normal controls.

6.4.1 Steps in Testing Statistical Hypotheses

A *statistical* hypothesis is a statement of belief about population parameters. Like the term *probability,* the term *hypothesis* has a more precise meaning in statistics than in its everyday use, as we will see shortly. The steps involved in testing a statistical hypothesis are outlined in the following subsections.

6.4.1.a Step 1. State the Research Question in Terms of Statistical Hypotheses:
The **null hypothesis,** H_0, is a statement claiming that there is no difference between the population mean μ and the assumed or hypothesized value; ie, *null* means "no difference." The **alternative hypothesis,** H_1, is a statement that disagrees with the null hypothesis. If the null hypothesis is rejected as a result of sample evidence, then the alternative hypothesis is the conclusion. If there is not sufficient evidence to reject the null hypothesis, it is retained but *not* "accepted." Scientists distinguish between "not rejecting" the null hypothesis and "accepting" it; they argue that a better study may be designed in which the null hypothesis will be rejected. Therefore, we do not "accept" the null hypothesis from current evidence; we merely state that it cannot be rejected.

For Presenting Problem 2, the null and alternative hypotheses are as follows:

H_0: μ = 9.9 mg/dL   **(The mean in the population is 9.9 mg/dL.)**

H_1: $\mu \neq$ 9.9 mg/dL   **(The mean in the population is not 9.9 mg/dL.)**

These hypotheses result in a **two-tailed (or nondirectional) test:** The null hypothesis will be rejected if the serum calcium level is sufficiently greater than 9.9 or if it is sufficiently less than 9.9. A two-tailed test is appropriate when the investigators do not have an a priori expectation regarding the value they expect to observe in the sample; they want to know if the sample mean is different from the population mean, in either direction.

A **one-tailed** (or **directional**) **test** occurs when investigators do have an a priori expectation about the size of the sample mean, and they want to test only whether it is larger or smaller than the population mean. A one-tailed alternative hypothesis is

H_1: $\mu <$ 9.9 mg/dL   **or** H_1: $\mu >$ 9.9 mg/dL

Note that it is the alternative hypothesis that indicates whether a test is one-tailed or two-tailed.

The advantage of a one-tailed test is that statistical significance can be obtained with a smaller departure from the hypothesized value, because the interest is in one direction only. Therefore, whenever a one-tailed test is used, the investigators should have been interested in a departure in only one direction before the data were examined. The disadvantage of a one-tailed test is that once investigators commit themselves to this approach, they are obligated to test only in the hypothesized direction. If, for some strange reason, the sample mean departs from the population mean in the opposite direction than the one hypothesized, they cannot claim the departure as significant.

6.4.1.b Step 2. Decide on the Appropriate Test Statistic for the Hypotheses:
Statistics, such as the critical ratio, whose primary use is in testing hypotheses are called **test statistics.** Choosing a test statistic is a major topic in statistics, and subsequent chapters focus on which test statistics are appropriate to answer specific research questions.

Each test statistic has a probability distribution. In this example, the appropriate test statistic is based on the standard normal distribution because the sample size is large and the population standard deviation is known. That is, we will assume that the mean is normally distributed in large samples, according to the central limit theorem, and use as the test statistic the critical ratio

$$z = \frac{\overline{X} - \mu}{\sigma/\sqrt{n}}$$

If the population standard deviation σ is not known—which is almost always the situation in research—the appropriate test statistic is based on the t distribution instead of the z distribution; the t statistic is discussed in Chapter 7.

6.4.1.c Step 3. Select the Level of Significance for the Statistical Test:
The **level of significance,** when chosen before the statistical test is performed, is called the **alpha value,** denoted by α (Greek letter alpha); it gives the probability of incorrectly rejecting the null hypothesis when it is actually true. This probability should be small, because we do not want to reject the null hypothesis when it is true. Traditional values used for α are .05, .01, and .001. Let us be very traditional in this example and choose α = .05.

Another concept related to significance is *P*-value. A *P*-**value** is always related to a hypothesis test; it is the probability of obtaining a result as extreme as or more extreme than the one observed, *if* the null hypothesis is true. Some people like to think of the *P*-value as the probability that the observed result is due to chance alone. The *P*-value is calculated *after* the statistical test has been performed; if the *P*-value is less than α, the null hypothesis is rejected.

6.4.1.d Step 4. Determine the Value the Test Statistic Must Attain to Be Declared Significant:
This "significant" value is called the **critical value** of the test statistic. Determining the critical value is quite a simple process: Each test statistic has a distribution; the distribution of the test statistic is divided into an area of acceptance and an area of rejection. For a one-tailed test, the area of rejection is in either the lower or the upper tail of the distribution. For a two-tailed test, there are two areas of rejection, one in each tail of the distribution. The areas of acceptance and rejection are determined by the value chosen for α. Fig 6–6 illustrates the areas of acceptance and rejection.

An illustration should make the process clear. The test statistic in the example we are considering follows the standard normal (*z*) distribution; α is .05; and a two-tailed test was specified by the alternative hypothesis. Thus, the area of acceptance is the central 95% of the *z* distribution, and the areas of rejection are the 2.5% of the area in each tail. (See Fig 6–6.) From Table A–2, the value of *z* that defines these areas is −1.96 for the lower tail and +1.96 for the upper tail. Therefore, the null hypothesis that the mean serum calcium level is equal to 9.9 mg/dL will be rejected if the critical value of the test statistic is less than −1.96 or greater than +1.96.

6.4.1.e Step 5. Perform the Calculations:
To summarize once more, the mean serum calcium level observed in the 43 men with rheumatoid arthritis was 9.5 mg/dL. This value is to be compared with the population value of 9.9 mg/dL, with standard deviation 0.66, in normal male controls. Substituting these values in the test statistic yields

$$z = \frac{\overline{X} - \mu}{\sigma/\sqrt{n}} = \frac{9.5 - 9.9}{0.66/\sqrt{43}} = \frac{-0.4}{0.10} = -4.00$$

6.4.1.f Step 6. Draw and State the Conclusion:
Draw the appropriate conclusion from the testing procedure, and state the conclusion in words. The latter point is important because, in our experience, readers sometimes learn the mechanics of hypothesis testing but are not able to apply the concepts. In the example, the observed value for *z* is −4.00. Locating this value for *z* in Fig 6–6A, we see that −4.00 falls in the rejection area in the lower tail of the distribution, to the left of −1.96. Therefore, the decision is to reject the null hypothesis that the mean serum calcium for men with rheumatoid arthritis is equal to 9.9 mg/dL. Another way to state this conslusion is: Reject the hypothesis that the sample of men with rheumatoid arthritis comes from a population with a mean serum calcium of 9.9 mg/dL. Thus, the con-

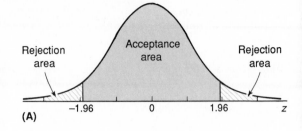

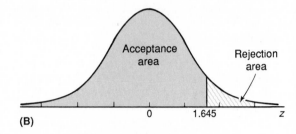

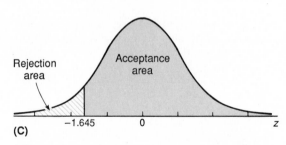

Figure 6-6 Defining areas of acceptance and rejection in standard normal distribution using *a* = .05. (*A*) Two-tailed or nondirectional. (*B*) One-tailed or directional upper tail. (*C*) One-tailed or directional lower tail.

clusion is that men with rheumatoid arthritis have serum calcium levels different from 9.9 mg/dL; they are lower. The probability of observing a mean level as extreme as 9.5 mg/dL in a random sample of 43 men, if the true mean level is actually 9.9 mg/dL, is less than .05, the alpha value chosen for the test.

The actual *P*-value for this test cannot be obtained from Table A–2 because this table lists values of *z* only between −3.00 and +3.00. For *z* = ±3.00, the two-tailed *P*-value is .003; for *z* = ±4.00, the *P*-value is .00006 (obtained from more complete tables published elsewhere). Some authors report actual *P*-values; others report them as less than some traditional value. In this example, the *P*-value can be reported as *P* = .00006 or, alternatively, as *P* is less than .0001. The practice of reporting values less than some traditional value was established prior to the availability of computers when statistical tables such as Table A–2 were the only source of probabilities. Today, most researchers use computers to analyze their data, and computer programs generally give the exact probability; so some authors present exact values. We prefer the practice of reporting exact

P-values when they are available; otherwise, the arbitrary traditional values may lead an investigator (or reader of a journal article) to conclude that a result is significant when $P = .05$ but is not significant when $P = .06$.

6.4.2 Errors of Hypothesis Testing & Power

Two errors can be made in a test of hypothesis. We referred in step 3 to one of these errors—rejecting the null hypothesis when it is true—as a consideration when one selects the significance level α for the test. There is another error as well: not rejecting the null hypothesis when it is actually false, ie, not accepting the alternative hypothesis when it is true. Table 6–3 summarizes these errors. The situations marked by the asterisk (*) are correct decisions: rejecting the null hypothesis when a difference exists, and not rejecting the null hypothesis when there is no difference.

The situation marked by I is rejecting the null hypothesis when it is really true and is called a **type I error;** α is the probability of a type I error. In the rheumatoid arthritis study, a type I error is concluding that serum calcium levels are different (rejecting the null hypothesis) if in fact they are the same as in normal men (when the null hypothesis is true).

A **type II error** occurs in the situation marked by II; this error is failing to reject the null hypothesis when it is false (or not rejecting the null hypothesis when the alternative hypothesis is true). The probability of a type II error is generally denoted by β (Greek letter beta). In the arthritis example, a type II error is concluding that the serum calcium level in men with rheumatoid arthritis is the same as in normal healthy men if it is actually different.

Another important concept related to hypothesis testing is power. **Power** is defined as the probability of rejecting the null hypothesis when it is false or of accepting the alternative hypothesis when it is true. Some people think of power as the ability of a study to detect a true difference. In general terms, power is the ability of a study to

detect a difference of a given size if the difference really exists. High power is a good thing to have in a study, because all investigators want their study to detect a significant result if it really is present. Power is calculated as $1 - \beta$.

6.4.2.a Analogies to Hypothesis Testing: An analogy may help you better understand the meaning of hypothesis testing. Certain measures of diagnostic tests—eg, sensitivity, specificity—provide a straightforward analogy to hypothesis testing. A type I error, incorrectly concluding significance when the result is not significant, is similar to a false-positive test that incorrectly indicates presence of a disease when it is not present. Similarly, a type II error, incorrectly concluding no significance when the result is significant, is analogous to a false-negative test that incorrectly indicates no presence of disease when it is present. The power of a statistical test, the ability to detect significance when a result is significant, is analogous to the sensitivity of a diagnostic test, ie, its ability to detect a disease that is present. We may say we want the statistical test to be "sensitive" to detecting significance when it should be detected. These concepts are discussed in detail in Chapters 13 and 14.

Another useful analogy is to the legal system in the USA. The null hypothesis is equivalent to the assumption that a person is innocent until proved guilty. Just as it is the responsibility of the prosecution to show evidence that the accused person is guilty, the researcher must provide evidence that the null hypothesis is false. In the legal system, in order to avoid a type I error of convicting an innocent person, the prosecution must provide evidence to convince jurors "beyond the shadow of a doubt" that the accused is guilty before the null hypothesis of innocence can be rejected. In hypothesis testing, the evidence for a false null hypothesis must be such that the probability of incorrectly rejecting the null hypothesis is very small, generally less than .05.

The legal system would rather err in the direction of setting a guilty person free than unfairly convicting an innocent person. In statistical test-

Table 6-3. Correct decisions and errors in hypothesis testing.

		True situation	
		Difference exists (H₁)	No difference (H₀)
Conclusion from hypothesis test	Difference exists (Reject H₀)	* (Power or 1 − ß)	I (Type I error, or α error)
	No difference (Do not reject H₀)	II (Type II error, or ß error)	*

ing, the tradition is to prefer the error of missing a significant difference (arguing, perhaps, that others will come along and design a better study) to the error of incorrectly concluding significance when a result is not significant. These two errors are, of course, related to each other. If a society decides to reduce the number of guilty people that go free, it must increase the likelihood of innocent people being convicted. Similarly, an investigator who wishes to decrease the probability of missing a significant difference by decreasing β must compensate by increasing the probability α of declaring a false difference true. The only way the legal system can simultaneously reduce both types of errors is by requiring more evidence for a decision; similarly, the only way both type I and type II errors can be simultaneously reduced is by increasing the sample size n. When increasing the sample size is not possible—because the study is exploratory, the problem studied is rare, or the costs are too high—the investigator must carefully evaluate the values of α and β and make a judicious decision.

6.4.2.b Specifying α and β and Power: Most journal editors now require authors to specify the value of α or the *P*-values for statistical tests, which is not difficult to do. Specifying probabilities of type II errors after statistical tests have been done is a more difficult task, however. A type I error occurs when there truly is no difference, and this error can happen in only one way. A type II error occurs when there truly is a difference, and this error can happen in many different ways. For example, if the null hypothesis is true—the mean serum calcium level in men with rheumatoid arthritis equals 9.9 mg/dL—then μ is equal to 9.9 mg/dL. If, however, the alternative hypothesis is true—the mean is not equal to 9.9 mg/dL—then there are an infinite number of possible values for μ, (9.8, 10.0, 10.5, 9.2, etc). Furthermore, rejecting a false null hypothesis is easier if the value in the true alternative hypothesis is quite different. In the serum calcium example, we are more likely to obtain a sample mean that leads to rejecting the null hypothesis (H_0: $\mu = 9.9$ mg/dL) if the true mean serum calcium in men with rheumatoid arthritis is really 9.5 than if it is 9.7.

Unfortunately, type II errors are generally not explicitly discussed in journal articles. However, the issue can be addressed by describing the *power* of the statistical test to detect a specified difference. The following statements are typical: "This study was designed to have 90% power of detecting a difference of 0.2 mg/dL between the mean for study patients and the mean for normal controls," or "The sample size was determined to provide 80% power to detect a difference of 0.2 mg/dL or greater." In the first case, the value for β is .10; in the second case, it is .20.

The process of determining the power of a study is sometimes called **power analysis.** It consists of determining how large a sample is required to detect an actual difference of some specified magnitude. The power of a study should be determined before the study begins; otherwise, the study may require more time and resources than are available. Alternatively, the study may use more subjects than needed, thus wasting resources in another way.

The ramifications of not performing power calculations can be serious. For instance, a study may have low power because the sample size was too small to detect the presence of reasonable differences; it is a "negative" study. As a consequence, a promising line of research may be abandoned. To illustrate our point, Freiman et al (1978) reviewed 71 negative clinical trials published in 20 different medical journals between 1968–1977. They found that the sample size was so small in 50 of the studies that the investigators would have missed a 50% therapeutic improvement. DerSimonian et al (1982) examined 67 clinical trials in four leading US and UK journals and found, on the average, that only 12% of the articles addressed issues of sample size or power. Similarly, Brown et al (1987) investigated 14 negative trials in emergency medicine and found that only one had addressed these issues.

Power calculations are also essential when an investigator wants to conclude that two drugs or two procedures are not significantly different. In principle, this can be accomplished easily by having a very small sample size. The lesson for practitioners who read the literature is that whenever differences are not significant, they must know the power of the study for detecting differences. Quite possibly, a difference exists but the sample size was too small to detect it. Without information about power, readers cannot arrive at a conclusion. When information on power is given, it most often is presented near the end of the Method section of a journal article, generally in the same paragraph where statistical methods used in the study are listed. In subsequent chapters we illustrate the process of determining the sample size needed to attain specific levels of power for some of the commonly used statistical procedures.

6.4.3 Confidence Intervals vs Hypothesis Tests

Readers of the medical literature generally encounter more hypothesis tests and *P*-values than confidence limits or intervals. However, the practice of using hypothesis tests is changing. One explanation for the change may be that some investigators prefer confidence intervals because they

Es

PRE

P

pho
can
Alth
cage
tory
mus
failu
(198
func
corr
tory
scol
had
pne:
pers
lung
were
com
have
max
mea
of ;
give
the
of 4
scol
for
nor

P
with
gra\
carl
care
cor
ran;
lon;
the
ratc
hyp

T
pre
higl
Tirl
car
the
hyr

remind the person who reads and uses the results that the estimates presented in the study have variability and the same results may not be obtained if the study is replicated. A second possible reason is that confidence intervals provide the same information that a statistical test provides and more. For instance, in the serum calcium example, the null hypothesis that mean serum calcium levels equal 9.9 mg/dL in rheumatoid arthritic men was rejected with $\alpha = .05$; the P-value was .00006. The reader knows from the size of the P-value that observing a mean serum calcium level of 9.5 mg/dL is very unlikely if the true mean is 9.9 mg/dL. However, the 95% confidence interval was 9.3 and 9.7 mg/dL; this interval can be interpreted as a rejection of *all* null hypotheses associated with values less than 9.3 and greater than 9.7. Therefore in one sense, confidence intervals can be viewed as a summary of many statistical tests. A third and perhaps more important explanation relates to the purpose of the study. Many times, the objective is to estimate a parameter rather than to test any particular hypothesis about a parameter. For example, investigators may be more interested in an estimate of the mean serum calcium level in patients with rheumatoid arthritis and how that mean may vary in other groups of men with this disease than in a test of the hypothesis that serum calcium levels are lower in arthritic patients than in normal men. In this study, using confidence intervals makes more sense than performing hypothesis tests.

Many statisticians prefer confidence intervals to hypothesis tests, because in confidence intervals the role played by sample size is clear. For example, in a study involving a large sample size, even a trivial difference will be statistically significant, although the clinical significance of the difference may be very small; confidence intervals clearly illustrate the magnitude of the difference. As another example, the results of a negative study may be more appropriately interpreted if confidence intervals are used rather than tests of hypothesis, because confidence intervals expose the high degree of uncertainty resulting from small sample sizes. Confidence intervals also avoid the issue of what to report when a hypothesis test is not significant at .05 but is significant at .06.

In summary, practitioners who read the medical literature should become familiar with interpreting confidence intervals and hypothesis tests. There is every indication that more and more results will be presented by using confidence intervals. For example, the *British Medical Journal* has established the policy of having its authors use confidence intervals instead of hypothesis tests if confidence intervals are appropriate to their study (Gardner and Altman, 1986). To give you practice with both approaches to statistical inference, we will use both hypothesis tests and con-

fidence intervals throughout the remaining chapters.

6.5 SUMMARY

The next several chapters will help clarify the ideas presented in this chapter, because we shall reiterate the concepts and illustrate the process of estimation and hypothesis testing in a variety of situations. Therefore, these concepts, while difficult, will become easier to understand and use with practice.

We began this chapter by discussing the sampling distribution of the mean. If we know the sampling distribution of the mean, we can observe and measure only one sample, draw conclusions from that sample, and generalize the conclusions to what would happen if we had observed many similar samples. Relying on theory saves time and effort and allows research to proceed.

Next, we presented the central limit theorem, which says that the distribution of the *mean* follows a normal distribution, regardless of the shape of the parent population, as long as the sample sizes are large enough. Generally, a sample of 30 observations or more is large enough. We used the values of Po_2 from Presenting Problem 1 and values of blood pressure from Presenting Problem 5 in Chapter 5 to illustrate use of the normal distribution as the sampling distribution of the mean.

Estimation and hypothesis testing are two methods for making inferences about a value in a population of subjects by using observations from a sample of subjects. They approach inference from different perspectives but are entirely consistent with each other. Estimation involves the calculation of confidence limits for some statistic, such as the mean or the proportion. The confidence intervals obtained tell us that a given percentage, say 95%, of such confidence intervals contain the true value of the mean (or the proportion) in the population. We interpret confidence intervals as indicating the degree of confidence we can have in their containing the true mean (or proportion). The current trend is toward increased use of estimation and confidence intervals.

Hypothesis tests involve a set of formal steps to determine whether the sample provides sufficient evidence to draw hypothesized conclusions. Note that the null hypothesis, which is assumed to be true, is frequently just the opposite of what the investigator really thinks is true. The object in this case is to reject the null hypothesis in favor of the alternative hypothesis. Because hypothesis testing leads to a yes-or-no decision, two errors are possible: declaring a difference when none exists (called a type I or α error) and failing to declare a difference that does exist (a type II or β error).

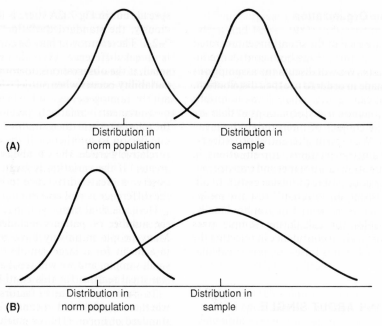

Figure 7-2. Comparison of distributions.

and statistical tests based on the t distribution are used to answer these questions.

The t **distribution** is sometimes called "Student's t," after the man who first noted the distribution of means from small samples in 1908. "Student" was really William Gosset, a mathematician who worked for the Guinness Brewery; he was forced to use the pseudonym Student because of company policy prohibiting publishing. He noted that, regardless of the sample size n, the sampling distribution of the mean of observations from a normally distributed population is also normally distributed when the population standard deviation σ is known. However, if σ is not known, it must be estimated by the sample stan-

Table 7-1. Inspiratory mouth pressure (Pimax) in kyphoscoliotic patients.[1]

Patient Number	Pimax (cm H_2O)
1	54.8
2	62.0
3	63.3
4	44.2
5	40.3
6	36.3
7	19.3
8	24.6
9	26.6
Mean	41.27
Standard deviation	16.23

[1]Adapted and reproduced, with permission, from Table 3 in Lisboa C et al: Inspiratory muscle function in patients with severe kyphoscoliosis. *Am Rev Respir Dis* 1985; **132**:48.

dard deviation s. This procedure presents no problem for large samples because s is a reliable estimate of σ with samples of size 30 or greater. When n is less than 30, however, the critical ratio (using s as an estimate of σ) is not normally distributed; Gosset named this distribution t.

The critical ratio or t statistic, defined as

$$\frac{\overline{X} - \mu}{s/\sqrt{n}} = \frac{\overline{X} - \mu}{SEM}$$

follows the t distribution. The distribution of t is similar to the standard normal (z) distribution in that it is symmetric with a mean of 0, but its standard deviation depends on a parameter called **degrees of freedom.** As the sample size increases, however, the standard deviation of the distribution approaches one, just as in the z distribution. Degrees of freedom is a complex concept beyond the scope of this book, but it is related to sample size. For example, for one sample, the degrees of freedom are $n - 1$. In a later chapter, we will attempt an intuitive description of the concept of degrees of freedom.

Fig 7-3 illustrates the standard normal distribution and the t distributions with one, five, and 25 degrees of freedom (df). Note that the t distribution is flatter than the standard normal distribution, and its tails are higher and wider, indicating that it has a greater standard deviation, especially for small sample sizes. As the sample size n increases, the degrees of freedom also increase; and the t distribution approaches the standard normal distribution. When n is 30 or more, the two curves are very close; either distribution can be used to

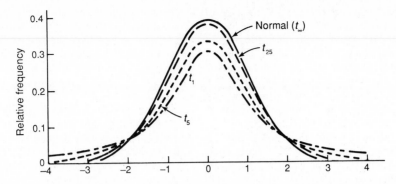

Figure 7-3. t distribution with 1, 5, and 25 degrees of freedom and standard normal (z) distribution.

answer statistical questions, although most researchers in medicine use the t distribution, even with large sample sizes. Therefore, in this book, we illustrate the common situation in which the sample standard deviation and the t distribution are used, regardless of the size of the sample. Because the t distribution has a greater standard deviation than the z distribution for n less than 30, using the t distribution requires investigators to observe a greater difference in order to declare the results significant.

As with the z distribution, we can integrate a mathematical function to find the area under the curve in the t distribution; and computers typically provide exact areas. However, we can also use tables (developed before the computers were readily available) to obtain critical values for the t distribution corresponding to commonly used areas under the curve. Table A–3 in Appendix A gives critical values for areas related to α values of .10, .05, .02, .01 and .001 for two-tailed tests (and for α values half that size for one-tailed tests). Note that for $n = \infty$ in Table A–3, the values are the same as in the z distribution in Table A–2. We will use Table A–3 in the examples to determine critical values related to specified levels of α in determining confidence limits, performing hypothesis tests, and estimating P-values.

The assumption needed for use of the t distribution with one sample is that the observations follow a normal (gaussian) distribution. There are two ways to evaluate this assumption: empirically or logically. Both methods are discussed, along with alternative approaches to use when data are not normally distributed, in Section 7.2.4.

7.2.2 CONFIDENCE INTERVALS FOR THE POPULATION MEAN

Confidence intervals for the mean were illustrated in Chapter 6, Section 6.3, using the z distribution. The general formula for 95% confidence limits

for the unknown population mean μ uses the sample mean $\overline{X}$. When the population standard deviation σ is known, the formula is as follows:

$$\overline{X} \pm 1.96 \frac{\sigma}{\sqrt{n}} = \overline{X} \pm 1.96(\text{SEM})$$

where 1.96 is the critical value from the z distribution corresponding to 95% confidence.

When σ is not known, it is replaced by s in the above formula. In addition, ± 1.96 is replaced by the critical value from the t distribution reflecting 95%. The formula for confidence limits when σ is unknown is

$$\overline{X} \pm t \frac{s}{\sqrt{n}} = \overline{X} \pm t(\text{SEM})$$

where t is the appropriate critical value from the t distribution. In Table A–3, the critical value corresponding to 95% confidence limits is the value for 5% of the area in two tails, the third column in Table A–3. The value of t, as indicated earlier, is also a function of the degrees of freedom, or $n - 1$ for a single sample. There are 9 observations in the Pimax study, so there are 8 degrees of freedom, and the value of t from Table A–3 is 2.306. Therefore, 2.306 replaces 1.96 in the above formula to reflect the larger standard deviation of the t distribution. Substituting 41.27 for the mean and 16.23 for the standard deviation gives

$$\overline{X} \pm 2.306(\text{SEM}) = 41.27 \pm (2.306)\left(\frac{16.23}{\sqrt{9}}\right)$$

$$= 41.27 \pm (2.306)\left(\frac{16.23}{3}\right)$$

or

28.79 and 53.75 cm H_2O

Thus, we have 95% confidence that the interval of 28.79 and 53.75 cm H_2O contains the actual *mean* maximal inspiratory pressure in the population of patients from which this sample was taken. Even though this interval is wide, it seems unlikely that the patients with kyphoscoliosis come from the same population as patients without kyphoscoliosis, because the latter group has a mean Pimax of 110 cm H_2O. If the investigators want to be more confident that the interval contains the true mean, they can use a 99% confidence interval, which, of course, will be wider than the 95% interval. To increase the precision with which the mean Pimax is measured in the patient group, the investigators would need to increase the sample size and thus obtain a narrower confidence interval.

An "eyeball" or graphic test follows directly from the concept of a confidence interval. Fig 7-4 is a plot of the mean Pimax *with 95% confidence limits*. The mean in the normal population, symbolized by the asterisk (*), is outside the confidence interval, so we have 95% confidence that 110 is not the mean Pimax of patients with kyphoscoliosis. Note that some studies in the literature present similar graphs, except that the lines on each side of the mean represent *one standard deviation* or *one standard error*. The only time the eyeball test can be used, however, is when confidence intervals are reported or when 2 standard errors are reported with sample sizes of 30 or more.

7.2.3 Test of Hypothesis for the Mean

If, instead of evaluating the estimate of mean Pimax by determining confidence limits, the re-searchers want to perform a test of hypothesis, the six-step procedure outlined in Chapter 6 can be used. A test of hypothesis evaluates whether the mean maximal inspiratory pressure of patients with kyphoscoliosis is equal to a known standard (norm). Of course, the graph in Fig 7-4 indicates that there is really no need to perform a hypothesis test. But for the sake of illustration, we assume that the authors have not created such a graph and thus wish to test the hypothesis that the mean Pimax in patients with kyphoscoliosis is equal to 110 cm H_2O and to obtain a *P*-value. The steps are illustrated below.

Step 1.   State the null and alternative hypotheses.

H_0: $\mu = 110$ (The true mean Pimax value is 110.)
H_1: $\mu \neq 110$ (The true mean Pimax value is *not* 110.)

Step 2.   Decide on the appropriate test statistic.
This problem involves one group measured on one numerical variable, and the question involves a difference from a standard. If we assume that the variable "maximal inspiratory pressure" is normally distributed and the value of the standard deviation is unknown, the appropriate test is the *t* test. The test statistic, which follows the *t* distribution with $n - 1$ degrees of freedom, is

$$ t = \frac{\overline{X} - \mu}{s/\sqrt{n}} = \frac{\overline{X} - \mu}{SEM} $$

with $n - 1$ df.

Step 3.   Select a value for α.
The value of α is the probability of a type I error, ie, the probability of concluding that Pimax is different for patients wth kyphoscoliosis when it is not. We assume that the investigators choose $\alpha = .05$.

Step 4.   Determine the critical value of the test.
For the *t* distribution with $9 - 1 = 8$ degrees of freedom, $\alpha = .05$, and a two-tailed test to detect a difference greater than 110 as well as a difference less than 110, the value from Table A-3 is 2.306. Because the test is two-tailed, the decision is to reject the null hypothesis if the value of the *t* statistic is less than -2.306 or greater than $+2.306$. The graphic form of this decision strategy is given in Fig 7-5.

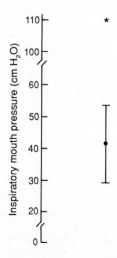

Figure 7-4. Visual assessment of difference between sample mean and norm, using 95% confidence limits. (Adapted, with permission, from Lisboa C et al: Inspiratory muscle function in patients with severe kyphoscoliosis. *Am Rev Respir Dis* 1985;**132**:48.)

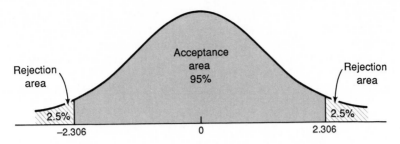

Figure 7-5. Areas of acceptance and rejection for testing hypothesis that mean Pimax = 100 cm H_2O (α = .05, two-tailed).

Step 5. Perform the calculations.

$$t = \frac{41.27 - 110}{16.23/\sqrt{9}}$$

$$= \frac{-68.73}{5.41}$$

$$= -12.70$$

Step 6. Draw and state the conclusion. The value of the test statistic, −12.70, is much less than −2.306, and it falls in the rejection area on the left-hand side of the curve in Fig 7-5. Therefore, we reject the null hypothesis that the true mean Pimax for patients with kyphoscoliosis is equal to 110. Our conclusion may be stated in one of several ways: (1) The difference is statistically significant; or (2) sampling variation is an unlikely explanation for the observed difference; or (3) it is likely that the mean Pimax for patients with kyphoscoliosis is different from (less than) the mean Pimax for patients without this condition.

Calculating the P-value is left until Section 7.3.3. As a final note, compare the calculations for the confidence limits and for the hypothesis test. Observe that the degrees of freedom are the same, the value of t from the t distribution is the same (and always will be for a two-tailed test using the same α level), and the conclusions are identical. Whether an investigator uses confidence limits or a hypothesis test is primarily a matter of preference, although confidence limits may be more versatile.

7.2.4 Single Groups With Data Not Normally Distributed

As we stated earlier, one assumption for use of the t distribution for one group or mean is that the observations are normally distributed. This assumption can be tested empirically by using the observed data and plotting the observations on a normal probability graph called a Lilliefors graph (Iman, 1982). Some computer programs will produce a plot of the data on a normal probability scale; for example, SPSS will plot a normal distribution overlay on a histogram, and SAS provides a normal probability plot as part of UNIVARIATE. (See Fig 3-18 in Chapter 3 for an illustration of a normal probability plot for the colorectal cancer data.) Often, you can simply examine the distribution plotted as a histogram or box and whisker plot for obvious skewness. In actual practice, investigators sometimes assume normality from prior experience with similar data.

What is the ramification of using the t distribution for observations that are not normally distributed? Little problem results for a reasonable sample size, such as 30 or more, because the central limit theorem (see Chapter 6) then ensures that the means are normally distributed, regardless of the distribution of the data. With sample sizes smaller than 30, however, using the t distribution with observations that are not normally distributed will generally result in a reported P-value that is smaller than its actual value or in a confidence interval that is too narrow. How much smaller or how much narrower depends on how much the data depart from the normal distribution. If the departure is small, the t distribution may still be used because it is somewhat *robust* for data that are not normally distributed. **Robustness** means that the conclusion is unchanged even when the assumptions are not met. However, if the departure is large, another procedure should be used. The two alternative procedures are transforming the scale of the observations or using different statistical methods, called nonparametric procedures, to analyze the observations.

7.2.4.a Transformations: A **transformation** expresses the values of observations on another scale. Transforming observations before they are analyzed may be desirable for several reasons. **Linear transformations,** for instance, which involve only a change in the origin (or mean) and a scaling factor, are often done for convenience; eg, the z transformation is used to reexpress the mean as 0 and the standard deviation as 1. In this chapter, we are discussing the need for observations to follow a normal distribution so that the t distribution applies. In this case, we want to change the shape of the distribution and thus need to use a **nonlinear transformation.** Nonlinear transformation can be used to "straighten out" the relationship between two measures and make it linear.

Consider the survival time of patients who are diagnosed as having cancer of the prostate. A graph of possible values of survival time (in years) for a group of patients with prostate cancer metastatic to the bone is given in Fig 7–6A. The distribution has a substantial positive skew, so methods that assume a normal distribution would not be appropriate. Fig 7–6B illustrates the distribution if the logarithm of survival time is used instead, ie, $X' = log(X)$. The log transformation spreads out the observations with smaller values and compresses (or "bunches up") the larger values. Another transformation that has much the same effect is the square root transformation, $X' = \sqrt{X}$, but this transformation is not used as frequently in medicine as the log transformation.

Another transformation was discussed in Chapter 4: the rank transformation. Observations are rank-ordered from lowest to highest; the lowest observation is assigned the number 1, the second lowest observation the number 2, etc. The rank transformation is especially appropriate when the objective is to use statistical methods to analyze skewed observations. All the nonparametric procedures illustrated in this chapter are based on ranks of observations.

Transformations can also be used to stabilize the variance (or standard deviation) so that it is independent of the value of the mean; these transformations are especially appropriate when obser- vations follow the binomial or Poisson distribution. Recall that for the binomial distribution, the mean is π and the variance is $n\pi(1 - \pi)$; we see from the formula that the size of the variance is related to the size of the mean. Therefore, researchers using the binomial distribution sometimes convert the binomial to an angle measured in either degrees or radians. The transformation is $X' = \sin^{-1}\sqrt{X/n}$ (or, equivalently, $X' = $ arcsin $\sqrt{X/n}$), and the resulting transformed data are approximately normally distributed. The logit transformation is sometimes used to analyze dose-response relationships or survival data when the response is a proportion. The logit transformation is the natural logarithm of the ratio of the proportion to 1 − the proportion, ie, $\ln[p/(1 - p)]$. When the Poisson distribution is of interest (recall that the mean and the variance of the Poisson distribution are the same), the square root transformation, $X' = \sqrt{X}$, can be used to obtain an approximately normal distribution. If you are interested in learning more about transformations, see the text by Murphy (1982; see Chapter 4).

7.2.4.b Nonparametric Procedures: The alternative to performing a transformation is to use statistical procedures called nonparametric methods. **Nonparametric methods** are based on weaker assumptions; they do not assume a normal distribution, and thus, they are also called distribution-free methods. There is no nonparametric procedure for analyzing the difference between a single group and a norm; but there are procedures for analyzing the difference between two groups, and we discuss them as their applications arise. A useful discussion of nonparametric methods is given in the article by Moses, Emerson, and Hosseini (1984).

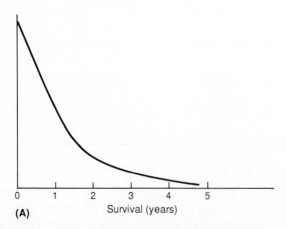

(A) Survival (years)

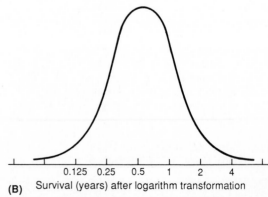

(B) Survival (years) after logarithm transformation

Figure 7–6. Example of logarithm transformation for survival in patients with cancer of prostate metastatic to bone.

7.3 DECISIONS ABOUT PAIRED GROUPS

Two research designs focus on the *difference* between two observed means (rather than the difference between the mean in one sample and a standard mean, as in the previous section). One design involves two totally separate (independent) groups: In one group, patients may receive a treatment; and in another independent group, patients may receive a placebo. The *t* test for two independent groups is appropriate for this design and is discussed in Section 7.4. The other situation is called a **matched, or paired, design.** The most common paired design results when one group is measured twice, as in Presenting Problem 2. Frequently, the first measurement occurs before treatment or intervention, and the second measurement occurs after treatment; the purpose is to decide whether the intervention makes a difference. In this design, each subject serves as his or her own "control." Less often, subjects are matched on one or more relevant characteristics and then treated as though they were the same subject. For example, this procedure is used in laboratory experiments: One litter mate is given the treatment, and another litter mate serves as the control. The goal of paired designs is to control for extraneous factors that might influence the result; then, any differences caused by the treatment will not be masked by the differences among the subjects themselves.

7.3.1 Reasons for Using a Paired Design

Paired designs can be very powerful in detecting differences, which makes them highly useful. Let us examine a simplified example. Suppose a researcher wants to evaluate the effect of a new diet for weight loss. The population consists of 6 patients who have used the diet for 2 months; their weights before and after the diet are as follows:

Patient	Weight Before (kg)	Weight After (kg)
1	100	95
2	89	84
3	83	78
4	98	93
5	108	103
6	95	90

To estimate the amount of weight loss, the researcher decides to select a random sample of 3 patients before the diet to determine their mean weight and an independent random sample of 3 patients after the diet to determine their mean weight. Just by chance, the random sample of patients in the before-diet sample includes patients 2, 3, and 6; their mean weight is $(89 + 83 + 95)/3 = 89$ kg. Also just by chance, the random sample of patients in the after-diet sample consists of patients 1, 4, and 5; their mean weight after the diet is $(95 + 93 + 103)/3 = 97$ kg. (Of course, we have contrived the makeup of the random samples in this example to make a point, but they really could occur by chance.)

The conclusion is that patients gained an average of 8 kg on the diet. What is the problem here? The means for the two independent random samples indicate that the patients gained weight; in fact, they each lost 5 kg on the diet. We know the conclusion based on random samples is incorrect because we can examine the entire population and determine the actual differences; but in real life, we can rarely observe the population. The problem is that the characteristic being studied—weight—is quite variable from one patient to another; in this small population of 6 patients, weight varies from 83 to 108 kg before the diet program. Furthermore, the amount of change, 5 kg, is relatively small compared with the variability among patients and is overwhelmed (or overshadowed) by this variability. The researcher needs a way to control for variability among patients. The solution, as you may have guessed, is to select a single random sample of patients and measure their weights both before and after the diet; because the measurements are taken on the same patients, the true change is more likely to be observed. The paired design allows the researcher to detect change more easily by controlling for extraneous variation among the observations. Many biologic measurements exhibit wide variation among individuals, and the use of the paired design is thus especially appropriate in medicine.

The paired design is analyzed by using the **paired *t* test** (or, synonymously, **matched-group *t* test** or correlated *t* test). Observations from this design are especially simple to analyze, because the difference between the two measurements on each subject is treated as a single measure. Then, the procedure for a single mean, illustrated in Section 7.2, is used. This approach works because the difference between two means is equal to the mean of the differences.

To illustrate, examine the mean weights of the 6 subjects in the weight loss example. Before the diet, the mean weight was 95.5 kg; after the diet, the mean was 90.5 kg. The difference between the means, $95.5 - 90.5 = 5$ kg, is exactly the same as the mean weight loss, 5 kg for each subject. The standard deviation of the differences, however, is not necessarily equal to the difference between the standard deviations in the after and before measurements. The differences themselves must be analyzed in order to obtain the standard deviation of the differences. Finally, as with the *t* test for one mean, use of the paired *t* test assumes that the observations are normally distributed.

7.3.2 Confidence Intervals for the Mean Difference in Paired Designs

To demonstrate the steps in calculating and interpreting the paired observations, we consider the study by Tirlapur and Mir (1984) on the effect of low-calorie intake on abnormal pulmonary physiology in patients with chronic hypercapneic respiratory failure (Presenting Problem 2). Measurements of patients' arterial oxygen tension and arterial carbon dioxide tension before and after the weight loss program are given in Table 7–2.

The mean and the standard deviation of arterial oxygen tension are 55.6 and 9.2, respectively, before the program and 69.1 and 7.9, respectively, after the program. To calculate the confidence limits for the difference, we must subtract arterial oxygen tension measures before the weight reduction program from measures after the program. (The subtraction can be done the other way as well, after from before, and the same conclusion will be reached; we prefer differences expressed as positive numbers.) The differences are given in Table 7–2.

The calculations for the mean and the standard deviation of the differences used the formulas given in Chapter 4, but we replace the X's in the formulas with d's (for the difference). So the mean difference is

$$\overline{d} = \frac{\Sigma d}{n}$$

and the standard deviation of the differences is

$$s_d = \sqrt{\frac{\Sigma(d - \overline{d})^2}{n - 1}}$$

where n refers to the number of patients (8 in our example). The mean difference and the standard deviation of differences are 13.5 and 8.2, respectively.

The procedure for determining 95% confidence interval for the mean difference is the same as the procedure presented in Section 7.2.2 (single mean); but we use $\overline{d}$ instead of $\overline{X}$ and the standard error of the differences ($SE_{\overline{d}}$) instead of SEM:

$$\overline{d} \pm (t \text{ value})(SE_{\overline{d}})$$

The use of 95% confidence limits is traditional, but other limits, such as 99% and 90%, can be employed as well. We will use 90% for this example.

For the 8 patients, there are $8 - 1 = 7$ degrees of freedom; and the value of t from Table A–3 for 90% confidence limits is 1.895. The standard error of the differences $SE_{\overline{d}}$), is the standard deviation of the differences divided by the square root of the sample size: $8.2/\sqrt{8} = 8.2/2.83 = 2.9$. Therefore, the 90% confidence limits are

$$13.5 \pm (1.895)(2.9) = 13.5 \pm 5.5$$

or

8 and 19 mm Hg

Thus, we can be 90% sure that the interval from 8 to 19 mm Hg contains the actual mean increase in arterial oxygen tension for patients after a weight reduction program. Note that zero is not contained in the interval; therefore, we can be 90% sure that the actual mean difference is *not* zero. Of course, if greater confidence is desired, 95% or 99% confidence limits can be used.

There is no accurate way to examine graphs of

Table 7-2. Patients with chronic hypercapneic respiratory failure.[1]

Patient	Arterial Oxygen Tension (mm Hg)			Arterial Carbon Dioxide Tension (mm Hg)		
	Before	**After**	**Difference**[2]	**Before**	**After**	**Difference**[3]
1	70	82	12	49	45	4
2	59	66	7	68	54	14
3	53	65	12	65	60	5
4	54	62	8	57	60	−3
5	44	74	30	76	59	17
6	58	77	19	62	54	8
7	64	68	4	49	47	2
8	43	59	16	53	50	3
Mean	55.6	69.1	13.5	59.9	53.6	6.3
Standard deviation	9.2	7.9	8.2	9.6	5.9	6.5

[1]Adapted and reproduced, with permission, from Table 1 in Tirlapur VG, Mir MA: Effect of low-calorie intake on abnormal pulmonary physiology in patients with chronic hypercapneic respiratory failure. *Am J Med* 1984; **77**:987.
[2]After minus before.
[3]Before minus after.

the 95% confidence intervals for the *two* means and tell whether the mean difference is significant. Accurate methods are not available because the standard error of the difference is a complicated function of the standard errors of the two means and, in addition, depends on the nature of the relationship between the two sets of measurements. Therefore, the graphic test for a mean difference between two paired sets of observations requires that we know the mean difference and the standard deviation (or the standard error) of the differences. Then, we graph the 95% confidence interval for the mean difference. If the graph does not include zero, we may be 95% confident that a change has occurred, ie, there is a difference. Fig 7–7 illustrates the 90% confidence interval for the mean arterial oxygen tension in our example.

Unfortunately, journal articles often do not include the information needed for the graphic test. They frequently give the means and the standard deviations (or standard errors) of the two groups but not the mean and the standard deviation (or standard error) of the differences. Therefore, you may not be able to perform this graphic test on published data.

7.3.3 Test of Hypothesis for the Mean Difference in Paired Designs

The test of hypothesis about the mean difference with paired observations is identical to the test about a single mean, with differences d replacing the X values. In Section 7.3.2, we concluded, with 90% confidence, that the mean change in arterial oxygen tension was between 8–19 mm Hg; therefore, the hypothesis that the difference is equal to

zero will be rejected. (In fact, if the confidence interval does not include zero, a hypothesis test is really not necessary.) Let's illustrate this hypothesis test and show that we reach the same conclusion as with the confidence interval.

Step 1. State the hypotheses.

H_0: $\delta = 0$ **(The mean difference in the population is zero.)**

H_1: $\delta = 0$ **(The mean difference is not zero.)**

Here, δ (Greek letter delta) is the parameter for the mean difference in the population.

Step 2. Decide on the test statistic. The test statistic for comparing the means between two paired groups, assuming normality, is the t statistic for the mean difference:

$$t = \frac{\overline{d}}{SE_{\overline{d}}}$$

with $n - 1$ degrees of freedom. Note that the formula assumes that the hypothesis being tested is "The difference is equal to zero," which is the most common case. However, we can also test the hypothesis that the difference is equal to a particular value, such as $\delta = 10$; then, the numerator of the t statistic is $\overline{d} - \delta$.

Step 3. Select a value for α. To be consistent with the confidence limits, we choose $\alpha = .10$.

Step 4. Determine the critical value. The alternative hypothesis specifies a two-tailed test. The value of the t distribution with 7 degrees of freedom that separates the central 90% of the distribution from the 5% in each tail, from Table A-3, is 1.895. Therefore, we will reject the null hypothesis of no difference if the calculated value of t is greater than 1.895 or less than -1.895. See Fig 7–8.

Step 5. Perform the calculations. The mean difference is 13.5, the standard deviation of the differences is 8.2, and the sample size is 8. Therefore, the standard error of the differences, $SE_{\overline{d}}$ is 2.9, as calculated earlier. Then,

$$t = \frac{\overline{d}}{SE_{\overline{d}}} = \frac{13.5}{2.9} = 4.66$$

Step 6. Draw and state the conclusion. The observed value of t, 4.66, is greater than 1.895. Therefore, we reject the null hypothesis of no difference and conclude, with $P < .10$, that there is a difference (increase) in arterial oxygen

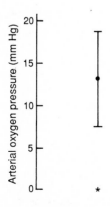

Figure 7–7. Visual assessment of mean increase in paired groups, using 90% confidence limits. (Adapted, with permission, from Tirlapur VG, Mir MA: Effect of low-calorie intake on abnormal pulmonary physiology in patients with chronic hypercapneic respiratory failure. *Am J Med* 1984;**77**:987.)

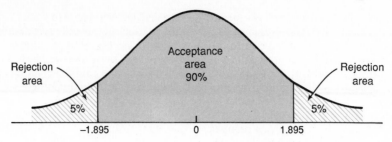

Figure 7–8. Areas of acceptance and rejection for testing hypothesis on mean change in arterial oxygen tension ($\alpha = .10$, two-tailed).

tension after the weight reduction program. (We hope you are becoming convinced of the correspondence between confidence limits and tests of hypothesis.)

To determine the *P*-value specifically, we note that the values of the *t* distribution, from Table A–3 for 7 degrees of freedom corresponding to 1% of the area in the tails, are ± 3.499. The values corresponding to .1% of the area are ± 5.408. Therefore, the *P*-value for 4.66 is between 1% and 0.1%, written as $0.001 < P < .01$. The *P*-value gives the probability of observing a difference this great if there actually is no difference in mean arterial oxygen tension following a weight reduction program. Computer programs generally give the exact *P*-value to three or four decimal places.

7.3.4 The Signed-Ranks Test for the Mean Difference

The paired *t* test requires that the observations being analyzed follow a normal (gaussian) distribution. This assumption is important with small sample sizes, *n* less than 30. If researchers suspect that the distribution of the variable of interest departs significantly from the normal distribution, either from theory or from the empirical Lilliefors test (Iman, 1982), they should either transform the observations or use a nonparametric statistical test that does not require the observations to be normally distributed.

The most commonly used nonparametric test for the difference between two paired samples is the **Wilcoxon signed-ranks test.** It actually tests the hypothesis that the medians, rather than the means, are equal in the two paired samples. It is an excellent alternative to the *t* test because it is almost as powerful (rejecting the null hypothesis when it is false) as the *t* test in detecting an actual difference when the observations follow a normal distribution. When the observations are not normally distributed, it is more powerful than the *t* test. For this reason, the Wilcoxon signed-ranks test is increasingly used in medical research.

The traditional calculations for the Wilcoxon signed-ranks test require either extensive counting (of all observations larger or smaller than all others) or the use of extensive tables. Perhaps for this reason, this method is not included in most introductory statistics texts. However, two statisticians (Iman, 1974; Conover and Iman, 1981) developed a simpler approach that is a very good approximation of the more cumbersome method. The procedure is to convert the observations to a rank order and then calculate the *t* statistic by using ranks instead of the original observations. As with the paired *t* test, the raw data must be available; the Wilcoxon signed-ranks test statistic cannot be calculated from the values of the means and the standard deviations.

Consider an example. We suspect that the differences in arterial oxygen tension are not normally distributed; a stem and leaf plot of the observations would indicate a positively skewed distribution because of the value 30 mm Hg for Patient 5. So let's perform the Wilcoxon test for these data. The differences in arterial oxygen tension are listed in Table 7–3 along with the steps needed to calculate the Wilcoxon signed-ranks test statistic.

The first step is to write the absolute value of the differences in a column in the table (column 3). Because none of the patients in this study had a decrease in arterial oxygen tension, there are no negative values; so the absolute values of the differences are the same as the original differences. The second step is to rank-order the absolute value of the differences, from smallest (for patient 7) to largest (for patient 5) (column 4). In the arterial oxygen tension data, there is one tie; patients 1 and 3 have a difference of 12. These patients rank as the fourth and fifth smallest values, and they are each assigned the mean rank of 4.5. For three tied observations—eg, 17th, 18th, and 19th—the mean rank of 18 is assigned to each; etc. The third step is to "sign the ranks," ie, give each rank a + or − sign according to the original difference (column 5). In this example, all the differences are positive; but generally they will not be (see Exercise 4).

To compute the Wilcoxon signed-ranks test statistic, we treat the signed ranks as though they were the original observations. Thus, the mean

Table 7-3. Arterial oxygen tension before and after weight loss.[1]

(1) Patient	(2) Difference ($X_2 - X_1$)	(3) Absolute Value of Difference	(4) Rank *R*	(5) Signed Rank *SR*
1	+12	12	4.5	+4.5
2	+7	7	2	+2
3	+12	12	4.5	+4.5
4	+8	8	3	+3
5	+30	30	8	+8
6	+19	19	7	+7
7	+4	4	1	+1
8	+16	16	6	+6

[1]Adapted and reproduced, with permission, from Table 1 in Tirlapur VG, Mir MA: Effect of low-calorie intake on abnormal pulmonary physiology in patients with chronic hypercapneic respiratory failure. *Am J Med* 1984; **77**:987.

and the standard deviation of the ranks are computed and are used to obtain the standard error of the ranks. The idea behind this approach is that positive and negative ranks will tend to cancel each other if some patients have a positive difference and others have a negative difference. Then, the *t* statistic is calculated from

$$t = \frac{\overline{R}}{SE_{\overline{R}}}$$

with $n-1$ degrees of freedom. The mean of the ranks, $\overline{R}$, is 36/8 = 4.5; the standard deviation is 2.43. Therefore, the standard error of the ranks is $2.43/\sqrt{8}$ = .86. The critical value of the *t* distribution, with $n - 1 = 7$ degrees of freedom and $\alpha = .10$, is 1.895, as we found earlier. The observed value of *t* is

$$t = \frac{4.5}{0.86} = 5.23$$

Because the observed value of *t* is greater than the critical value—ie, 5.23 is greater than 1.895—we reject the null hypothesis of no difference. Therefore, even if the data are not normally distributed, the differences are all in the same direction (all patients experienced an increase in arterial oxygen tension). With 8 patients, this result is sufficient to indicate, with *P* less than .10, that an actual difference has occurred. In this example, the conclusion is the same as the conclusion for the *t* test on the original observations.

To summarize, the Wilcoxon signed-ranks test is appropriate for small samples that are not normally distributed. It does not require the assumption of normality as does the *t* test. In this test the *rank* of the difference is given the same *sign* as the original difference (thus, the name *signed-ranks* test). The method we used to calculate the Wilcoxon test statistic (converting the observations to ranks and using the ranks in a *t* test) is an approximate procedure; however, the approximation is acceptable.

Two final comments are worth mentioning.

First, because the test is performed on the ranks of the observations instead of the original observations, it is not very informative to determine confidence limits. Second, the appropriate use of nonparametric statistics is extremely important in drawing correct conclusions from data; and using these procedures is quite easy, because almost all statistical packages for computers perform these tests.

7.4 DECISIONS ABOUT TWO INDEPENDENT GROUPS

When observations come from two separate or independent groups, such as in Presenting Problem 3, and the research question is whether there is a difference between the means of the two groups, the appropriate test is the **two-sample independent-groups *t* test.** As with one sample, the *z* distribution can be used if the population standard deviations are known for each group; however, this condition rarely applies. The *z* distribution can also be used if sample sizes are 30 or more; but in this situation, the *z* and the *t* distributions are very similar. For these reasons and because of its more frequent use in medicine, only the *t* test is illustrated.

The assumptions needed for use of the two-sample *t* test for independent groups are more complicated than those for one-sample or paired-sample tests. The assumption that the observations be normally distributed in both groups remains, and the ramifications of violating this assumption are as discussed earlier: Reported *P*-values are lower than they should be, and confidence intervals are narrower than they should be. If the sample size is 30 or more in each group, the need for normality is reduced. With smaller samples from populations that are not normally distributed, however, another nonparametric procedure, discussed in Section 7.4.4, should be used.

The second assumption of the two-sample *t* test is that the variances or standard deviations be equal in the two groups (homogeneous variances).

This assumption arises from the null hypothesis, which says that the means are equal, or, equivalently, that the observations in the two groups come from the same population. Thus, under the null hypothesis, the variances are the same as well. Fortunately, statistical research has shown that the *t* test is robust with respect to unequal variances, and this assumption may be ignored if the sample sizes are equal (Box, 1953). Satisfying this second assumption is one of the reasons investigators strive for fairly equal numbers in each group. If the sample sizes are not equal, the assumption of equal variances should be tested before a *t* test is performed, and many computer programs do this test as a matter of course. An illustration of the test, called the *F* test for equal variances, and the steps to take if the variances are not equal are covered in Section 7.4.3.

The third and last assumption is that the observations occur independently. Operationally, independence means that knowing the value of an observation in one group does not provide any information about the value of an observation in the second group. Unfortunately, this assumption is not generally given much attention and there is no statistical test to determine whether it has been violated. Independence is a consequence of how the study is designed and carried out; this topic is addressed in Chapter 15.

7.4.1 Confidence Intervals for the Difference Between Independent Means

We will use data from Presenting Problem 3—measures of the highest urinary excretion of 5-HIAA in the study of carcinoid heart disease, to illustrate the calculation of confidence limits for the differences between two means. The data are given in Table 7–4. Recall that the general form for a confidence interval is

Estimate ± (Factor related to level of confidence)
× (Standard error of estimate)

For the difference between two means, the estimate is simply the difference. That is, if $\overline{X}_1$ is the mean of sample 1, and $\overline{X}_2$ is the mean of sample 2, then the estimate of the difference is $\overline{X}_1 - \overline{X}_2$. The factor related to the level of confidence is a critical value from the *t* distribution, with degrees of freedom equal to the sum of the two sample sizes minus 2. That is, if n_1 is the number of subjects in sample 1, and n_2 is the number of subjects in sample 2, then the degrees of freedom are $n_1 + n_2 - 2$ or $n_1 - 1 + n_2 - 1$).

Calculating the standard error is somewhat complicated. Under the assumption that the variances (or standard deviations) are equal, we can combine, or "pool," the standard deviations to have a single, more reliable estimate based on a

Table 7-4. Highest urinary excretions of 5-HIAA (milligrams per 24 hours).[1]

Subjects With Carcinoid Heart Disease		Subjects Without Carcinoid Heart Disease	
Subject	5-HIAA	Subject	5-HIAA
1	263	1	60
2	288	2	119
3	432	3	153
4	890	4	588
5	450	5	124
6	1270	6	196
7	220	7	14
8	350	8	23
9	283	9	43
10	274	10	854
11	580	11	400
12	285	12	73
13	524		
14	135		
15	500		
16	120		
Mean	429.0	220.6	
Standard deviation	294.7	261.8	

[1]Adapted and reproduced, with permission, from Tables 1 and 2 in Ross EM, Roberts WC: The carcinoid syndrome: Comparison of 21 necropsy subjects with carcinoid heart disease to 15 necropsy subjects without carcinoid heart disease. *Am J Med* 1985; **79**:339.

larger, combined sample size. The formula for the **pooled standard deviation** follows; it looks complex but is really just a **weighted average** of the two sample variances.

$$\text{Pooled } s = s_p = \sqrt{\frac{(n_1 - 1)s_1^2 + (n_2 - 1)s_2^2}{n_1 + n_2 - 2}}$$

Where

$s_1 = $ **Standard deviation in sample 1**
$s_2 = $ **Standard deviation in sample 2**

Then the standard error of the difference is:

$$\text{SE}_{(\overline{x}_1 - \overline{x}_2)} = s_p\sqrt{\frac{1}{n_1} + \frac{1}{n_2}}$$

The calculation for the pooled standard deviation in this example, using the observations in Table 7–4, is as follows:

$$s_p = \sqrt{\frac{(16 - 1)(294.7)^2 + (12 - 1)(261.8)^2}{16 + 12 - 2}}$$

$$= \sqrt{\frac{(15)(86,848.09) + (11)(68,539.24)}{26}}$$

$$= 281.25$$

Note that the pooled standard deviation falls between the two separate sample standard deviations of 294.7 and 261.8.

The standard error of the difference is

$$SE_{(\bar{x}_1 - \bar{x}_2)} = 281.25 \sqrt{\frac{1}{16} + \frac{1}{12}}$$

$$= 281.25 \sqrt{0.14583}$$

$$= 107.40$$

From Table 7-4, the mean highest urinary excretion of 5-HIAA for the 16 subjects with carcinoid heart disease is 429.0; and for the 12 subjects without carcinoid heart disease it is 220.6. To determine 95% confidence limits for the difference $429.0 - 220.6 = 208.4$, we use the two-tailed value of t from Table A-3 (for $16 + 12 - 2 = 26$ degrees of freedom) that separates the central 95% of the distribution from the 5% in the two tails. This value is 2.056.

Now, we use these results in the formula for the confidence limits. The 95% confidence limits for the difference in mean highest urinary excretion of 5-HIAA between subjects with and without carcinoid heart disease are

$$208.4 \pm (2.056)(107.4) = 208.4 \pm 220.8$$

or

$$-12.4 \text{ and } 429.2 \text{ mg/24 h}$$

Thus, we are 95% confident that the interval from -12.4 to 429.2 contains the true mean difference in highest urinary excretion of 5-HIAA. Note that this interval contains zero, indicating that zero is a possible value; in other words, it is possible that the mean difference is zero.

The graphic test for the difference between means of two independent groups is easy to perform and is quite accurate for sample sizes of 10 or more (Browne, 1979). The 95% confidence interval for each group is calculated, and a graph is prepared. One of following three results must occur.

1. The 95% confidence intervals for means do not overlap, as in Fig 7-9A. In this case, we are 95% *sure that a difference exists.*

2. The 95% confidence intervals for the means overlap so much that the mean of one group is contained within the interval of the other group, as in Fig 7-9B. Here, we conclude that *there is no difference between the means.*

3. The 95% confidence intervals for the means overlap but not so much that the mean of one group is contained within the interval of the other group, as in Fig 7-9C. In this case, *we cannot make a decision;* the hypothesis test outlined in

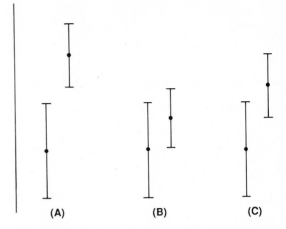

Figure 7-9. Visual assessment of differences between 2 independent groups, using 95% confidence limits.

the next section must be performed before we can say whether the means are different.

To illustrate the graphic procedure with the 5-HIAA data, we determine the 95% confidence interval for the mean in each individual group. For subjects with carcinoid heart disease, the 95% confidence limits are

$$429 \pm (2.131)\left(\frac{294.7}{\sqrt{16}}\right)$$

or

$$272.0 \text{ and } 586.0 \text{ mg/24 h}$$

For subjects without carcinoid heart disease, the 95% confidence limits are

$$220.6 \pm (2.201)\left(\frac{261.8}{\sqrt{12}}\right)$$

or

$$54.3 \text{ and } 386.9 \text{ mg/24 h}$$

A graph of the 95% confidence interval for each mean in the two patient groups is shown in Fig 7-10 and illustrates the third result: The intervals overlap but not enough to include one of the means. Therefore, a statistical test of the difference must be performed, or a confidence interval for the *difference* between means must be determined (as in the previous example), if we are to decide whether the means differ.

Before leaving this section, we make two comments. First the graphic procedure using 95% confidence limits for the mean (or mean difference) is fairly accurate, especially if sample sizes are 10 or more; however, journal editors may require in-

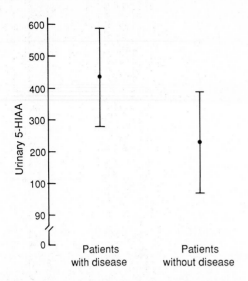

Figure 7-10. Ninety-five percent confidence limits for patients with and without carcinoid heart disease. (Adapted, with permission, from Ross EM, Roberts WC: The carcinoid syndrome: Comparison of 21 necropsy subjects with carcinoid heart disease to 15 necropsy subjects without carcinoid heart disease. *Am J Med* 1985;**79**:339.)

Table 7-5. Information on hypertensive patients: means ± standard deviation.[1]

	Hyper-adrenergic	Normo-adrenergic
Number of patients	13	9
Serum catecholamine (mg/mL)	0.484 ± 0.133	0.206 ± 0.060
Heart rate (beats/min)	90.7 ± 11.5	77.8 ± 13.2
Systolic BP (mm Hg)	171.3 ± 13.7	147.4 ± 9.9
Diastolic BP (mm Hg)	103.0 ± 8.3	95.6 ± 12.9

[1]Adapted and reproduced, with permission, from the American Heart Association and from Table 4 in de Champlain J et al: Circulating catecholamine levels in human and experimental hypertension. *Circ Res* 1976; **38**:109.

vestigators to perform the appropriate statistical test anyway. Second, to repeat our word of caution regarding published graphs, authors frequently display graphs of the mean plus or minus one standard error of the mean, ie, approximately 68% confidence intervals. (Recall from the sampling distribution of the mean in Chapter 6 that $\overline{X} \pm 1$ SEM contains 68% of the means of samples of size *n.)* In other words, you should examine the author's graph labels and legends before applying the eyeball test to published graphs.

7.4.2 Hypothesis Test for the Difference Between Independent Means

In the study described in Presenting Problem 4, the investigators wanted to determine whether mean heart rate was higher in hyperadrenergic hypertensive patients than in normoadrenergic hypertensive patients. They were not interested in testing whether mean heart rate was lower; thus, they were interested in a directional or one-tailed test, looking only for a positive difference. Information from the study is given in Table 7-5. If the investigators can assume (or verify from the data) that heart rates are normally distributed, a *t* test will answer this research question. Let the hyperadrenergic patients be designated as group 1 and the normoadrenergic patients as group 2. The six steps in testing this hypothesis follow.

Step 1. Hypotheses:

H_0: $\mu_1 \leq \mu_2$ **(The mean heart rate is lower or the same in hyperadrenergic patients.)**

H_1: $\mu_1 > \mu_2$ **(The mean heart rate is higher in hyperadrenergic patients.)**

If you find it difficult to decide which statement should be the null hypothesis and which should be the alternative hypothesis when specifying a one-tailed test, here is a hint: The alternative hypothesis often reflects the suspected conclusion (that rates are higher in hyperadrenergic patients, in this example); the null hypothesis is then the converse.

Step 2. Test statistic: The test statistic for comparing the means from two independent groups (assuming normality, equal variances if sample sizes are unequal, and independent observations) is the *t* statistic:

$$t = \frac{(\overline{X}_1 - \overline{X}_2) - (\mu_1 - \mu_2)}{s_p \ \sqrt{(1/n_1) + (1/n_2)}}$$

with $n_1 + n_2 - 2$ degrees of freedom.

Step 3. Value of α: Let's use α equal to .05 for this example.

Step 4. Critical value: The alternative hypothesis calls for a one-tailed test, and we are looking for a difference in the positive direction. The critical value of the *t* distribution with 13 + 9 − 2 = 20 df, from Table A-3, is +1.725. Therefore, we will reject the null hypothesis if the observed t is +1.725 or greater. See Fig 7–11.

Step 5. Calculations: We know $\overline{X}_1 = 90.7$, $\overline{X}_2 = 77.8$, $s_1 = 11.5$, and $s_2 = 13.2$. We now find the pooled standard deviation s_p:

$$s_p = \sqrt{\frac{(13-1)(11.5)^2 + (9-1)(13.2)^2}{13 + 9 - 2}}$$

$$= \sqrt{\frac{2980.92}{20}}$$

$$= 12.21$$

Then, the standard error of the mean difference is

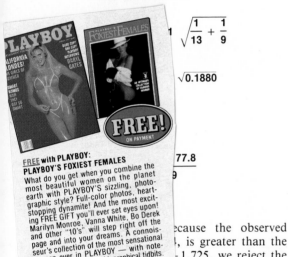

$$\sqrt{\frac{1}{13} + \frac{1}{9}}$$

$$\sqrt{0.1880}$$

...ecause the observed ..., is greater than the ...-1.725, we reject the ... We conclude that the ...othesis is true—ie, ... rate is higher in hyperadrenergic hypertensive patients than in normoadrenergic hypertensive patients.

If the investigators want to draw a graph before performing the t test, they should use 90% instead of 95% confidence limits to be equivalent to $\alpha = .05$ for a one-tailed test. In general, for a two-tailed test at a level of significance equal to α, the $(1.00 - \alpha)$ 100% confidence limits are determined. A one-tailed test, however, places the area defined by α at one end of the distribution only. Therefore, $(1.00 - 2\alpha)$ 100% confidence limits are necessary.

7.4.3 Equality of Population Variances

The t test for independent groups assumes that the variances of the observations are equal. Statistical research (Box, 1953) has shown that this assumption can be violated when sample sizes are equal without having a major impact on the significance level of the test; ie, the t test is robust for unequal variances if the sample sizes are equal. If the sample sizes are different, however, one should perform a test of the equality of variances before proceeding with a t test. If the variances are significantly different, a downward adjustment of the degrees of freedom is made, and separate variance estimates are used instead of the pooled variance.

The statistical test to compare two variances is called the **F test** and is briefly described here because of its other applications. To perform the F test, we form the ratio of the two variances, with the larger variance in the numerator, and compare the ratio with the critical value of the F probability distribution. If the ratio is significantly greater than 1, the variances are declared unequal.

We will use the observations on highest urinary excretion of 5-HIAA to illustrate the procedure. From Table 7–4, the standard deviation of the highest urinary excretions of 5-HIAA for the 16 subjects with carcinoid heart disease is 294.7; for the 12 patients without carcinoid heart disease, the standard deviation is 261.8.

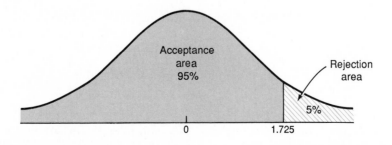

Acceptance area 95%

Rejection area

5%

0   1.725

Figure 7–11. Areas of acceptance and rejection for testing hypothesis on mean heart rate in hyperadrenergic patients (α = .05, one-tailed).

Step 1. Hypotheses:

H_0: $\sigma_1^2 = \sigma_2^2$ **(The variances in groups 1 and 2 are equal.)**
H_1: $\sigma_1^2 \neq \sigma_2^2$ **(The variances in groups 1 and 2 are not equal.)**

Step 2. Test statistic: The test statistic to compare two variances (assuming normality) is the F statistic, ie, the ratio of the largest variance to the smallest variance. The F statistic ha two sets of degrees of freedom: on less than the sample size for the vari ance in the numerator, and one less than the sample size for the variance in the denominator. Thus, if s_1^2 is the larger variance and s_2^2 is the smaller variance, the test statistic is

$$F = \frac{s_1^2}{s_2^2}$$

with $n_1 - 1$ (numerator) and $n_2 - 1$ (denominator) degrees of freedom.

Step 3. Value of α: Use an α of .05.
Step 4. Critical value: The larger variance is the one for subjects with carcinoid heart disease, and it will be in the numerator; therefore, the degrees of freedom are $16 - 1 = 15$ for the numerator and $12 - 1 = 11$ for the denominator. The critical values for the F distribution are given in Table A–4 for $\alpha = .05$ and .01. With 15 and 11 degrees of freedom, the value of F that divides the distribution into 95% and 5% is between 2.85 and 2.62; interpolation gives 2.74. Because of the way the F statistic is formed, when the larger variance is placed in the numerator, the entire 5% of the area is in the right end of the distribution. Therefore, we will reject the null hypothesis of equal variances if the observed value of F is greater than 2.74. Fig 7–12 is a graph of the F distribution illustrating the areas of acceptance and rejection.
Step 5. Calculations: The F statistic is

$$F = \sqrt{\frac{(294.7)^2}{(261.8)^2}} = 1.27$$

Step 6. Conclusion: Because the observed value of F, 1.27, is less than 2.74, the null hypothesis of equal variances is not rejected. So the t test to compare means may be performed, assuming other assumptions are met.

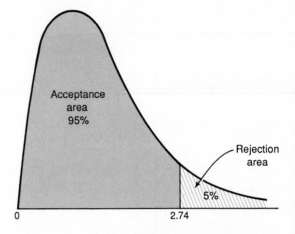

Figure 7-12. F distribution with 15 and 11 degrees of freedom (with $\alpha = .05$ critical value, one-tailed).

If the F test is significant and the hypothesis of equal variances is rejected, the standard deviations from the two samples cannot be pooled, because pooling assumes they are equal. Instead, the separate variances are used, and the degrees of freedom for the t test are reduced by a formula known as the Satterthwaite correction. The effect of reducing the degrees of freedom is to increase the size of the observed t statistic needed to reject the null hypothesis and conclude a difference between means. This procedure thus requires a larger difference between means in order to declare significance. The reduction may be thought of as a kind of penalty for violating the assumption of equal variances with unequal sample sizes. If you are interested in the details of this procedure, consult Winer (1971; see Chapter 1.9).

Computer programs perform the F test for equality of variances, often in the same program that performs the t test, as we shall see in Fig 7–15. If results are given for both equal and unequal variances, the results for equal variances (sometimes called results for pooled degrees of freedom) are used if the sample sizes are equal. If the sample sizes are not equal, the results of the F test are consulted. If the F test results are significant, the results for unequal variances with adjusted degrees of freedom are used. If the F test is not significant, the equal-variances or pooled-degrees-of-freedom results are used.

7.4.4 The Wilcoxon Rank-Sum Test (Mann-Whitney U)

If investigators wish to compare the means of two independent groups, and one or more of the assumptions for the t test are seriously violated, they can employ an excellent nonparametric alternative to the t test. The test is called by various names:

Mann-Whitney U test, **Wilcoxon rank-sum test,** or Mann-Whitney-Wilcoxon rank-sum test. It is a test of equality of medians rather than means. Similar to the Wilcoxon signed-ranks test for paired groups, the Wilcoxon rank-sum test is computationally arduous and time-consuming; an approximate test based on a rank order of the observations also exists (Conover and Iman, 1981). Again, as with the Wilcoxon signed-ranks test, raw data are required to perform the Wilcoxon rank-sum text. Of course, this test is available in most computer packages.

We will use the data on highest urinary excretions of 5-HIAA for patients with and without carcinoid heart disease to illustrate the steps of the Wilcoxon rank-sum test. The first step is to rank the observations from lowest to highest (or vice versa), ignoring the group an observation belongs to. Table 7-6 lists the data and ranks; eg, the smallest 5-HIAA value was observed in Subject 7 in the group without carcinoid heart disease and is given a rank of 1. There are no duplicate or tied values in this set of data; when ties are present, they are handled as in the Wilcoxon signed-ranks test, by taking the mean rank for the tied observations.

The next step, after the observations have been ranked, is to calculate the mean and the standard deviation of the ranks separately for each group of patients. The pooled standard deviation of the ranks is then computed. Finally, the t statistic for two independent groups is determined, using the calculations for ranks, not original observations.

The hypothesis is that the mean ranks are equal in the two groups. The idea behind the test is as follows: If there is no significant difference between the two groups, then there should be some low rank-ordered observations in each group and some high rank-ordered observations in each group—ie, the rank orders should be fairly evenly distributed across groups (within sampling variation, of course). Therefore, if the mean ranks are equal, the groups are similar. In contrast, if the mean rank is substantially higher in one group than the other, there must be a number of individuals with higher observations in that group.

The mean rank in the group with carcinoid heart disease, $\overline{R}_1$, is 17.875, with a standard deviation of 6.13. For patients without the disease, the mean rank $\overline{R}_2$ is 10, with a standard deviation of 8.73. The formula for the pooled standard deviation of ranks is the same as the formula given in Section 7.4.1:

$$s_{R_p} = \sqrt{\frac{(15)(6.13)^2 + (11)(8.73)^2}{16 + 12 - 2}}$$

$$= 7.34$$

The degrees of freedom are $n_1 + n_2 - 2 = 26$ in this example, where n_1 is the sample size for patients with carcinoid heart disease and n_2 is the

Table 7-6. Ranked values of highest urinary excretions of 5-HIAA (milligrams per 24 hours).[1]

Subjects With Carcinoid Heart Disease			Subjects Without Carcinoid Heart Disease		
Subject	5-HIAA	Rank	Subject	5-HIAA	Rank
1	263	13	1	60	4
2	288	17	2	119	6
3	432	20	3	153	10
4	890	27	4	588	25
5	450	21	5	124	8
6	1270	28	6	196	11
7	220	12	7	14	1
8	350	18	8	23	2
9	283	15	9	43	3
10	274	14	10	854	26
11	580	24	11	400	19
12	285	16	12	73	5
13	524	23			
14	135	9			
15	500	22			
16	120	7			
Mean		$\overline{R}_1 = $ 17.875	$\overline{R}_2 = 10.00$		
Standard deviation		$s_{\overline{R}1} = 6.13$	$s_{\overline{R}2} = 8.73$		

[1]Adapted and reproduced, with permission, from Tables 1 and 2 in Ross EM, Roberts, WC: The carcinoid syndrome: Comparison of 21 necropsy subjects with carcinoid heart disease to 15 necropsy subjects without carcinoid heart disease. *Am J Med* 1985; **79**:339.

sample size for patients without the disease. The critical value of t, for $\alpha = .05$ (to correspond to the 95% confidence limits in Section 7.4.1), is 2.056. The null hypothesis of equal ranks will be rejected if the observed value of t computed for ranks is either greater than 2.056 or less than -2.056.

The t statistic calculated for ranks is

$$t = \frac{\overline{R}_1 - \overline{R}_2}{s_{R_p}\sqrt{(1/n_1) + (1/n_2)}}$$

$$= \frac{17.875 - 10}{7.34\sqrt{(1/16 + 1/12)}}$$

$$= \frac{7.875}{2.80}$$

$$= 2.81$$

Because 2.81 is greater than 2.056, we reject the null hypothesis of equal ranks ($P < .05$). We conclude that there is a difference in mean highest urinary excretions of 5-HIAA for patients with and without carcinoid heart disease, with diseased patients having higher levels.

Note that this conclusion is not the same one we reached by using confidence limits in Section 7.4.1; the confidence limits were -12.4 and 429.2 mg/24 h, and the conclusion was that the mean difference could be zero. Why are the conclusions different? Examination of the values for 5-HIAA in Table 7–4 reveals that they are not normally distributed; the distributions for both sets of patients are skewed positively. Therefore, the t distribution should not have been used to calculate confidence limits or to perform a statistical test for these data. The Wilcoxon rank-sum test was able to detect the difference because it is a more powerful test when the assumptions of the t test are violated. In this example, inappropriately using the t distribution with data that are not normally distributed leads to the wrong conclusion.

7.5 DETERMINATION OF SAMPLE SIZE

Prior determination of the sample size needed to demonstrate an important difference is essential in a well-designed study. Recall from Chapter 6 that missing a significant difference is called a type II error, and this error occurs when the sample size is too small. For example, Freiman et al (1978) reported on 71 negative clinical trials published in the literature and found that 50 of the trials would have missed a 50% improvement because the sample size was too small. Authors are beginning to provide some information about the sample size needed to detect a given difference,

and we expect this practice to be followed with increasing frequency. For proposers of grants for funding from the National Institutes of Health and other agencies, it is almost a requirement.

Several different formulas may be used to estimate sample sizes. The formulas presented here take both type I and type II errors into consideration. If you are interested in more detail about the procedures, consult the book by Cohen (1977), which is devoted entirely to this subject.

7.5.1 Sample Sizes for a Single Mean

This section presents the formula needed to estimate the sample size for a "study comparing the mean in a single sample with a standard value. In Section 7.5.2, the formula to calculate sample sizes for comparing two means is given.

To estimate sample size for a research study involving a single mean, the researcher must answer four questions:

1. What is the desired level of significance (α level) related to the null hypothesis involving μ_0?

2. What chance should there be of detecting an actual difference; ie, what is the power (equal to $1 - \beta$) associated with the alternative hypothesis involving μ_1?

3. What difference between the means must be detected; ie, what is $\mu_1 - \mu_0$?

4. What is a good estimate of the standard deviation σ in the population?

Specifications of α for a null hypothesis and β for an alternative hypothesis permit us to solve for the sample size. These specifications lead to the two critical ratios given below, where z_α is two-tailed z_β denotes the lower tail, and $\overline{X}$ is the mean in the sample.

$$z_\alpha = \frac{\overline{X} - \mu_0}{\sigma/\sqrt{n}} \quad \text{and} \quad z_\beta = \frac{\overline{X} - \mu_1}{\sigma/\sqrt{n}}$$

Solving these two critical ratios for the sample size n gives

$$n = \left[\frac{(z_\alpha - z_\beta)\sigma}{\mu_1 - \mu_0} \right]^2$$

(see Exercise 9).

Let us illustrate the procedure by using the data of Presenting Problem 1; recall that the investigators want to know whether Pimax (maximal inspiratory mouth pressure) is the same in patients with kyphoscoliosis and in normal patients without kyphoscoliosis. Suppose the investigators want the type I error (concluding that there is a difference in Pimax when there is not) to be .05, and they want a .90 probability of detecting a true difference. Assume that the mean Pimax is 110 cm H_2O in normal patients, and the standard deviation in the population is 20 cm H_2O. Suppose that the investigators want to be able to say that a mean Pimax of 80 cm H_2O or less in kyphoscoliotic patients is significantly different from the norm. From the given information, what is the sample size needed to detect a difference?

The two-tailed z value related to α is ± 1.96 (from Table A–2; this is the critical value that divides the central 95% of the z distribution from the 5% in the tails). The lower one-tailed z value related to β is -1.28 (the critical value that separates the lower 10% of the z distribution from the upper 90%). For a standard deviation of 20 and the difference the investigators want to be able to detect $(110 - 80)$, the sample size is

$$n = \left[\frac{(1.96 + 1.28)(20)}{110 - 80} \right]^2 = \left(\frac{64.8}{30} \right)^2$$
$$= 4.67 \quad \text{or} \quad 5$$

In the determination of sample sizes, the convention is to always round up. Because the investigators were interested in detecting only a rather large difference between patients with kyphoscoliosis and normal patients, they do not need a very large sample. In Exercise 8 you are asked to calculate how large a sample would be needed if they wanted to detect a difference of 10 cm H_2O.

7.5.2 Sample Sizes for Two Means

A similar formula can be used to calculate sample sizes for comparing the means from two independent groups. For this formula, two simplifying assumptions must be met: The standard deviations in the two populations are equal, and the sample sizes are equal in the two groups. Then, if $\mu_1 - \mu_2$ is the magnitude of the difference to be detected between the two groups, the sample size needed in *each* group is

$$n = 2 \left[\frac{(z_\alpha - z_\beta)\sigma}{\mu_1 - \mu_2} \right]^2$$

where z_α and z_β are defined as in the formula for the sample size for a single mean.

To illustrate, we consider Presenting Problem 4, which examined the role of circulating catecholamines in the serum of patients with essential hypertension. Suppose the investigators want to compare the heart rate in patients with essential hypertension and high catecholamine levels with the heart rate in patients with essential hypertension and low catecholamine levels. They are willing to accept a type I error (incorrectly concluding that there is a difference in heart rate) of .05, and they want a probability of .80 of detecting a true difference. Further assume that the investigators decide a difference of 10 or more beats per minute is clinically significant, and that an estimate of the standard deviation in heart rate is 15 beats per

minute. For this information, what is the sample size needed to detect a difference?

The two-tailed z value related to α again is ± 1.96, and the lower one-tailed z value related to β is $-.84$ (the critical value separating the lower 20% of the z distribution from the upper 80%). From the given estimates, the sample size for each group is

$$n = 2 \left[\frac{(1.96 + 0.84)(15)}{10} \right]^2 = 2 \left(\frac{42}{10} \right)^2$$
$$= 2(17.64) \quad \text{or} \quad 36$$

Thus, 36 patients are needed in each group if the investigators want to have an 80% chance (or 80% power) of detecting a difference of 10 or more beats per minute.

We developed a rule of thumb for quickly determining the sample sizes needed for comparing the means of two groups. First, determine the ratio of the standard deviation of the measure of interest to the difference to be detected between the means $(\sigma/(\mu_1 - \mu_2))$; then, square this ratio. For a study with a P-value of .05, an experiment will have a 90% chance of detecting an actual difference between the two groups if the sample size in each group is approximately 20 times the squared ratio. For a study with the same P-value but only an 80% chance of detecting an actual difference, a sample size of approximately 15 times the squared ratio is required. To see how this rule of thumb was obtained, see Exercise 10.

7.6 COMPUTER PROGRAMS ILLUSTRATING COMPARISON OF MEANS

We will illustrate output from MINITAB, SPSS, STATISTIX, and SYSTAT to analyze numerical data when the questions involve means. We use the observations from Presenting Problem 1 on maximum inspiratory mouth pressure (Pimax) in nine patients with kyphoscoliosis to illustrate the confidence interval program for one mean in MINITAB. Fig 7–13 illustrates the brief output from this program. The results agree with our calculations in Section 7.2.2.

The study described in Presenting Problem 2 involved eight patients with chronic obstructive lung disease, and measurements were made on their arterial oxygen and carbon dioxide tension before and after a low carbohydrate diet. The SPSS program in Fig 7–14 illustrates analysis of the observations on arterial oxygen tension. As an exploratory step, the SPSS command to compute a new variable was used to calculate the difference in arterial oxygen tension (called OXYDIFF); the DESCRIPTIVES program was used to produce statistics on the change. The mean and the standard deviation of the difference in oxygen tension are

	N	MEAN	STDEV	SE MEAN	95.0 PERCENT C.I.
pimax	9	41.27	16.23	5.41	(28.79, 53.75)

Figure 7–13. MINITAB program to find confidence limits for one mean, using observations on Pimax from Presenting Problem 1. (Adapted, with permission, from Lisboa C et al: Inspiratory muscle function in patients with severe kyphoscoliosis. *Am Rev Respir Dis* 1985;**132**:48. MINITAB is a registered trademark of Minitab, Inc. Used with permission.)

13.5 and 8.246, respectively, the same values we calculated. The TTEST program for paired samples provides summary statistics for the values of arterial oxygen tension before (OXYBEF) and after (OXYAFT) as well as for the difference. Note the agreement between the mean and the standard deviation of the difference (13.5 and 8.246) in the TTEST and DESCRIPTIVES procedures. The t value from the program is -4.63, but we calculated 4.66; the difference in value is simply round-off error on our part, and the difference in sign occurs because we subtracted arterial oxygen tension values before the diet from those after the diet in order to have positive numbers for calculations. Note that SPSS provides the actual P-value (.002), accurate to three decimal places.

The STATISTIX computer printout in Fig 7–15 illustrates the use of the two-sample t test for the observations on 5-HIAA in patients with and without carcinoid heart disease. Most computer programs will not accept variable names beginning with a number, so 5-HIAA is denoted simply HIAA in the printout. STATISTIX gives mean, sample size, standard deviation (S.D.), and standard error (S.E.) in each sample. Note that the F statistic for equal variances is 1.27, as in our calculations. The F test is not significant, so the t test corresponding to equal variances, $t = 1.94$ with 26 df, is used. Although STATISTIX does not print the value of the pooled standard deviation, it is used in the calculation of the equal-variance t test. We did not perform the two-sample t test but calculated the 95% confidence interval of

```
Number of Valid Observations (Listwise) =          8.00

Variable   OXYDIFF

Mean              13.500          S.E. Mean          2.915
Std Dev            8.246          Variance          68.000
Kurtosis           1.471          S.E. Kurt          1.481
Skewness           1.144          S.E. Skew           .752
Range             26.000          Minimum            4.00
Maximum           30.00           Sum              108.000

Valid Observations -        8      Missing Observations -         0
Paired samples t-test:   OXYBEF
                         OXYAFT

Variable       Number                Standard    Standard
             of Cases      Mean      Deviation     Error

OXYBEF           8       55.6250       9.242       3.267
OXYAFT           8       69.1250       7.864       2.780


(Difference) Standard    Standard  |      2-Tail   |    t      Degrees of   2-Tail
     Mean    Deviation     Error   |  Corr.  Prob. | Value     Freedom      Prob.

  -13.5000     8.246       2.915   |  .545    .162 |  -4.63       7           .002
```

Figure 7–14. SPSS program to produce descriptive statistics and perform paired t test using observations on arterial oxygen and carbon dioxide tension from Presenting Problem 2. (Adapted, with permission, from Tirlapur VG, Mir MA: Effect of low-calorie intake on abnormal pulmonary physiology in patients with chronic hypercapneic respiratory failure. *Am J Med* 1984;**77**:987. SPSS is a registered trademark of SPSS, Inc. Used with permission.)

```
TWO SAMPLE T TESTS FOR HIAA = CARCINOID

                      SAMPLE
CARCINOID    MEAN      SIZE      S.D.        S.E.
---------  ---------  ------  ---------   ---------
    1        429.0       16     294.7       73.67
    2        220.6       12     261.8       75.58

                 T        DF        P
              ------   ------   ------
EQUAL VARIANCES  1.94     26     0.0632
UNEQUAL VARIANCES 1.97   25.2    0.0594

                 F      NUM DF   DEN DF       P
              -------  -------  ------    -------
TESTS FOR EQUALITY
     OF VARIANCES 1.27    15       11      0.3516

CASES INCLUDED 28   MISSING CASES 0
```

Figure 7-15. STATISTIX program to calculate two-sample t test using observations on 5-HIAA in patients with and without carcinoid heart disease from Presenting Problem 3. (Adapted, with permission, from Ross EM, Roberts WC: The carcinoid syndrome: Comparison of 21 necropsy subjects with carcinoid heart disease to 15 necropsy subjects without carcinoid heart disease. *Am J Med* 1985;**79**:339. STATISTIX is a registered trademark of NH Analytical Software. Used with permission.)

−15.31 to 428.11. This confidence interval contains zero, indicating that the null hypothesis of no difference would be rejected at $\alpha = .05$, and the *P*-value of .0632 associated with the t test is consistent with this conclusion.

In Chapter 3, we used the computer program SYSTAT to make box plots of percentage saturation of bile in men and women (Fig 3–19). The box plots contain some marks that look like parentheses ( ) called notches; these notches are similar to the eyeball test for the difference between two independent means illustrated in Fig 7–10. They actually constitute a test of equality of two medians. The SYSTAT box plot procedure for patients with and without carcinoid heart disease is reproduced in Fig 7–16. These plots clearly show

how positively skewed the distributions of the observations are. If the notches do not overlap, you can be 95% confident that the medians are different. Although the precision of the computer graphics is fairly low, the notches do not overlap; in fact, they are printed at the same place in the distributions. The graphic test thus substantiates the conclusion that the groups do differ when medians rather than means are compared.

7.7 SUMMARY

This chapter illustrates a variety of methods for estimating and computing means or differences in means for observations measured on a numerical

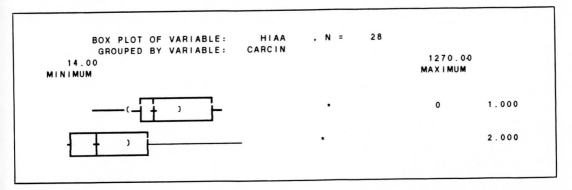

Figure 7-16. SYSTAT box plot program to compare two medians using observations on 5-HIAA in patients with and without carcinoid heart disease from Presenting Problem 3. (Adapted, with permission, from Ross EM, Roberts WC: The carcinoid syndrome: Comparison of 21 necropsy subjects with carcinoid heart disease to 15 necropsy subjects without carcinoid heart disease. *Am J Med* 1985;**79**:339. SYSTAT is a registered trademark of SYSTAT, Inc. Used with permission.)

scale. The next chapter extends this discussion to research questions involving three or more means. In this summary, we first discuss the Presenting Problems; then we give an outline useful in selecting the appropriate statistical method.

In the study described in Presenting Problem 1, the authors measured the maximal inspiratory mouth pressure (Pimax) in patients with kyphoscoliosis. They found a significant relationship between Pimax and Pa_{CO_2} and between Pimax and Pa_{O_2}; they concluded that the impairment of inspiratory muscle function is related to the development of respiratory failure in patients with severe kyphoscoliosis. Using measurements of inspiratory mouth pressure for nine patients, we calculated the observed mean value and the 95% confidence limits for this observed mean. We also used the one-sample t test to determine whether the observed mean was different from the norm. The 95% confidence interval was approximately 29 to 54 cm H_2O; thus, clinicians can conclude, with 95% confidence, that this interval contains the actual mean maximal inspiratory pressure in patients with kyphoscoliosis. In the test of hypothesis, the mean was found to be significantly different from that in normal patients.

Presenting Problem 2 discussed a study in which a low-carbohydrate, low-calorie diet given to a group of mostly obese patients with chronic hypercapneic respiratory failure resulted in a significant improvement in mean arterial oxygen tension after weight reduction had occurred and reached a steady state. We used the data to illustrate how we can determine whether there was a significant change in arterial oxygen tension as the result of the weight loss program. Both the confidence interval for the mean difference and the paired t test for the hypothesis of a difference equal to zero were performed. The 90% confidence interval for the change of arterial oxygen tension was 8–19 mm Hg, indicating that the investigators can be 90% sure that this interval contains the actual mean increase in arterial oxygen tension for patients on a weight reduction program like this one. Because zero is not included in this interval, the t test rejects the null hypothesis of no change. In other words, the investigators can be sure that a significant increase in arterial oxygen tension did occur. Even though the observations were not normally distributed, the Wilcoxon signed-ranks test confirmed this conclusion. However, because there was no control group, the authors cannot state unequivocally that this change in arterial oxygen tension occurred solely as a result of the diet.

Presenting Problem 3 looked at two independent groups of patients with carcinoid syndrome, those with and without carcinoid heart disease. The authors sought clues that might help them distinguish clinically subjects with carcinoid heart disease from those without carcinoid heart dis-

ease. They compared the levels of urinary excretions of 5-HIAA in the two groups of patients. Because these two groups are independent, the two-sample (independent groups) t test is appropriate for comparing the mean measures of 5-HIAA. The mean value in the 16 subjects with carcinoid heart disease was 429 mg/24 h and was 220 mg/24 h in the 12 subjects without carcinoid heart disease. These values were found to be not significantly different. The 95% confidence interval for the difference in highest urinary excretions of 5-HIAA was from -12 to $+429$ mg/24 h. These limits are wide and indicate that the measure has considerable variability. We are 95% confident that this interval contains the true difference. But because the interval contains zero, we cannot say that there is a difference between patients with and without carcinoid heart disease. However, further analysis we performed showed that 5-HIAA levels are not normally distributed. When the appropriate nonparametric test, the Wilcoxon rank-sum test, is used, there is evidence to indicate differences between patients with and without carcinoid heart disease.

Patients with essential hypertension were separated into two groups in Presenting Problem 4, hyperadrenergic and normoadrenergic hypertensive patients. The authors found the heart rate and systolic blood pressure were significantly higher in patients with elevated circulating catecholamine levels than in those with normal catecholamine levels. From these findings, they concluded that the sympathetic system may play an important role in maintaining elevated blood pressure in a significant portion of patients with essential hypertension. As in Presenting Problem 3, two independent groups were studied, and the two-sample t test is the appropriate method for determining whether mean heart rates and systolic blood pressures are the same in the two samples. The statistical conclusion was to reject the null hypothesis of no difference in mean heart rates and systolic blood pressures and to claim that there are significant differences in these values.

The flowchart for Chapter 7 in Appendix C (Flowchart C-1) will help both readers and researchers to determine which statistical procedure is appropriate for comparing means.

EXERCISES

1. Calculate the mean and the standard deviation of the Pimax data in Table 7–1.
2. Calculate the P-value for the t test on the Pimax data in Table 7–1, assuming that the investigators want to know whether the mean Pimax is less than 60 cm H_2O.
3. **a.** Recalculate the 95% confidence interval for the true mean Pimax in patients with kyphoscoliosis, using the z distribution. Com-

pare this interval with the correct interval obtained by using the t distribution.

b. What effect on the confidence interval is observed if n is increased to 36 patients?

c. Why do you think the investigators did not use more patients?

4. For the calculations in Table 7–2, test whether the mean difference in arterial carbon dioxide tension is significant, using $\alpha = .05$.

5. Using the information in Table 7–5, calculate a 90% confidence interval for the mean difference in heart rate. Compare the results with the hypothesis test in Section 7.4.2.

6. Perform the eyeball test for the serum catecholamine data in Table 7–5 (95% confidence intervals are 0.16 to 0.25 and 0.40 to 0.57, respectively).

7. Compute the 95% and 99% confidence intervals for the difference in arterial oxygen tension from the data in Table 7–2. Compare these intervals with the 90% interval obtained

in Section 7.3.2. What is the effect of higher confidence on the width of the interval?

8. Compute the sample size needed in the Pimax example to provide a 90% chance of detecting a difference of 10 cm Hg, using $\alpha = .05$.

9. Show that solving

$$z_\alpha = \frac{\overline{X} - \mu_0}{\sigma/\sqrt{n}} \quad \text{and} \quad z_\beta = \frac{\overline{X} - \mu_1}{\sigma/\sqrt{n}}$$

for the sample size n gives

$$n = \left[\frac{(z_\alpha - z_\beta)\sigma}{\mu_1 - \mu_0} \right]^2$$

10. How was the rule of thumb for calculating the sample size for two independent groups found?

8

Comparing Three or More Means

PRESENTING PROBLEMS

Presenting Problem 1. In recent years researchers have become increasingly interested in the neuromodulation of immune function. Irwin and colleagues (1987) published the results of a cross-sectional study of the relationship between major life events and immune function. Subjects in the study included women whose husbands were undergoing treatment for metastatic lung cancer, women whose husbands had died of lung cancer in the preceding 1–6 months, and women whose husbands were in good health. Scores for the Social Readjustment Rating Scale and the Hamilton Rating Scale for Depression were recorded for each of the women; in addition, two measures of immune system function, natural killer (NK) cell activity and T cell subpopulation numbers (T helper cells and T suppressor cells), were determined.

According to their scores on the Social Readjustment Rating Scale, the women were divided into three groups: those with low scores (54 or less), those with moderate scores (55–99), and those with high scores (100 or more). The relationships between scores on the Social Readjustment Scale and alterations in immune function were analyzed. How can we determine whether NK cell activity and T cell numbers are the same for women with low, moderate, and high scores? (The data from the study are given in Table 8–1).

Presenting Problem 2. Alzheimer's disease, the most common cause of dementia affecting millions of Americans, has a broad social and economic impact on our society. An intense effort to learn more about this disease is under way. Recent research has focused on neurotransmitter deficits in patients with Alzheimer's disease. Francis et al (1985) studied the relationship between acetylcholine synthesis in brain biopsy specimens in patients with Alzheimer's disease and a rating of their cognitive impairment. They also compared the differences in neurochemical markers of various neurotransmitters in brain tissue taken during autopsy from 48 patients with either early onset (age < 80 years) or late onset (age ≥ 80 years) of Alzheimer's disease and 34 controls of similar ages. The investigators found that acetylcholine synthesis measured in temporal biopsy specimens decreases

with increasing cognitive impairment. (Their data from brain tissue samples obtained at autopsy are shown in Table 8–4.) The objective is to determine whether there are differences in choline acetyltransferase levels between Alzheimer patients and controls or between those with early-onset and late-onset disease.

8.1 PURPOSE OF THE CHAPTER

Many research projects in medicine employ more than two groups, but the t tests discussed in Chapter 7 are applicable only for the comparison of two groups. For example, in Presenting Problem 1, the mean natural killer cell activity (measured in lytic units) is examined for three groups of women—those with low, moderate, and high scores on the Social Readjustment Rating Scale. Other studies examine the influence of more than one factor, such as the study described in Presenting Problem 2, in which the researchers were interested in differences in neurochemical markers in subjects with and without Alzheimer's disease and also in younger and older subjects in these two groups. These situations call for a global, or omnibus, test to see whether there are any differences in the data prior to testing various combinations of means to determine individual group differences.

If a global test is not performed, multiple tests between different pairs of means will alter the α level, not for each comparison but for the experiment as a whole. For example, in a study of drug use, physicians, pharmacists, medical students, and pharmacy students were asked how frequently and recently they had used drugs (McAuliffe et al, 1986). These researchers may wish to compare drug use between (1) physicians and pharmacists, (2) physicians and medical students, (3) pharmacists and pharmacy students, and (4) medical students and pharmacy students. They may also wish to compare (5) those in medicine (physicians and medical students) with those in pharmacy (pharmacists and pharmacy students) and (6) practitioners (physicians and pharmacists) with students (medical and pharmacy). If each comparison is made by using $\alpha = .05$, there is a 5% chance that each comparison will falsely be called significant; ie, a type I error may occur six different times. Overall, therefore, there is a 30% chance ($6 \times 5\%$) of declaring one of the comparisons incorrectly significant.*

The recommended approach for analyzing data in this situation is called the **analysis of variance,** abbreviated **ANOVA**; it protects the researcher against error "inflation" by first asking if there are any differences at all among means of the groups. If the result of the ANOVA is significant, the answer is yes; and the investigator is then free to make comparisons between pairs or combinations of groups.

The topic of analysis of variance is complex, and there are many textbooks devoted to the subject. However, its use in the medical literature is somewhat limited, ranging from a low of about 2% of the articles in oncology (Hokanson, Luttman, and Weiss, 1986) and pathology (Hokanson et al, 1987b) journals, to about 3–4% in *The New England Journal of Medicine* (Emerson and Colditz, 1983) and various surgery journals (Reznick, Dawson-Saunders, and Folse, 1987), to a high of almost 10% in the psychiatric literature (Hokanson et al, 1986). The purpose of introducing ANOVA in this text is to familiarize you with the terms used and to give you an idea of how actual analyses are performed. As with other procedures, our goal is to provide enough discussion so that students and medical professionals are able to identify situations in which ANOVA is appropriate and to interpret the results. If you are interested in learning more about analysis of variance, consult Armitage (1971; Chapters 8 and 12) and Dunn and Clark (1974; Chapters 5–7). Except for very simple study designs, the best approach investigators can take when their study involves more than two groups or two or more variables is to consult with a statistician to determine how best to analyze the data.

Our approach to the topic will be as follows: In Section 8.2, we attempt an intuitive overview of the logic involved in ANOVA; the results from Presenting Problem 1 are then used to illustrate the computations in an intuitive manner. Section 8.3 presents the traditional approach and the formulas used in ANOVA; they are illustrated for Presenting Problems 1 and 2. Section 8.4 discusses some of the more commonly used methods for comparing means by using ANOVA; these methods are called **multiple-comparison procedures.** The published findings from the Presenting Problems are interpreted, without computations, and other ANOVA designs are briefly discussed in Section 8.5. Finally, in Section 8.6, the data from

the Presenting Problems are used to illustrate computer programs that perform ANOVA.

8.2 INTUITIVE OVERVIEW TO ANOVA

8.2.1 The Logic of ANOVA

In Presenting Problem 1, the natural killer cell activity was measured for three groups of subjects—those who had low, medium, and high scores on the Social Readjustment Rating Scale. The original observations, sample sizes, means, and standard deviations provided by Irwin and his colleagues are given in Table 8–1.

ANOVA provides a way to divide the total variation in natural killer cell activity for each subject into two parts. For example, let us denote a given subject's natural killer cell activity as X and consider how much X differs from the mean natural killer cell activity for all the subjects in the study, abbreviated $\overline{X}.$, where the . denotes the grand (overall) mean of all subjects, regardless of which group they are in. This difference, symbolized $X - \overline{X}.$, can be divided into two parts: the difference between X and the mean of the group this subject is in, $\overline{X}_j$, and the difference between the group mean and the grand mean. In symbols, we write

$$(X - \overline{X}.) = (X - \overline{X}_j) + (\overline{X}_j - \overline{X}.)$$

For example, Subject 1 in the low-score group has a natural killer cell activity of 22.2 lytic units.

Table 8-1. Natural killer cell activity[1] (lytic units)[2]. The grouping is based on scores from the Social Readjustment Rating Scale.

	Low Score ($n = 13$)	Moderate Score ($n = 12$)	High Score ($n = 12$)
	22.2	15.1	10.2
	97.8	23.2	11.3
	29.1	10.5	11.4
	37.0	13.9	5.3
	35.8	9.7	14.5
	44.2	19.0	11.0
	82.0	19.8	13.6
	56.0	9.1	33.4
	9.3	30.1	25.0
	19.9	15.5	27.0
	39.5	10.3	36.3
	12.8	11.0	17.7
	37.4		
Mean	40.23	15.60	18.06
Standard deviation	25.71	6.42	9.97

[1]Observations used, with permission, from Irwin M et al: Life events, depressive symptoms, and immune function. *Am J Psychiatry* 1987; **144:**437.
[2]One lytic unit is defined as the number of effector cells killing 20% of the target cells.

---

*Actually, 30% is only approximately correct; it does not reflect that all the comparisons are not independent. See Exercise 3.

The grand mean for all patients is 25.05, so Subject 1 differs from the grand mean by 22.2 − 25.05, or −2.85. This difference can be divided into two parts: the difference between 22.2 and the mean for the low group, 40.23; and the difference between the mean for the low group and the grand mean. Thus,

$$(22.2 - 40.23) + (40.23 - 25.05) = -18.03 + 15.18$$
$$= -2.85$$

Subject 1 appears to be "more different" from the mean of the low group than the mean of the low group is from the grand mean.

While our example does not show exactly how ANOVA works, it is helpful for understanding the concept of dividing the variation into different parts. Here, we have been looking at simple differences related to just one observation; ANOVA considers the variation in all observations and divides it into (1) the variation between each subject and the subject's group mean and (2) the variation between each group mean and the grand mean. If the group means are quite different from one another, there will be considerable variation between them and the grand mean, compared with the variation within each group. If the group means are not very different, however, the variation between them and the grand mean will not be much more than the variation among the subjects within each group. Therefore, the *F* test for two variances (discussed in Chapter 7) can be used to test the ratio of the variance among means to the variance within each group.

The null hypothesis for the *F* test is that the two variances are equal; if they are, the variation among means is not much greater than the variation among individual observations within any given group. Therefore, there is not sufficient evidence to conclude that the means are different from one another. Thus, we think of ANOVA as a test of the equality of means, even though it is the variances that are tested in the process. If the null hypothesis is rejected, we conclude that not all the means are equal; however, we do not know which ones are not equal, which is why post hoc comparison procedures are necessary.

8.2.2 Illustration of Intuitive Calculations for ANOVA

Recall that the formula for the variance of the observations (or squared standard deviation; see Chapter 4) involves the sum of the squared deviations of each X from the mean $\overline{X}$:

$$s^2 = \frac{\Sigma(X - \overline{X})^2}{n - 1}$$

A similar formula can be used to find the variance of means from the grand mean:

$$\text{Estimate of variance of means} = \frac{\Sigma\, n_j(\overline{X}_j - \overline{X}.)^2}{j - 1}$$

where n is the number of observations in each group and j is the number of groups. This estimate is called the **mean square among groups**, abbreviated MS_A, and it has $j - 1$ degrees of freedom.

To obtain the variance within groups, we use a pooled variance like the one for the *t* test for two independent groups:

$$\text{Estimate of variance within groups} = \frac{\Sigma(n_j - 1)s_i^2}{\Sigma(n_j - 1)}$$

This estimate is called the **error mean square** (or **mean square within**), abbreviated MS_E. It has $\Sigma\,(n_j - 1)$ degrees of freedom, or, if the sum of the number of observations is denoted by N, $N - j$ degrees of freedom. The F ratio is formed by dividing the estimate of the variance of means (mean square among groups) by the estimate of the variance within groups (error mean square), and it has $j - 1$ and $N - j$ degrees of freedom.

We will use the data on the natural killer cell activity to illustrate the calculations. In this example, the variable of interest (natural killer cell activity) is the **dependent variable,** and the grouping variable (score on Social Readjustment Rating Scale) is the **independent variable.** The data in Table 8–1 indicate that the mean natural killer cell activity for the low-score group is higher than the means for the moderate and high groups. If these three groups of women are viewed as coming from a larger population, then the question is whether mean natural killer cell activity levels are different in the population. Although there are differences in the means in Table 8–1, some differences in the samples would occur simply by chance, even when there are no differences in the population. The question is therefore reduced to whether the observed differences are large enough to convince us that they did not occur merely by chance but reflect real differences in the population.

The statistical hypothesis being tested, the null hypothesis, is that the mean natural killer cell activity is equal among women with low, moderate, and high scores. The alternative hypothesis is that there is a difference; ie, not all the means are equal. The steps in testing the null hypothesis follow.

Step 1.

H_0: $\mu_1 = \mu_2 = \mu_3$ (The mean natural killer cell activity is equal in the three groups.)
H_1: $\mu_1 \neq \mu_2$ or $\mu_2 \neq \mu_3$ or $\mu_1 \neq \mu_3$ (The means are not all equal.)

Step 2. The test statistic in the test of equality of means in ANOVA is the F ratio, F

$= MS_A/MS_E$, with $j - 1$ and $\Sigma (n_j - 1)$ degrees of freedom.

Step 3. Let us use $\alpha = .01$ for this statistical test.

Step 4. The value of the F distribution from Table A–4 with $j - 1 = 2$ degrees of freedom in the numerator and $\Sigma (n_j - 1) = 34$ degrees of freedom in the denominator, for $\alpha = .01$, is between 5.39 and 5.18; interpolation gives 5.31. Therefore, the decision will be to reject the null hypothesis of equal means if the observed value of F is greater than 5.31 and falls in the rejection area (see Fig 8–1).

Step 5. We begin by calculating the grand mean. Because we already know the means for the three groups, we can form a weighted average of these means, ie, the grand mean:

$$\overline{X}. = \frac{(13)(40.23) + (12)(15.6) + (12)(18.06)}{13 + 12 + 12}$$

$$= \frac{926.91}{37}$$

$$= 25.05$$

The numerator of the among-groups mean square MS_A is then

$$\Sigma\, n_j(\overline{X}_j - \overline{X}.)^2 = 13(40.23 - 25.05)^2 + 12(15.60 - 25.05)^2 + 12(18.06 - 25.05)^2$$
$$= 13(230.43) + 12(89.30) + 12(48.86)$$
$$= 4653.57$$

The term MS_A is found by dividing the numerator by the number of groups minus 1 ($j - 1$), which is 2 in this example. Therefore,

$$MS_A = \frac{4653.57}{2}$$

$$= 2326.79$$

The individual group variances are used to calculate the pooled estimate of the error mean square MS_E:

$$MS_E = \frac{\Sigma(n_j - 1)s_j^2}{\Sigma(n_j - 1)}$$

$$= \frac{12(25.71)^2 + 11(6.42)^2 + 11(9.97)^2}{12 + 11 + 11}$$

$$= \frac{9478.84}{34}$$

$$= 278.79.$$

Finally, the F ratio is found by dividing the mean square among groups by the error mean square:

$$F = \frac{MS_A}{MS_E}$$

$$= \frac{2326.79}{278.79}$$

$$= 8.35$$

Step 6. The observed value of the F ratio is 8.35, which is larger than 5.31 (the critical value from Table A–4). Therefore, the null hypothesis of equal means is rejected. We conclude that there is a difference in mean natural killer cell activity among patients with low, moderate, and high scores on the Social Readjustment Rating Scale. Note that rejecting the null hypothesis does not tell us *which* group means differ, only that a difference exists; in Section 8.4, we will discuss how to determine which specific groups differ.

The results of ANOVA traditionally are presented in an ANOVA table similar to Table 8–2. The terms "sums of squares" and "mean squares"

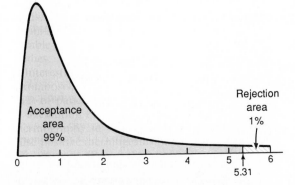

Figure 8–1. Illustration of critical values in F distribution.

Table 8–2. ANOVA table for number of killer cells.[1,2]

Source of Variation	Sums of Squares	Degrees of Freedom	Mean Squares	F Ratio
Among groups	4,653.57	2	2326.79	8.35
Error	9,478.84	34	278.79	
Total	14,132.41	36		

[1]Observations used, with permission, from Irwin M et al: Life events, depressive symptoms, and immune function. *Am J Psychiatry* 1987; **144**:437.
[2]Figures are based on calculations from the intuitive overview and vary slightly from those found in Section 8.3.

are discussed in the next section. When the results are given at this level of detail, we can easily determine exactly how the data were analyzed. Not all the authors, however, present this amount of detail; some simply list means and standard deviations along with F values or P- values.

8.3 TRADITIONAL APPROACH TO ANOVA

In the preceding section, we presented a simple illustration of ANOVA by using formulas to estimate the variance among individual group means and the grand mean, called the among-groups mean square (MS_A), and the variance within the groups, called the error mean square (MS_E). Traditionally in ANOVA, formulas are given for **sums of squares,** which are generally equivalent to the numerators of the formulas used in the preceding section; then, sums of squares are divided by appropriate degrees of freedom to obtain mean squares. Before illustrating the calculations for the data on number of killer cells (Presenting Problem 1), we define some terms and give the traditional formulas.

8.3.1 Terms & Formulas for ANOVA

In ANOVA, the term **factor** refers to the variable by which groups are formed, the **independent variable.** For example, in Presenting Problem 1, there was only one variable by which subjects were grouped, the Social Readjustment Rating Scale; therefore, this study is an example of a one-factor ANOVA, called one-way ANOVA. The number of groups defined by a given factor is referred to as the number of *levels* of the factor; the factor in Presenting Problem 1 has three levels: low, moderate, and high scores. In experimental studies in medicine, levels are frequently referred to as **treatments.**

Some textbooks approach analysis of variance from the perspective of models. The **model** for one-way ANOVA states that an individual observation can be decomposed into three components related to (1) the grand mean, (2) the group to which the individual belongs, and (3) the individual observation itself. To write this model in symbols, we let i stand for a given individual observation and j stand for the group to which this individual belongs. Then, X_{ij} denotes the observation of individual i in group j; eg, X_{11} is the first observation in the first group, and X_{53} is the fifth observation in the third group. The grand mean in the model is denoted by μ. The *effect* of being a member of group j may be thought of as the difference between the mean of group j and the grand mean; the effect associated with being in group j is written α_j. Finally, the difference between the individual observation and the mean of the

group to which the observation belongs is written e_{ij} and is called the **error,** or **residual.** Putting these symbols together, we can write the model for one-way ANOVA as

$$X_{ij} = \mu + \alpha_j + e_{ij}$$

which states that the ith observation in the jth group, X_{ij}, is the sum of three components: the grand mean μ; the effect associated with group j, α_j; and an error (residual) e_{ij}.

The size of an effect is, of course, related to the size of the difference between a given group mean and the grand mean. When inferences involve only specific levels of the factor included in the study, the model is called a **fixed-effects model.** The fixed-effects model assumes we are interested in making inferences only to the populations represented in the study. For example, if investigators wish to draw conclusions about three dosage levels of a drug, the model is a fixed-effects model. If, in contrast, the dosage levels included in the study are viewed as being randomly selected from all different possible dosage levels of the drug, the model is called a **random-effects model,** and inferences can be made to other levels of the factor not represented in the study. (The calculations of sums of squares and mean squares are the same in both models, but the type of model determines the way the F ratio is formed when there are two or more factors in a study.) Both models are used to test hypotheses about the equality of group means. However, the random-effects model can also be used to test hypotheses and form confidence intervals about group variances, and it is also referred to as the components-of-variance model for this reason.

8.3.1.a Definitional Formulas: In Section 8.2.1, we showed that the variation of 22.2 lytic units from the grand mean of 25.05 (Subject 1 in the low-score group) can be expressed as a sum of two differences: (1) the difference between the observation and the mean of the group it is in, and (2) the difference between its group mean and the grand mean. This result is also true when the differences are squared and the squared deviations added to form the sum of squares.

For example, for one factor with j groups, we use the following definitions:

> X_{ij} is the ith observation in the jth group.
> $\overline{X}_j$ is the mean of all observations in the jth group.
> $\overline{X}.$ is the grand mean of the observations.

Then, $\Sigma(X_{ij} - \overline{X}.)^2$, the total sum of squares, or SS_T, can be expressed as the sum of

> $\Sigma(X_{ij} - \overline{X}_j)^2$, the error sum of squares, or SS_E.
> $\Sigma(\overline{X}_j - \overline{X}.)^2$, the among-groups sum of squares, or SS_A.

That is,

$$\Sigma(X_{ij} - \overline{X}.)^2 = \Sigma(X_{ij} - \overline{X}_j)^2 + \Sigma(\overline{X}_j - \overline{X}.)^2$$

or

$$SS_T = SS_E + SS_A$$

A proof of this equality is beyond the scope of this text but may be found in many statistical references (eg, Armitage, 1971, p 191; Daniel, 1974, pp 186–187; Hays, 1973, pp. 465–475).

8.3.1.b Computational Formulas:
Computational formulas are more convenient than definitional formulas when sums of squares are calculated manually or by using small calculators. Additionally, computational formulas are preferred because they reduce round-off errors. Computational formulas can be derived from definitional ones, but the algebra is complex and is not given here. If you are interested in the details, consult the texts listed above for derivations.

The symbols in ANOVA vary somewhat from one text to another; the following formulas are similar to those used in many books and are the ones we will use to illustrate calculations for ANOVA. Let N be the total number of observations in all the groups, ie, $N = \Sigma n_j$. Then, the computational formulas for the sums of squares are

$$SS_T = \Sigma(X_{ij} - \overline{X}.)^2 = \Sigma X_{ij}^2 - \frac{(\Sigma X_{ij})^2}{N}$$

$$SS_A = \Sigma(\overline{X}_j - \overline{X}.)^2 = \Sigma n_j\overline{X}_j^2 - \frac{(\Sigma X_{ij})^2}{N}$$

and SS_E is found by subtraction:

$$SS_E = SS_T - SS_A$$

The sums of squares are divided by the degrees of freedom to obtain the mean squares:

$$MS_A = \frac{SS_A}{k - 1}$$

where k is the number of groups or levels of the factor; and

$$MS_E = \frac{SS_E}{N - k}$$

8.3.2 One-Way ANOVA

Presenting Problem 1 is an example of a one-way ANOVA model in which there is only one independent variable, the score on the Social Readjustment Rating Scale. There are three levels of scores—low, moderate, and high; and the mean

number of lytic units is examined for women at each level. (See Table 8–1.)

8.3.2.a Illustration of Traditional Calculations:
To calculate sums of squares by using traditional ANOVA formulas, we must obtain three terms:

1. We square each observation (X_{ij}) and add, to obtain ΣX_{ij}^2.
2. We add the observations, square the sum, and divide by N, to obtain $(\Sigma X_{ij})^2 / N$.
3. We square each mean ($\overline{X}_j$), multiply by n_j, and add, to obtain $\Sigma n_j\overline{X}_j^2$.

These three terms for the data in Table 8–1 are

$$\Sigma X_{ij}^2 = 22.2^2 + 97.8^2 + 29.1^2 + \ldots + 36.3^2 + 17.7^2$$
$$= 37,353.65$$

$$\frac{(\Sigma X_{ij})^2}{N} = \frac{(22.2 + 97.8 + 29.1 + \ldots + 36.3 + 17.7)^2}{13 + 12 + 12}$$

$$= \frac{(926.9)^2}{37}$$
$$= 23,220.10$$

$$\Sigma n_j\overline{X}_j^2 = (13)(40.23)^2 + (12)(15.60)^2 + (12)(18.06)^2$$
$$= 27,874.28$$

Then, the sums of squares are

$$SS_T = \Sigma X_{ij}^2 - \frac{(\Sigma X_{ij})^2}{N} \quad = 37,353.65 - 23,220.10$$
$$= 14,133.55$$

$$SS_A = \Sigma n_j\overline{X}_j^2 - \frac{(\Sigma X_{ij})^2}{N} \quad = 27,874.28 - 23,220.10$$
$$= 4654.18$$

$$SS_E = SS_T - SS_A = 14,133.55 - 4654.18 = 9479.37$$

Next, the mean squares are calculated:

$$MS_A = \frac{SS_A}{k - 1} = \frac{4654.18}{2} = 2327.09$$

$$MS_E = \frac{SS_E}{N - k} = \frac{9479.37}{34} = 278.81$$

Slight differences between these results and the results for the mean squares calculated in Section 8.2.2 are due to round-off error. Otherwise, the results are the same regardless of which formulas are used.

Finally, the F ratio is determined:

$$F = \frac{MS_A}{MS_E} = \frac{2327.18}{278.81} = 8.35$$

The calculated F ratio is compared with the value from the F distribution with 2 and 34 degrees of freedom at the desired level of significance. As we found in Section 8.2.2, for $\alpha = .01$, the value of $F(2, 34)$ is 5.31. Because 8.35 is greater than 5.31, the null hypothesis is rejected; and we conclude that mean killer cell activity is not the same for patients who score low, moderate, and high on the Social Readjustment Rating Scale.

The formulas for one-way ANOVA are summarized in Table 8–3.

8.3.2.b Assumptions in ANOVA: Before leaving this topic, we will discuss the three assumptions made in ANOVA:

1. The values of the dependent variable are assumed to be normally distributed within each group, ie, at each level of the factor or independent variable.

2. The population variance is the same in each group, ie, $\sigma_1^2 = \sigma_2^2 = \sigma_3^2$.

3. The observations are a random sample, and they are independent; ie, the value of one observation is not related in any way to the value of another observation.

Not all the above assumptions are equally important. For example, the results of the F test are not affected by moderate departures from normality, especially for a large number of observations in each group or sample—ie, the F test is **robust** with respect to violations of the assumption of normality. However, if the observations are extremely skewed, especially for small samples, the Kruskal-Wallis nonparametric procedure, discussed later in this chapter, should be used.

The F test is more sensitive to the second assumption of equal variances, also called **homogeneity** of variances. However, concern about this assumption is eliminated if sample sizes are equal (or close to equal) in each group (Box, 1953, 1954). For this reason, investigators try to design studies with equal sample sizes. If they cannot, as is sometimes the case in observational studies, there are two possible solutions. The first is to transform data within each group to obtain equal variances, using one of the transformations discussed in Chapter 7. The second solution is to se-

lect samples of equal sizes randomly from each group, although many investigators do not like this solution because perfectly good observations are thrown away. Investigators should consult a statistician for studies resulting in greatly unequal variances and unequal sample sizes.

The last assumption is particularly important. In general, investigators should be certain that they have **independent observations.** Independence is a problem primarily with studies involving repeated measurements of the same subjects, and they must be handled in a special way, as we briefly discuss later in this chapter.

As a final comment, recall that the fixed-effects model assumes that each observation is really a sum, consisting of the grand mean, the effect of being a member of the particular group, and the error (residual) representing any unexplained variation. Some studies involve observations that are proportions, rates, or ratios; and for these data, the assumption about sums does not hold.

8.3.3 Two-Way ANOVA

Two-way ANOVA is similar to one-way ANOVA except that there are two factors (or two independent variables). For example, in the study described in Presenting Problem 2, the investigators were interested in comparing acetylcholine synthesis in patients with Alzheimer's disease and in controls without Alzheimer's disease. Therefore, disease status—Alzheimer's versus none—is one factor. The investigators were also interested in comparing acetylcholine synthesis in younger and older subjects; thus, the second factor is age at onset—younger (< 80 years) versus older (80 years or more). In this example, both factors are measured at two levels and are said to be *crossed.* Data from the study are given in Table 8–4.

Because there are two factors in this study (disease status and age), each measured at two levels (patients vs controls and younger vs older), there are $2 \times 2 = 4$ treatment combinations: younger controls, younger patients, older controls, and older patients. Three questions may be asked in this two-way ANOVA:

Table 8-3. Formulas for one-way ANOVA.

Source of Variation	Sums of Squares	Degrees of Freedom	Mean Squares	F Ratio
Among groups	$SS_A = \Sigma\, n_i \overline{X}_j^2 - \dfrac{(\Sigma\, X_{ij})^2}{N}$	$k - 1$	$MS_A = \dfrac{SS_A}{k - 1}$	$F = \dfrac{MS_A}{MS_E}$
Error	$SS_E = SS_T - SS_A$	$N - k$	$MS_E = \dfrac{SS_E}{N - k}$	
Total	$SS_T = \Sigma\, X_{ij}^2 - \dfrac{(\Sigma\, X_{ij})^2}{N}$	$N - 1$		

Table 8–4. Neurochemical changes of temporal cortex in Alzheimer and matched control patients.[1]

	Means and Standard Deviations			
	Early Onset (< 80 years)		Late Onset (≥ 80 years)	
	Controls	Patients	Controls	Patients
n	14	19	20	29
Choline acetyltransferase[2]	99.7 ± 18.7	39.3 ± 33.1[3]	101.9 ± 41.6	42.1 ± 25.8[3]
Serotonin	1392 ± 868	504 ± 397[3]	915 ± 474	625 ± 420[3]
Norepinephrine	646 ± 385	230 ± 214[3]	549 ± 264	340 ± 280[3]
3-Methoxy-4-hydroxy-phenylglycol	1451 ± 935	2187 ± 1268[3]	2126 ± 962	1768 ± 953
5-Hydroxyindoleacetic acid	6714 ± 3079	5423 ± 3308	8673 ± 4244	7094 ± 5649

[1]Adapted and reproduced, with permission, from Table 1 in Francis PT et al: Neurochemical studies of early-onset Alzheimer's disease: Possible influence on treatment. *N Engl J Med* 1985; **313**:7.
[2]Choline acetyltransferase values are expressed as picomoles per hour per milligram of protein; all other values are expressed as femtomoles per milligram of protein.
[3]$P < .05$ between patients and controls in a least significant difference test.

1. Are there differences between patients and controls? If so, the means for each treatment combination might be as given in Table 8–5A, and we say there is a difference in the *main effect* for disease status. The null hypothesis for this question is that choline acetyltransferase is the same in patients and in controls ($\mu_P = \mu_C$).

2. Are there differences between younger and older subjects? If so, the means for each treatment combination might be as given in Table 8–5B, and we say there is a difference in the *main effect* for age. The null hypothesis for this question is that choline acetyltransferase is the same in younger subjects and in older subjects ($\mu_Y = \mu_O$).

3. Are there differences owing to neither disease state nor age alone but owing to the combination

Table 8–5. Possible results for hypothetical data in two-way ANOVA: Means for each treatment combination.

A. Difference Between Patients and Controls

	Age Group	
Subjects	Younger	Older
Patients	50	50
Controls	100	100

B. Differences Between Younger and Older Subjects

	Age Group	
Subjects	Younger	Older
Patients	100	50
Controls	100	50

C. Differences Owing Only to Combinations of Factors

	Age Group	
Subjects	Younger	Older
Patients	50	100
Controls	100	50

of factors? If so, the means for each treatment combination might be as given in Table 8–5C, and we say there is an **interaction** effect between the two factors. The null hypothesis for this question is that any difference in choline acetyltransferase between young patients and young controls is the same as the difference between older patients and older controls ($\mu_{YP} - \mu_{YC} = \mu_{OP} - \mu_{OC}$).

This study can be viewed as two separate experiments on the same set of subjects for each of the first two questions. The third question can only be answered, however, in a single experiment in which both factors are measured and there is more than one observation at each treatment combination of the factors (ie, in each cell).

The calculations in two-way ANOVA are tedious and will not be illustrated in this book. However, they are similar to the calculations in the simpler one-way situation: The total variation in observations is divided, and sums of squares are determined for the first factor, the second factor, the interaction of the factors, and the error (residual), which is analogous to the within-group sums of squares in one-way ANOVA. A summary of the ANOVA for Presenting Problem 2 is discussed in Section 8.5. Equal sample sizes are assumed, as in one-way ANOVA. In Presenting Problem 2, however, the sample sizes are not equal; they vary from 14 in younger controls to 29 in older patients. If sample sizes are not equal, the analysis must be modified in ways too complicated to discuss in this text.

The topic of interactions is important and worth pursuing a bit further. Fig 8–2A is a graph of the mean choline acetyltransferase levels from Table 8–5A for younger and older patients and controls. When lines connecting means are parallel, there is no interaction between the factors of disease status and age, and the effects are said to

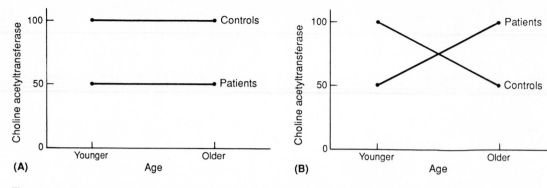

Figure 8–2. Graphs of interaction. (A) No interaction; effects are additive. (B) Significant interaction; effects are multiplicative.

be *additive*. However, if there is significant interaction, as in Table 8–5C, the lines intersect and the effects are called *multiplicative*. Fig 8–2B illustrates this situation and shows that main effects, such as disease status and age at onset, are difficult to interpret when there are significant interactions. For example, if the interaction is significant, any conclusions regarding decreased levels of choline acetyltransferase depend on both disease status and age; any comparison between patients with Alzheimer's disease and controls depends on age of the subject. Although this example illustrates extreme interaction, some statisticians recommend that interaction be tested first and, if significant, that main effects *not* be tested.

8.4 MULTIPLE-COMPARISON PROCEDURES

The discussion thus far has ignored studies in which the investigator has a limited number of specific comparisons in mind prior to designing the study, called planned, or a priori, comparisons. In this special case, comparisons can be made without performing an ANOVA first, although in actual practice, most investigators prefer to perform an ANOVA anyway. Typically, investigators want the freedom and flexibility of making comparisons afforded by the posteriori, or post hoc, methods. However, before discussing these types of comparisons, we need two definitions.

A **comparison** or **contrast** between two means is the difference between the means, such as $\mu_1 - \mu_2$. Comparisons or contrasts can also involve more than two means. For example, suppose a study undertaken to compare a new drug with a placebo uses two dosage levels. The investigators may wish to compare the mean response for dosage 1 with the mean response for placebo, $\mu_1 - \mu_P$, as well as the mean response for dosage 2 with

the mean response for placebo, $\mu_2 - \mu_P$. In addition, they may wish to compare the overall effect of the drug with the effect of placebo, $[(\mu_1 + \mu_2)/2] - \mu_P$. Note that in all examples, the *coefficients* of the means add to zero; ie, rewriting the first comparison slightly gives $(1)\mu_1 + (-1)\mu_2$ and $(1) + (-1) = 0$; rewriting the last comparison gives $(1/2)\mu_1 + (1/2)\mu_2 + (-1)\mu_P$ and $(1/2) + (1/2) + (-1)$ again is 0.

The second definition involves the distinction between two different kinds of comparisons or contrasts. Two comparisons or contrasts are *orthogonal* if they do not use the same information. For example, suppose a study involves four different therapies, 1, 2, 3, and 4. Then, comparisons between $\mu_1 - \mu_2$ and between $\mu_3 - \mu_4$ are orthogonal because information used to compare groups 1 and 2 is not the same information used to compare groups 3 and 4. In a sense, the questions asked by two orthogonal comparisons may be considered independent from each other. Conversely, comparisons of $\mu_1 - \mu_2$ and $\mu_1 - \mu_3$ are not orthogonal because they utilize redundant information; ie, observations in group 1 are used in both comparisons.

8.4.1 A Priori, or Planned, Comparisons

When comparisons are planned, they may be undertaken without first performing ANOVA. When the comparisons are all orthogonal, the *t* test for independent groups (see Chapter 7) may be used with the following modification: Instead of using the pooled standard deviation s_P in the denominator of the *t* ratio, we use the error mean square MS_E. When sample sizes are equal, denoted by n, the *t* ratio becomes

$$t = \frac{(\overline{X}_i - \overline{X}_j)}{\sqrt{2MS_E/n}}$$

with $N - k$ degrees of freedom, where N is the total number of observations in the study.

8.4.1.a Adjusting the Level Downward:
When several planned comparisons are made, the probability of obtaining significance by chance is increased; ie, the probability of a type I error is increased. For example, for four independent comparisons, all at $\alpha = .05$, the probability of one or more significant results is $4 \times .05 = .20$. One way to compensate for multiple comparisons is to decrease the α level, and one way to decrease the α level is to divide α by the number of comparisons made. For example, if four independent comparisons are made and the investigator wants to maintain the overall probability of a type I error at .05, this value is divided by 4 to obtain a comparisonwise α of $.05/4 = .0125$. For this method, each orthogonal comparison must be significant at the .0125 level in order to be declared statistically significant.

8.4.1.b The Bonferroni t Procedure:
Another approach for planned comparisons is the **Bonferroni t method,** also called Dunn's multiple-comparison procedure. This approach is more versatile because it is applicable for both orthogonal and nonorthogonal comparisons. The Bonferroni t method increases the critical F value needed for the comparison to be declared significant. The amount of increase depends on the number of comparisons and the sample size.

To illustrate, we will consider the three groups of patients defined by scores on the Social Readjustment Rating Scale (Presenting Problem 1). The pairwise differences for mean number of killer cells for these three groups are listed in Table 8-6. These data will be used to illustrate all multiple comparisons in this section so that results of the different procedures can be compared. To simplify the comparisons, we will assume that there are 12 subjects in the group with low scores, although there actually were 13 in that group.

In the Bonferroni t procedure, $\sqrt{2MS_E/n}$ is multiplied by a factor related to the number of comparisons made and the degrees of freedom for error mean square. In this example, there are three possible pairwise comparisons, and 34 degrees of freedom. For $\alpha = .01$, the multiplier is 3.17. (If

you are interested in more detail, consult Kirk, 1982, Chapter 3 and Table E16.) Therefore,

$$3.17 \times \sqrt{\frac{2MS_E}{n}} = 3.17 \times \sqrt{\frac{2 \times 278.81}{12}}$$
$$= 3.17 \times 6.82$$
$$= 21.62$$

(We have used the value for MS_E found in Section 8.3.) All pairwise differences are compared with 21.62, and if they exceed this value, the differences are significantly different from zero with $\alpha = .01$. Two of the pairwise differences in Table 8-6 are greater than 21.62, those between $\overline{X}_{low}$ and $\overline{X}_{high}$ and between $\overline{X}_{low}$ and $\overline{X}_{mod}$. Therefore, the conclusion is that persons with low scores on the Social Readjustment Rating Scale have a higher mean number of killer cells than persons with high or moderate scores on the scale; there are no differences, however, between the high and moderate groups.

8.4.2 Posteriori, or Post Hoc, Comparisons

Post hoc comparisons (the Latin term means "after this") are made after an ANOVA has resulted in a significant F test. The t test introduced in Chapter 7 should *not* be used for these comparisons. Although t tests allow investigators to easily obtain significant differences for post hoc comparisons, they do not take into consideration the number of comparisons being made, the possible lack of independence (nonorthogonality) of the comparisons, and the post hoc (unplanned) nature of the comparisons.

Several procedures are available for making post hoc comparisons. Four of them are recommended by statisticians, depending on the particular research design, and two others are not recommended but are commonly used nonetheless. We discuss all six procedures below. The data from Presenting Problem 1, as summarized in Table 8–6, are again used to illustrate the procedures. As with the Bonferroni t, special tables are needed to find appropriate multipliers for these tests when computer programs are not used for the analysis; excerpts from the tables are reproduced in Table 8–7 corresponding to $\alpha = .01$.

Table 8-6. Differences between means of number of killer cells for groups based on Social Readjustment Rating Scale.

	$\overline{X}_{low}$	$\overline{X}_{high}$	$\overline{X}_{mod}$
$\overline{X}_{low}$	—	22.17	24.63
$\overline{X}_{high}$		—	2.46
$\overline{X}_{mod}$			—

Table 8-7. Excerpts from tables for use with multiple-comparison procedures for $\alpha = .01$.

Error df	Number of Means or Steps for Studentized Range		Number of Means for Dunnett's Test	
	2	3	2	3
30	3.89	4.45	2.46	2.72
34[1]	3.87	4.42	2.45	2.71
40	3.82	4.37	2.42	2.68

[1]Found by interpolation.

8.4.2.a Tukey's HSD Procedure: The first procedure we discuss was developed by the same statistician who gave us stem and leaf plots and box and whisker plots, and he obviously has a sense of humor. The procedure is called the Tukey test or **Tukey's HSD** (honestly significant difference) test, so named because some post hoc procedures make significance too easy to obtain. It is applicable only for pairwise comparisons, but it permits the researcher to compare all pairs of means. A study by Stoline (1981) found it to be the most accurate and powerful procedure to use in this situation. Recall that power is the ability to detect a difference if one actually exists, so high power means that the null hypothesis will be correctly rejected more often.

The HSD statistic, like the Bonferroni t, has a multiplier, which is based on the number of treatment levels and the degrees of freedom for error mean square (in our example, 3 and 34, respectively). In Table 8-7, under the column for 3 and studentized range,* we find the multiplier 4.42. The HSD statistic is

$$\text{HSD} = \text{multiplier} \times \frac{\text{MS}_E}{n}$$

where n is the sample size in each group. Thus,

$$\text{HSD} = 4.42 \times \sqrt{\frac{278.81}{12}}$$

$$= 21.31$$

The differences in Table 8–6 are now compared with 21.31 and declared to be significantly different if they exceed this value. From the table, we see that the differences between the low and high groups and between the low and moderate groups are greater than 21.31. Therefore, the conclusion is the same as for the Bonferroni t—that persons with low scores on the Social Readjustment Rating Scale have a higher mean number of killer cells than persons with high or moderate scores on the scale; there are no differences, however, between the high and moderate groups. Tukey's procedure can also be used to form confidence intervals about the mean difference (as can the Bonferroni t method). For instance, 99% confidence interval for the mean difference in number of killer cells between low and moderate groups is

$$(\overline{X}_{\text{low}} - \overline{X}_{\text{mod}}) \pm 21.31 \quad \text{or} \quad 3.32 \text{ to } 45.94$$

8.4.2.b Scheffe's Procedure: Scheffe's **procedure** is the most versatile of all the post hoc procedures because it permits the researcher to make all types of comparisons, not simply pairwise comparisons. For example, Scheffe's procedure allows one to compare the overall mean of two or more dosage levels with a placebo, as discussed earlier. However, a price is extracted for this flexibility: A higher critical value is used to determine significance. Thus, Scheffe's procedure is also the most conservative of the multiple-comparison procedures.

The formula, which looks somewhat complicated, is

$$S = \sqrt{(k - 1)F_{\alpha,\text{df}}}\sqrt{\text{MS}_E \frac{\Sigma\ C_j^2}{n_j}}$$

where k is the number of groups, $F_{\alpha,\text{df}}$ is the critical value of F used in ANOVA, MS_E is the error mean square, and $\Sigma\ C_j^2/n_j$ is the sum of squared coefficients divided by sample sizes in the contrast of interest. For example, in the contrast defined by $\overline{X}_{\text{low}}$ - $\overline{X}_{\text{high}}$, $\Sigma\ C_j^2/n_j = (1)^2/12 + (-1)^2/12 = 0.167$; this value will be the same for all pairwise comparisons listed in Table 8–6.*

The critical value for F at $\alpha = .01$ with 2 and 34 degrees of freedom was found to be 5.31 in Section 8.2. Substituting values in S yields

$$S = \sqrt{(k - 1)F_{\alpha,\text{df}}}\sqrt{\text{MS}_E \frac{\Sigma\ C_j^2}{n_j}}$$

$$= \sqrt{2 \times 5.31}\ \sqrt{278.81 \times 0.167}$$
$$= 3.259 \times 6.824$$
$$= 22.24$$

Therefore, any contrast greater than 22.24 is significant. As we see from Table 8–6, the difference in means between low and high groups is 22.17 and just barely misses being significant by Scheffe's test. Because Scheffe's test is the most conservative of all post hoc tests, 22.24 is the largest critical value required by any of the multiple-comparison procedures. Confidence intervals can also be formed by using Scheffe's procedure.

8.4.2.c The Newman-Keuls Procedure: The **Newman-Keuls procedure** uses a stepwise approach to testing differences between means and can be used only to make pairwise comparisons. The procedure rank-orders means from lowest to highest, and the number of steps that separate pairs of means is noted. For our example, the rank orders and the number of steps are given in Fig. 8–3. In this procedure, the mean differences are compared with a critical value that depends on the number of steps between the two means, the

---

*The studentized range distribution was developed for a test of the largest observed difference in a set of means, essentially a test of significance of the range of means. It is used as the basis for several post hoc comparisons.

*However, this term is changed if, say, the mean of the low group is compared with the mean of the moderate and high groups combined, ie, $\overline{X}_{\text{low}}$ - $[(\overline{X}_{\text{mod}} + \overline{X}_{\text{high}})/2]$. In this case, $\Sigma\ C_j^2/n_j = (1)^2/12 + (-1/2)^2/12 + (-1/2)^2/12 = 0.125$.

Figure 8-3. Ranking of means and steps for Newman-Keuls test.

sample size, and the number of groups being compared. In addition, the testing must be done in a prescribed sequence.

The critical value also uses the studentized range, but the value corresponds to the number of steps between the means (instead of the number of means, as in the Tukey test); this value is multiplied by $\sqrt{MS_E/n}$, as in the Tukey test. The value from Table 8-7 corresponding to two steps with 34 degrees of freedom is 3.87; the value for three steps is 4.42. Therefore, the critical values for this example are

$$\text{Newman-Keuls} = \text{Multiplier} \times \sqrt{\frac{MS_E}{n}}$$

$$\text{Newman-Keuls} = 3.87 \times \sqrt{\frac{MS_E}{n}}$$

$$= 3.87 \times 4.82$$
$$= 18.65$$

$$\text{Newman-Keuls} = 4.42 \times \sqrt{\frac{MS_E}{n}}$$

$$= 4.42 \times 4.82$$
$$= 21.31$$

The critical value for three steps is the same as the critical value for Tukey's HSD test, but the critical value for two steps is less. Therefore, although the conclusions are the same as in Tukey's test, we can see that using the Newman-Keuls procedure with several groups may permit the investigator to declare a difference between two means significant when the difference would not be significant in Tukey's HSD test. The primary disadvantage of the Newman-Keuls procedure is that one cannot form confidence intervals for mean differences.

8.4.2.d Dunnett's Procedure: The fourth procedure recommended by statisticians is called **Dunnett's procedure,** and it is applicable *only* in situations in which the investigator wants to compare several treatment means with a single control mean. No comparisons are permitted between the treatment means themselves, so this test has a very specialized application. However, in this special situation, Dunnett's test is convenient to use because it has a relatively low critical value. The size of the multiplier depends on the number of groups, including the control group, and the degrees of freedom for error mean square. The formula is

$$\text{Dunnett's test} = \text{Multiplier} \times \sqrt{\frac{2MS_E}{n}}$$

Even though Dunnett's test is not applicable to our example, we will determine the critical value for the sake of comparison. From Table 8-7, under the column for Dunnett's test and three groups, we find the multiplier 2.71. Multiplying it by $\sqrt{2MS_E/n}$ gives 2.71 × 6.82, or 18.49, a value almost 2 units lower than Tukey's value and almost 4 units lower than Scheffe's value.

8.4.2.e Other Tests: Two procedures that appear in the medical literature but are *not* recommended by statisticians are Duncan's new multiple-range test and the least significant difference test. Duncan's new multiple-range test uses the same principle as the Newman-Keuls test. However, the multipliers in the formula are smaller, so statistically significant differences are found with smaller mean differences. Duncan argued that there is more likelihood of finding differences when there is a larger number of groups, and he increased the power of his test by using smaller multipliers. But his test, as a result, rejects the null hypothesis too often. Thus, it is not recommended by statisticians.

The least significant difference (LSD) test is one of the oldest multiple-comparisons procedures. As with the other post hoc procedures, it requires a significant F ratio from the ANOVA in order to make pairwise comparisons. However, instead of utilizing an adjustment to make the critical value larger, as the other tests have done, the LSD test uses the t distribution corresponding to the number of degrees of freedom for error mean square. The reason statisticians do not recommend this test is that with a large number of comparisons, the α levels of each comparison are inflated and differences that are too small may be incorrectly declared significant.

8.5 ADDITIONAL ILLUSTRATIONS OF THE USE OF ANOVA

In this section, we use the Presenting Problems to show you how to interpret results from ANOVA as they are typically presented in the literature. Then we briefly discuss additional ANOVA designs that are useful in medicine.

8.5.1 Interpretation of ANOVA Using the Presenting Problems

8.5.1.a Summary of Presenting Problem 1: The ANOVA results for Presenting Problem 1 are given in Table 8-8. The table provides much useful information, including sample sizes, means, and standard deviations (SD) for each group on

Table 8-8. Age, race, depression score, and immune variables of women with low, moderate, and high Social Readjustment Rating Scale scores.[1]

| | Group | | | | | | ANOVA | |
| | Low (n = 13)[2] | | Moderate (n = 12)[2] | | High (n = 12)[2] | | F | |
Measure	Mean	SD	Mean	SD	Mean	SD	(df = 2, 34)	P
Social Readjustment Rating Scale Score	31.9	16.4	81.7	11.1	126.6	24.7	84.4	< .001
Age (years)	54.8	9.5	55.3	6.3	57.8	9.4	—	n.s.
Race[3]	0.7	0.2	0.8	0.2	0.8	0.2	—	n.s.
No. of blood samples	3.6	0.6	3.8	0.9	3.8	0.9	—	n.s.
Hamilton depression score	5.3	5.2	14.7	7.5	12.0	6.8	6.8	< .003
Natural killer cell activity (lytic units)	40.2	25.7	15.6	6.4	18.1	10.0	8.3	< .001
Absolute no. of lymphocytes (cells/mL × 10)	1.8	0.5	2.2	0.5	2.5	0.8	—	n.s.
Percent Th cells[4]	34.0	6.5	38.0	9.8	37.0	8.4	—	n.s.
Percent Ts cells[4]	27.0	8.7	23.0	8.4	26.0	8.5	—	n.s.
Th-to-Ts ratio[4]	1.6	0.6	2.0	0.6	1.6	0.4	—	n.s.

[1]Reproduced, with permission, from Table 1 in Irwin M et al: Life events, depressive symptoms, and immune function. *Am J Psychiatry* 1987; **144**:437.
[2]Low = 54 or less; moderate = 55–99; high = 100 or more.
[3]0 = black; 1 = white.
[4] n = 9 in low-score group; n = 9 in moderate-score group; n = 10 in high-score group.

each dependent measure and results from the ANOVA.

The first measure in the table, Social Readjustment Rating Scale score, was actually the independent variable used to form the groups for ANOVA. These results are given to assure the reader that the division of the subjects into three groups did, in fact, result in groups that were significantly different.

Why was ANOVA performed on the next three measures—age, race, and blood samples? Generally, this type of information is presented in order to demonstrate that groups are similar with respect to baseline measures that could have a possible confounding effect. The designation "n.s." in the column headed by P stands for "not significant."

The authors chose not to present F ratios for measures that were not significant, although there is no reason to omit this information; it would not require additional space. In fact, when F ratios are given, they help the reader determine the chances that a type II error (not rejecting a false hypothesis) occurred. That is, if the F ratios that are not significant have a value close to 1, we can be more confident that the measures do not differ than if the F ratios are larger than 1.

The two measures that are significant in the analysis of Presenting Problem 1 are the Hamilton depression score and the natural killer cell ac-

tivity, the first with P < .003 and the second with P < .001. Thus, the probability of observing a difference this large merely by chance (ie, if there really is no difference between the low, moderate, and high groups) is less than 3 in 1000 and less than 1 in 1000, respectively. Natural killer cell activity was used to illustrate computations in this chapter, and the value of F ratio (8.3) agrees with the results from our analysis. The authors used the Newman-Keuls comparison procedure and reported, as with the analysis of natural killer cell activity, that the low group differed significantly from the high and moderate groups on mean Hamilton depression score, but the moderate and high groups did not differ. There were no differences between the groups on the other immune function variables analyzed, total lymphocyte count and T cell subpopulations.

The investigators controlled for possible **confounding variables** by excluding women who had a history of any chronic medical disorder associated with altered immune function or the abuse of alcohol or other substances. They concluded that women who were undergoing major life changes, such as bereavement, had lower natural killer cell activity than women who were not experiencing major life events. Additional analyses indicated that the severity of depressive symptoms, as measured by the Hamilton Rating Scale for Depression, was associated with impaired natural killer cell activity, a decrease in T suppressor cells, and an increase in the ratio of T helper to T suppressor cells.

8.5.1.b Summary of Presenting Problem 2:

The investigators for the study described in Presenting Problem 2 were interested in the differences in neurochemical markers between patients with Alzheimer's dementia and controls and between younger and older subjects. The results of their analyses are reported in Table 8–4 (except that the reported standard errors of the mean have been converted to standard deviations for our convenience in interpreting the findings). The authors do not provide F ratios for ANOVA; they simply indicate the comparisons between patients and controls that were significant at $P < .05$. Apparently, the authors actually performed a one-way ANOVA on the four groups of patients in Table 8–4. The shortcoming of this approach is that it is impossible to evaluate an interaction effect between disease status and age. A two-way ANOVA is a reasonable analysis (see Section 8.3.3). The authors appropriately used a logarithmic transformation on the data for serotonin and 5-hydroxyindoleacetic acid in order to obtain observations that were normally distributed before they performed the ANOVA.

The results of the ANOVA performed by the authors indicate differences between younger patients and controls on choline acetyltransferase, serotonin, norepinephrine, and 3-methoxy-4-hydroxyphenylglycol levels. Comparisons between older patients and controls were significant for the first three of these variables. Because older patients and controls did not differ in levels of 3-methoxy-4-hydroxyphenylglycol, whereas younger patients and controls did, it is likely that the interaction effect would be significant. Note that mean values for this measure are lower in younger controls than in younger patients, but they are higher in older controls than in older patients. The authors used the least significant difference test to make pairwise comparisons and concluded that the results of this study are consistent with the view that deficiency in the presynaptic cholinergic system is a relatively early change in the development of clinical features of Alzheimer's disease.

8.5.2 Other Designs & Methods in ANOVA

Many experimental designs for ANOVA are possible. Most designs, however, are combinations of a relatively small number of designs: randomized factorial designs, randomized block designs, and Latin square designs. The principle of randomized assignment resulted from the work of two statisticians in the early 20th century, Ronald Fisher and Karl Pearson, who had considerable influence on the development and the direction of modern statistical methods. For this reason, the term *randomized* occurs in the names of many designs; and, of course, one of the assumptions is that a random sample has been assigned to the different treatment levels. In this section, we will describe several designs sometimes used for an ANOVA.

8.5.2.a Randomized Factorial Designs:

The studies discussed earlier in this chapter are examples of randomized **factorial designs** with one or two factors. Randomized factorial designs with three or more factors are possible as well, and the ideas introduced in Section 8.3 generalize in a logical way. For example, a study with factors A, B, and C (with more than one observation per treatment combination) has sums of squares and mean squares for factor A, factor B, and factor C; for the interactions between A and B, between B and C, and between A and C; and, finally, for the interaction between A and B and C. Studies that employ more than three or four factors are rare in medicine because of the large number of subjects needed for such studies. For example, a study with three factors, each with two levels, has $2 \times 2 \times 2 = 8$ treatment combinations; if one factor has two levels, another has three levels, and the third has four levels, then there are $2 \times 3 \times 4 = 24$ different treatment combinations. Finding an equal number of subjects for each treatment combination can be difficult.

Randomized Block Designs:

A factor is said to be **confounded** with another factor if it is impossible to determine which factor is responsible for the observed effect. Age is frequently a confounding factor in medical studies, so investigators often age-match control subjects with treatment subjects. Randomized **block designs** are useful when there is a confounding factor that contributes to variation.

In the **randomized block design,** subjects are first subdivided into homogeneous blocks; one or more subjects from each block is then randomly assigned to each level of the experimental factor. This type of study is especially useful in laboratory experiments in which investigators are concerned about genetic variation and its effect on the outcome being studied. Litters are defined as the blocks, and littermates are then randomly assigned to the different levels of treatment. In this experiment, blocking is said to "control for" genetic differences. In studies involving humans, blocking on age or severity of a condition is often useful. Investigators may also want to have more than one subject from each block assigned to each treatment level.

Sometimes, investigators cannot control for possible confounding factors in the design of a study. The procedure called **analysis of covariance** allows investigators to control statistically for such factors in the analysis; this procedure is discussed in Chapter 12.

8.5.2.c Latin Square Designs:

Latin square designs employ the blocking principle for two confounding (or nuisance) factors. The levels of the confounding factors are assigned to the rows and the columns of a square; then, the cells of the square identify treatment levels. For example, suppose that both age and body weight are important blocking factors in an experiment that has as treatment 3 dosage levels of a drug. Then, 3 blocks of age and 3 blocks of body weight form a Latin square with 9 cells, and each dosage level appears one or more times for each age-weight combination. Table 8–9 illustrates this design. The Latin square design can be powerful, because only 9 subjects are needed and yet two possible confounding factors have been controlled.

8.5.2.d Nested Designs:

In the factorial designs described above, the factors are *crossed;* ie, all levels of one factor occur within all levels of the other factors. Thus, all possible combinations are included in the experiment, as in the study described in Presenting Problem 2 with younger and older patients and controls.

However, in some situations, crossed factors cannot be employed; so *hierarchical,* or *nested, designs* are used instead. In nested designs, one or more of the treatments is nested within levels of another factor. Nesting is needed when entire sets of subjects must be given the same treatment, often because of administrative concerns. For example, educational experimentation is difficult to perform in medical schools because classes are often small, and medical students communicate with one another about their educational experiences. Thus, an investigator who wants to compare two methods of teaching physical examination skills, say, may opt for designing a cooperative study among six medical schools. Medical schools are randomly assigned to one of the two treatment methods, three schools for each method. Medical schools are then said to be nested within teaching conditions, and it is impossible to determine interaction effects between medical school and teaching condition.

8.5.2.e Repeated-Measures Designs:

Recall from Chapter 7 that the paired t test should be used when the same sample of subjects is observed on two occasions. This design is powerful because it controls for individual variation; and for both biologic and psychologic measurements, individual variation can be large. The counterpart to the paired t test in ANOVA is the **repeated-measures** (or split-plot) **design.** In this design, subjects serve as their own controls, so that the variability owing to individual differences is eliminated from the error (residual) term, increasing the chances of observing significant differences between levels of treatment. In this sense, repeated-measures designs have the same goal as designs using blocks. The repeated measurements may be the same treatment level measured at more than one time, or they may involve the same subjects measured under different levels of a treatment at more than one time.

Recall that one of the assumptions in ANOVA is that the observations be independent of one another. This assumption is frequently not met in repeated-measures ANOVA; therefore, certain other assumptions concerning the dependent nature of the observations must be made and tested. We will not discuss these assumptions here. From an interpretation perspective, readers of the medical literature merely need to know that such designs are possible and should be used in studies with repeated measurements of the same subjects.

8.5.2.f Nonparametric ANOVA:

Nonparametric ANOVA is not a different design but a different method of analysis. Recall from Chapter 7 that if the assumptions for the t tests are seriously violated, then nonparametric methods such as the Wilcoxon rank-sum test or the signed-ranks test should be used instead. A similar situation holds in ANOVA. Even though the F test is robust with respect to violating the assumption of normality and, if the sample sizes are equal, the assumption of equal variances, there are situations in which transforming observations to a logarithm scale or using nonparametric procedures is advisable. For example, when observations from small samples greatly depart from the normal distribution or when markedly unequal variances occur along with different sample sizes, nonparametric ANOVA should be considered.

Like the nonparametric procedures discussed in Chapter 7, the nonparametric methods in ANOVA are based on the analysis of ranks of observations rather than on original observations. For one-way ANOVA, the nonparametric procedure is

Table 8–9. Latin square design with three dosage levels (D1, D2, D3).

Body weight (kg)	Age level (years)		
	<50	50–70	>70
<60	D1	D2	D3
65–90	D2	D3	D1
≥90	D3	D1	D2

phy was the same as the proportion diagnosed from venography.

9.1 PURPOSE OF THE CHAPTER

In the previous two chapters, we examined the different kinds of research questions that occur when the characteristic of interest is measured on a numerical scale. In those situations, means are calculated and the *t* **distribution,** the **analysis of variance** (ANOVA), or the appropriate **nonparametric method** is used to form confidence intervals or to test hypotheses. This chapter takes the same approach, except that the characteristic of interest is nominal, and thus, the research questions involve comparing proportions. Four situations are examined.

1. When the research question concerns inferences about the observed proportion in a single group of patients, an approximation to the binomial distribution based on the *z* distribution can be used to determine confidence limits or to compare the observed proportion with a standard.

2. When the research question compares the proportions from two independent groups, the *z* distribution procedure can be extended; but the chi-square test is also an alternative.

3. When the research question compares the proportions from three or more independent groups, the chi-square test is appropriate.

4. When the research question compares the proportions from two dependent or correlated groups, a modification of the chi-square test, known as the McNemar test, can be used.

Articles in the medical literature often report use of the *z* tests and chi-square tests for proportions. Recent literature reviews quoted in earlier chapters indicate their use ranges from 3% in pathology articles (Hokanson, Ladoulis, Quinn and Bienkowski, 1987) to 25% of the articles in family practice journals (Fromm and Snyder, 1986).

9.2 PROPORTIONS IN SINGLE GROUPS

Occasionally in medicine, a group is examined or observed and the proportion of subjects who respond favorably is compared with a well-accepted norm. For example, suppose clinical investigators wanted to examine the efficacy of a new drug in healing benign gastric ulcers, as in Presenting Problem 1. One approach is to give the drug to a sample of patients and determine the proportion of patients with healed ulcers after 6 weeks of treatment. In this type of study, the binomial distribution introduced in Chapter 5 can be used to determine confidence limits or to test hypotheses about the observed proportion. Recall that the **binomial distribution** is appropriate when there are

n independent trials, each with the same probability of success, π. Here, each patient who is given the new drug is considered a trial, and the probability of success is equal to the proportion of patients whose ulcers are healed at the end of 6 weeks.

Fig 9–1 illustrates the shape of the binomial distribution for $\pi = .2$ and $.4$ for sample sizes of 5, 10, and 25. Notice how the distribution becomes more bell-shaped as the sample size increases and as the population proportion π approaches .5. This result should not be surprising, because a proportion is actually a special case of a mean in which successes equal 1 and failures equal 0; and the central limit theorem discussed in Chapter 6 states that the sampling distribution of means for large samples resembles the normal distribution. These observations lead naturally to the idea of using the normal distribution as an approximation to the binomial distribution. Just as the sampling distribution of the mean has a mean μ and a standard error $\sigma/\sqrt{n}$, the sampling distribution of the proportion has a mean π and a standard error $\sqrt{\pi(1 - \pi)/n}$. The normal approximation to the binomial distribution, also called the *z* **approximation,** is appropriate as long as both $n\pi$ and $n(1 - \pi)$ exceed 5. As in Fig 9–1, when $n\pi$ is close to 5, the binomial distribution becomes bell-shaped, similar to the normal (gaussian) distribution.

9.2.1 Confidence Intervals for a Proportion

Since proportions for large samples [both $n\pi$ and $n(1-\pi)$ are greater than 5] have a sampling distribution that is normally distributed with mean π and standard error $\sqrt{\pi(1-\pi)/n}$, these properties can be used to form confidence intervals for the population proportion. Because the true proportion π is not known, the estimate of π, denoted p, is used *both* for the estimate of π and in the formula for the standard error. Therefore, the 95% confidence limits are given by

$$\text{Observed proportion} \pm 1.96 \times \text{Standard error of proportion}$$

$$= p \pm 1.96 \sqrt{\frac{p(1 - p)}{n}}$$

The value 1.96 comes from the standard normal (*z*) distribution and corresponds to the point that separates the central 95% of the distribution from the 2.5% in each tail. Of course, other values of the *z* distribution may be used if other confidence intervals, such as 90% or 99%, are desired for the proportion.

In the study described in Presenting Problem 1, the investigators found that 43 of the 66 patients, or 65%, who received the drug cimetidine and were examined after 6 weeks showed healing of

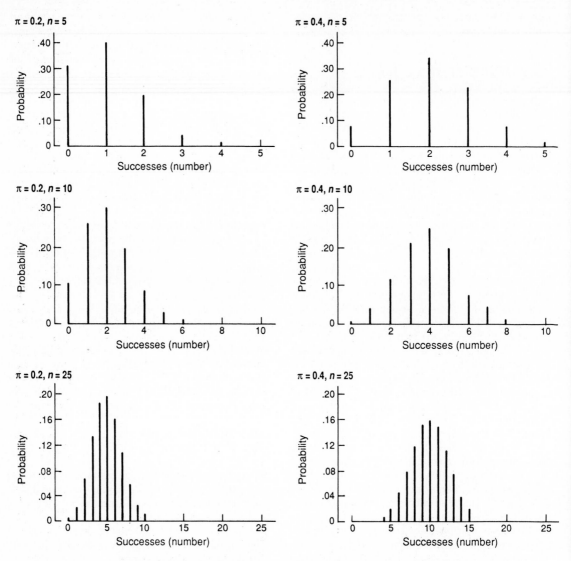

Figure 9-1. Probability distributions for binomial when $\pi = .2$ and .4.

their gastric ulcer. (See Table 9-1.) From the formula above, the 95% confidence limits for the true proportion of patients healed with this treatment are

$$0.65 \pm 1.96 \sqrt{\frac{0.65(1 - 0.65)}{66}} = 0.65 \pm 0.12$$

or

0.53 to 0.77

Therefore, the investigators may be 95% confident that the interval 0.53 to 0.77 (or 53% to 77%) contains the true proportion of patients in the population with healed ulcers following 6 weeks of treatment with cimetidine.

Various graphs and pictorial aids have been de-veloped to help researchers who test hypotheses about proportions. Fig 9–2 shows a graph that can be used to obtain estimates of 95% confidence intervals for the proportion. To use the graph, we locate the observed proportion on the horizontal

Table 9-1. Effect of treatment after 6 weeks with cimetidine or placebo in patients with benign gastric ulcers.[1]

Treatment	Number of Patients	Number of Ulcers Healed
Cimetidine	66	43
Placebo	67	30

[1]Adapted and reproduced, with permission, from Table 1 in Graham DY et al: Healing of benign gastric ulcer: Comparison of cimetidine and placebo in the United States. *Ann Intern Med* 1985: **102**:573–576.

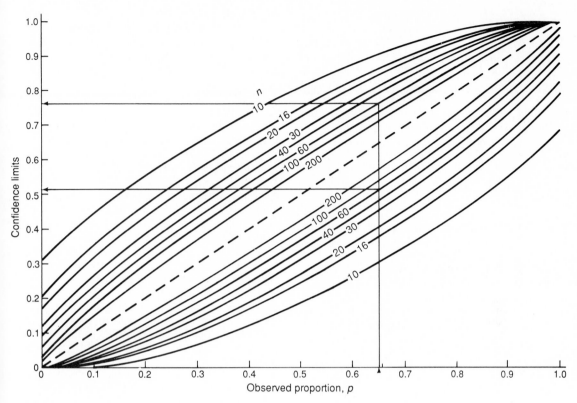

Figure 9–2. 95% confidence intervals for population proportion. (Adapted and reproduced, with permission, from Pearson ES, Hartley HO [editors]: *Biometrika Tables for Statisticians*, 3rd ed. Vol 1. Cambridge, Univ Press, 1966.)

axis and determine the points at which a vertical line extended upward from the observed proportion intersects the sample size lines. These intersection points are then projected on the vertical axis to determine the 95% confidence limits.

For our example, the observed proportion is 0.65 and the sample size is 66. Since there is no sample size of 66 on the graph, we will use the next smallest sample size, 60, which results in a more **conservative** estimate. A vertical line drawn at $p = 0.65$ in Fig 9–2 intersects the lines for a sample size of 60 so that the projections onto the vertical axis are approximately 0.52 and 0.77. This estimate results in a confidence interval only slightly wider than the one found above, close enough for a reasonable approximation.

9.2.2 Hypothesis Test for a Proportion

Suppose, continuing our example, that the investigators want to test whether the observed proportion of patients whose ulcer heals following treatment with cimetidine is greater than the proportion observed for patients who receive no treatment. The best way to test this research hypothesis is with a clinical trial in which patients are randomized to treatment with cimetidine and compared with patients randomized to a placebo treatment, as was actually done in Presenting Problem 1. However, not all studies have concurrent control groups; so the observed proportion is sometimes compared with a standard or known value.

To illustrate this approach, let's suppose that there is only one group of patients, those who received cimetidine; and after 6 weeks of treatment, 65% of the 66 patients examined showed healing of their gastric ulcer. Furthermore, suppose that we want to know whether the observed proportion of 0.65 is significantly larger than 0.50, because a rule of thumb from clinical experience is that 50% of patients who have symptoms of gastric ulcer improve within 6 weeks, with or without treatment. The z distribution is used below in the hypothesis test for this research question.

Step 1. H_0: $\pi \leq 0.50$ **(The population proportion is 0.5 or less.)**

H_1: $\pi > 0.50$ **(The population proportion is greater than 0.5.)**

In this example, we are interested only in testing whether the proportion of ulcers healed with cimetidine is greater than the commonly accepted 50%; therefore, a **one-tailed test** to detect only a positive difference is appropriate.

Step 2. In this example, $n\pi = n(1 - \pi)$, because π is equal to 0.50. That is, both $= 66 \times 0.50 = 33$; 33 exceeds 5, so the z approximation to the binomial can be used. The population proportion π is used instead of p in the formula for the standard error:

$$z = \frac{p - \pi}{\sqrt{\pi(1 - \pi)/n}}$$

Note the similarity of this formula to $z = (\overline{X} - \mu)/(\sigma/\sqrt{n})$ in the hypothesis test for a mean.

Step 3. For this example, let us use $\alpha = .01$.

Step 4. We are performing a one-tailed test, placing all the area in the upper part of the distribution. The value of z divides the normal distribution into the lower 99% and the upper 1% is 2.326, from Table A–2. Therefore, the hypothesis that the true population proportion is equal to 0.50 will be rejected if the observed value of z exceeds 2.326.

Step 5. The calculation is as follows:

$$z = \frac{0.65 - 0.50}{\sqrt{0.50(0.50)/66}}$$

$$= \frac{0.15}{0.0616}$$

$$= 2.44$$

Step 6. Since the observed value of z, 2.44, is larger than the critical value, 2.326, the decision is to reject the null hypothesis that the proportion of patients treated with cimetidine who have healed ulcers is less than or equal to 0.50. The probability of observing a proportion as large as 0.65 if the true healing rate is 50% is less than 1 in 100. The conclusion is that cimetidine appears to result in a proportion of healed ulcers larger than that occurring without treatment with $P < .01$.

Exercise 7 discusses the actual analysis performed by the investigators in this study, comparing the proportion of patients receiving cimetidine with healed ulcers with the proportion receiving placebo with healed ulcers.

9.3 COMPARING TWO INDEPENDENT PROPORTIONS

Presenting Problem 2 compares the proportion of gram-negative episodes, including those leading to shock and lethal shock, in patients who were immunized with J5 vaccine and in control patients. Clearly, these patients represent two **independent groups** of patients, and the investigators need a method for answering the research question "Is there a difference in the proportion of patients who have shock episodes under the two treatment regimens?" One of three methods can be used to answer questions about differences in proportions in two independent groups: approximate confidence limits for the difference in proportions from the z distribution, an extention of the z approximation to the binomial distribution (illustrated in Section 9.2.2), and a test based on the chi-square distribution. Each method, while resulting in different numbers, leads to the same conclusions about any differences between the groups of patients. All three methods are illustrated frequently in the medical literature. We will use Presenting Problem 2 to illustrate each method in turn.

9.3.1 Confidence Intervals for the Difference Between Two Independent Proportions

Table 9–2 gives the number of episodes of shock that occurred in all patients included in the study by Baumgartner et al (1985). Shock occurred in 6 of the 126 patients (4.8%) given J5 vaccine and in 15 of the 136 control patients (11.0%). The method for determining confidence limits for the difference between the two proportions is similar to the method for determining confidence limits

Table 9-2. Episodes of systemic gram-negative infection in surgical patients treated with J5 versus controls.[1]

Episode	Abdominal Surgery		All Patients	
	J5	Control	J5	Control
Shock	2	13	6	15
Lethal shock	1	9	2	9
Total number with infection	8	15	16	23
Number of patients	71	83	126	136

[1]Adapted and reproduced, with permission, from Table 4 in Baumgartner J et al: Prevention of gram-negative shock and death in surgical patients by antibody to endotoxin core glycolipid. *Lancet* 1985; 2:59.

for the differences between two means (illustrated in Chapter 7).

Let p_t be the proportion of patients in the treatment group who suffered shock and p_c be the proportion in the control group. Then, the difference between the two proportions is estimated by $p_t - p_c$. The standard error of the difference is actually

$$SE(\pi_t - \pi_c) = \sqrt{\frac{\pi_t (1 - \pi_t)}{n_t} + \frac{\pi_c (1 - \pi_c)}{n_c}}$$

However, the true proportions are unknown, and p_t and p_c must be used to estimate the standard error of the difference as well as the difference itself.

Just as the two-sample standard deviations are pooled when two means are compared, the two sample proportions are pooled by forming a weighted average, with the sample sizes as weights. The pooled proportion provides a better estimate to use in the standard error; it is designated simply as p without any subscripts and is calculated as follows:

$$p = \frac{n_t p_t + n_c p_c}{n_t + n_c}$$

Then, the estimate of the standard error for the difference between two proportions is

$$SE(p_t - p_c) = \sqrt{\frac{p(1 - p)}{n_t} + \frac{p(1 - p)}{n_c}}$$

$$= \sqrt{p(1 - p)\left(\frac{1}{n_t} + \frac{1}{n_c}\right)}$$

Therefore, the formula for 99% confidence limits is

$$(p_t - p_c) \pm 2.575 \sqrt{p(1 - p)\left(\frac{1}{n_t} + \frac{1}{n_c}\right)}$$

where 2.575 is from Table A-2.

This method is appropriate whenever the product of sample size and observed proportion exceeds 5 in each group, ie, whenever $n_t p_t$ and $n_c p_c$ are 5 or greater, which is true for this example. From the data in Table 9-2, $p_t = 6/126 = 0.048$ and $p_c = 15/136 = 0.110$; the pooled proportion of patients with shock is

$$p = \frac{(126)(0.048) + (136)(0.110)}{126 + 136}$$

$$= 0.080$$

The standard error of the difference between the two proportions is

$$SE(p_t - p_c) = \sqrt{0.080(1 - 0.080)\left(\frac{1}{126} + \frac{1}{136}\right)}$$

$$= (0.080)(0.920)(0.008 + 0.007)$$

$$= 0.034$$

The 99% confidence interval is therefore

$$(p_t - p_c) \pm 2.575 \sqrt{p(1 - p)\left(\frac{1}{n_t} + \frac{1}{n_c}\right)}$$

$$= (0.048 - 0.110) \pm (2.575)(0.034)$$

$$= -0.062 \pm 0.088$$

or

$$-0.150 \text{ to } 0.026$$

We have 99% confidence that the interval from -0.150 to $+0.026$ contains the true difference in the proportion of patients with episodes of shock. Since zero is in this interval, the difference between the proportions may be equal to zero; and we cannot conclude that there is a difference in the proportion of episodes of shock for patients who received J5 vaccine or placebo vaccine.

9.3.2 Using the z-Approximation to Compare Two Independent Proportions

We will use the same data on gram-negative infection from Presenting Problem 2 to illustrate the hypothesis test of the difference between two independent proportions using the z approximation to the binomial distribution.

Step 1. $H_0: \pi_t \geq \pi_c$ **(The proportion with shock is greater with J5 treatment.)**
$H_1: \pi_t < \pi_c$ **(The proportion with shock is smaller with J5 treatment.)**

π_t and π_c are the proportions of treated and control patients experiencing shock in the population. Here, a one-tailed test is employed because the investigators are interested in testing whether the J5 vaccine is superior to the placebo, in which case the proportion of patients with lethal shock will be less with the J5 vaccine than with the placebo vaccine.

Step 2. The z test for the approximation to the binomial distribution (introduced in Section 9.2.2) can be extended to compare two independent proportions. The test statistic is

$$z = \frac{p_t - p_c}{\sqrt{p(1 - p)(1/n_t + 1/n_c)}}$$

where p is the pooled proportion defined in Section 9.3.1. The z approximation may be used whenever both np_t and np_c are greater than 5. In our example, $np_t = (126)(0.048) = 6$ and $np_c = (136)(0.110) = 15$. Since both values are greater than 5, the z approximation may be used.

Step 3. Let us use $\alpha = .05$.

Step 4. For a one-tailed test at $\alpha = .05$, the value of the z distribution that separates the lower 5% of the area under the curve from the upper 95% is -1.645 (from Table A–2). Therefore, we will reject the null hypothesis of equal proportions if the observed value of the z statistic is less than -1.645.

Step 5. The calculation is

$$z = \frac{0.048 - 0.110}{\sqrt{0.080(1 - 0.080)(1/126 + 1/136)}}$$

$$= -1.824$$

(See Section 9.3.1 for details on calculating the pooled proportion.)

Step 6. Since the observed value of z, -1.824, is less than -1.645, the null hypothesis—that the proportion of patients in the treatment groups experiencing shock is greater than or equal to the proportion of patients in the control group experiencing shock—is rejected. The conclusion, therefore, is that the study cited in Presenting Problem 2 provides sufficient evidence to show that the proportion of shock episodes in patients who receive J5 vaccine is less than the proportion of shock episodes in patients who receive a placebo vaccine ($P < .05$).

Is this conclusion consistent with the confidence limits found earlier? See Exercise 1.

9.3.3 Using Chi-Square to Compare Two Independent Proportions

The **chi-square test** is the most commonly used method for comparing proportions, because it can be used to compare two or more independent proportions in place of the z approximation. Like the z approximation, the chi-square test is an approximate test and should be used in accordance with some general guidelines, which will be discussed shortly. Chi-square also has a variety of other applications and thus is a very versatile test. In addition, the calculations are relatively easy.

We will discuss the use of the chi-square test for comparing two independent proportions first, using Presenting Problem 2 once more. Later, we will illustrate chi-square for more than two groups. Recall that the investigators want to know whether vaccinating patients with J5 has an effect

on the number of episodes of shock. This research question can be phrased in two different ways:

1. Is there a *difference* in the proportion of gram-negative shock episodes in patients who receive J5 vaccine versus those who receive a placebo vaccine?

2. Is there an *association* (or relationship or dependency) between receiving the J5 vaccine and the occurrence of gram-negative shock episodes?

The z approximation test approached the research question from the perspective of a difference; ie, the test involved the difference between the two proportions. The chi-square test, in contrast, can approach the research question from either perspective, but it is the second perspective of an association (or relationship) that makes the test applicable in so many research situations. The test illustrated below is commonly called the chi-square test for proportions (or the chi-square test of independence).

9.3.3.a Intuitive Overview to the Chi-Square Test: Before illustrating the computation of the chi-square test for the data in Presenting Problem 2, we give an intuitive description of the test to help you understand the logic behind it. Consider the hypothetical data in the contingency table in Table 9–3A. These **marginal frequencies** represent a hypothetical study in which 100 patients are given an experimental treatment and 100 patients receive a control treatment. Fifty patients respond positively; the remaining 150 patients respond negatively.

Now, if there is no relationship between treatment and outcome—ie, if treatment and outcome are independent—we would expect approximately half of the 50 patients who respond positively to be in the treatment group and approximately half to be in the control group (with equal numbers in treatment and control groups). Similarly, we would expect approximately half of the 150 patients who respond negatively to be in the treatment group and approximately half in the control

Table 9–3. Hypothetical data for chi-square.

A. Marginal Frequencies			
	Treatment	Control	Total
Positive			50
Negative			150
Total	100	100	200
B. Expected Frequencies			
	Treatment	Control	Total
Positive	25	25	50
Negative	75	75	150
Total	100	100	200

group. Thus, if there is not a relationship between treatment and outcome, the frequencies should be as listed in Table 9–3B. The numbers in the cells of Table 9–3B are called **expected frequencies.**

The logic of the chi-square test is as follows:

1. The number of observations in each column (treatment and control) and the number of observations in each row (positive and negative response) are considered to be given; as we mentioned in Chapter 5, they are called **marginal frequencies.**

2. From the given column and row totals, the number of observations *expected* to occur in each cell of the table, assuming that columns and rows are independent, is calculated; they are called expected cell frequencies, or, simply, expected frequencies.

3. The expected frequency in each cell is compared with the frequency that actually occurred (ie, **observed frequencies**). The differences between the observed frequencies and the expected frequencies are combined to form the chi-square statistic.

4. If there is no relationship between the column and row variables—ie, if they are independent from each other—the observed frequencies will differ from the expected frequencies only by a small amount, owing to sampling variability; and the value of the chi-square statistic will be small. If there is a relationship (or dependency), the observed frequencies will vary substantially from the expected frequencies; and the value of the chi-square statistic will be large.

We let O denote the observed frequency in a cell and E denote the expected frequency in a cell. Then, the chi-square statistic is

$$X^2(df) = \sum_{\substack{\text{all} \\ \text{cells}}} \frac{(O - E)^2}{E}$$

where $X^2(df)$ designates the chi-square statistic with degrees of freedom df.

9.3.3.b The Chi-Square Distribution:

The **chi-square (χ^2) distribution** (χ is the Greek letter chi), like the t distribution, has degrees of freedom. In the chi-square test for independence, the degrees of freedom equal the number of rows minus 1 multiplied by the number of columns minus 1, or

$$df = (r - 1)(c - 1)$$

where r denotes the number of rows and c the number of columns. Fig 9–3 illustrates the chi-square distribution for one degree of freedom. The chi-square distribution has nonnegative values only, and the center (or mean) of the distribution is equal to the degrees of freedom. (The standard deviation of the chi-square distribution is

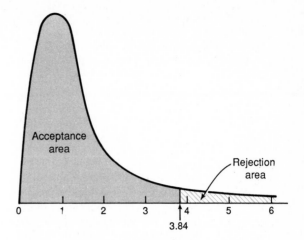

Figure 9–3. χ^2 distribution (with 1 df and α = .05 critical value).

also equal to the degrees of freedom. Therefore, only one parameter, degrees of freedom, is needed to specify any chi-square distribution.)

To use the chi-square distribution for hypothesis testing, we find the critical value in Table A–5 that separates the area defined by α (in the upper tail) from that defined by $1 - \alpha$. Table A–5 contains only one-tailed values for χ^2 because they are the values generally used in hypothesis testing. For example, the critical value for $X^2(1)$ with α = .05 is 3.841.

9.3.3.c The Chi-Square Test:

Let us now apply the chi- square test to the observations from the gram-negative infection study in Table 9–2. This time, we will look at the occurrence of lethal shock. The data are arranged as a 2 × 2 contingency table in Table 9–4A, with occurrence (yes

Table 9–4. 2 × 2 table for presenting problem 2.[1]

A. Observed Frequencies			
Lethal Shock Episodes	J5 Vaccine	Control Vaccine	Total
Yes	2	9	11
No	124	127	251
Total	126	136	262

B. Expected Frequencies			
Lethal Shock Episodes	J5 Vaccine	Control Vaccine	
Yes	5.29	5.71	
No	120.71	130.29	

[1]Adapted and reproduced, with permission, from Table 4 in Baumgartner J et al: Prevention of gram-negative shock and death in surgical patients by antibody to endotoxin core glycolipid. *Lancet* 1985; **2**:59.

or no) of lethal shock as the rows of the table and treatment (J5 or placebo) as the columns. The steps of the hypothesis test follow.

Step 1. H_0: Treatment and outcome (ie, columns and rows) are independent.

H_1: Treatment and outcome are not independent (ie, they are associated or related).

Step 2. The chi-square test is appropriate for this research question because the data are nominal.

Step 3. Let us use an α of .05.

Step 4. The contingency table has 2 rows and 2 columns; therefore, df $= (2 - 1)(2 - 1) = 1$. The critical value in Table A–5 that separates the upper 5% of the χ^2 distribution from the remaining 95% is 3.841. The chi-square test is performed as a one-tailed test, detecting whether the observed frequencies depart from the expected frequencies by *more* than the amount that would occur by chance. Therefore, the decision is to reject the null hypothesis of independence if the observed value of X^2 is greater than 3.841.

Step 5. The procedure for calculating chi-square requires that expected frequencies be found for each cell. The illustration using hypothetical data (Table 9–3) indicates how expected frequencies are calculated. For instance, for cell (1, 1) (meaning row 1 and column 1, corresponding to treatment and positive), the proportion of patients in the treatment group, 100/200, is multiplied by the number of patients who responded positively, 50; thus, $(100)(50)/200 = 25$ is the expected frequency. This calculation uses the following rule:

$$E = \frac{\text{Row total} \times \text{Column total}}{\text{Grand total}}$$

Using this rule, we can find expected frequencies for the data in Table 9–4*A* as follows:

$$E(1, 1) = \frac{11 \times 126}{262} = 5.29 \quad \text{(row 1, column 1)}$$

$$E(1, 2) = \frac{11 \times 136}{262} = 5.71 \quad \text{(row 1, column 2)}$$

$$E(2, 1) = \frac{251 \times 126}{262} = 120.71 \quad \text{(row 2, column 1)}$$

$$E(2, 2) = \frac{251 \times 136}{262} = 130.29 \quad \text{(row 2, column 2)}$$

These expected frequencies are listed in Table 9–4*B*. The calculation for X^2 is

$$X^2(1) = \Sigma \frac{(O - E)^2}{E}$$

$$= \frac{(2 - 5.29)^2}{5.29} + \frac{(9 - 5.71)^2}{5.71}$$

$$+ \frac{(124 - 120.71)^2}{120.71} + \frac{(127 - 130.29)^2}{130.29}$$

$$= 2.05 + 1.90 + 0.09 + 0.08$$
$$= 4.12$$

Step 6. The observed value of $X^2(1)$, 4.12, is greater than 3.841; therefore, we reject the null hypothesis of independence. We conclude that there is a relationship between the vaccine and occurrence of lethal shock. Note that this conclusion is the same as the one regarding shock in Section 9.3.2.

9.3.4 Using Chi-Square Tests

Because of the widespread use of chi-square tests in the literature, we will discuss several aspects of these tests.

9.3.4.a Small Expected Frequencies and Fisher's Exact Test: The chi-square procedure, like the test based on the z approximation, is an approximate method. Just as the z test should not be used unless both np and $n(1 - p)$ are greater than 5, the chi-square test should not be used when there are small *expected frequencies*. Looking at the formula for chi-square

$$X^2(\text{df}) = \Sigma \frac{(O - E)^2}{E}$$

we can see what happens when expected frequencies are small: A small expected frequency in the denominator of one of the terms in the equation causes that term to be large, which, in turn, inflates the value of chi-square.

How small can expected frequencies be before we must worry about them? Although there is no absolute rule, most statisticians agree that an expected frequency of 2 or less means that the chi-square test should not be used; and many argue that chi-square should not be used if an expected frequency is less than 5. We suggest that if any expected frequency is less than 2 or if more than half the expected frequencies are less than 5, then an alternative procedure called **Fisher's exact test** should be performed. (We emphasize that it is the *expected* values that one must worry about and not the *observed* values. This point is often misunderstood, and researchers sometimes mistakenly think that the chi-square test cannot be performed if there is a zero or a very small observed value in one of the cells.) If the contingency table of observations is larger than 2 × 2, categories should be combined to eliminate most of the expected values less than 5; this procedure is discussed further in the Section 9.4.

For Fisher's exact test, the exact probability of the occurrence of the observed frequencies, given the assumption of independence and the size of the marginal frequencies (row and column totals), is computed for the 2 × 2 table. For example, assume that the cells of a 2 × 2 table are labeled *a, b, c,* and *d,* as in Table 9–5. Then, the probability P of obtaining the observed frequencies in the table is

$$P = \frac{(a + b)!(c + d)!(a + c)!(b + d)!}{a\,!\,b\,!\,c\,!\,d\,!\,n\,!}$$

where $a + b$ and $c + d$ are the row totals, $a + c$ and $b + d$ are the column totals, n is the grand total, and ! is the symbol for factorial (introduced in Chapter 5); ie, $n! = n(n - 1)(n - 2) \ldots (4)(3)(2)(1)$. Calculating the probability of the observed frequencies for Presenting Problem 2 in Table 9–4*A* gives

$$P = \frac{11!\ 251!\ 126!\ 136!}{2!\ 9!\ 124!\ 127!\ 262!} = .0326$$

The null hypothesis tested with both the chi-square test and Fisher's exact test is that the observed frequencies or frequencies more extreme could occur by chance, given the fixed values of the row and column totals. Therefore, for Fisher's exact test, the probabilities of frequencies more extreme than each observed frequency must also be calculated, and the probabilities of all the more extreme sets are added to the probability of the observed set. For our example, then, we need to compute P for two additional situations: for frequencies more extreme than the 2 in row 1, column 1 of Table 9–4*A*—ie, for frequencies of 1 and of 0. Thus, we calculate

$$P = \frac{11!\ 251!\ 126!\ 136!}{1!\ 10!\ 125!\ 126!\ 262!} = .0066$$

for a 1 in row 1, column 1, and

$$P = \frac{11!\ 251!\ 126!\ 136!}{0!\ 11!\ 126!\ 125!\ 262!} = .0006$$

for a 0 in row 1, column 1.

Adding the probabilities for the observed frequencies plus the more extreme frequencies gives .0326 + .0066 + .0006 = .0398. When the approximate chi-square statistic was calculated for this example in Section 9.3.3, we rejected the null hypothesis at $P < .05$; and here, we see the exact probability given by Fisher's exact test is .0398.

As we can see from the above example, Fisher's exact test is quite arduous to compute by hand. Some statistics books present tables for calculating the probabilities, and most computer programs will calculate Fisher's exact test. However, the reader of medical journals need only have a basic understanding of the purpose of this statistic—ie, you need only remember that Fisher's exact test is used as an alternative to the chi-square test to examine association in 2 × 2 tables when expected frequencies are small.

9.3.4.b Shortcut Chi-Square Formula for 2 × 2 Tables:

There is a shortcut formula that makes it easy to calculate X^2 for 2 × 2 tables, because expected frequencies do not need to be computed. Table 9–5 gives the setup of the table for the shortcut formula.

The shortcut formula for calculating X^2 from a 2 × 2 contingency table is

$$X^2(1) = \frac{n(ad - bc)^2}{(a + c)(b + d)(a + b)(c + d)}$$

Using this formula with the lethal shock data, we get

$$X^2(1) = \frac{262[(2)(127) - (9)(124)]^2}{(11)(251)(126)(136)}$$
$$= 4.11$$

This value for $X^2(1)$ agrees (within rounding error) with the X value obtained in Section 9.3.3. In fact, the two approaches are equivalent for 2 × 2 tables.

9.3.4.c Continuity Correction:

Some investigators report "corrected" chi-square values, called

Table 9–5. Standard notation for chi-square 2 × 2 table.

	Treatment	Control	Total
Positive	a	b	a + b
Negative	c	d	c + d
Total	a + c	b + d	a + b + c + d = n

Fisher's exact test

$$P = \frac{(a + b)!(c + d)!(a + c)!(b + d)!}{a!\ b!\ c!\ d!\ n!}$$

Shortcut formula

$$X^2(1) = \frac{n(ad - bc)^2}{(a + c)(b + d)(a + b)(c + d)}$$

chi-square with **continuity correction** or chi-square with **Yates' correction.** This correction involves subtracting 0.5 from the difference between observed and expected frequencies in the numerator of X^2 before squaring; it has the effect of making the value for X^2 smaller. (In the short-cut formula, 0.5 is subtracted from the absolute value of $ad - bc$ prior to squaring.)

A smaller X^2 means that the null hypothesis will not be rejected as often as it is with the larger, uncorrected chi-square; ie, it is more conservative. Thus, there is a smaller risk of a type I error (rejecting the null hypothesis when it is true); however, there is then an increased risk of a type II error (not rejecting the null hypothesis when it is false and should be rejected). Some statisticians recommend the use of the continuity correction for all 2 × 2 tables (Yates, 1984); others caution against its use (Grizzle, 1967). Both corrected and uncorrected chi-square statistics are commonly encountered in the medical literature.

9.3.4.d Risk Ratios Versus Chi-Square: Both the chi-square test and the z approximation test allow investigators to test a hypothesis about equal proportions or about a relationship between two nominal measures, depending on how the research hypothesis is articulated. It may have occurred to you that the **relative risk** or the **odds ratio** introduced in Chapter 4 (Section 4.5.4) could also be used with 2 × 2 tables, and you are correct. The statistic selected depends on the purpose of the analysis. If the objective is to estimate the relationship between two nominal measures, then the relative risk or the odds ratio is appropriate. Furthermore, investigators can form confidence intervals about the relative risk or the odds ratio (illustrated in Chapter 10), which, for all practical purposes, accomplishes the same end as a significance test; and this approach is being used with increasing frequency in medical journals.

9.3.4.e. Overuse of Chi-Square: Because the chi-square test is so easy to understand and calculate, it is sometimes used when another method is more appropriate. A common misuse of chi-square tests occurs when there are two groups and the characteristic of interest is measured on a numerical scale. Instead of correctly using the t test, researchers convert the numerical scale to an ordinal or even dichotomous scale and then use chi-square. As an example, investigators brought the following problem to one of the authors:

> Some patients who undergo a surgical procedure are more likely to have complications than other patients. The investigators had collected data on one group of patients who had complications following surgery and on another group of patients who did not have complications, and they wanted to know whether there was a relationship between the patient's age and whether the patient had a complica-tion. The investigators had formed a 2 × 2 contingency table, with columns being complication–no complication, and rows being patient age 45 years or greater–age less than 45 years; and they had performed a chi-square test for independence. The results, much to their surprise, indicated no relationship between age and complication.

The problem is the arbitrary selection of 45 years as a cutoff point for age. When a t test was performed, the mean age of patients who had complications was significantly greater than the mean age of patients who did not. Forty-five years of age, although meaningful perhaps from a clinical point of view related to other factors, was not the age sensitive to the occurrence of complications.

When numerical variables are analyzed with methods designed for ordinal or categorical variables, the greater specificity of the numerical variables is wasted. Investigations may wish to categorize a numerical variable, such as age, for graphic or tabular presentation; however, they should analyze the variable correctly and not throw away the additional information reflected in the numerical scale.

9.3.4.f Illustration of Degrees of Freedom: The chi-square test provides a nice illustration of the concept of degrees of freedom. Suppose we have a contingency table with 3 rows and 4 columns and marginal frequencies, such as Table 9-6. How many cells in the table are "free to vary" (within the constraints imposed by the marginal frequencies)?

In column 1, the frequencies for rows 1 and 2 can be any value at all, as long as neither is greater than their row total and their sum does not exceed 100, the column 1 total. The frequency for row 3, however, is determined once the frequencies for rows 1 and 2 are known; ie, it is 100 minus the values in rows 1 and 2. The same reasoning applies for columns 2 and 3. At this point, however, the frequencies in row 3 as well as all the frequencies in column 4 are determined. Thus, there are $(3 - 1)(4 - 1) = 2 \times 3 = 6$ degrees of freedom. In general, the degrees of freedom for a contingency table are equal to the number of rows minus 1 times the number of columns minus 1, or, symbolically, $df = (r - 1)(c - 1)$.

Table 9-6. Illustration of degrees of freedom in chi-square.

Rows	Columns				Total
	1	2	3	4	
1	*	*	*		75
2	*	*	*		100
3					225
Total	100	100	100	100	

9.4 COMPARING PROPORTIONS IN MORE THAN TWO GROUPS

The chi-square test presented in Section 9.3 can be extended to include any number of rows and columns. A common application in medicine is for a comparison of several different independent groups. For example, in Presenting Problem 3, three different antibiotics were compared in the treatment of cystitis. The frequencies from the study are presented in Table 9–7.

Let us use the frequencies for overall cure rate to illustrate the use of chi-square in comparing three independent groups of patients. The frequencies have been rearranged in Table 9–8 in the form of a 2 × 3 contingency table with cure (yes or no) as rows and treatment group (trimethoprim-sulfamethoxazole, amoxicillin, cyclacillin) as columns.

The chi-square statistic for this example is calculated from the formula given in Section 9.3, with degrees of freedom equal to $(r - 1)(c - 1)$:

$$X^2[(r - 1)(c - 1)] = \Sigma \frac{(O - E)^2}{E}$$

However, for this calculation, there will be 6 terms in the formula corresponding to the 6 cells of the 2 × 3 table, instead of 4 terms as in the 2 × 2 table.

In the formulation of the null and alternative hypotheses, the research question may be stated in terms of either equal proportions or an association. The hypothesis-testing procedure follows.

Step 1. H_0: **The proportions are all equal (or, equivalently, the occurence of cure is independent from the antibiotic used).**

H_1: **The proportions are not all equal (or, equivalently, the occurrence of cure is dependent on the antibiotic used).**

Step 2. The chi-square statistic is appropriate for analyzing characteristics measured on nominal scales, ie, data displayed in a contingency table.

Step 3. Let us use $\alpha = .01$ for this example.

Step 4. The degrees of freedom for this example are $(2 - 1)(3 - 1) = 2$. The value of the chi-square distribution with 2 degrees of freedom that divides the area into the lower 99% and the upper 1% is 9.210 (from Table A–5). Therefore, we will reject the null hypothesis of equal proportions (or independence) if the observed value of chi-square is greater than 9.210.

Step 5. The expected frequencies are calculated according to the formula

$$\frac{\textbf{Row total} \times \textbf{Column total}}{\textbf{Grand total}}$$

and are as follows:

$$E(1, 1) = \frac{20 \times 13}{35} = 7.43 \quad \textbf{(row 1, column 1)}$$

$$E(1, 2) = \frac{20 \times 12}{35} = 6.86 \quad \textbf{(row 1, column 2)}$$

Similarly, $E(1, 3) = 5.71$, $E(2, 1) = 5.57$, $E(2, 2) = 5.14$, and $E(2, 3) = 4.29$. None of the expected frequencies are less than 2, and only one is less than 5; therefore, we can use the chi-square test. The calculation for X^2 is

$$X^2(2) = \Sigma \frac{(O - E)^2}{E}$$

$$= \frac{(11 - 7.43)^2}{7.43} + \ldots + \frac{(7 - 4.29)^2}{4.29}$$

$$= 1.72 + 0.11 + 1.29 + 2.29 + 0.14 + 1.71$$

$$= 7.26$$

Step 6. The observed value of $X^2(2)$ is 7.26 and is not greater than 9.210; therefore, the null hypothesis of equal proportions cured of cystitis is not rejected. We conclude that the proportions cured do not differ for the different antibiotics. Equivalently,

Table 9-7. Outcome of treatment for cystitis.[1]

Outcome	Treatment[2] Trimethoprim-sulfamethoxazole	Amoxicillin	Cyclacillin	Total
Initial care	13/13	9/13	8/12	30/38
Relapse[3]	2/13	2/12	3/10	7/35
Overall cure rate[3]	11/13	6/12	3/10	20/35

[1]Adapted and reproduced, with permission, from Table 2 in Hooten TM, Running K, Stamm WE: Single-dose therapy for cystitis in women. *JAMA* 1985; **253**:387.
[2]Numbers of patients are given in the denominators.
[3]Patients are not indicated in the denominator if they were initially cured but missed a follow-up visit.

Table 9-8. Contingency table for study of treatment of cystitis.[1]

| Outcome | Treatment | | | Total |
	Trimethoprim-sulfamethoxazole	Amoxicillin	Cyclacillin	
Cured	11	6	3	20
Not cured	2	6	7	15
Total	13	12	10	35

[1]Adapted and reproduced, with permission, from Table 2 in Hooten TM, Running K, Stamm WE: Single-dose therapy for cystitis in women. *JAMA* 1985; **253**:387.

there does not appear to be a relationship between cure of cystitis and antibiotic used.

Two comments about this study deserve mention. First, as we discussed in Chapter 6, there is a relationship—namely, power—between sample size and the ability to detect an actual difference in treatment methods. As the sample size increases, the power to detect an actual difference also increases. In the analysis illustrated above, a larger sample size might well have led to significance if the proportions of patients cured remained the same. (See Exercise 3.) Whenever a study fails to detect a significant difference, investigators should examine the sample size before accepting the results without question. Section 9.7 discusses guidelines for estimating sample sizes.

Second, in this particular study, the authors did not analyze the data as we have done. Instead of performing a chi-square analysis of a 2 × 3 contingency table, they compared each antibiotic with every other antibiotic; ie, they performed three 2 × 2 table analyses. They concluded that the cure rate with trimethoprim-sulfamethoxazole was greater than that with cyclacillin ($P = .01$). However, we should note that the same situation can occur in studies using proportions as occurs in studies performing multiple *t* tests, where it is advisable to perform an analysis of variance prior to comparing individual groups. Although there is no analogue to analysis of variance with nominal data, some statisticians recommend that an overall test of hypothesis, as illustrated above, be done prior to performing paired comparisons. Otherwise, the strength of the conclusions drawn in the study are weakened owing to the increased chances of making a Type I error, ie, of concluding a difference when there is none.

The example we used to illustrate chi-square for more than two proportions had only three groups. But contingency tables can have more than two rows and any number of columns. Of course, the sample size must increase as the number of categories within a variable increases. If some expected frequencies are less than 2, or if more than half the expected frequencies are less than 5, the only solution is to collapse some of the categories.

For example, Table 9-9 is a contingency table with four rows and two columns. These data come from a randomized, double-blind clinical trial (Feldman et al 1985) designed to determine the effect on recurrent abdominal pain, if any, of the addition of fiber to children's diets. The authors examined differences in the educational background of parents of the children given fiber and those given placebo. Two of the expected frequencies in Table 9-9 are 3 or less, so the investigators might consider collapsing the "Elementary or less" and "Secondary" categories into one category with 18 and 23 subjects in fiber and placebo groups, respectively.

9.5 COMPARING PROPORTIONS IN PAIRED GROUPS

Sometimes, subjects are measured or observed twice, and situations analogous to those using the paired-groups *t* test arise. If the measured characteristic is nominal, a test for paired proportions that analyzes the number of disagreements, called the **McNemar test,** can be used. This application arises in medicine most often when two procedures purporting to measure the same characteris-

Table 9-9. Contingency table from fiber study.[1]

Parental Education	Fiber	Placebo	Total
Elementary or less	2	3	5
Secondary	16	20	36
Apprenticeship, college, university	24	21	45
Postgraduate, professional	8	8	16
Total	50	55	105

[1]Adapted and reproduced, with permission, from Table 1 in Feldman W et al: The use of dietary fiber in the management of simple, childhood, idiopathic, recurrent, abdominal pain. *Am J Dis Child* 1985; **139**:1216.

tic are compared or when the opinions of two experts are compared. An example of the first application is given in Presenting Problem 4, in which the accuracy of the diagnosis of acute deep vein thrombosis is compared for thermography and venography. An example of the second application occurs when two radiologists are asked to read and classify the same set of x-rays. In both cases, because the items or subjects analyzed are measured two times, we would expect to see a relationship (or correlation) between the two measurements.

The data for Presenting Problem 4 are reproduced in Table 9–10. They are used to illustrate the McNemar test for correlated proportions testing whether the proportions of positive diagnosis are the same for the two imaging methods. The steps in the hypothesis test follow.

Step 1. H_0: $\pi_1 = \pi_2$ **(The paired proportions are equal.)**

H_1: $\pi_1 \neq \pi_2$ **(The paired proportions are not equal.)**

Step 2. The rows represent the results of one measurement (thermography), and the columns represent the results of another measurement (venography); ie, the measurements have been made on the same patients. In this situation, the McNemar test for paired proportions is appropriate. The McNemar test follows a chi-square distribution with 1 degree of freedom; its formula is

$$X^2(1) = \frac{(|b - c| - 1)^2}{b + c}$$

where b and c are the frequencies in the cells corresponding to disagreement (also see the notation in Table 9–5). Thus, the McNemar test does not consider the items on which the two methods agreed; the analysis is based only on the number of disagreements.

Step 3. Let us use an α of .05 in this example.

Step 4. The critical value that divides the chi-square distribution with 1 degree of freedom into the lower 95% and the upper 5% is 3.841 (from Table A–5).

Step 5. The computation for the data in Table 9–10 is

$$X^2(1) = \frac{(|8 - 1| - 1)^2}{8 + 1}$$

$$= \frac{6^2}{9}$$

$$= 4.00$$

Step 6. The observed value of chi-square, 4.00, is larger than the critical value, 3.841; therefore, we reject the null hypothesis of equal proportions. We conclude that there is a difference in the proportion of diagnoses of acute deep vein thrombosis when thermography or venography is used to make the diagnosis.

We should note that another method is frequently used to compare two diagnostic procedures if one of the procedures can be declared to be the gold standard. This method uses the concepts of sensitivity and specificity of the diagnostic procedure and is presented in Chapter 13. When neither procedure can be designated as the gold standard, the κ statistic (discussed in Chapter 4, Section 4.6) can be used to describe the magnitude of agreement, or the McNemar test just illustrated can be used to test the significance of the relationship.

9.6 OTHER APPLICATIONS OF CHI-SQUARE

We have described several situations in which chi-square is appropriate. There are other applications of the chi-square statistic in medicine as well. Variations of the chi-square statistic are illustrated in Chapter 11 for analyzing survival data and in Chapter 12 for analyzing multiple variables. In addition, the chi-square statistic can be used in a test generally referred to as a goodness-of-fit test.

The chi-square test for goodness of fit is performed when the question is whether an observed distribution of observations follows a theoretical distribution of some kind. For example, this approach can be used to test whether a set of observations follows a binomial or Poisson distribution. The procedure is to divide the distribution into several intervals and then, on the basis of sample size, to determine the numbers of observations that should fall into each interval. These values serve as the expected frequencies. The actual numbers of observations that fall in each in-

Table 9–10. Diagnosis of acute deep vein thrombosis.[1]

Thermography	Venography	
	Positive	**Negative**
Positive	19	8[2]
Negative	1	27

[1]Reproduced, with permission, from Table 1 in Watz R, Ek I, Bygdeman S: Noninvasive diagnosis of acute deep vein thrombosis. *Acta Med Scand* 1979; **206**:463.

[2]Among patients who were negative on venography, 3 were positive and 5 were suspected for DVT using thermography.

terval are the observed frequencies, and the chi-square statistic is calculated as above, with one change. The degrees of freedom are equal to the number of intervals minus one; ie, when k is the number of intervals, $X^2 (k - 1) = \Sigma(O - E)^2/E$ is the goodness-of-fit chi-square statistic.

9.7 SAMPLE SIZES FOR PROPORTIONS

As we discussed in Chapter 7, investigators should determine the sample size required to demonstrate a difference prior to beginning a study, and authors should also provide readers with this information by specifying the power of a study. Knowing the sample sizes needed is helpful in readers' determining whether a negative study is negative because the sample size was too small. Some journal editors now require authors to provide this key information.

9.7.1 Sample Size for a Single Proportion

This section presents the formula for estimating the sample size needed in a study comparing the proportion in a single sample with a standard value. As in estimating the sample size for a mean (illustrated in Chapter 7, Section 7.5), the researcher must answer four questions in order to estimate the sample size needed for a single proportion.

1. What is the desired level of significance (the α level) related to the null hypothesis π_0?
2. What chance should there be of detecting an actual difference, ie, what power $(1 - \beta)$ is desired?
3. What difference between the proportions must be detected; ie, what is $\pi_1 - \pi_0$?
4. What is a good estimate of the standard deviation in the population? For a proportion, the value in the null hypothesis, π_0, determines the estimated standard deviation.

The formula to determine the sample size is

$$n = \left[\frac{z_\alpha\sqrt{\pi_0(1 - \pi_0)} - z_\beta\sqrt{\pi_1(1 - \pi_1)}}{\pi_1 - \pi_0} \right]^2$$

where z_α is the two-tailed z value related to the null hypothesis and z_β is the lower one-tailed z value related to the alternative hypothesis.

To illustrate, we consider a problem in which a group of physicians wants to know whether sending patients a postcard reminder prior to their next annual physical examination will increase the proportion of patients who actually return for the exam. In the past, only about 40% have come to the office for the exam within 2 months of the time they were scheduled to do so. The physicians are willing to make a Type 1 error (concluding that

the postcards have made a difference when they really have not) only 2% of the time, and they want a .90 probability of detecting a true difference. They want to be able to detect a 25% improvement, ie, an increase from the previous 40% to 50%. What is the sample size they need to detect this change?

The two-tailed z value related to α is $+2.326$ (from Table A–2, the critical value dividing the central 98% of the z distribution from the 2% in the tails). The lower one-tailed z value related to β is -1.28 (the critical value separating the lower 10% of the z distribution from the upper 90%). Then, the estimated sample size is

$$n = \left[\frac{2.326\sqrt{(0.40 \times 0.60)} - (-1.28)\sqrt{(0.50 \times 0.50)}}{0.40 - 0.50} \right]^2$$

$$= \left(\frac{1.78}{-0.10} \right)^2 = -17.79^2$$

which, squaring and rounding up, gives 317. Thus, this group of physicians must use the postcard reminders with a random sample of 317 patients in order to determine whether this procedure will increase to 50% the percentage who keep their follow-up appointments for physical examinations. From what they learn by using the sample of 317, they can decide whether to implement the postcard reminders in their entire practice.

9.7.2 Sample Sizes for Comparing Two Proportions

A formula similar to the one for determining sample sizes for two means (Chapter 7, Section 7.5.2) can be used for two proportions. To simply matter, we assume that the sample sizes will be the same in the two groups. The symbol π_t denotes the proportion in the treatment group, and π_c denotes the proportion in the control group. Then, the formula for n is

$$n =$$
$$\left[\frac{z_a\sqrt{2\pi_c(1 - \pi_c)} - z_\beta\sqrt{\pi_t(1 - \pi_t) + \pi_c(1 - \pi_c)}}{\pi_t - \pi_c} \right]^2$$

where n again is the size needed in *each* group and z_α and z_β are defined as in the formula for the sample size for a single proportion.

Let us use Presenting Problem 2 to illustrate calculating the sample size for comparing proportions in two groups. This study involved a trial of J5 antiserum in surgical patients to determine whether it is effective in preventing gram-negative infections. The actual study utilized 126 patients in the treatment group and 136 in the control group. Let us suppose that the investigators, prior to doing the study, want to estimate the sample size needed to detect a reduction in the proportion of patients who experience shock from the 10% level according to the investigators' previous experience (π_c) to 5% or less if patients are given transfusions from donors treated with J5. They are willing to accept a Type I error (of falsely concluding that there is a difference when there really is none) of .05, and they want a .90 probability of detecting a true difference.

The two-tailed z value related to α is ± 1.96, and the lower one-tailed z value related to β is -1.28. Then, the estimated sample size is

$$n = \left[\frac{1.96\sqrt{2 \times 0.10 \times 0.90} - (-1.28)\sqrt{(0.05 \times 0.95) + (0.10 \times 0.90)}}{0.05 - 0.10} \right]^2$$

$$= \left(\frac{1.306}{0.05} \right)^2 = 26.12^2 = 682.46$$

or 683 patients are needed in each group. This sample size is very large, and chances are that the investigators will compromise and recalculate the sample size with less power or a larger difference (see Exercise 8).

9.8 COMPUTER PROGRAMS THAT COMPARE PROPORTIONS

We use output from MINITAB, STATISTIX, and SYSTAT to illustrate the analysis of contingency table data for observations from Presenting Problems 2 and 3.

In Presenting Problem 2, a randomized trial was used to evaluate the efficacy of J5 antiserum in preventing the serious consequences of gram-negative infection. In Section 9.3.3, we used the data from Table 9–4 to compare the proportion of patients receiving J5 who experienced lethal shock

with the proportion receiving control vaccine who experienced lethal shock; the calculated value of the chi-square statistic was 4.12. We used the MINITAB program for calculating chi-square for this problem; the output is given in Fig 9–4. The columns represent the experimental and control groups, respectively; and the rows are lethal shock versus no lethal shock. The words "J5" and "Control" at the tops of the columns are user-provided labels. MINITAB prints the reproduced data and calculates the expected values for each cell. The details from the calculation of the chi-square statistic are given as well; however, MINITAB does not provide a P-value, so users must consult a chi-square table, such as Table A–5, in order to determine whether the result is significant.

We also used the STATISTIX program to calculate chi-square for Presenting Problem 2; the output is reproduced in Fig 9–5. The rows and columns are defined as in MINITAB. The value of X^2, called Pearson chi-square by the STATISTIX program, is 4.11, and the P-value associated with this value is .0425. STATISTIX also provides the Yates, or corrected, chi-square for researchers who prefer to use this statistic. As a brief review, recall (from Chapter 4) that the odds ratio is also called the cross-product ratio; the value of the odds ratio calculated by STATISTIX is given in scientific notation, $2.276E-01$; ie, the risk of lethal shock among the patients who were given J5 is 0.2276, indicating that patients given J5 were only 23% as likely to have lethal shock as those given the control plasma. If desired, the reciprocal of the odds ratio can be taken ($1/0.2276 = 4.3917$) so that the risk can be stated in terms of the unprotected group; ie, the patients not given J5 were 4.39 times more likely to develop lethal shock than those given J5.

```
Expected counts are printed below observed counts

              J5   Control   Total
      1        2         9      11
            5.29      5.71

      2      124       127     251
          120.71    130.29

Total      126       136     262

ChiSq =   2.046 +   1.896 +
          0.090 +   0.083 =  4.115

df = 1
```

Figure 9–4. MINITAB program illustrating chi-square, using observations on lethal shock in patients given J5 or a control vaccine from Presenting Problem 2. (Adapted, with permission, from Baumgartner J et al: Prevention of gram-negative shock and death in surgical patients by antibody to endotoxin core glycolipid. *Lancet* 1985;**2**:59. MINITAB is a registered trademark of Minitab, Inc. Used with permission.)

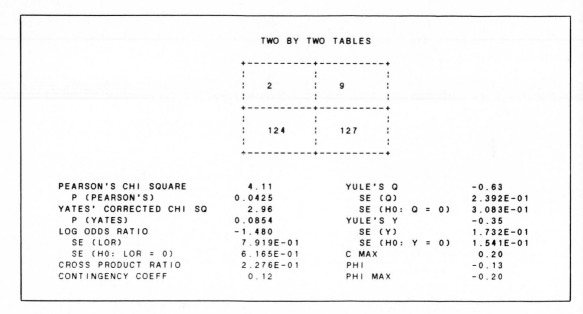

Figure 9-5. STATISTIX program illustrating chi-square, using observations on lethal shock in patients given J5 or control vaccine from Presenting Problem 2. (Adapted, with permission, from Baumgartner J et al: Prevention of gram-negative shock and death in surgical patients by antibody to endotoxin core glycolipid. *Lancet* 1985;**2**:59. STATISTIX is a registered trademark of NH Analytical Software. Used with permission.)

The SYSTAT program for calculating chi-square is used for the data from Presenting Problem 3 (Table 9–8) on outcome of treatment with three different antibiotics for cystitis. The SYSTAT TABLES program permits users to label the rows and columns (with eight or less characters or letters), as illustrated in Fig 9–6; we labeled the rows "Cure" and the columns "Treatment." The value for the Pearson chi-square is 7.265, with P = .026, and agrees with our calculations in Sec-

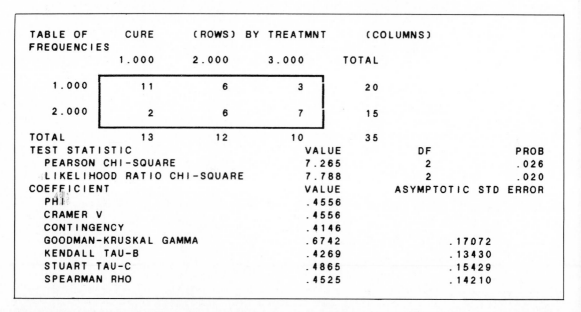

Figure 9-6. SYSTAT program illustrating chi-square, using observations on outcome of treatment of cystitis with three different antibiotics from Presenting Problem 3. (Adapted, with permission, from Hooten TM, Running K, Stamm WE: Single-dose therapy for cystitis in women: A comparison of trimethoprim-sulfamethoxazole, amoxicillin, and cyclacillin. *JAMA* 1985;**253**:387. SYSTAT is a registered trademark of SYSTAT, Inc. Used with permission.)

tion 9.4. Note that both SYSTAT and STATISTIX specify chi-square as the Pearson chi-square to distinguish it from other methods of calculating the chi-square statistic. SYSTAT provides several other statistics; one of them, Spearman's rho, was discussed in Chapter 4 and is further illustrated in Chapter 10.

9.9 SUMMARY

In the study described in Presenting Problem 1, significantly more patients were healed at both 2 and 6 weeks after cimetidine treatment than after placebo treatment. Confidence limits for a proportion were used to illustrate the proportion of patients expected to heal following 6 weeks of treatment. As expected, large ulcers heal more slowly than small ones, and the authors suggest that 8 weeks is a logical time for the first endoscopic or radiographic follow-up of benign gastric ulcers.

In the study described in Presenting Problem 2, J5 prophylaxis did not prevent focal gram-negative infection, but it did reduce the frequency of gram-negative shock in these high-risk patients, especially those who had undergone abdominal surgery. There was a significant decrease in the proportion of patients suffering shock and lethal shock when given J5 vaccine. This result was illustrated by a confidence interval for the difference in the proportion of patients receiving J5 vaccine who had shock and the proportion receiving placebo vaccine who had shock, as well as by statistical tests (z approximation and chi-square) of the difference in the proportion of patients suffering lethal shock. Vaccine J5 can protect against septic shock, but its protective effect seems to be specific for gram-negative infection. J5 antibodies probably act primarily against the lipopolysaccharide in the outer membrane of the bacteria.

The trial reported in Presenting Problem 3 was prematurely stopped because of frequent treatment failures, although the chi-square analysis we performed failed to show a difference at $P = .01$. These failures turned out to be occurring primarily in the amoxicillin- and cyclacillin-treated patients. The overall cure rate of 85% in the trimethoprim-sulfamethoxazole-treated group was comparable with the cure rate reported in previous studies.

Interestingly, the clinical suspicion of deep vein thrombosis is confirmed by venography in only 36% of patients, according to the study reported in Presenting Problem 4. The main problem with thermography in the study was the incidence of false-positive thermograms; however, only 1 of 20 patients had a false-negative thermogram. (False-positives, false-negatives, and other terminology related to diagnostic procedures are discussed in Chapter 13.) The relationship between thermography and venography was highly significant, as would be expected, and was tested by using the McNemar chi-square test for paired groups.

A summary of the statistical methods discussed in this chapter is given in Appendix C. Flowchart C-3 indicates the appropriate methods to use for different research questions involving nominal variables.

EXERCISES

1. In Section 9.3.1, data from Presenting Problem 2 were used to determine 99% confidence limits for the difference between the proportion of patients who suffered shock on J5 vaccine and the proportion who suffered shock on placebo; the 99% confidence interval was -0.150 to $+0.028$, indicating that a difference of zero is possible. In Section 9.3.2, the null hypothesis of no difference between the proportions was rejected for $\alpha = .05$ (for a one-tailed test). What level of confidence must be used to form a confidence interval consistent with the hypothesis test? Calculate the appropriate confidence limits.

2. Use the data in Table 9–2 to compare the total number of all patients with a gram-negative infection receiving J5 vaccine and placebo vaccine.
   a. What conclusion is drawn if a chi-square test is performed, using $\alpha = .05$?
   b. What conclusion is drawn if the relative risk of infection with placebo to that with J5 vaccine is determined?

3. Using Presenting Problem 3, and referring to Table 9–8, double the sample size but keep all the proportions the same, and recalculate $X^2(2)$. Compare your answer with the value obtained in Section 9.4. What does this result indicate about the power of the study in Presenting Problem 3?

4. Leveno et al (1986) studied the effects of electronic fetal monitoring during labor by comparing a group of women who universally had monitoring with another group of women who had monitoring only when there was a risk to the fetus. An alternate-month study design was used in which monitoring was provided in all pregnancies during one month (universal monitoring) and then provided only for high-risk pregnancies during the next month (selective monitoring); the study was performed over a 36-month period. The investigators were interested in determining whether universal electronic monitoring would improve perinatal results. Data from this study are given in Table 9–11. Perform an appropriate statistical procedure to answer the following questions.
   a. Is there a difference in the cesarean rate?
   b. Among low-risk pregnancies, is there a dif-

Table 9–11. Observations on electronic fetal monitoring.[1]

Observation	Selective Monitoring (N = 17, 409)	Universal Monitoring (N = 17, 586)
Number of cesarean sections	1777	1933
Number of low-risk pregnancies[2]	7330	7288
Number of neonatal deaths in low-risk pregnancies	5	4
Abnormal fetal heart rate in low-risk pregnancies	196	551

[1]Adapted and reproduced, with permission, from Tables 2 and 6 in Leveno KJ et al: A prospective comparison of selective and universal electronic fetal monitoring in 34,995 pregnancies. *N Engl J Med* 1986; **315**:615.

[2]Defined as a single fetus in a cephalic presentation; spontaneous, uncomplicated labor; and a birth weight greater than 2500 g.

ference in the proportion with abnormal fetal heart rate?

c. Among low-risk pregnancies, is there a difference in neonatal deaths?

d. The authors concluded that low-risk pregnancies do not need continuous electronic fetal monitoring during labor. Do you agree?

5. Table 9–12 presents some of the findings from a randomized, double-blind, placebo-con-

Table 9–12. Information on patients with Lyme arthritis.[1]

Patient Characteristics	Responders	Non-responders
No. of patients	n = 18	n = 22
History of ECM	16	11
Antibiotic therapy for ECM	8	1
Antibiotic therapy for arthritis	4	2
Mean ± standard deviation Months from ECM to treatment of arthritis	32 ± 15	34 ± 19
Months of active arthritis	12 ± 10	11 ± 7
ESR (mm/h) after treatment	13 ± 9	18 ± 10

[1]Adapted and reproduced, with permission, from Table 3 in Steer AC et al: Successful parenteral penicillin therapy of established Lyme arthritis. *N Engl J Med* 1985; **312**:869.

Table 9–13. Recurrence of acute alcohol-induced pancreatitis.[1,2]

	Total Patients	Pancreatitis Recurred[3]	Recurrence per Patient[3]
Declined participation	61	49 (80%)	1.9 (93)
Encouraged to abstain[4]	33	24 (73%)	1.7 (41)
Allocated to operation	31	2 (6%)	1.0 (2)

[1]Reproduced, with permission, from Table 4 in Stone HH, Mullins RJ, Scovill WA: Vagotomy plus Bilroth II gastrectomy for the prevention of recurrent alcohol-induced pancreatitis. *Ann Surg* 1985; **201**:686.

[2]Follow-up of 2 to 26 months (average 14 months).

[3]Recurrences based upon subsequent hospitalization for documented acute alcohol-associated pancreatitis.

[4]Including the three patients who did abstain; none of these three experienced a recurrent episode.

trolled study of the use of parenteral penicillin in patients with Lyme arthritis (Steere et al, 1985). Identify the appropriate statistical method for determining whether there was a difference between responders and nonresponders for the following conditions.

a. History of erythema chronicum migrans (ECM).

b. Previous use of antibiotic for ECM.

c. Months from ECM to treatment of arthritis.

d. Erythrocyte sedimentation rate (ESR) after treatment.

6. A prospective nonrandomized study reported by Stone, Mullins, and Scovill (1985) was used to determine whether truncal vagotomy with Bilroth II gastrectomy would prevent recurrent alcohol-induced pancreatitis. Of 125 patients who met the entry criteria for the study, 61 refused to participate, 33 were encouraged to abstain from alcohol, and 31 had surgery. The results are given in Table 9–13. Is there a difference in the rate of pancreatitis recurrence in the three groups?

7. Replicate the analysis done by the authors of the study described in Presenting Problem 1 by testing the hypothesis that the proportion of patients with healed ulcers is the same with cimetidine and placebo. See Table 9–1. Would you prescribe cimetidine for a patient with a benign gastric ulcer?

8. Refer to the study on evaluating the use of J5 vaccine in preventing gram-negative infections in surgery patients (Presenting Problem 2). How large a sample is needed to detect a drop in infection rate from 10% to 3%, with 70% power?

Correlation and Regression

10

PRESENTING PROBLEMS

Presenting Problem 1. The causes and pathogenesis of steroid-responsive nephrotic syndrome (also known as minimal-change disease) are unknown. Levinsky et al (1978) postulated that this disease may have an immunological basis because it may be associated with atopy, recent immunizations, or a recent upper respiratory infection. It is also responsive to corticosteroid treatment. They analyzed the serum from 39 children with steroid-responsive nephrotic syndrome (SRNS) for presence of IgG-containing immune complexes and, in a subset of these patients, the complement-binding properties (C1q-binding) of these complexes. They found elevated levels of IgG immune complexes in 17 of 18 children who had had a relapse of their steroid responsive nephrotic syndrome. These raised levels of immune complexes differed significantly from those of controls and of children whose disease was in remission.

These investigators also studied IgG immune complexes and C1q binding in 25 adults and 4 children with systemic lupus erythematosus (SLE). The IgG complexes from 5 of the children with SRNS in relapse did not bind complement, but complexes in serum of patients with SLE had complement-binding properties. The observations on complement binding and IgG-containing immune complexes in patients with SLE are given in Table 10–1, and a scatterplot is shown in Fig 10–2; the question is whether there is a significant correlation between these two measures.

Presenting Problem 2. Measurement of protein excretion in a 24-hour urine collection is commonly used in the assessment of renal disease and the effectiveness of appropriate therapies. Ginsberg et al (1983) contend that the 24-hour urine collection is often inaccurate, cumbersome, and time-consuming. They reasoned that if creatinine and protein excretions were fairly constant in the presence of stable renal function, then a simple ratio of the excreted protein and creatinine in a single voided urine specimen (expressed in milligrams per deciliter) would reflect the 24-hour excretion rate of protein. They collected 24-hour urine samples for measurement of protein excretion from 46 ambulatory patients attending their renal clinic. They also obtained a random urine specimen for creatinine and protein measurement from each of these patients and from 30 normal controls. Fig 10–4 shows the graph of protein/creatinine ratio in the single specimen as a function of the 24-hour protein excretion per 1.73 m^2 of body surface area. The investigators wish to examine the correlation between these measurements.

Presenting Problem 3. A major route of hepatitis B virus (HBV) transmission is percutaneous. Many groups of health care workers who are repeatedly exposed to blood have a higher risk of HBV infection than the general population. Approximately 1–3% of people with HBV infection may eventually develop chronic active hepatitis. Immunization against HBV is recommended for individuals at high risk. Therefore, Berry et al (1984) wanted to determine the prevalence of HBV infection among individuals who administer anesthesia and commonly handle blood samples or other body fluids in order to make appropriate recommendations regarding immunization to the group. Serum samples were collected from 86 anesthesiologists, nurse anesthetists, and anesthesia assistants at four major hospitals. Each sample was tested for hepatitis B surface antigen (HBsAg), surface antibody (anti-HBs), and core antibody (Anti-HBc). The authors stated that "the prevalence of hepatitis B markers showed a statistically significant increase with time since graduation from medical school, nursing school or college." The results are shown in Table 10–4. What statistical method can be used to evaluate the stated relationship?

Presenting Problem 4. Academic counselors must be able to predict whether a student will succeed in his or her educational aspirations so that the counselor may suggest remedial course work or alternative careers if necessary. Dawson-Saunders, Paiva, and Doolen (1986) presented information about predicting medical school applicants' MCAT examination scores from their ACT scores. A random sample of their data is given in Table 10–6. Can one predict MCAT science problem scores if one knows the ACT composite score? How good is the prediction; ie, how much error can be expected when one makes this prediction?

10.1 PURPOSE OF THE CHAPTER

This chapter reviews the concept of association between two characteristics (which was introduced in Chapter 4) and extends this idea to predicting the value of one characteristic from the other. Statistical tests are given for determining whether a relationship between two characteristics is significant. Some typical applications in medicine are used to illustrate the concepts of correlation and regression. These methods, although used less often than *t* tests and chi-square tests, are seen in approximately 10% of the published journal articles. They are used in from 6–7% of articles in surgery journals (Reznick, Dawson-Saunders, and Folse, 1987) to 12% of articles in *The New England Journal of Medicine* (Emerson and Colditz, 1983). Two probability distributions introduced previously—namely, the *t* distribution and the chi-square distribution—can be used for statistical tests in correlation and regression. Thus, many of the topics we will be discussing in this chapter have been introduced earlier.

10.2 AN OVERVIEW OF CORRELATION & REGRESSION

In Presenting Problem 1, the investigators wanted to know the relationship between IgG-containing immune complexes and C1q-binding properties of these complexes in their patients. In Presenting Problem 3, interest was in whether the prevalence of hepatitis B viral markers is increased in people working in anesthesia, and whether the risk increases with time on the job. In neither case was the prediction of one measure from another a specific goal, such as predicting C1q-binding properties from the presence of IgG-containing complexes, or predicting presence versus absence of hepatitis B markers from the number of years a person has been working in anesthesiology. When the goal is merely to establish a relationship (or association) between two measures, as in these studies, the correlation coefficient (Pearson product moment correlation coefficient) introduced in Chapter 4 is the statistic most often used. Recall that the correlation coefficient is a measure of the linear relationship between two variables measured on a numerical scale.

In Presenting Problem 4, however, the goal is clearly prediction; the investigators wanted to be able to predict how well a premedical student would perform on the MCAT examination from knowledge of his or her ACT test performance prior to entering college. In this situation, one of the variables, student score on the ACT examination, is considered to be the **independent,** or **explanatory, variable;** score on the MCAT examination is the **dependent** or **response, variable.**

Generally, the explanatory characteristic is the one that occurs first or is easier or less costly to measure. The statistical method of **linear regression** is used; this technique involves determining an equation for predicting the value of the response variable from knowledge of the value of the explanatory variable. Therefore, one of the major differences between correlation and regression is the purpose of the analysis—whether it is to describe a relationship or to predict a value. There are also several important similarities, however. As we will see in this chapter, if the means and the standard deviations of both characteristics are known, we can calculate a correlation coefficient from the regression equation and vice versa. In addition, if the correlation coefficient is statistically significant, the regression equation will also be statistically significant. Many of the same assumptions are required for correlation and regression, and both measure the extent of a linear relationship between the two characteristics.

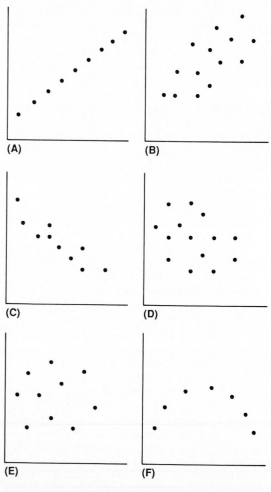

Figure 10-1. Scatterplots and correlations. *(A)* $r = +1.0$. *(B)* $r \approx .7$. *(C)* $r \approx -.9$. *(D)* $r \approx -.4$. *(E)* $r \approx 0.0$. *(F)* $r \approx 0.0$.

10.3 CORRELATION

Fig 10–1 illustrates several hypothetical **scatterplots** of data to demonstrate the relationship between size of the correlation coefficient r and shape of the scatterplot. When the correlation is near zero, as in Fig 10–1E, the pattern of plotted points is somewhat circular. When there is a small degree of relationship, the pattern is more like an oval, as in Figs 10–1D and 10–1B. As the value of the correlation gets closer to either $+1$ or -1, as in Fig 10–1C, the plot has a long, narrow shape; at $+1$ and -1, the observations fall directly on a line, as in Fig 10–1A.

The scatterplot in Fig 10–1F illustrates a situation in which there is a strong but nonlinear relationship. For example, with temperatures less than 10–15°C, a cold nerve fiber discharges few impulses; as the temperature increases, so do numbers of impulses per second until the temperature reaches about 25°C. As the temperature increases beyond 25°C, the numbers of impulses per second decrease once again until they cease at 40–45°C. The correlation coefficient, however, measures only a linear relationship, and it has a value close to zero in this situation.

One of the reasons for drawing scatterplots of data as part of the initial analysis is to identify nonlinear relationships when they occur. Otherwise, if the correlation coefficient is calculated without examining the data, one can miss a strong, but nonlinear, relationship, such as the one between temperature and number of cold nerve fiber impulses.

10.3.1 Calculating the Correlation Coefficient

We use Presenting Problem 1 to give a brief review of the concepts involved in correlation; for more specific details on calculating the **correlation coefficient,** refer to Chapter 4, Section 4.5. The levels of IgG complexes and C1q-binding complexes (expressed as a percentage of inhibition of agglutination of IgG-coated latex particles) for the 29 patients with SLE in this study are given in Table 10–1.

As we discussed above, a very valuable first step when we look at the relationship between two characteristics is to examine the relationship graphically. Fig 10–2 is a scatterplot of the data from Presenting Problem 1, with IgG complexes on the X-axis and C1q-binding complexes on the Y-axis. We see from Fig 10–2 that there is indeed a relationship between these two characteristics: As the percentage inhibition of IgG complex increases, so does the percentage inhibition of C1q-binding complex.

The extent of the relationship can be found by calculating the correlation coefficient. In Present-

Table 10-1. Complexes and their relationship to C1q-binding complexes.

Subject	IgG Complex	C1q-Binding Complex
1	34	35
2	15	16
3	95	98
4	92	99
5	4	1
6	58	31
7	36	37
8	69	38
9	78	72
10	9	11
11	30	25
12	61	75
13	1	2
14	8	17
15	11	2
16	35	20
17	90	72
18	5	5
19	30	39
20	30	20
21	48	63
22	60	59
23	36	27
24	16	5
25	31	28
26	59	53
27	66	72
28	50	31
29	42	15

[1]Observations inferred and reproduced, with permission, from Fig 3 in Levinsky RJ et al: Circulating immune complexes in steroid-responsive nephrotic syndrome. *N Engl J Med* 1978; **298**:126.

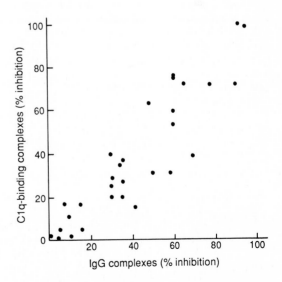

Figure 10-2. Scatterplot of C1q-binding complexes and IgG-containing complexes (expressed as a percentage of inhibition of agglutination of IgG-coated latex particles). (Adapted and reproduced, with permission, from Levinsky RJ et al: Circulating immune complexes in steroid-responsive nephrotic syndrome. *N Engl J Med* 1978;**298**:126.)

ing Problem 1, the correlation coefficient between IgG complexes and C1q-binding complex is

$$r_{x,y} = \frac{\Sigma(X - \bar{X})(Y - \bar{Y})}{\sqrt{\Sigma(X - \bar{X})^2}\sqrt{\Sigma(Y - \bar{Y})^2}}$$

$$= \frac{20{,}168}{\sqrt{(21{,}374)}\sqrt{(22{,}879}}$$

$$= .91$$

Thus, there is a correlation of .91 between IgG-containing complexes and C1q-binding complex. This relationship is fairly strong, since the largest value the correlation can attain is 1 (or −1 if it is a negative relationship). (See Chapter 4, Section 4.5, for a review of the properties of the correlation coefficient.)

10.3.2 Interpreting the Size of r

10.3.2.a The Coefficient of Determination: The correlation coefficient can be squared to form the statistic called the **coefficient of determination.** For the immune complex data, the coefficient of determination is $(.91)^2$, or .83; this result means that 83% of the variation in the values for one of the measures, such as C1q-binding complex, may be accounted for by knowing the values for the other characteristic, IgG-containing complex inhibition. This concept is demonstrated by the Venn diagrams in Fig 10–3. For the left diagram, $r^2 = .25$; so 25% of the variation in A is accounted for by knowing B (or vice versa). The middle diagram illustrates $r^2 = .50$, and the diagram on the right represents $r^2 = .83$ for the IgG–C1q example. (Also see Exercise 1).

The coefficient of determination tells us how strong the relationship really is. However, in the medical literature, confidence limits or results of a statistical test for significance of the correlation coefficient are usually presented.

10.3.2.b The t Test for Correlation: The symbol for the correlation coefficient in the population

(the population parameter) is ρ (Greek letter rho). In a random sample, ρ is estimated by r. If several random samples of the same size are selected from a given population and the correlation coefficient r calculated for each, we expect the rs to vary from each other but to follow some sort of distribution about the population value ρ. And, in fact, they do; unfortunately, the sampling distribution of the correlation does not behave as nicely as the sampling distribution of the mean, which is normally distributed for large samples.

Part of the problem is what we might describe as a "ceiling" effect when the correlation approaches either −1 or +1. If the value of the population parameter is, say, .8, the sample values can exceed .8 only up to 1.0, but they can be less than .8 all the way to −1.0. The maximum value of 1.0 acts as a "ceiling" in keeping the sample values from varying very far above .8 and results in a **skewed distribution.** However, when the population parameter is zero, the ceiling effects are equal, and the sample values are approximately distributed according to the t distribution.

Therefore, the t distribution can be used to test the hypothesis that the true value of the population parameter ρ is equal to zero. The following mathematical expression involving the correlation coefficient has been found to have a **t distribution** with $n - 2$ degrees of freedom:

$$t = \frac{r\sqrt{n - 2}}{\sqrt{1 - r^2}}$$

Let us use the above t ratio to test whether the observed value of $r = .91$ is sufficient evidence with 29 observations to conclude that the true population value of the correlation ρ is different from 0.

Step 1. H_0: $\rho = 0$ **(The true correlation is zero.)**
H^1: $\rho \neq 0$ **(The true correlation is not zero.)**

Step 2. Since the null hypothesis is a test of whether ρ is 0, the t ratio may be used

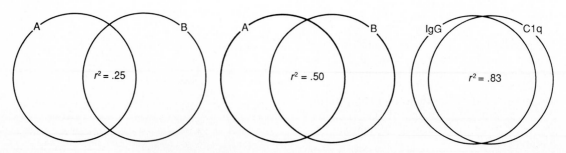

Figure 10–3. Illustration of r^2, proportion of explained variance.

when the assumptions for correlation (discussed below) are met.

Step 3. Suppose the investigators chose $\alpha = .01$ for this example.

Step 4. The degrees of freedom are $n - 2 = 29 - 2 = 27$. The value of t distribution with 27 degrees of freedom that divides the area into the central 99% and the upper and lower 1% is 2.771 (Table A–3). Therefore, we will reject the null hypothesis of no correlation if (the absolute value of) the observed value of t is greater than 2.771.

Step 5. The calculation is

$$t = \frac{.91 \sqrt{27}}{\sqrt{1 - .91^2}}$$

$$= 11.40$$

Step 6. The observed value of the t ratio with 27 degrees of freedom is 11.40, which is greater than 2.771. Therefore, the null hypothesis of zero correlation is rejected. We conclude that there is a significant correlation between IgG-containing and C1q-binding complexes in patients with SLE.

10.3.2.c Fisher's z Transformation to Test the Correlation:

To test the statistical significance of the correlation coefficient when the population parameter is not equal to zero, we must transform the correlation and then use the standard normal (z) distribution. We need a transformation because, as we described above, the distribution of sample values of the correlation is skewed when $\rho \neq 0$. Although this method is a bit complicated, it is much more flexible than the t test. The transformation is called **Fisher's z transformation**, and it was proposed by the same statistician who developed Fisher's exact test for 2×2 contingency tables (discussed in Chapter 9).

Although investigators frequently use a significance test to determine whether $\rho = 0$, there are numerous occasions in medicine when interest is in whether the correlation is equal to another value. For example, consider a diagnostic test that gives accurate numerical values but that is also invasive and somewhat risky for the patient. Suppose that researchers have developed an alternative testing procedure, and they want to demonstrate that the new procedure is as accurate as the standard test. To do so, they must select a sample of patients and perform both the standard test and the new procedure on each patient. An appropriate statistic for demonstrating the relationship between the standard test and the new procedure is the correlation coefficient, so the

next step is to calculate this statistic. Finally, the researchers will want to show that the correlation measures more than a chance relationship.

Would it be reasonable in this illustration to determine whether the observed correlation was significantly greater than zero, or would it be more appropriate to establish that the correlation exceeded some minimum value acceptable to physicians who use the diagnostic test? If the latter approach is preferable, either a test of hypothesis can be performed to show that the correlation is greater than some value, such as .90, or a confidence interval about the observed correlation can be calculated. In either case, we can use Fisher's z transformation, which allows us to test any null hypothesis—not just $\rho = 0$, as with the t test—and also to form confidence intervals.

Fisher's z transformation is as follows:

$$z(r) = \frac{1}{2} \ln \left(\frac{1 + r}{1 - r} \right)$$

where ln represents the natural logarithm. Table A–6 gives the z transformation for different values of r, so we do not need to do the calculation of the formula. With moderate-sized samples, this transformation follows a normal distribution, and the following expression for the z test can be used:

$$z = \frac{z(r) - z(\rho)}{\sqrt{1/(n - 3)}}$$

To illustrate Fisher's z transformation for testing the significance of ρ, let us use Presenting Problem 1 again, but let us assume the investigators want to know whether the relationship between IgG and C1q is greater than .70.

Step 1. H_0: $\rho \leq .70$ **(The true correlation is .70 or less.)**

H_1: $\rho > .70$ **(The true correlation is greater than .70.)**

Step 2. Fisher's z transformation may be used with the correlation coefficient to test any hypothesis.

Step 3. Let us again use $\alpha = .01$ for this example.

Step 4. The alternative hypothesis specifies a one-tailed test. The value of the z distribution that divides the area into the lower 99% and the upper 1% is 2.326 (from Table A–2). Therefore, we will reject the null hypothesis that the correlation is .70 or less if the observed value of z is greater than 2.326.

Step 5. The first step is to find the transformed values for $r = .91$ and $\rho = .70$ from Table A–6; these values are +1.528 and +0.867, respectively. Then, the calculation for the z test is

$$z = \frac{1.528 - 0.867}{\sqrt{1/(29 - 3)}}$$

$$= 3.370$$

Step 6. The observed value of the z statistic, 3.37, is greater than 2.326. Therefore, the null hypothesis that the correlation is .70 or less is rejected, and the conclusion is that the correlation between IgG-containing and C1q-binding complexes exceeds .70, with $P \leq .01$.

10.3.2.d Confidence Interval for the Correlation: A major advantage of Fisher's z transformation is that **confidence intervals** can be formed. The transformed value of the correlation is used to calculate confidence intervals in a manner similar to the calculations presented in previous chapters; however, after the limits have been determined, they must be transformed back to values corresponding to the correlation coefficient.

To illustrate, we will calculate a 99% confidence interval, using the z distribution in Table A–2 to find the critical value, for the correlation coefficient .91 in Presenting Problem 1. The confidence interval is

Transformed correlation $\pm$ **2.575** $\times$ **Standard error**

or

Transformed correlation $\pm$ **2.575** $\times \sqrt{\dfrac{1}{n - 3}}$

$$= 1.5238 \pm (2.575)(0.196)$$
$$= 1.528 \pm 0.505$$

or **1.025 to 2.035**

Transforming the values 1.025 and 2.035 back to correlations by using Table A–6 gives, approximately, $r = .77$ and $r = .97$. therefore, the 99% confidence interval for the observed correlation of .91 is .77 to .97; ie, we are 99% confident that the true value of the correlation in the population is contained within this interval. Note that the interval is not symmetric about the sample value of .91, because of the skewness in the distribution of the correlation coefficient.

10.3.3 Assumptions in Correlation

The assumption needed to make inferences about the correlation coefficient is that the two variables, X and Y, vary together in a joint distribution that is normally distributed, called the bivariate normal distribution. However, just because each is normally distributed when examined separately does not guarantee that, jointly, they have a bivariate normal distribution. Some guidance is available, however. The rule of thumb for deciding whether use of correlation is appropriate is as

follows: If either of the two variables is *not* normally distributed, Pearson's product moment correlation coefficient is *not* the most appropriate method. Instead, either one or both of the variables may be transformed so that they more closely follow a normal distribution, as discussed in Chapter 7, Section 7.2; or the observations may be converted into ranks and the Spearman rank correlation calculated. This topic is discussed in the next section.

10.4 OTHER MEASURES OF CORRELATION

Several other measures of correlation are often found in the medical literature. Spearman's rho, the rank correlation introduced in Chapter 4, is used with ordinal data or in situations in which the numerical variables are not normally distributed. With nominal data, the risk ratio or the kappa statistic (κ) discussed in Chapter 4 can be used. We discuss confidence intervals and significance tests for these statistics in this section.

10.4.1 Spearman's Rho (Rank Correlation)

Recall that the value of the correlation coefficient is markedly influenced by extreme values and thus does not provide a good description of the relationship between two variables when the distributions of the variables are skewed or contain outlying values. For example, consider the data for Presenting Problem 2, listed in Table 10–2. Several of the observations for protein excretion and ratio of urine protein to creatinine clearance are extremely large. Five of the subjects have protein excretion of 9.9 g per 24 hours or greater along with a ratio of 11 or more.

When the data are plotted on a graph, as in Fig 10–4, the nature of these extreme observations becomes even more apparent. (Notice that the authors have broken the regression line as well as the X - and Y - axes with two short lines to call the break in scale to the reader's attention.) These five outlying individuals have an undue effect on the size of the correlation coefficient. If the Pearson product moment correlation coefficient is calculated with these five individuals included in the sample, it has a value of .969.

As we indicated in Chapter 4, a simple method for dealing with the problem of extreme observations in correlation is to transform the data to **ranks** and then recalculate the correlation on ranks to obtain the nonparametric correlation called **Spearman's rho** or **rank correlation.** To illustrate this procedure, we have converted the observations from Presenting Problem 2 to ranks in Table 10–3. Ranking was done separately for each of the variables, protein excretion and ratio; when

Table 10-2. Protein excretion and ratio of urinary protein to creatinine for 43 patients.[1]

Subjects	Protein Excretion (g/24 h)	Ratio
1	1.3	1.0
2	0.5	0.6
3	3.0	5.5
4	10.0	11.0
5	3..0	2.7
6	4.4	3.9
7	0.3	0.1
8	2.1	2.5
9	0.8	1.7
10	2.5	2.1
11	2.9	3.2
12	0.3	0.3
13	6.1	5.1
14	3.5	3.5
15	0.8	0.8
16	2.3	1.8
17	0.1	0.1
18	8.0	4.5
19	0.1	0.3
20	0.6	0.3
21	1.2	1.2
22	2.9	4.5
23	20.0	17.0
24	3.6	4.4
25	3.4	5.3
26	0.2	0.2
27	27.0	31.0
28	1.0	1.0
29	0.4	1.0
30	1.8	2.0
31	2.2	4.5
32	4.3	3.8
33	9.9	12.0
34	1.2	2.4
35	3.5	3.1
36	0.5	0.8
37	1.2	0.9
38	16.0	13.0
39	4.6	3.5
40	2.2	1.3
41	2.3	2.1
42	2.1	3.7
43	1.5	1.9

[1]Observations inferred and reproduced, with permission, from Fig 1 in Ginsberg JM et al: Use of single voided urine samples to estimate quantitative proteinuria. *N Engl J Med* 1983; **309**:1543.

Ratio of urinary protein to creatinine concentration ([PR/CR]u) of random single voided urine samples expressed as a function of protein excretion per 24 hours per 1.73 m². The concentrations of protein and creatinine in the urine were measured in milligrams per deciliter. The five open circles denote values for patients who had protein:creatinine ratio of more than 3.5 and a protein excretion rate of less than 3.5 per 24 hours per 1.73 m².

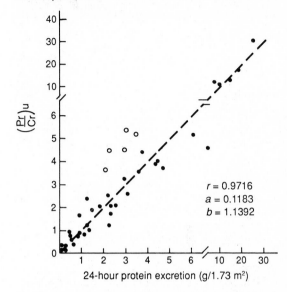

$r = 0.9716$
$a = 0.1183$
$b = 1.1392$

Figure 10-4. Scatterplot of observations on ratio of urine protein to creatinine and 24-hour protein excretion from Presenting Problem 2. (Reproduced, with permission, from Ginsberg JM et al: Use of single voided urine samples to estimate quantitative proteinuria. *N Engl J Med* 1983;**309**:1543.)

where R_X is the rank of the X variable, R_Y is the rank of the Y variable, and $\overline{R}_X$ and $\overline{R}_Y$ are the mean ranks for the X and Y variables, respectively. The rank correlation r_s may also be calculated by using other formulas, but the approximate procedure for ranks outlined here is quite good (Conover and Iman, 1981).

Calculating r_s for the ranked observations in Table 10-3 gives

$$r_s = \frac{6142.5}{\sqrt{6642}\ \sqrt{6614}}$$

$$= .927 \text{ or } .93$$

The value of r_s is slightly smaller than the value of Pearson's correlation and indicates a strong relationship between the ranked observations. To see whether it is significantly different from zero, we can use the t test, just as we did for the Pearson correlation. For example, the following procedure tests whether the value of Spearman's rho in the population, symbolized ρ_s (Greek letter rho with subscript denoting Spearman) differs from zero.

Step 1. H_0: $\rho_s = 0$ **(The population value of Spearman's rho is zero).**

ties occurred, we used the average ranks of the tied values.

The ranks of the X variable, protein excretion, are used for values of the X variable in the equation for the correlation coefficient; the ranks of ratio are used for the Y variable. Then, Spearman's rank correlation in a sample of observations, denoted r_s, is

$$r_s = \frac{\Sigma(R_X - \overline{R}_X)(R_Y - \overline{R}_Y)}{\sqrt{\Sigma(R_X - \overline{R}_X)^2\ \ \Sigma(R_Y - \overline{R}_Y)^2}}$$

Table 10-3. Ranks of protein excretion (grams per 24 hours) and ratio of urinary protein to creatinine.

Subjects	Protein Excretion Rank	Ratio Rank
1	16	12
2	7.5	7
3	28.5	38
4	40	39
5	28.5	24
6	35	31
7	4.5	1.5
8	19.5	23
9	10.5	16
10	25	20.5
11	26.5	26
12	4.5	5
13	37	36
14	31.5	27.5
15	10.5	8.5
16	23.5	17
17	1.5	1.5
18	38	34
19	1.5	5
20	8	5
21	14	14
22	26.5	34
23	42	42
24	33	32
25	30	37
26	3	3
27	43	43
28	12	12
29	6	12
30	18	19
31	21.5	34
32	34	30
33	39	40
34	14	22
35	31.5	25
36	7.5	8.5
37	14	10
38	41	41
39	36	27.5
40	21.5	15
41	23.5	20.5
42	19.5	29
43	17	18

H_1: $\rho_s \neq 0$ (The population value of Spearman's rho is not zero.)

Step 2. Since the null hypothesis is a test of whether or not ρ_s is zero the t ratio may be used.

Step 3. Let us use $\alpha = .01$ for this example.

Step 4. The degrees of freedom are $n - 2 = 43 - 2 = 41$. The value of the t distribution with 40 degrees of freedom that divides the area into the central 99% and the upper and lower 1% is 2.704 (Table A–3) and is used as an approximation to 41 df. Therefore, we will reject the null hypothesis of no correlation if (the absolute value of)

the observed value of t is greater than 2.704.

Step 5. The calculation is

$$t = \frac{r\sqrt{n - 2}}{\sqrt{1 - r^2}}$$

$$= \frac{.927\sqrt{41}}{\sqrt{1 - .927^2}}$$

$$= 15.83$$

Step 6. The observed value of the t ratio with 41 degrees of freedom is 15.83, which is much larger than 2.704. Therefore, and not surprisingly, we reject the null hypothesis of no correlation and conclude that there is a significant nonparametric correlation between protein excretion and ratio of urine protein to creatinine clearance.

To summarize, Spearman's rho (rank correlation) is used when investigators want to measure the relationship between two ordinal variables. It may also be used with numerical variables if one or both are not normally distributed. Alternatively, a numerical measure may be transformed so that they are normally distributed, and Pearson's correlation can then be used.

10.4.2 Confidence Interval for the Relative Risk & the Odds Ratio

Chapter 4 introduced the **relative risk** and the **odds ratio** as measures of relationship between two nominal characteristics. These statistics were developed for epidemiology and are used in medicine for studies examining risk of exposure to conditions that may result in a disease. Presenting Problem 3, for instance, examines prevalence of hepatitis B infection in anesthesiology workers and the relationship between the number of years a person has worked in anesthesiology and presence or absence of the hepatitis B viral marker. The data from that study are given in Table 10–4.

Table 10-4. Relationship between presence of hepatitis B viral markers in anesthesia personnel and time since graduation.[1]

Hepatitis B	Years Since Graduation				
	≤ 9	10-19	20-19	30-39	
Present	6	4	5	3	18
Absent	40	15	9	2	66
	46	19	14	5	84

[1]Adapted and reproduced, with permission, from Fig 2 in Berry AJ et al: The prevalence of hepatitis B viral markers in anesthesia personnel. *Anesthesiology* 1984; **60**:6.

To calculate a single odds ratio for length of time working in anesthesiology, we must rearrange the data into a contingency table with two columns and two rows (as in Table 4–11). For example, suppose that 20 years is taken as the dividing point. The data can then be rearranged as in Table 10–5, with working in anesthesiology 20 or more years playing the role of the presence of the risk factor and having the hepatitis B viral marker playing the role of the disease.

The odds ratio is

$$OR = \frac{AD}{BC}$$

where A, B, C, and D are the cell frequencies (see Table 4–11). Then,

$$OR = \frac{(8)(55)}{(11)(10)}$$
$$= 4.00$$

In many studies, the investigators want to determine whether the estimate of the relative risk (the risk ratio in cohort studies or the odds ratio in case-control studies) is statistically significant. There are several ways to determine significance. For instance, to test the significance of the relationship between working 20 or more years and having the hepatitis B marker, investigators may use the chi-square test introduced in Chapter 9. In this case, the degrees of freedom are equal to 1. The chi-square test for this example is left as an exercise (see Exercise 2). An alternative chi-square test, based on the natural logarithm of the odds ratio, is also available, but it results in values close to the chi-square test illustrated in Chapter 9 (Fleiss, 1981).

More often, articles in the medical literature use confidence intervals for risk ratios or odds ratios. The procedure for obtaining confidence intervals is a bit more complicated than usual because it requires finding natural logarithms and antilogarithms. Since many inexpensive calculators have these functions available, however, we briefly il-

lustrate the procedure. If a calculator with these functions is not available, an elementary mathematics textbook containing tables of the natural logarithm can be consulted. The formula for a 95% confidence interval for the true relative risk is

$$\exp\left[\ln(OR) \pm 1.96 \sqrt{\left(\frac{1}{A}\right) + \left(\frac{1}{B}\right) + \left(\frac{1}{C}\right) + \left(\frac{1}{D}\right)} \right]$$

where exp denotes the exponential function, or antilogarithm, of the natural logarithm, and ln is the natural logarithm. The calculation is

$$\exp\left[\ln(4.00) \pm 1.96 \sqrt{\left(\frac{1}{7}\right) + \left(\frac{1}{12}\right) + \left(\frac{1}{10}\right) + \left(\frac{1}{55}\right)} \right]$$

$$= \exp(1.39 \pm 1.96\sqrt{0.334})$$
$$= \exp(0.25, 2.52)$$
or $$= 1.29 \text{ to } 12.42$$

This interval contains the value of the true relative risk with 95% confidence. Since the interval does not include 1, we may be 95% confident that the relative risk is not 1, ie, that there is an elevated risk of infection with hepatitis B to anesthesia workers who have been employed 20 years or more. Of course, 90% or 99% confidence intervals can be formed by using 1.645 or 2.575 instead of 1.96 in the above equation. (For a 99% confidence interval, see Exercise 12.) For a detailed and insightful discussion of the odds ratio and its advantages and disadvantages, see Feinstein (1985, Chap 20) and Fleiss (1981, Chap 5).

10.4.3 Measuring Relationships in Other Situations

We have discussed how to measure and test the significance of relationships by using Pearson's product moment correlation coefficient, Spearman's nonparametric procedure based on ranks, and risk or odds ratios. However, not all situations are covered by these procedures, such as when one variable is measured on a nominal scale and the other is numerical but has been classified into categories, as in Table 10–4 (in which "years" has been collapsed into four categories); or when one variable is nominal and the other is ordinal; or when both are ordinal but there are only a few categories. In these cases, a contingency table is formed and the chi-square test is used, as illustrated in Chapter 9, Section 9.3.4.

Table 10–5. Data rearrangement for odds ratio for relationship between hepatitis B in anesthesia personnel and time since graduation.[1]

Years of Work	Hepatitis B Marker		
	Present	Absent	
≥ 20	8	11	19
19 or less	10	55	65
	18	66	84

[1]Adapted and reproduced, with permission, from Fig 2 in Berry AJ et al: The prevalence of hepatitis B viral markers in anesthesia personnel. *Anesthesiology* 1984; **60**:6.

10.5 LINEAR REGRESSION

As we stated in the introduction to this chapter, when the goal is to predict the value of one characteristic from knowledge of another, one statistical method used is **regression** analysis. This method is also called linear regression, simple linear regression, or least squares regression. Let us take a moment to review the history of these terms, for it sheds some light on the nature of regression analysis.

The concepts of correlation and regression were developed by Sir Francis Galton, a cousin of Charles Darwin, who studied both mathematics and medicine in the mid nineteenth century (Walker, 1931). Galton was interested in heredity and wanted to understand why a population remains more or less the same over many generations if the "average" offspring resemble their parents, ie, why it does not become more diverse. By growing sweet peas and observing the average size of seeds from parent plants of different sizes, he discovered regression, which he termed as the "tendency of the ideal mean filial type to depart from the parental type, reverting to what may be roughly and perhaps fairly described as the average ancestral type." This phenomenon is more typically known as regression toward the mean. The term *correlation* was used by Galton in his work on inheritance in terms of the "co-relation" between such characteristics as heights of fathers and sons. It was the mathematician Karl Pearson who went on to work out the theory of correlation and regression, and the correlation coefficient is named after him for this reason.

The term *linear regression* refers to the fact that correlation and regression measure only a straight-line, or linear, relationship between two variables. When the term *simple* is used with regression, it refers to the situation in which only one explanatory (independent) variable is used to predict another. In **multiple regression,** more than one explanatory variable is included in the prediction equation (a method discussed in Chapter 12).

Least squares regression describes the mathematical method for obtaining the statistical estimators in the regression equation. The important thing to remember is that when the term *regression* is used alone, it generally means simple linear regression based on the least squares method. The conceptual idea behind least squares regression is described in the next section. The application of regression is discussed in Section 10.5.2.

10.5.1 Least Squares Method

Several times previously in this text, we have mentioned the linear nature of the pattern of points in a scatterplot. For example, in Figure 10–2, a straight line can be drawn through the points representing the values of IgG complexes and C1q-

binding complexes to indicate the direction of the relationship. The least squares method is a way to determine the equation of the line that provides a "good fit" to the points.

To illustrate the method, we consider the straight line in Fig 10–5. Elementary geometry can be used to determine the equation for the straight line. If the point where the line crosses, or **intercept,** the Y-axis is denoted by *a* and the **slope** of the line is *b,* then the equation is

$$Y = a + bX$$

The slope of the line measures the amount of change in Y for each one-unit change in X. If the slope is positive, Y increases as X increases; if the slope is negative, Y decreases as X increases. In the regression model, the slope in the population is generally symbolized by β_1, called the **regression coefficient;** and β_0 denotes the **intercept** of the regression line—ie, β_1 and β_0 are the population parameters in regression. Furthermore, the values of the Y's obtained from the regression equation are predicted values, rather than actual values of Y. For this reason, predicted values of Y are usually distinguished from Y values, often by using the symbol Y'. Therefore, the regression equation is given by

$$Y' = \beta_0 + \beta_1 X$$

Of course, not all predictions for Y will be right on target, because in most applications, the points do not fall exactly along a straight line. For this reason, the regression model contains an *error term, e,* which is the amount the actual values of Y depart from the predicted values based on the regression line. Thus,

$$Y' = \beta_0 + \beta_1 X + e$$

Fig 10–6 illustrates a straight line drawn through a set of data points that represent the sort of variability often seen in medical applications.

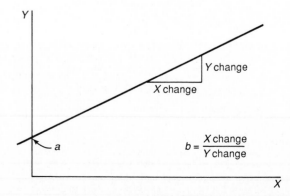

Figure 10–5. Geometric interpretation of regression line.

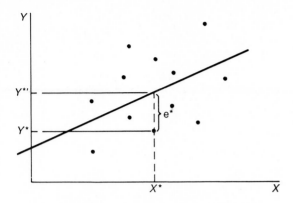

Figure 10-6. Least squares regression line.

For a given value of X, say X^*, the predicted value of Y^* is found by extending a horizontal line from the regression line to the Y-axis. The difference between the actual value for Y^* and the predicted value, $e^* = Y^* - Y^{*'}$, provides a criterion for judging how good the line fits the data points. The least squares method determines the line that minimizes the sum of the squared vertical differences between the actual and predicted values of the Y variable; ie, β_0 and β_1 are determined so that

$$\Sigma(Y - Y')^2$$

is minimized. The formulas for β_0 and β_1 are found,* and in terms of the sample estimates b and a, these formulas are

$$b = \frac{\Sigma(X - \overline{X})(Y - \overline{Y})}{\Sigma(X - \overline{X})^2}$$

$$a = \overline{Y} - b\overline{X}$$

10.5.2 Calculating the Regression Equation

There are two approaches for calculating the regression equation. The first approach uses raw data and the formulas for a and b given above. This approach requires many calculations, regardless of whether the formulas given in Section 10.5.1 are used or alternative computational formulas are used. The second method takes advantage of the relationship between r and b. When the correlation has a positive value, the slope of

---

*The procedure for finding β_0 and β_1 involves the use of differential calculus; the idea is as follows: The partial derivatives of the above equations are found with respect to β_0 and β_1; the two resulting equations are set equal to 0 to locate the minimum values; these two equations in two unknowns, β_0 and β_1, are solved simultaneously to obtain the formulas for β_0 and β_1.

the regression line drawn through the observations is also positive. Similarly, a negative correlation is associated with a negative slope. If the correlation is zero, the regression line is horizontal with a slope of zero. Thus, the formulas for the correlation coefficient and the regression coefficient are closely related. If r has already been calculated, it can be multiplied by the ratio of the standard deviation of Y to the standard deviation of X, s_Y/s_X, to obtain b (see Exercise 5). Thus,

$$b = r \frac{s_Y}{s_X}$$

Similarly, if the regression coefficient is known, r can be found by

$$r = b \frac{s_X}{s_Y}$$

Of course, a far more practical way to obtain the regression equation is by using a computer, and except for illustration purposes, we will assume anyone doing regression analysis uses a computer.

In the study described in Presenting Problem 4, the investigators want to be able to predict a potential medical school applicant's MCAT scores from his or her previous ACT examination scores. The ACT composite score and the MCAT Science Problems score for 42 undergraduate students are given in Table 10-6.

Before calculating the regression equation for these data, let's create a scatterplot and practice "guesstimating" the value of the correlation coefficient from the plot [although it is difficult to estimate the size of r accurately when it is between $-.5$ and $+.5$ unless the sample size is quite large (Sokal and Rohlf, 1981)]. Fig 10-7 is a scatterplot of ACT score as the explanatory X variable and MCAT score as the response Y variable. How large do you think the correlation is? (See Exercise 4 to check your estimate.)

If we knew the correlation between ACT and MCAT scores, we could use it to calculate the regression equation. Since we do not, we will assume the needed terms have been calculated; they are $\Sigma(X - \overline{X})(Y - \overline{Y}) = 197.78$, $\Sigma(X - \overline{X})^2 = 487.49$, $\overline{X} = 24.667$, and $\overline{Y} = 8.405$. Then,

$$b = \frac{\Sigma(X - \overline{X})(Y - \overline{Y})}{\Sigma(X - \overline{X})^2} = \frac{197.78}{487.49} = 0.406$$

$$a = \overline{Y} - b\overline{X} = 8.405 - (0.406)(24.667)$$

$$= 8.405 - 10.015 = -1.61$$

In this example, the MCAT scores are said to be regressed on the ACT scores, and the regression equation is written as $Y' = -1.61 + 0.406X$, where Y' is the predicted MCAT score and X is

Table 10-6. ACT and MCAT test scores for medical school applicants.[1]

Subject	ACT Composite	MCAT Science Problems
1	28	8
2	25	7
3	24	7
4	28	9
5	25	10
6	25	8
7	24	9
8	21	5
9	25	9
10	29	9
11	27	10
12	18	6
13	24	10
14	29	10
15	27	8
16	21	6
17	27	6
18	17	3
19	28	8
20	20	6
21	24	9
22	27	6
23	27	10
24	31	12
25	25	10
26	27	12
27	21	4
28	24	9
29	26	11
30	22	10
31	24	12
32	27	7
33	23	9
34	28	10
35	23	11
36	28	11
37	24	9
38	16	3
39	25	7
40	26	9
41	18	8
42	28	10
Mean	24.667	8.405
Standard deviation	3.448	2.275

[1]Observations used, with permission from Dawson-Saunders B, Paiva RE, Doolen DR: Using ACT scores and grade-point averages to predict students' MCAT scores. *J Med Educ* 1986; **61**:681.

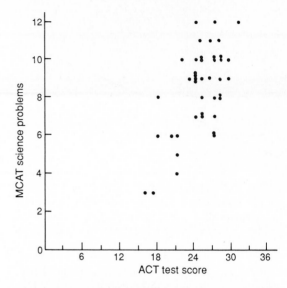

Figure 10-7. Scatterplot of ACT and MCAT test scores. (Adapted, with permission, from Dawson-Saunders B, Paiva RE, Doolen DR: Using ACT scores and grade-point averages to predict students' MCAT scores. *J Med Educ* 1986;**61**:681.)

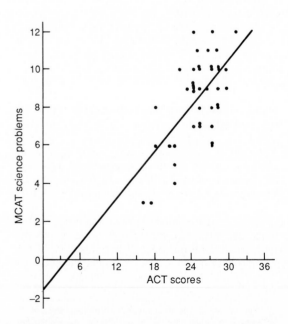

Figure 10-8. Regression line for regression of MCAT on ACT. (Adapted, with permission, from Dawson-Saunders B, Paiva RE, Doolen DR: Using ACT scores and grade-point averages to predict students' MCAT scores. *J Med Educ* 1986;**61**:681.)

the ACT score. Fig 10–8 illustrates the regression line drawn through the observations. The regression equation has a negative intercept of −1.61, so that a student who scores zero on the ACT composite is predicted to score −1.61 on the MCAT Science Problems subtest (even though one cannot obtain a negative score on this examination). The slope of 0.406 indicates that each time a student's ACT score increases by 1 point, his or her predicted MCAT score increases by approximately 0.4 point. Whether there is a *significant* relationship between ACT scores and MCAT scores is discussed in the next section.

10.5.3 Assumptions & Inferences in Regression

In the previous section, we worked with a sample of observations instead of the population of observations. Just as the sample mean $\overline{X}$ is an estimate of the population mean μ, the regression line determined from the formulas for a and b in the previous section is an estimate of the regression equation for the underlying population—ie, in the population,

$$Y = \beta_0 + \beta_1 X + e$$

As in Chapter 7, in which we performed statistical tests to determine how likely it was that the observed differences between two means, $\overline{X}_1 - \overline{X}_2$, occurred by chance, in regression analysis, we must perform statistical tests to determine how likely any observed relationship between X and Y variables is. Again, there are two ways to approach the question—using hypothesis tests or forming confidence intervals. Before discussing these approaches, however, we present the assumptions required in regression analysis.

If we are to use a regression equation, the observations must have certain properties. Thus, for each value of the X variable, the Y variable is assumed to have a normal distribution, and the mean of the distribution is assumed to be the predicted value, Y'. In addition, no matter what the value of the X variable is, the standard deviation of Y is assumed to be the same. These assumptions are rather like imagining a large number of individual normal distributions of the Y variable, all of the same size, one for each value of X. The assumption of this equal variation in the Y's across the entire range of the X's is called **homogeneity,** or **homoscedasticity.** It is analogous to the assumption of equal variances (homogeneous variances) in the t test for independent groups, as discussed in Chapter 7. If the variability of the Y's differs by a large amount, transformations or other modifications must be made before regression can be performed. Fig 10–9 is a schematic drawing of these assumptions.

Another assumption of regression is the straight-line, or linear, assumption. It requires that the mean values of Y corresponding to various values of X fall on a straight line, as indicated in Fig 10–9. Finally, the values of Y are assumed to be independent from each other. This assumption is not met when repeated measurements are made on the same subjects; ie, a subject's measure at one time is not independent from the measure at another time.

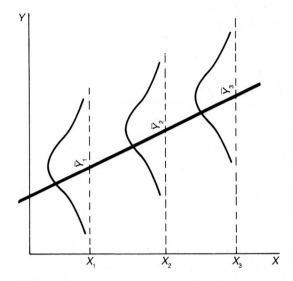

Figure 10-9. Assumption of equal variances of Y's for all values of X.

As with many other statistical methods, regression is generally robust and may still be used in most situations in which the assumptions are not met, as long as the measurements are fairly reliable and the correct regression model is used. (Other regression models are discussed in Chapter 12). Meeting the regression assumptions generally causes fewer problems in experiments or clinical trials than in observational studies because reliability of the measurements tends to be greater in experimental studies. However, special procedures can be used when the assumptions are seriously violated; and as in ANOVA, researchers should seek a statistician's advice before using regression if there is any question about its applicability.

10.5.3.a The Standard Error of the Estimate: Regression lines, like other statistics, can vary. After all, the regression equation computed for any one sample of observations is only an estimate of the true population regression equation. If other samples are chosen from the population and regression equations calculated for each sample, these equations will vary from one sample to another with respect to both their slopes and their intercepts. An estimate of this variation is symbolized $s_{Y.X}$ and is called the standard error of regression, or the **standard error of the estimate.** It is based on the squared deviations of the predicted Y's from the actual Y's and is found as follows:

$$s_{Y.X} = \sqrt{\frac{\Sigma(Y - Y')^2}{n - 2}}$$

The computation of this formula is quite tedious; and although more user-friendly computational forms exist, we assume that a computer program is used to calculate the standard error of the estimate. Because both the slope and the intercept can vary, it makes sense to perform a statistical test on each one. In each situation, a t test can be used, and the standard error of the estimate is part of the formula. It is also used in determining confidence limits. To present these formulas and the logic involved in testing the slope and the intercept, we illustrate the test of hypothesis for the intercept and the calculation of a confidence interval for the slope, using the ACT–MCAT regression equation determined above.

10.5.3.b Inferences About the Intercept: To test the hypothesis that the intercept departs significantly from 0, we use the following procedure.

Step 1. H_0: $\beta_0 = 0$ **(The intercepts are equal.)**

H_1: $\beta_0 \neq 0$ **(The intercepts are not equal.)**

Step 2. Since the null hypothesis is a test of whether the intercept is zero, the t ratio may be used if the assumptions are met. The t ratio is

$$t = \frac{a - \beta_0}{\sqrt{s_{Y\cdot x}^2 [1/n + \overline{X}^2/\Sigma(X - \overline{X})^2]}}$$

Step 3. Let us use α equal to the traditional value .05 for this example.

Step 4. The degrees of freedom are $n - 2 = 42 - 2 = 40$. The value of the t distribution with 40 degrees of freedom that divides the area into the central 95% and the combined upper and lower 5% is 2.021 (from Table A–3). Therefore, we will reject the null hypothesis of a zero intercept if (the absolute value of) the observed value of t is greater than 2.021.

Step 5. The calculation follows; a computer program has been used to calculate $s_{Y\cdot X} = 1.82$.

$$t = \frac{-1.61 - 0}{\sqrt{1.82^2(1/42 + 24.667^2/487.4)}}$$

$$= -0.784$$

Step 6. The absolute value of the observed t ratio is 0.784, which is not greater than 2.021. Therefore, the null hypothesis of a zero intercept cannot be rejected. We conclude that there is not sufficient evidence to show that the intercept is nonzero for the regression of MCAT scores on ACT scores; ie, the observed value of -1.61 could occur by chance when the intercept is zero.

10.5.3.c Inferences About the Regression Coefficient: Next, let us determine a 95% confidence interval for the population regression coefficient β_1. The interval is given by

$$b \pm t(n - 2) \sqrt{s_{Y\cdot x}^2 \left[\frac{1}{\Sigma(X - \overline{X})^2} \right]}$$

$$= 0.406 \pm 2.021 \sqrt{(1.82)^2 \left(\frac{1}{487.4} \right)}$$

or

$$= 0.239 \text{ to } 0.573$$

Because the interval excludes zero, we can be 95% confident that the regression coefficient is not zero but that it is between 0.239 and 0.573. Since the regression coefficient is significantly greater than zero, can the correlation coefficient be equal to zero (see Exercise 4)? The relationship between b and r illustrated earlier should convince you of the equivalence of the results obtained with testing the significance of correlation and the regression coefficient. In fact, many authors in the medical literature perform a regression analysis and then report the P-values for the correlation coefficient to indicate significance.

10.5.3.d. Predicting With the Regression Equation: One of the primary reasons for obtaining a regression equation is to predict future scores for a groups of subjects (or for individual subjects). For example, a premed advisor may want to predict the mean MCAT Science Problems score for a group of students who took special course work. Or the advisor may wish to predict the score for a particular student. In either case, the variability associated with the regression line must be reflected in the prediction. The 95% confidence interval for a *predicted mean Y* in a group of subjects is

$$Y' \pm t(n - 2) \sqrt{s_{Y\cdot x}^2 \left[\frac{1}{n} + \frac{(X - \overline{X})^2}{\Sigma(X - \overline{X})^2} \right]}$$

The 95% confidence interval for *predicting an individual subject's score* is

$$Y' \pm t(n - 2) \sqrt{s_{Y\cdot x}^2 \left[1 + \frac{1}{n} + \frac{(X - \overline{X})^2}{\Sigma(X - \overline{X})^2} \right]}$$

Comparing these two formulas, we see that the confidence interval for an individual prediction is wider than the interval for the mean of a group of individuals; 1 is added to the standard error term for the individual case. This result makes sense, because for a given value of X, there will be greater variation in the scores of individuals than in the mean scores of groups of individuals. Note also that the numerator of the third term in the standard error is the squared deviation of X from $\overline{X}$. Therefore, the size of the standard error depends on how close the X score is to $\overline{X}$; and the closer X is to its mean, the more accurate the prediction of Y is. For values of X quite far from $\overline{X}$, there is considerable variability in predicting the Y score.

Table 10–7 gives 95% confidence intervals associated with predicted mean MCAT scores and predicted MCAT scores for an individual corresponding to several different ACT scores (and for the mean ACT score in this sample of students). Several insights about regression analysis can be gained by examining this table. First, note the differences in magnitude of the standard errors associated with the predicted mean MCAT and associated with individual MCAT scores: The standard errors are much larger when we predict individual scores than when we predict the mean score. In fact, the standard error for individuals can never be less than one because of the additional 1 in the formula. Also note that the standard errors and confidence intervals take on their smallest values when $X = 24.667$, the mean of the ACT scores. As X departs in either direction from 24.667, the standard errors and confidence intervals become increasingly larger, reflecting the squared difference between X and $\overline{X}$. If the confidence intervals are plotted as **confidence bands** about the regression line, they are closest to the line at the mean of X and curve away from it in both directions on each side of $\overline{X}$. Fig 10–10 shows the graph of the confidence bands.

Table 10–7 illustrates another interesting feature of the regression equation. When the mean of X is used in the regression equation, the predicted Y' is the mean of Y. Therefore, the regression line goes through the point $(\overline{X}, \overline{Y})$. Some elementary statistics texts give guidelines on fitting the least squares regression line by hand, and one of the points the line must go through is $(\overline{X}, \overline{Y})$.

Now we can see why confidence bands about the regression line are curved. The error in the intercept means that the true regression line can be either above or below the line calculated for the

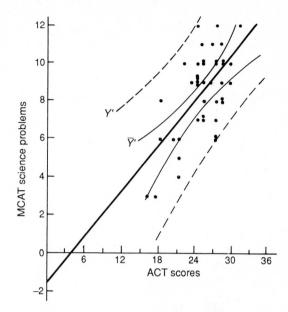

Figure 10–10. Confidence bands for regression line for MCAT on ACT scores. (Adapted, with permission, from Dawson-Saunders B, Paiva RE, Doolen DR: Using ACT scores and grade-point averages to predict students' MCAT scores. *J Med Educ* 1986;**61**:681.)

sample observations, although it will maintain the same orientation (slope). And the error in measuring the slope means that the true regression line can rotate about the point $(\overline{X}, \overline{Y})$ to a certain degree. The combination of these two errors results in the *concave* confidence bands illustrated in Fig 10–10. It is not uncommon in the medical literature, however, to see regression lines with confidence bands that are parallel rather than curved. These confidence bands are incorrect, although they may correspond to standard errors or to confidence intervals at their narrowest distance from the regression line. Readers cannot always tell what the parallel lines mean, in which case they indicate that the authors are less than expert in their application of statistical methods.

Table 10–7. 95% Confidence intervals for predicted mean MCAT scores and predicted individual MCAT scores.

Act X	Predicting Mean MCAT			Predicting Individual MCAT	
	Y'	SE[1]	Confidence Intervals	SE[1]	Confidence Intervals
16	4.9	0.77	3.3 to 6.5	1.98	0.9 to 8.9
20	6.5	0.48	5.5 to 7.5	1.88	2.7 to 10.3
22	7.3	0.36	6.6 to 8.0	1.86	3.5 to 11.1
24.667	8.4	0.28	7.8 to 9.0	1.84	4.7 to 12.1
27	9.4	0.34	8.7 to 10.1	1.85	5.7 to 13.1
29	10.2	0.45	9.3 to 11.1	1.88	6.4 to 14.0
33	11.8	0.74	10.3 to 13.3	1.97	7.8 to 15.8

[1]SE denotes standard error.

10.5.4 Comparing Two Regression Lines

Sometimes investigators wish to compare two regression lines to see whether they are the same. For example, the investigators in Presenting Problem 1 could have compared the regression line predicting C1q-binding complex from IgG complex in patients with systemic lupus erythematosus (SLE) with the regression line for children with steroid-responsive nephrotic syndrome (SRNS). Although they did not do this analysis, they presented a scatterplot with observations from both sets of patients, so the statistical question is reasonable. The scatterplot is shown in Fig 10–11; the two sets of observations appear to be quite different. (Also note that parallel lines are drawn about the regression line instead of the correct, curved lines.) The regression lines for these two groups (omitting the calculations) follow.

For SLE patients: $Y' = 2.17 + 0.94X$
For SRNS patients: $Y' = 7.62 + 0.10X$

However, as we have seen from the standard error of the estimate, the regression lines in samples of

Correlation between C1q binding and IgG-containing complexes in children and adults with systematic lupus erythematosus (SLE) and nephritis (o) and in children with steroid-responsive nephrotic syndrome (SRNS) (•). The regression line and 95% confidence limits for the patients with lupus are shown. The limit of normal for both systems is indicated.

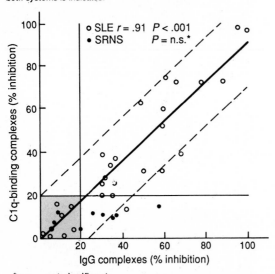

*n.s. = not significant.

Figure 10–11. Scatterplot of C1q-binding complexes and IgG complexes in patients with SLE (open circles) and SRNS (dark circles), illustrating possible differences in regression lines for SLE and SRNS patients. (Reproduced, with permission, from Levinsky RJ et al: Circulating immune complexes in steroid-responsive nephrotic syndrome. *N Engl J Med* 1978;**298**:126.)

SLE and SRNS patients can be expected to vary from one sample to another, and a statistical test must be done in order to conclude that the observed regression lines differ significantly.

When we compare two regression lines, four situations can occur, as illustrated in Fig 10–12. In Fig 10–12*A,* the slopes of the regression lines are the same and the intercepts differ. This situation occurs, for instance, in blood pressure measurements regressed on age in men and women; ie, the relationship between blood pressure and age is similar for men and women (equal slopes), but men tend to have higher blood pressure levels at all ages than women have (higher intercept for men).

Fig 10–12*B* illustrates just the opposite situation: The intercepts are equal but the slopes are different. This pattern may describe, say, the regression of platelet count on number of days following bone marrow transplantation in two groups of patients: those for whom adjuvant therapy results in remission of the underlying disease and those for whom the disease remains active. That is, prior to and immediately after transplantation, the platelet count is similar for both groups (equal intercepts); but at some time after transplantation, the platelet count remains steady for patients in remission and begins to decrease for patients not in remission (more negative slope for patients with active disease).

The apparent situation illustrated by the observations on SLE and SRNS patients from Presenting Problem 1 is similar to Fig 10–12*C,* in which

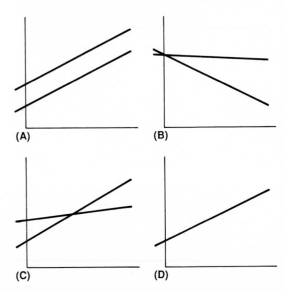

Figure 10–12. Illustration of ways regression lines can differ. (*A*) Equal slopes and different intercepts. (*B*) Equal intercepts and different slopes. (*C*) Different slopes and different intercepts. (*D*) Equal slopes and equal intercepts.

both the intercept and the slope of the regression lines differ. Finally, in Fig 10–12*D,* the regression lines are coincident; both intercepts and slopes are equal. This situation occurs in many situations in medicine and is considered to be the expected pattern (the null hypothesis), until it is shown not to apply.

From the situations illustrated in Fig 10–12, we can see that there are three possible statistical questions that can be asked:

1. Are the slopes equal?
2. Are the intercepts equal?
3. Are both the slopes and the intercepts equal?

Statistical tests based on the *t* distribution can be used for the first two questions; these tests are illustrated in Kleinbaum and Kupper (1978, Chap 8). However, as Kleinbaum and Kupper point out, the preferred approach is to use regression models for more than one independent variable—a procedure called multiple regression—to answer these questions. The procedure consists of pooling observations from both samples of subjects (eg, observations on both SLE and SRNS patients) and computing one regression line for the combined data. In addition, other regression coefficients are determined, coefficients that indicate whether it matters to which group the observations belong. This approach uses a technique called dummy coding to indicate the group to which an observation belongs, and it is discussed in greater length in Chapter 12 in the presentation of multiple regression.

10.6 USE OF CORRELATION & REGRESSION

Some of the characteristics of correlation and regression have been noted throughout the discussions in this chapter, and we will recapitulate and reemphasize them here and also mention other features as well. An important point to note is that correlation and regression describe *only* linear relationships. If correlation coefficients or regression equations are calculated blindly, without examining plots of the data, investigators can miss very strong, but nonlinear, relationships.

10.6.1 Analysis of Residuals

A procedure useful in evaluating the fit of the regression equation is the analysis of residuals (Pedhazur, 1982). A **residual** is the difference between actual value of Y and the predicted value of Y', or $Y - Y'$. It is the part of Y that is not predicted by X (the part left over, or the residual). The residual values, $Y - Y'$, on the Y-axis are plotted against the X values on the X-axis. The mean of the residuals is zero and because the slope has been "subtracted" in the process of calculating the residuals, the correlation between them and the X values should also be zero.

Stated another way, if the regression model provides a good fit to the data, as in Fig 10–13*A,* the values of the residuals are not related to the values of X; therefore, a plot of the residuals and the

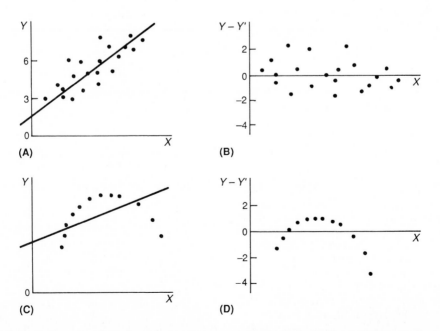

Figure 10–13. Illustration of analysis of residuals. (*A*) Linear relationship between X and Y. (*B*) Residuals versus values of X for relation in (*A*). (*C*) Curvilinear relationship between X and Y. (*D*) Residuals versus values of X for relation in (*C*).

X values should resemble a patternless scatter of points corresponding to Fig 10-13*B,* in which there is no correlation between the residuals and the values of X. If, in contrast, there is a curvilinear relationship between Y and X, such as in Fig 10-13*C,* the residuals are negative for both small values and large values of X, because the corresponding values of Y fall below a regression line drawn through the data; but they are positive for mid-sized values of X because the corresponding values of Y fall above the regression line. In this case, instead of obtaining a random scatter, we obtain a plot like the curve in Fig 10-13*D,* with the values of the residuals related to the values of X. Other patterns can be used by statisticians to help diagnose problems, such as a lack of homoscedasticity or various types of nonlinearity.

10.6.2 Dealing With Nonlinear Observations

Several alternative actions can be taken if there are serious problems with nonlinearity of data. As we discussed previously, a **transformation** of the data may make the relationship linear, and regular regression methods can then be used on the transformed data. Another possibility, especially for a curve, is to fit a straight line to one part of the curve and a second straight line to another part of the curve, a procedure called *piecewise linear regression.* In this case, one regression equation is used with all values of X less than a given value, and the second equation is used with all values of X greater than the given value. A third strategy, also useful for curves, is to perform polynomial regression; this technique is briefly discussed in Chapter 12. Finally, more complex approaches called nonlinear regression may be used (Snedecor and Cochran, 1980).

10.6.3 Confidence Bands

The confidence bands for the regression line, calculated in Section 10.5.3.d, demonstrate that the error of prediction becomes increasingly large as the value of X departs from the mean of the X observations. Predictions are most stable when observations near the middle of the distribution are used; therefore, it is risky to try to predict Y for extremely large or small values of X. This feature of regression illustrates why economists and others who try to predict future events have such difficulty; they are trying to use information based on past performance to predict values beyond the range of X and into the future.

The problem inherent in trying to predict beyond the range of X was illustrated with the ACT–MCAT data. The intercept corresponds to the predicted Y when X is zero. The value of the intercept for the ACT–MCAT data was -1.61, corresponding to a prediction of a negative score on the MCAT examination if zero is scored on ACT. However, zero is an extreme ACT score—one that rarely happens, if ever—and trying to predict beyond the range of observed ACT scores leads to an impossible result. Consistent with this interpretation, the statistical test for the intercept indicates that a value of -1.61 can easily occur by chance when the intercept is actually zero.

10.6.4 Regression Toward the Mean

The phenomenon called *regression toward the mean* is a problem that often occurs in applied research and often goes unrecognized. A good illustration of regression toward the mean occurred in the MRFIT study (Multiple Risk Factor Intervention Trial Research Group, 1982) designed to evaluate the effect of diet and exercise on blood pressure in men with mild hypertension. To be eligible to participate in the study, men had to have a diastolic blood pressure of 90 mm Hg or higher. The eligible subjects were then assigned to either the treatment arm of the study, consisting of programs to encourage appropriate diet and exercise, or to the control arm, consisting of typical care.

To illustrate the concept of regression toward the mean, we consider the hypothetical data in Table 10-8 for diastolic blood pressure in 12 men. If these men were being screened for the MRFIT study, only subjects 7 through 12 would be accepted; subjects 1 through 6 would not be eligible because their diastolic pressure is less than 90 mm Hg. Suppose all subjects had another blood pressure measurement some time later. Since there is considerable variation in a person's blood pressure from one reading to another, about half the men would be expected to have higher blood pressures and about half would have lower blood pressures, owing to random variation. Regression toward the mean tells us that those men who had

Table 10-8. Hypothetical data on diastolic blood pressure to illustrate regression toward the mean.

Subject	Baseline	Repeat
1	78	80
2	80	81
3	82	82
4	84	86
5	86	85
6	88	90
7	90	88
8	92	91
9	94	95
10	96	95
11	98	97
12	100	98

lower pressures on the first reading are more likely to have higher pressures on the second reading. Similarly, men who had a diastolic blood pressure of 90 or more on the first reading are more likely to have lower pressures on the second reading. If the entire sample of men is measured, the increases and decreases tend to cancel each other. However, if only a subset of the subjects is examined again—eg, the men with initial diastolic pressures greater than 90—the mean blood pressure will appear to have dropped, when in fact it has not.

Regression towards the mean can result in a treatment or procedure appearing to be of value when it has had no actual effect; as we discuss in Chapter 15, the use of a control group will guard against this effect. We should point out the investigators in the MRFIT study were aware of the problem of regression toward the mean and discussed precautions they took to reduce its effect.

10.6.5 Common Errors in Regression

A common error in regression analysis occurs when multiple observations on the same subject are treated as though they were independent. As a simple example, consider 10 patients who have their weight and skinfold measurements recorded prior to a low-calorie diet. We may reasonably expect a moderately positive relationship between weight and skinfold thickness. Suppose, however, that the same 10 patients are weighed and measured again after 6 weeks on the diet. If all 20 observations are treated as though they were independent, there will be several effects. First, the sample size will appear to be increased, which means that we are more likely to obtain a significant correlation. Second, the relationship between weight and skinfold thickness in the same person is somewhat stable across minor shifts in weight. Therefore, using both observations has the same effect as using duplicate measures, spuriously increasing the size of the correlation.

The magnitude of the correlation can also be spuriously increased by combining two different groups. For example, consider the relationship between height and weight. Suppose the heights and weights of 10 men and 10 women are recorded, and the correlation between height and weight is calculated for the combined samples. Fig 10–14 illustrates what the scatterplot might look like and indicates the problem that results from combining men and women in one sample. The relationship between height and weight appears to be more significant in the combined sample than it is when measured in men and women separately. Much of the "significance" results because men tend both to weigh more and to be taller than women. Inappropriate conclusions often result from mixing two different populations, and this is another

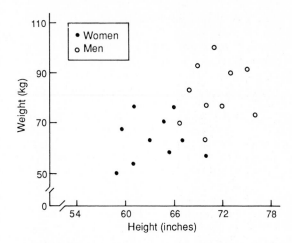

Figure 10-14. Hypothetical data illustrating spurious correlation.

common error to watch for in the medical literature.

10.6.6 Comparing Correlation & Regression

Correlation and regression have some similarities and some differences. First, correlation is scale-independent, but regression is not. That is, the correlation between two characteristics, such as height and weight, is the same whether height is measured in centimeters or inches and weight in kilograms or pounds. The regression equation resulting from regressing weight on height, however, depends on which scales are being used; ie, predicting weight measured in kilograms from height measured in centimeters will give different values for a and b than predicting weight in pounds from height in inches.

Second, an important consequence of scale independence in correlation is the equivalence between the correlation between X and Y and the correlation between Y' and Y. These correlations are equal because the regression equation itself, $Y' = a + bX,$ is a simple rescaling of the X variable; ie, X is multiplied by a constant value b and then a constant amount a is added. Therefore, it does not matter whether a significance test is performed on the correlation r or on the regression coefficient b. In addition, the fact that the correlation between the original variables X and Y is equal to the correlation between Y and the values of Y' obtained from the regression equation (ie, $r_{X,Y} = r_{Y',Y}$) provides a useful alternative method for testing the significance of the regression, as we will see in Chapter 12. Finally, the slope of the regression line has the same sign (+ or −) as the correlation coefficient (see Exercise 5).

10.6.7 Multiple Regression

Multiple regression analysis is a straightforward generalization of simple regression for two or more independent (explanatory) variables used in the prediction equation. For example, in the study described in Presenting Problem 4, the investigators wanted to predict a medical student's score on the MCAT Science Problems subtest. Although they found that the composite score on the ACT test was significantly related to performance on MCAT, better prediction might be achieved if other performance measures, such as the student's grade point average (GPA), is considered as well. In this study, the investigators evaluated undergraduate GPA in science courses, undergraduate GPA in nonscience courses, and gender of the student, in addition to ACT score. The prediction equation obtained in the study was

Predicted Science
Problems score $=$ **0.18** × **ACT composite score**
+ 1.62 × science GPA
+ 1.02 if male −2.17

Multiple regression and other statistical methods based on regression are discussed in detail in Chapter 12.

10.7 COMPUTER PROGRAMS USING CORRELATION & REGRESSION

We will use output from several different programs in SPSS, SYSTAT, and MINITAB to illustrate analysis of the Presenting Problems in this chapter. A scatterplot of C1q-binding complex versus IgG-containing immune complex produced by SPSS is given in Fig 10–15. The "2" in the plot

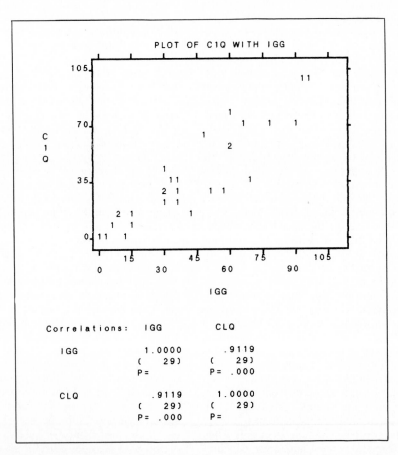

Figure 10–15. SPSS program illustrating a scatterplot and correlation coefficient, using observations on C1q-binding and IgG-containing complexes from Presenting Problem 1. (Adapted, with permission from Levinsky RJ et al: Circulating immune complexes in steroid-responsive nephrotic syndrome *N Engl J Med* 1978;**298**:126. SPSS is a registered trademark of SPSS, Inc. Used with permission.)

indicates that two observations have identical values for the plotted variables. The output from the SPSS program computing the correlation coefficient is presented in Fig 10–15 as well. The value obtained, .9119, agrees with our findings in Section 10.3.1. The SPSS program also has options to request a *P*-value for the statistical test comparing the correlation with zero. The output value $P = 0.000$ does not mean that the probability of observing $r = .9119$ is 0 but that it is less than .001, the smallest *P*-value SPSS prints. Thus, we would report these findings as $P < .001$.

The scatterplot produced by SYSTAT displays the relationship between the ratio of excreted protein and creatinine, labeled RATIO, and 24-hour excretion rate of protein, labeled PROTEXCR (see Fig 10–16). The plot from SYSTAT is much wider than the one from SPSS. The plot clearly shows the nature of the skewed distributions of both variables. SYSTAT also performs the calculations for the Spearman correlation coefficient. The value in the printout is .926, very close to the value .927 we obtained in Section 10.4.1. SYSTAT does not perform a statistical test or provide a *P*-

value for the Spearman correlation; so if they are desired, they must be done by hand.

Output from the MINITAB program illustrating the regression of MCAT scores on ACT scores (Presenting Problem 4) is given in Fig 10–17. We see that MINITAB uses the term Predictor to refer to both the Constant and the ACT score; it does so because the program can also be used to perform multiple regression with more than one predictor variable. The values for the slope and the intercept are given under the column heading "Coef" and agree, within round-off error, with the results we obtained in Section 10.3.2. MINITAB also gives the results of a *t* test for both the slope and the intercept, along with the *P*-values The *P*-value of .439 for the intercept agrees with our conclusion that the intercept was not different from zero, and the *p*-value of 0.000 is consistent with the confidence limits we found for the regression coefficient. As we noted above, the *P*-value is not really 0; it is simply smaller than .001, the smallest value printed by the computer program. MINI-TAB finds the value of the coefficient of determination, r^2, denoted "R = sq" in

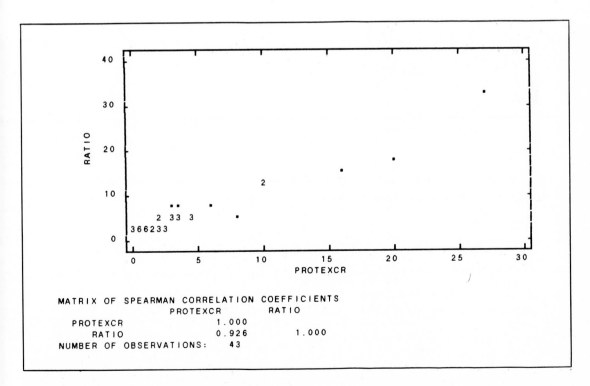

Figure 10–16. SYSTAT program illustrating scatterplot and Spearman correlation coefficient, using observations on ratio of protein and creatine in single-voided sample and excreted protein over 24-hour period from Presenting Problem 2. (Adapted, with permission, from Ginsberg JM et al: Use of single voided urine samples to estimate quantitative proteinuria. *N Engl J Med* 1983;**309**:1543. SYSTAT is a registered trademark of SYSTAT, Inc. Used with permission.)

```
The regression equation is
mcat = - 1.60 + 0.406 act

Predictor        Coef        Stdev       t-ratio        p
Constant       -1.600        2.049         -0.78      0.439
act            0.40561       0.08227        4.93      0.000

s = 1.816        R-sq = 37.8%       R-sq(adj) = 36.2%

Analysis of Variance

SOURCE          DF          SS          MS          F          p
Regression       1        80.175      80.175      24.31      0.000
Error           40       131.944       3.299
Total           41       212.119

Unusual Observations
Obs        act        mcat        Fit    Stdev.Fit    Residual    St.Resid
 18       17.0       3.000       5.295      0.690       -2.295      -1.37 X
 31       24.0      12.000       8.134      0.286        3.866       2.16R
 38       16.0       3.000       4.889      0.766       -1.889      -1.15 X

R denotes an obs. with a large st. resid.

X denotes an obs. whose X value gives it large influence.
```

Figure 10-17. MINITAB program illustrating calculation of regression equation and analysis of residuals, using observations on predicting MCAT performance from ACT scores from Presenting Problem 4. (Adapted, with permission, from Dawson-Saunders B, Paiva RE, Doolen DR: Using ACT scores and grade-point averages to predict students' MCAT scores. *J Med Educ* 1986;**61**:681. MINITAB is a registered trademark of Minitab, Inc. Used with permission.)

the output, as a percentage rather than a proportion. "R-sq(adj)" is the size of the coefficient of determination one would expect to see in another group of applicants; it is an indication of the stability of the relationship between ACT and MCAT scores.

An interesting component of the regression analysis produced by MINITAB is an analysis of variance. When ANOVA is done within the context of regression, the between sum of squares and the mean square are called regression sum of squares and mean square; the error term has the same name as in ANOVA. Note that the *F* value is significant, with a *P*-value equal to 0.000, the same as the *P*-value for the regression coefficient for ACT score. As it turns out, regression analysis and analysis of variance always lead to the same conclusion.

Another interesting component of the MINITAB printout is a listing of unusual observations. MINITAB lists three observations, subjects 18, 31, and 38, as being "unusual." Observation 31 is unusual because it has a large residual value, indicating that this observation does not fit the model very well. Subject 31 has a score of 24 on ACT, a value close to the middle of the distribution, and a score of 12 on MCAT, one of the highest in the group. MINITAB indicates that the other two observations have a large influence, as is clear from Fig 10–8; these two points are quite far from the majority of observations. A message like this one tells us to check and see whether observed values were incorrectly transcribed. We transcribed the correct values, so we merely note the influence of the unusual values and decide whether or not we wish to reanalyze the data excluding them from the sample.

10.8 SUMMARY

Four Presenting Problems were used in this chapter to illustrate the application of correlation and regression in medical studies. The findings from the study described in Presenting Problem 1 suggest that immune complexes may have a role in the pathogenesis of this steroid-responsive form of nephrotic syndrome, although the mechanisms of injury are not known. The authors caution us, however, that these immune complexes may be merely an epiphenomenon. Even though the correlation (.91) between these two immune complexes is high in patients with systemic lupus erythematosus, this result in and of itself does not establish a cause-and-effect relationship.

The investigators in the study described in Presenting Problem 2 found a high correlation between the protein creatinine ratio measured in a single daytime urine sample and protein excretion measured in a 24-hour urine collection. They found that the protein/creatinine ratio tended to be lower when measured during a period of recumbency and that proper interpretation of the ratio must take into account that the rate of urinary protein excretion is modified by the rate of creatinine excretion. They concluded that determination of this ratio can, in most clinical settings, replace the measurement of protein excretion in a 24-hour period. They did not, however, acknowledge the extreme values in their samples or show how these values might affect the relationship. Also, they did not present the results of a regression analysis and give the magnitude of the standard error of the estimate, two procedures that would seem to be required if the protein/creatinine ratio is recommended as a replacement of 24-hour protein excretion.

In the study described in Presenting Problem 3, the investigators determined that the 23.3% prevalence of serologic markers of hepatitis B in the anesthesiology workers was much greater than the 3–5% reported for the general population. Approximately 20% of the individuals without a history of hepatitis had positive markers for a subclinical infection. The authors used a chi-square test to evaluate the relationship between prevalence of positive markers and length of time since the worker graduated from school, an appropriate method for analyzing these data. They did not calculate odds ratios, as we did, although this approach leads to conclusions consistent with those from a chi-square test. From their results, the authors recommend that individuals administering anesthesia receive immunization against hepatitis B virus.

In the study described in Presenting Problem 4, the investigators were able to develop a regression equation to predict a medical school applicant's MCAT Science Problems score from his or her performance on the ACT examination 3 or 4 years earlier. In the study, the authors used a sample size of 197, and they found that a student's grade point average in science subjects, in addition to the ACT score, played a significant role in predicting all the MCAT subtest scores. In addition, they performed an analysis that demonstrated the stability of the regression equation in a different sample of students.

The flowchart for Chapter 10 in Appendix C summarizes the methods for measuring an association between two characteristics measured on the same subjects. Flowchart C–4 indicates how the methods depend upon the scale of measurement for the variables.

EXERCISES

1. This exercise, although complex, leads to an understanding of the coefficient of determination. (Because of the computations involved, answers but not complete solutions are given in Appendix B.)
   a. Compute the mean and the variance (squared standard deviation) of the C1q-binding complex data for Presenting Problem 1.
   b. Compute the regression equation to predict C1q-binding complex from IgG-containing complex. (*Hint:* Use $r = .91$ and the numerators of the standard deviations calculated in the process of calculating r in Section 10.3.1)
   c. Predict each C1q-binding complex value, using the regression equation.
   d. Find the residual—the difference between the actual C1q-binding complex value and the predicted value—for each patient in the study.
   e. Compute the mean predicted C1q, and confirm that it is the same value as the mean C1q found in part A.
   f. Compute the mean and the variance of the residuals. Confirm that the mean residual is zero.
   g. Verify that the coefficient of determination is equal to the proportion of variance accounted for by the prediction; ie, show that

$$r^2 = \frac{\text{Original variance} - \text{Residual variance}}{\text{Original variance}}$$

2. a. Perform a chi-square test of the significance of the relationship between length of employment and presence of hepatitis B marker, using the data in Table 10–5.
   b. Can a chi-square test be performed for the data in Table 10–4?

3. Show that the regression coefficient b can be obtained by multiplying the correlation by the ratio of the standard deviation of Y to the standard deviation of X.

4. Calculate the correlation between ACT composite scores and MCAT Science Problems scores, using the results in Section 10.5.2 to find b and s_X. The value of s_Y is 2.275.

5. Develop an intuitive argument to explain why the sign of the correlation coefficient and the sign of the slope of the regression line are the same.

6. In the study described in Presenting Problem 1, the investigators found that the raised levels of IgG immune complexes in 18 children who had had a relapse of their steroid-responsive nephrotic syndrome differed significantly from levels in 12 controls and in 23 children whose disease was in remission (both $P < .001$).

What is the most appropriate statistical method for analyzing these differences?

7. In the study described in Presenting Problem 2, the investigators obtained creatinine and protein measurements for 30 normal control patients as well as for the 46 patients attending their renal clinic. What was the reason for collecting information on the control patients, and what is the most appropriate statistical method to make comparisons between the two groups?

8. Fig 10–18 is from the study described in Presenting Problem 2. What is the authors' purpose in displaying this graph? Do you detect any errors?

9. Explain why the mean $\overline{Y}'$ of the predicted values of Y is equal to $\overline{Y}$.

10. Goldsmith et al (1985) examined 35 patients with hemophilia to determine whether there is a relationship between impaired cell-mediated immunity and the amount of factor concentrate used. In one of their studies, the ratio of OKT4 (helper T cells) to OKT8 (suppressor-cytotoxic T cells) was formed, and the logarithm of this ratio was regressed on the logarithm of lifetime concentrate use. See Fig 10–19.

   a. Why were logarithm scales used for both variables?

Linear regression of log OKT4/OKT8 on log lifetime concentrate usage (in units of 10^3) with 95% confidence bands (dashed lines) for single observation ($P = .006$, $r = -.453$).

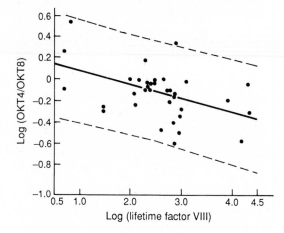

Figure 10-19. Regression of logarithm of OKT4:OKT8 on logarithm of factor concentrate use (Reproduced, with permission, from Goldsmith JM et al: Sequential clinical and immunologic abnormalities in hemophiliacs. *Arch Intern Med* 1985;**145**:431.)

b. Interpret the correlation.

c. What do the confidence bands mean?

11. Helmrich et al (1987) conducted a study to assess the risk of deep vein thrombosis and pulmonary embolism in relation to the use of oral contraceptives. They were especially interested in the risk associated with low dosage (less than 50 μg estrogen) and confined their study to women under the age of 50 years. They administered standard questionnaires to women admitted to the hospital for deep vein thrombosis or pulmonary embolism as well as to a control set of women admitted for trauma and upper respiratory infections to determine their history and current use of oral contraceptives. Twenty of the 61 cases and 121 of the 1278 controls had used oral contraceptives in the previous month.

   a. What was the research design used in this study?

   b. Find 95% confidence limits for the odds ratio for these data.

   c. The authors reported an age-adjusted odds ratio of 8.1 with 95% confidence limits of 3.7 and 18. Interpret these results.

12. Determine 99% confidence limits for the odds ratio of 4.00 for the risk of hepatitis B in workers with employment in anesthesiology 20 or more years. Compare the answer with that found in Section 10.4.2.

13. The graphs in Fig 10–20 were published in the

Time course throughout the day (abscissa) for the slopes expressing the relation between the protein/creatinine ratio ([Pr/Cr]$_u$) and 24-hour protein excretion per 1.73 m^2 (ordinate). Urine samples were grouped as follows: A, first sample obtained after initial voiding upon arising; B, all samples voided between 12:00 noon and 6:00 p.m.; D, samples voided between 6:00 p.m. and 12:00 midnight; E, samples voided from midnight up to sample A. Bars encompass 95% confidence limits. Slopes of the A and E samples were significantly different ($P < .05$) from those of the B and C samples.

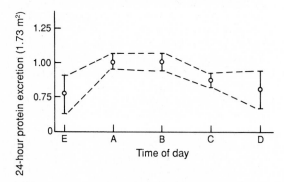

Figure 10-18. Graph of time course throughout day (*X*-axis) versus ratio of urine protein to creatinine and 24-hour protein excretion (*Y*-axis) from Presenting Problem 2. (Reproduced, with permission, from Ginsberg JM et al: Use of single voided urine samples to estimate quantitative proteinuria. *N Engl J Med* 1983;**309**:1543.)

Relation between age and hepatic secretion of cholesterol (*n* = 22), total bile acid synthesis (*n* = 18), and size of cholic acid pool (*n* = 18). Open circles denote women and closed circles men. To convert values for bile acids and cholesterol to milligrams, multiply by 400 and 0.39, respectively.

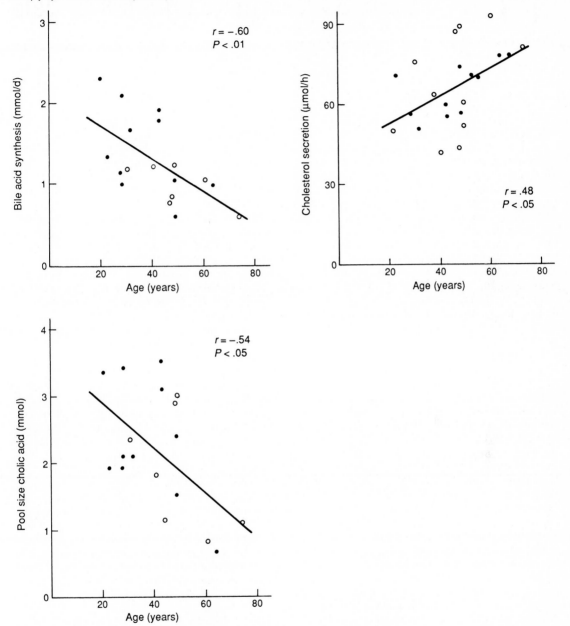

Figure 10-20. Scatterplots and regression lines for relation between age and hepatic secretion of cholesterol, total bile acid synthesis, and size of cholic acid pool for women (open circles) and men (dark circles) from Presenting Problem 3 in Chapter 3. (Reproduced, with permission, from Einarsson K et al: Influence of age on secretion of cholesterol and synthesis of bile acids by the liver. *N Engl J Med* 1985;**313**:277.)

study by Einarsson et al (1985) and summarized in Chapter 3, Presenting Problem 3.

a. Which graph exhibits the strongest relationship with age?

b. Which variable would be best predicted from a patient's age?

c. Do the relationships between the variables and age appear to be the same for men and women; ie, was it appropriate to combine the observations for men and women in the same figure?

11 Methods for Analyzing Survival Data

PRESENTING PROBLEMS

Presenting Problem 1. A renal transplant surgeon collected data on two groups of patients who received kidney transplants. The two groups are similar in all respects except that the 31 patients undergoing transplantation in 1978 and 1979 received azothioprine but the 21 patients undergoing transplantation in 1984 were treated with cyclosporine, a newer immunomodulatory substance used to prevent rejection of the transplanted organ. Data for 1978–1979 are given in Table 11–1; data for 1984 are given in Table 11–6. The surgeon wants to analyze and compare survival of the transplanted kidneys in the two groups of patients.

Presenting Problem 2. Chronic lymphocytic leukemia (CLL) is a disease usually affecting older people and is characterized by neoplastic transformation of the lymphoid system. Different types of lymphocytes can be involved, but B cell CLL is the most common form. The disease may have a variable course. The prognosis appears to be related to the extent of organ involvement at the time of diagnosis and presence or absence of thrombocytopenia and anemia. Several studies have shown cytogenic abnormalities of lymphocytes stimulated with B cell mitogen. An extra chromosome 12 (trisomy 12) is the most frequent abnormal karyotype in B cell CLL.

Han et al (1984) studied the occurrence of B lymphocyte chromosomal abnormalities and their relationship with clinical stage, progression of disease, and survival. Among the 53 patients in the cohort study, 21 (40%) had abnormal karyotypes. Trisomy 12 was the only abnormality found in 8; trisomy 12 in combination with other abnormalities occurred in 5; and other abnormalities without trisomy 12 were found in 8. There was no relationship between chromosomal abnormalities and duration of disease or treatment status. The investigators did find, however, that patients with more advanced stages of disease (Stage III or Stage IV) had a significantly higher prevalence of abnormal karyotypes. They also studied the survival distributions of these patients according to karyotype. Survival from time of diagnosis was significantly longer in the group with normal karyotypes than in those with abnormalities. Patients with trisomy 12 only appeared to have a survival similar to that of patients with normal karyotypes. Patients with other types of chromosomal abnormalities had a significantly shorter survival. Survival curves from their study are shown in Fig 11–7 and must be interpreted.

Presenting Problem 3. Many experts consider radical mastectomy to be the appropriate therapy for nonmetastatic breast cancer. However, other treatment protocols involving the use of less extreme surgical procedures—including modified radical mastectomy and total, or simple, mastectomy—have demonstrated survival times equal to that of radical mastectomy. Fisher et al (1985) published their data from a randomized trial to evaluate the use of segmental mastectomy (removal of primary tissue and a minimal amount of surrounding tissue to ensure that specimen margins are tumor-free) in the treatment of Stage I and Stage II breast tumors less than 4 cm. In this study, women were randomly assigned to one of three treatment groups: (1) total mastectomy, (2) segmental mastectomy, or (3) segmental mastectomy followed by breast irradiation. Axillary dissection was done in all treatment groups. Patients with one or more positive axillary nodes received systemic adjuvant chemotherapy regardless of treatment group. The mean duration of follow-up was 39 months (5–99 months). Data were available on 1843 women for analysis to determine whether length of survival depends on treatment given; the results are presented in Fig 11–8. What conclusions can we draw from these results?

11.1 PURPOSE OF THE CHAPTER

Many studies in medicine are designed to determine whether a new medication, a new treatment, or a new procedure will perform better than the one in current use. Although measures of short-term effects are of interest, long-term outcomes, such as mortality and major morbidity, are also important. For example, the CASS study discussed in Chapters 2 and 5 (Cass, 1983) was designed to assess survival and several indicators of quality of life among patients who had surgical treatment versus those who had medical treatment for stable ischemic heart disease. In Presenting

Problem 1, the investigator with the data on kidney transplants wants to examine survival of the kidney in patients who have had transplants. In both studies, there is a dichotomous outcome, survival or death of the patient and retention or loss of the kidney, and the desire is to estimate the length of time patients survive or retain their kidneys with specific types of treatment. Sometimes, the study focuses on comparing survival times for two or more groups of patients, as in the CASS study.

The methods of data analysis discussed in previous chapters are not appropriate for measuring length of survival time for two reasons. First, investigators frequently must analyze data before all patients have died; otherwise, it may be many years before they know which treatment is better. When analysis of survival is done while some patients in the study are still living, the observations on these patients are called **censored observations,** because we do not know how long these patients will remain alive. Fig 11–1 illustrates a situation in which observations on patients B and E are censored.

The second reason special methods are needed to analyze survival data is that patients do not typically begin treatment or enter the study at the same time, as they did for Fig 11–1. For example, in the kidney transplant study, not all patients had surgery on the same day; patients entered the study at different times. When the entry time for patients is not simultaneous and when some patients are still in the study when the analysis is done, the data are called *progressively* censored. Fig 11–2 shows results for a study with progressively censored observations. The study began at time 0 months with patient A; then, patient B entered the study at time 7 months; patient C entered at time 8 months; etc. Patients B and E are still alive at the time the data are analyzed at 40 months.

Analysis of survival times is sometimes called **actuarial,** or **life table, analysis.** Historically, life tables were first used by astronomer Edmund Halley (of Halley's comet fame) in the seventeenth

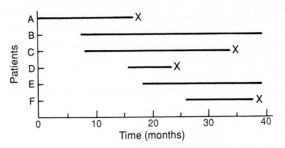

Figure 11-2. Example of progressively censored observations (X means patient died).

century to describe survival times of residents of a town. Since then, these methods have been used in various ways. For example, life insurance companies use life tables to determine the life expectancy of individuals, and this information is subsequently used to establish premium schedules. Insurers generally use cross-sectional data about how long people of different age groups are expected to live, and this information is used to develop a *current* life table. In medicine, however, most studies of survival use *cohort* life tables, in which the same group of subjects is followed for a given period of time. The data for life tables may come from cohort studies (either prospective or historical) or from clinical trials; the key feature is that the same group of subjects is followed for a prescribed amount of time.

Analysis of survival data occurs in varying degrees in the medical literature, depending on the specialty of the journal. For example, these methods are used in only about 1% of articles published in psychiatry (Hokanson et al, 1986) and pathology (Hokanson, Ladoulis, et al, 1987); they are used in 12% of surgical articles (Reznick, Dawson-Saunders, & Folse, 1987) and more than 14% of oncology articles (Hokanson, Luttman, & Weiss, 1986).

In this chapter, we examine two commonly used methods to determine survival curves, life (actuarial) curves, and Kaplan-Meier curves as well as a method to determine confidence limits about the curves. After discussing the concept of hazard rate, we examine ways survival curves may be compared with one another. Two methods used extensively in the medical literature are illustrated, the generalized Wilcoxon test and the logrank test, using data from Presenting Problem 1. Then, survival analyses published by the authors of the studies described in Presenting Problems 2 and 3 are examined and interpreted. Finally, computer programs for survival analysis are illustrated. As with many statistical procedures, the computations involved in estimating survival curves and comparing them are somewhat tedious; we employ small samples to illustrate the logic of the

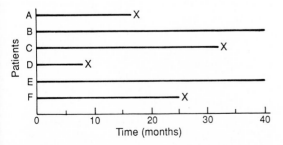

Figure 11-1. Example of censored observations (X means patient died).

method, assuming that computer programs will always be used to perform these analyses. Before illustrating the methods for analyzing survival data, we consider briefly why some intuitive methods are not very useful or appropriate. To illustrate these points, we use the data for kidney retention and loss from rejection in Table 11–1.

11.2 WHY SPECIAL METHODS ARE NEEDED TO ANALYZE SURVIVAL DATA

Colton (1974, 238–241) gives a creative analysis of some simple methods to analyze survival data; the arguments presented in this section are modeled on his discussion. Some methods appear at first glance to be appropriate for analyzing survival data, but closer inspection shows that they are incorrect.

Suppose someone suggests calculating the mean length of time patients have retained their kidney graft in the kidney transplant study. Using the data in Table 11–1 to calculate a mean retention time presents some problems, and we must make some decisions about these data (the reasons will be made clear later in the chapter). First, for patients whose kidney was retained less than 1 month, let us use a value for survival of 0.5 month. Second, regarding patient 1 who was lost to follow-up 2 months after transplantation, let us remove this patient from analysis for the moment. Then, the mean survival time of kidneys for all patients is 9.4 months. This value may be further broken down into mean survival time of kidneys among patients who retained their kidney, 11.3 months, and mean survival time of kidneys among patients who did not retain it, 4.1 months—with the obvious conclusion that kidneys are retained longer by patients who keep their kidney than by those who reject their kidney! Just as inappropriate is the comparison of kidney retention time for those who had surgery in 1978 versus those who received transplants in 1979; those who received transplants in 1978 have, of course, an average kidney retention time almost 12 months longer.

Table 11–1. Survival of kidney in patients having a transplant; 1978–1979 data.[1]

Patient	Date of Transplant	Date Lost to Follow-Up	Date of Kidney Failure	Months in Study
1	1-11-1978	4-8-1978		2
2	1-18-1978			23
3	1-29-1978			23
4	4-4-1978		4-24-1978	< 1[2]
5	4-19-1978			20
6	5-10-1978			19
7	5-14-1978		8-28-1978	3[2]
8	5-21-1978		11-2-1978	5[2]
9	6-6-1978		11-15-1978	17[2]
10	6-17-1978			18
11	6-21-1978			18
12	7-22-1978		11-7-1978	3[2]
13	9-27-1978			15
14	10-5-1978		1-20-1979	3[2]
15	10-22-1978			14
16	11-15-1978			13
17	12-6-1978			12
18	12-12-1978			12
19	2-1-1979			10
20	2-16-1979			10
21	4-8-1979			8
22	4-11-1979			8
23	4-18-1979			8
24	6-26-1979		8-4-1979	1[2]
25	7-3-1979			5
26	7-12-1979			5
27	7-18-1979		8-1-1979	< 1[2]
28	8-23-1979			4
29	10-16-1979			2
30	12-12-1979			< 1
31	12-24-1979			< 1

[1]Data courtesy of Dr. A Birtch.
[2]Patients who lost their kidney from rejection.

The problem is that mean survival time depends on when the data are analyzed; it will change with each passing month until the point when all the subjects have either lost their kidney or died. Therefore, mean survival estimates calculated in this way are useful only when all the subjects have lost their kidney or have died—or whenever the event being analyzed has occurred for all subjects. Almost always, however, investigators wish to analyze their data prior to that time. In Section 11.5, we present a method for obtaining an estimate of mean survival by using the hazard function.

An estimate of median length of survival time is also possible, and it can be calculated after only one-half the subjects have died (or lost their transplanted kidney, in the present example). However, investigators often wish to evaluate the outcome of a new treatment regimen prior to that time.

A concept frequently used in epidemiology is *person-years of observations*. Using this concept, one can determine a rate, such as number of deaths per 100 person-years of observation. To illustrate the concept, we use the observations in Table 11-1 to determine the number of person-months of kidney survival. Regardless of whether patients have lost their kidney or still retain it at the end of the study, they contribute to the calculation for however long they have been in the study. Therefore, patient 1 contributes 2 months, patients 2 and 3 contribute 23 months, Patient 4 contributes 0.5 month, etc. The total number of months patients have been observed is 283 months; converting to years by dividing by 12 months gives 23.6 person-years. The 8 kidneys lost during the period give 8/23.6 = 0.339, or 33.9 losses per 100 person-years of observation. This number is useful in comparing the results during this time period with results during another time period or results obtained by another investigator.

One problem with using person-years of observation, however, is that the same number is obtained by observing 1000 patients for a period of 1 year or by observing 100 patients for 10 years. Therefore, although the number of subjects is involved in the calculation of person-years, it is not evident as an explicit part of the result; and no statistical methods are available to compare these numbers. Another problem with this approach is the inherent assumption that the risk of the event, eg, death or rejection, during any one unit of time is constant throughout the study, an assumption rarely met in applied clinical research.

Mortality rates (defined in Chapter 4, Section 4.4.2) are a familiar way to deal with survival data, and they are used (especially in oncology) to estimate 3- and 5-year survival with various types of cancer. However, one cannot determine a mortality rate using data on all patients until a given length of time has passed. To illustrate, again using the data in Table 11-1, we can estimate mortality or loss of kidney from rejection as the number of observed losses from rejection divided by the number of patients in the study, or 8/31 = 0.26, or 26%. But again, we do not know what to do with patient 1, who dropped out of the study 2 months after surgery; if this patient is ignored, the mortality rate is 8/30 = 0.27, or 27%. Furthermore, we still have the problem noted for mean length of survival: The mortality rate depends on when the data are analyzed. For example, on November 1, 1978, 15 patients have entered the study and patients 4 and 7 have lost their kidney; if we ignore patient 1, the loss rate from rejection is 2 of 14 patients, or 14%. However, as of January 1, 1979, 18 patients have entered the study; and in addition to patients 4 and 7, patients 8 and 12 have now lost their kidney. Thus, the loss rate is 4 of 17 patients, or 24% (again, we ignore patient 1). The mortality or rejection loss rate may be quite small if it is calculated early enough in a study; but it will be 100% if one waits long enough to analyze the data.

We can also calculate 1-year rejection rates for the data on kidney transplants. Among the 8 patients who rejected their kidney, 7 did so within 1 year of entry into the study (all but patient 9). Among the 22 who had functioning grafts on January 1, 1980, at the time of analysis (ignoring patient 1), only the first 18 patients had been followed at least a year; and we do not know whether patients who entered the study since January 1, 1979, will retain their kidney for an entire year. One solution is to divide the number who rejected their kidney in the first 12 months, 7, by the total number in the study, 31, for an estimate of 0.226, or 22.6%. This estimate, however, is probably too low, because it assumes that all 13 patients who enter the study in the second year retain their kidney at least 12 months, and we know that patients 24 and 27 do not.

An alternative solution is to subtract the number of patients who have not yet been in the study for 12 months—ie, all those who entered in 1979—to obtain 7/(31 − 13) = 0.389, or 38.9%. This technique is similar to the approach used in cancer research in which 3- and 5-year mortality rates are based on only those patients who have been in the study at least 3 or 5 years. The shortcoming of this approach is that it ignores completely the contribution of those who entered the study in 1979. For example, 2 of these patients have had functioning grafts for 10 months, 3 more for 8 months, etc, and they are not counted at all. This approach is acceptable when a large number of patients are being followed; however, for small samples sizes—as often occurs early in the study of a new drug or procedure—one should find a way to utilize information gained from all patients who have entered the study. A reasonable approach should produce an estimate between

22.6% and 38.9%, which is exactly what actuarial life table analysis and Kaplan-Meier product limit methods do. They give credit for the amount of time subjects have survived (or kidneys have been retained, in this example) up to the time when the data are analyzed.

11.3 ACTUARIAL, OR LIFE TABLE, ANALYSIS

Actuarial, or life table, analysis is also sometimes referred to in the medical literature as the Cutler-Ederer method (1958). To illustrate the calculations involved in actuarial analysis, we arranged length of time patients have been in the study in a frequency table (see Table 11–2A). In this example, we assume analysis occurs 2 years after the study began, ie, on January 1, 1980. (In real life, up-to-date data at the time of analysis would be used; we have assumed a January 1, 1980, analysis date to keep the number of calculations small.) The length of time intervals used in the analysis is somewhat arbitrary but should be selected so that the number of censored observations in any interval is small; we group by 2-month intervals for the first 6 months and then by 3-month intervals.

The column headed n_i is the number of patients in the study at the beginning of the interval; all patients (31) began the study, so n_1 is 31. During

the first time interval, from the time they entered the study up to 2 months later, 3 patients rejected their kidneys (patients 4, 24, and 27), so $d_1 = 3$. (We use the symbol d, which stands for "death," because many survival studies analyze survival of people rather than retention of organs, as in our example.) Patients 30 and 31, however, although they still have functioning grafts, have been in the study less than 2 months; they are considered to have withdrawn alive for the purposes of the analysis, even though they are still in the study. During the first 2 months, no patients were lost to follow-up, so the number of withdrawals in interval 1, w_1, is 2. Therefore, there are $31 - 3 - 2 = 26$ patients in the study at the end of the second month following entry into the study.

Looking at the second line in Table 11–2A, during the period of time from 2 up to 4 months after entry, we see that 3 more patients lost their kidney (patients 7, 12, and 14). Patient 29 has been in the study less than 4 months and is counted as a withdrawal. Patient 1 was lost to follow-up between 2 and 4 months and, in the life table method, is given credit for being in the study for that period of time and is also treated as a withdrawal. Thus, there are $26 - 3 - 2 = 21$ patients to begin the interval from 4 up to 6 months. This computation procedure continues until the table is completed.

To determine the **survival curve**, we calculate

Table 11–2. Data for actuarial (life table) analysis of rejection (death) of kidneys.

A. Arrangement of Survival Data			
Months Since Entry into Study	Alive at Beginning of Interval n_i	Rejections During Interval d_i	Withdrawn Alive or Lost to Follow-up w_i
0 up to 2	31	3	2
2 up to 4	26	3	2
4 up to 6	21	1	3
6 up to 9	17	0	3
9 up to 12	14	0	2
12 up to 15	12	0	4
15 up to 18	8	1	1
18 up to 21	6	0	4
21 up to 24	2	0	2

B. Actuarial Calculations			
Months Since Entry into Study	Probability of Rejection or Death $q_i = d_i / [n_i - (w_i/2)]$	Probability of Kidney Retention $p_i = 1 - q_i$	Cumulative Probability of Kidney Retention $S_i = p_i p_{i-1} p_{i-2} \ldots p_1$
0 up to 2	$3/[31 - (2/2)] = .10$	.90	.90
2 up to 4	$3/[26 - (2/2)] = .12$	.88	.79
4 up to 6	$1/[21 - (3/2)] = .05$	.95	.75
6 up to 9	$0/[17 - (3/2)] = 0$	1.00	.75
9 up to 12	$0/[14 - (2/2)] = 0$	1.00	.75
12 up to 15	$0/[12 - (4/2)] = 0$	1.00	.75
15 up to 18	$1/[8 - (1/2)] = .13$	.87	.65
18 up to 21	$0/[6 - (4/2)] = 0$	1.00	.65
21 up to 24	$0/[2 - (2/2)] = 0$	1.00	.65

the probability of kidney loss for each time period; for time interval i, the probability is symbolized by q_i. The ordinary calculation of q_i is to divide the number of losses in the interval by the number of patients who were in the study at the beginning of the interval, ie, d_i/n_i. The actuarial method, however, also counts patients who were in the study at the beginning of the interval but not at the end of it, either because they have not been in the study that long or because they have been lost to follow-up. This method assumes that patients withdraw randomly throughout the interval; therefore, on the average, they are in the study for only half the period of time represented by the interval. It also assumes that these patients will lose their kidney at the same rate as the rest of the patients in the interval; therefore, half the number who were censored is multiplied by the probability of rejection and added to the number of rejections. Thus,

$$q_i = \frac{d_i + [(1/2)w_i]q_i}{n_i}$$

which, when we solve for q_i, becomes

$$q_i = \frac{d_i}{n_i - (1/2)w_i}$$

The second step is to compute the probability of survival (or retaining one's kidney) during each time interval, $p_i = 1 - q_i$. Note that p_i is the probability of surviving interval i only; and in order to survive interval i, a patient (or organ) must have survived all previous intervals as well. Thus, p_i is an example of a conditional probability; ie, the probability of surviving interval i is dependent, or conditional, on surviving until that point, or $p_i =$ probability (surviving interval i, given survival of previous intervals). This probability is sometimes called the **hazard function** or the hazard rate. Recall from Section 5.3 in Chapter 5 that if one event is conditional on a previous event, the probability of their joint occurrence is found by multiplying the probability of the conditional event by the probability of the previous event. Therefore, the cumulative probability of surviving interval i plus all previous intervals is found by multiplying:

$$S_i = p_i p_{i-1} p_{i-2} \cdots p_2 p_1$$

For example, the probability of retaining a functioning graft up to 2 months is .90; given that a patient retained his or her kidney the first 2 months, the probability of retention from 2 months up to 4 months is .88. So the probability of retention up to 4 months is .88 × .90 = .79. The calculations for the actuarial analysis are given in Table 11–2B.

The results from an actuarial analysis help an-

swer useful questions about the observations. For example, we might ask, If X is the length of time a functioning kidney graft is retained by a patient selected at random from the population represented by these kidney transplant patients, what is the probability that X is 12 months or greater? From Table 11–2B, the probability is .75 that a patient will retain a transplanted kidney for at least 12 months.

Results from life table analysis are usually presented in a survival curve rather than in tables. The solid line in Fig 11–3 is a survival curve for the kidney transplant data. The dashed lines on either side of the survival curve represent 95% **confidence bands** for the curve. Although confidence bands are not always presented in medical journals, they should be included because they help readers interpret error in the study results. Typically, as the time interval from entry into the study becomes longer, the number of patients who have been in the study that long becomes increasingly smaller. The confidence limits become wider, reflecting the decreased confidence in the estimate of the proportion as the sample size decreases.

The formula for obtaining confidence bands assumes that under mild censoring and for sufficiently large sample sizes, the proportion surviving at any interval is approximately normally distributed. It uses *Greenwood's formula* for the **standard error** (Greenwood, 1926) and is as follows:

$$SE(S_i) = S_i \sqrt{\Sigma \left[\frac{q_i}{n_i - d_i - (1/2)w_i} \right]}$$

Table 11–3 illustrates the computations for 95% confidence bands, using 1.96 × SE(S_i). The dashed lines in Fig 11–3 illustrate these bands.

The actuarial method involves two assumptions about the data. The first assumption is that all

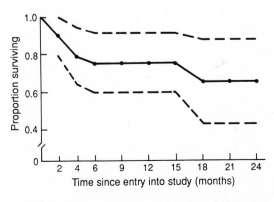

Figure 11-3. Actuarial curve (solid line) with 95% confidence limits (dashed lines) for kidney transplants 1978–1979.

Table 11-3. Calculations for confidence bands for actuarial curve.

Interval	q_i	n_i	d_i	w_i	$\dfrac{q_i}{n_i - d_i - (1/2)w_i}$	$\Sigma\left[\dfrac{q_i}{n_i - d_i - (1/2)w_i}\right]$	S_i	$S_i\,\Sigma\left[\dfrac{q_i}{n_i - d_i - (1/2)w_i}\right]$	1.96 SE
0 to 2	0.10	31	3	2	0.0037	0.0037	.90	0.0548	0.107
2 to 4	0.12	26	3	2	0.0055	0.0092	.79	0.0756	0.148
4 to 6	0.05	21	1	3	0.0027	0.0119	.75	0.0817	0.160
6 to 9	0	17	0	3	0	0.0119	.75	0.0817	0.160
9 to 12	0	14	0	2	0	0.0119	.75	0.0817	0.160
12 to 15	0	12	0	4	0	0.0119	.75	0.0817	0.160
15 to 18	0.13	8	1	1	0.0200	0.0319	.65	0.1160	0.227
18 to 21	0	6	0	4	0	0.0319	.65	0.1160	0.227
21 to 24	0	2	0	2	0	0.0319	.65	0.1160	0.227

withdrawals during a given interval occur randomly throughout the interval. This assumption is of less consequence when short time intervals are analyzed; however, there can be considerable bias if the time intervals are large, if there are many withdrawals, and if withdrawals do not occur midway in the interval. The Kaplan-Meier method introduced in the next section overcomes this problem. The second assumption is that although survival in a given time period depends on survival in all previous time periods, the probability of survival at one time period is independent from the probability of survival at other time periods. This condition, although probably violated somewhat in much applied research, does not appear to cause major concern to statisticians.

11.4 KAPLAN-MEIER PRODUCT LIMIT ESTIMATES OF SURVIVAL

The Kaplan-Meier method of estimating survival is similar to actuarial analysis except that time since entry in the study is not divided into intervals for analysis. For this reason, it is especially appropriate in studies involving a small number of patients. The **Kaplan-Meier product limit method** involves fewer calculations than the actuarial method, primarily because survival is estimated each time a patient dies, so withdrawals are ignored. An illustration should clarify this statement. Again, we use data for Presenting Problem 1 on kidney transplants to illustrate the method,

with loss of transplanted kidney rather than death being the event of interest.

The first step is to list the times at which loss of kidney occurred; we list them in the column "Month of Failure" in Table 11-4. In the kidney transplant study, 2 patients lost their kidney before the first month passed, 1 patient after 1 month, 3 patients after 3 months, 1 patient after 5 months, and 1 patient after 17 months. Then, for each time period, the number of patients who have retained their kidney and remain in the study is entered under the column "No. of Patients." For example, there are 31 patients who retained their kidney and were in the study at least 0 months, but patients 4 and 27 lost their kidney prior to 1 month. By the time patient 24 lost the kidney after 1 month, there were only 27 patients in the study; etc. Patients who are lost to follow-up and those who have retained their kidney but not for the given length of time merely drop out of the calculations by no longer being considered, as patients 30 and 31 do at month 1. This process continues as illustrated in Table 11-4.

The probability of kidney loss (or death) at each month of failure is

$$q_i = \frac{d_i}{n_i}$$

where i is the month of failure. As in the actuarial method, the probability of survival at month i is

$$p_i = 1 - q_i$$

Table 11-4. Kaplan-Meier product limit estimates for 1978–1979 data on rejection of kidneys.

Month of Failure	No. of Patients n_i	No. of Kidney Losses d_i	$q_i = d_i/n_i$	$p_i = 1 - q_i$	$S_i = p_i p_{i-1} \dots p_1$
0	31	2	.06	.94	.94
1	27	1	.04	.96	.90
3	24	3	.13	.87	.79
5	20	1	.05	.95	.75
17	7	1	.14	.86	.64

and the cumulative survival is

$$S_i = p_i p_{i-1} \ldots p_1$$

For example, at month 0, the probability of kidney loss is $2/31 = .06$, and the probability of retention (or survival) is $1 - .06 = .94$. At month 1, the probability of rejection is $1/27 = .04$, and the probability of survival is $.96$; however, the cumulative probability of survival is $.94 \times .96 = .90$. The complete calculations are given in Table 11–4. Note that the Kaplan-Meier procedure gives exact survival proportions because it uses exact survival times; the actuarial method gives approximations because it groups survival times into intervals. Prior to the widespread use of computers, the actuarial method was easier to use for a very large number of observations.

The standard error of the cumulative survival estimate S_i is similar to the standard error for an actuarial curve. For the Kaplan-Meier product limit estimate, the standard error is

$$SE(S_i) = S_i \sqrt{\Sigma \left[\frac{d_i}{n_i(n_i - d_i)} \right]}$$

For example, at month 0, there are 31 patients and 2 rejections; so

$$\frac{d_i}{n_i(n_i - d_i)} = \frac{2}{31 \times 29} = 0.0022$$

And the standard error is

$$S_i \sqrt{\Sigma \left[\frac{d_i}{n_i(n_i - d_i)} \right]} = .94 \sqrt{.0022} = 0.044$$

The results of the calculations are given in Table 11–5.

Figure 11–4 is a graph of the Kaplan-Meier product limit curve, along with 95% confidence limits for the curve. In this graph, the curve is step-like because the proportion of patients retaining their kidney changes precisely at the point when a subject dies.

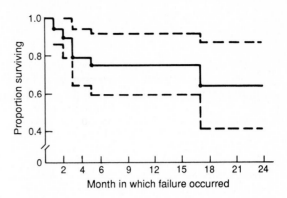

Figure 11–4. Kaplan-Meier curve (solid line) with 95% confidence limits (dashed lines) for kidney transplants 1978–1979.

11.5 THE HAZARD FUNCTION IN SURVIVAL ANALYSIS

In the introduction to this chapter, we stated that calculating mean survival is generally not useful, and we subsequently illustrated how its value depends on the time when the data are analyzed. However, there is a way to obtain estimates of mean survival that are reasonable when the sample size is fairly large. This procedure depends on the **hazard function,** which is the probability that a person dies in time interval i to $i + 1$, given that the person has survived until time i. The hazard function is also called the *conditional failure rate;* and in epidemiology, the term *force of mortality* is used.

Although the exponential probability distribution was not discussed in Chapter 5, many survival curves follow an exponential distribution. It is a continuous distribution that involves the natural logarithm, ln, and it depends on a constant rate (which determines the shape of the curve) and on time. It provides a model for describing, say, radioactive decay.

If an exponential distribution is a reasonable assumption for the shape of a survival curve, then the following formula can be used to estimate the hazard rate, symbolized by the letter H, when there are censored observations:

Table 11–5. Calculations for confidence bands for Kaplan-Meier curve.

Months to Failure	n_i	d_i	$\dfrac{d_i}{n_i(n_i - d_i)}$	$\Sigma \left[\dfrac{d_i}{n_i(n_i - d_i)} \right]$	s_i	$s_i \sqrt{\Sigma \left[\dfrac{d_i}{n_i(n_i - d_i)} \right]}$	1.96 SE
0	31	2	0.0022	0.0022	.94	0.044	0.087
1	27	1	0.0014	0.0036	.90	0.054	0.106
3	24	3	0.0060	0.0096	.79	0.077	0.151
5	20	1	0.0026	0.0122	.75	0.083	0.162
17	7	1	0.0238	0.0360	.64	0.121	0.238

$$H = \frac{d}{\Sigma f + \Sigma c}$$

where d is the number of failures (or in the kidney transplantation example, the number of kidneys lost from rejection), Σf is the sum of failure times, and Σc is the sum of censored times.

For example, the times to kidney loss for patients transplanted in 1978 as of January 1, 1980, are < 1, 3, 3, 3, 5, and 17 months. Let us again use 0.5 for < 1, assuming that the loss occurred midway through the first month. Thus, there are 6 failures, and the sum of the failure times is (0.5 + 3 + 3 + 3 + 5 + 17) = 31.5. The sum of the times for the censored observations is (12 + 12 + 13 + 14 + 15 + 18 + 18 + 19 + 20 + 23 + 23 + 2) = 189. Therefore, the estimate of the hazard rate is

$$\frac{6}{31.5 + 189} = 0.0272$$

and it is interpreted as a rate of 0.027 transplant failure per month.

One reason the hazard rate is of interest is that its reciprocal is an estimate of mean survival time, ie,

$$\overline{X} = \frac{1}{H}$$

where $\overline{X}$ is the mean survival time. In this example, the mean survival time is estimated to be l/0.0272 = 36.76 months. In addition, we can determine confidence limits for both hazard rate and mean survival time. The 95% confidence limits for the true mean survival time are

$$\overline{X} \pm 1.96 \times SE\,(\overline{X})$$

because the distribution of H with a large sample size is approximately normal. Although the formula for the standard error used in the calculation of confidence limits is quite complex, a reasonable approximation is

$$\sqrt{\frac{\overline{X}^2}{d}} = \sqrt{\frac{36.76^2}{6}}$$

$$= 15.01$$

Therefore, a 95% confidence interval for the mean number of months patients can be expected to retain a transplanted kidney is

$$36.76 \pm (1.96)(15.01)$$

or

7.34 to 66.18 months

When the assumption of an exponential distribution with a constant failure rate is not tenable, other forms of the hazard function based on different probability distributions are used. Details on using the hazard function are given in the comprehensive text on survival analysis by Lee (1980).

11.6 COMPARING TWO SURVIVAL CURVES

Although some journal articles report survival statistics for only one group, on many occasions investigators wish to compare survival between two samples of patients. To illustrate the computations involved in comparing two survival distributions, we will compare kidney retention times for patients who entered the study in 1978 with kidney retention times for patients who entered the study in 1984. Beginning in 1984, the drug cyclosporin was routinely used in this group of patients to reduce rejection of the kidney; prior to that time, the drug azathioprine was used. Investigators want to know whether the use of cyclosporin resulted in fewer cases of kidney rejection. Of course, there is no reason to limit the comparison to these two years; we have selected them to ease the burden of computations.

In the following examples, we again assume that analysis occurs 1 year after the end of the year in which the patient had transplant surgery, ie, December 31, 1979, and December 31, 1985. The retention times for patients who received a kidney transplant in 1984 are given in Table 11–6.

Calculations needed for drawing the survival curves for the actuarial and the Kaplan-Meier methods are left as exercises (see Exercises 1 and 2). However, the completed calculations and survival curves are given in Tables 11–7 and 11–8 and in Figs. 11–5 and 11–6.

In either method, the survival curve for patients receiving a kidney in 1984 is above the curve for patients receiving a kidney in 1978, indicating a higher proportion of patients retaining a functioning graft at any one point in time. However, variation in samples may be expected to occur simply by chance, and a reasonable question is whether the differences between the two years is greater than expected by chance. To test this hypothesis, we need methods to compare survival distributions. If there are no censored observations, the **Wilcoxon rank-sum test** introduced in Chapter 7 is an appropriate method for comparing length of survival for two independent samples. In this approach, survival times are ranked from shortest to longest, ignoring the group of patients to which the observation belongs, and the t test is performed on the ranks of survival time. The independent-groups t test of the original values is not appropriate because survival times are not normally distributed; they tend to be positively skewed

Table 11-6. Survival of kidneys in patients having transplant; 1984 data.[1]

Patient	Date of Transplant	Date Lost to Follow-up	Date of Kidney Failure	Months in Study
1	2-8-1984			22
2	2-22-1984			22
3	2-25-1984			22
4	2-29-1984		11-27-1984	8
5	3-12-1984			21
6	3-22-1984		4-15-1984	<1
7	4-26-1984			20
8	5-2-1984			19
9	5-9-1984			19
10	6-6-1984			18
11	7-11-1984			17
12	7-20-1984			17
13	8-18-1984			16
14	9-5-1984			15
15	9-15-1984			15
16	10-3-1984			14
17	11-9-1984			13
18	11-27-1984		6-11-1985	6
19	12-5-1984			12
20	12-6-1984			12
21	12-19-1984			12

[1]Data courtesy of Dr. A Birtch.

(extremely so, in some cases). However, if some observations are censored, other procedures must be used.

Unfortunately, there is no single method used by all investigators. As with post hoc multiple-comparison procedures in the analysis of variance, the most appropriate method for comparing survival distributions depends on the nature of the observations. Two procedures frequently used in medical research are illustrated in this section. We

caution you that the computations for these methods are time-consuming. We present them in detail because textbooks that discuss these methods are for advanced students of statistics, and few elementary texts illustrate more than one. Thus, medical practitioners may have difficulty finding accessible information about these tests. We advise you to go through the calculations in order to obtain an idea of the logic behind the procedures, recognizing that computer programs are certainly

Table 11-7. Life table analyses comparing kidney transplants in 1978 and 1984, analyzed January 1, 1980 and January 1, 1986, respectively.

A. Arrangement of Survival Data						
	1978			1984		
Months in Study	n_i	d_i	$w_i + l_i$	n_i	d_i	$w_i + l_i$
0 up to 2	18	1	0	21	1	0
2 up to 4	17	3	1	20	0	0
4 up to 6	13	1	0	20	0	0
6 up to 12	12	0	0	20	2	0
12 up to 18	12	1	5	18	0	10
18 up to 24	6	0	6	8	0	8

B. Calculations for Life Table						
	1978			1984		
Months in Study	q_i	p_i	S_i	q_i	p_i	S_i
0 up to 2	.06	.94	.94	.05	.95	.95
2 up to 4	.18	.82	.77	0	1.00	.95
4 up to 6	.08	.92	.71	0	1.00	.95
6 up to 12	0	1.00	.71	.10	.90	.86
12 up to 18	.11	.89	.63	0	1.00	.86
18 up to 24	0	1.00	.63	0	1.00	.86

Table 11-8. Kaplan-Meier product limit estimates for 1979 and 1984 data.

A. 1978 Kidney Transplants Analyzed December 31, 1979					
Month of Failure	No. of Patients n_i	No. of Kidney Losses d_i	q_i	p_i	S_i
0	18	1	.06	.94	.94
3	16	3	.19	.81	.77
5	13	1	.08	.92	.71
17	7	1	.14	.86	.61
B. 1984 Kidney Transplants Analyzed December 31, 1985					
Month of Failure	No. of Patients n_i	No. of Kidney Losses d_i	q_i	p_i	S_i
0	21	1	.05	.95	.95
6	20	1	.05	.95	.90
8	19	1	.05	.95	.86

the method to use for computing the statistics. After presenting the two methods, we summarize recommendations for using them.

11.6.1 The Gehan, or Generalized Wilcoxon, Test

The **Gehan,** or **generalized Wilcoxon, test** is an extension of the Wilcoxon rank-sum test illustrated in Chapter 7; it can be used with censored observations (Gehan, 1965). This test is also referred to in the literature as the Breslow test or the generalized Kruskal-Wallis test for comparison of more than two samples (Kalbfleisch and Prentice, 1980). Unfortunately, no one has yet developed a simple approximation, such as converting original observations to ranks and calculating the *t* statistic for ranks.

In Gehan's test, every observation in one sample is compared with every observation in the second sample. Since there are 18 patients in the 1978 sample and 21 in the 1984 sample, there will be 18 × 21 = 378 comparisons. A mathematical ex-

pression involving the ranks is calculated; this function follows an approximately **normal distribution,** and so the *z* **distribution** can be used to determine statistical significance. To illustrate the calculation of the Gehan statistic, we rearrange the data in a table with one sample across the top and the other down the side (see Table 11-9).

We next designate one sample as the reference sample, and each observation in the reference sample is compared one by one with each observation in the other sample. Using 1984 patients as the reference sample, we give each comparison a score of +1 if the 1984 patient in the pair *definitely* retained his or her kidney for a longer period of time; −1 if the 1984 patient in the pair *definitely* retained the kidney for a shorter period of time; and 0 if the patients in the two samples retained their kidneys the same length of time or if we *cannot tell* because both are censored. The steps in calculating the Gehan statistic follow.

1. Compare each pair of observations. Score them as follows: +1 if the reference sample observation is larger than the observation in the other

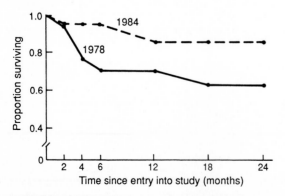

Figure 11-5. Actuarial curves comparing kidney transplants 1978 and 1984.

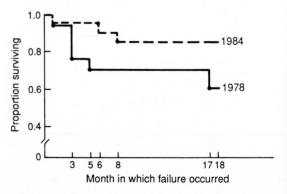

Figure 11-6. Kaplan-Meier curves comparing kidney transplants 1978 and 1984.

Wilcoxon) test and the logrank test, again using kidney retention data. We noted the confusion that results because these statistics are approximations and can be calculated in several different ways; they are also called by several different names in the literature. By using the observations on kidney retention to illustrate survival analysis, we hoped to show that survival analysis methods are applicable in a variety of situations, not simply in studies examining mortality.

We discussed two examples of survival analysis from the published literature. In the study described in Presenting Problem 2, the investigators compared survival in chronic lymphocytic leukemia patients who had normal karyotypes with those who had abnormal karyotypes. They found that patients with chromosomal abnormalities other than trisomy 12 have a poorer prognosis and suggested that cytogenetic analysis provides useful prognostic information. Published results on patient survival for three different methods of treating breast cancer in women were examined in Presenting Problem 3. This important study found that patients who received segmental mastectomy plus radiation had higher survival than patients who received segmental mastectomy alone or total mastectomy. There was no difference in survival between the latter two groups.

EXERCISES

1. Perform the calculations for drawing survival curves one year after the end of the year for the patients who had kidney transplants in 1979 and for those who had transplants in 1984. Use the actuarial method with 2-month intervals for the first 6 months and 6-month intervals thereafter. Draw the survival curves.
2. Perform the calculations for drawing survival curves one year after the end of the year for the patients who had kidney transplants in 1979 and for those who had transplants in 1984. Use the Kaplan-Meier method. Draw the survival curves.
3. In Presenting Problem 2, the investigators performed a survival analysis on the number of months of survival since the patients had had the karyotype study performed. What is a potential bias in this analysis?
4. Camitta et al (1979) studied 110 patients with severe aplastic anemia. Patients ($n = 47$) who had HLA-matched siblings entered the bone marrow transplantation arm of the study. Patients who did not have marrow donors were randomly assigned to one of three treatments to evaluate the role of androgens in marrow transplantation: oral androgen (PO, $n = 27$), intramuscular androgen (IM, $n = 23$), and no androgen ($n = 13$). Follow-up of the patients ranged from 9 to 45 months. Survival

distributions were evaluated by using the Kaplan-Meier method and the logrank test. Survival curves are given in Figs. 11–11 and 11–12.
a. What conclusion can you draw about the use of androgen from Fig 11–11?
b. Fig 11–12 illustrates survival in transplanted versus nontransplanted patients, the latter being all androgen groups combined. What conclusion can you draw?
c. From Fig 11–12, what is the median survival in patients given a marrow transplant? In patients not given a marrow transplant?
d. In interpreting the results shown in Fig 11–12, the authors stated:

> The estimated probability of surviving 6 months is 0.70 for transplanted patients (95% confidence limits 0.57–0.83) and 0.35 for nontransplanted patients (95% confidence limits 0.24–0.48). Thirty-three patients in the transplantation arm and 22 in the nontransplantation arm survived at least 6 months. The distributions of survival times beyond 6 months did not significantly differ between these two groups ($P = 0.27$), which indicates that the variation in overall survival occurs primarily during the first 6 months.

How do you interpret the 95% confidence limits at 6 months? What is an alternative explanation for why the survival distributions do not differ significantly beyond 6 months?
e. What do the small dots in Fig 11–12 designate?
5. Moertel et al (1985) performed a double-blind,

Kaplan-Meier product limit estimates of percentage surviving for patients given IM androgen, PO androgen, or no androgen.

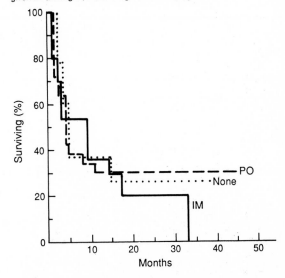

Figure 11–11. Comparison of survival in patients receiving androgen. (Reproduced, with permission, from Camitta BM et al: A prospective study of androgens and bone marrow transplantation for treatment of severe aplastic anemia. *Blood* 1979;**53**:504.)

Kaplan-Meier product limit estimates of percentage surviving for the pooled three nontransplantation groups of patients and for the patients given marrow transplants.

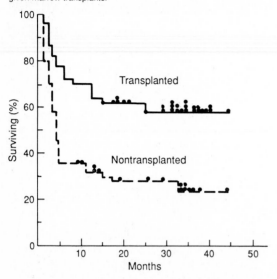

Figure 11-12. Comparison of patients with and without bone marrow transplantations. (Reproduced, with permission, from Camitta BM et al: A prospective study of androgens and bone marrow transplantation for treatment of severe aplastic anemia. *Blood* 1979;**53**:504.)

Time from the beginning of therapy to disease progression, according to treatment assignment in 100 patients with advanced cancer.

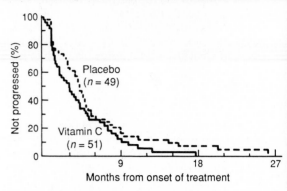

Figure 11-13. Comparison of disease progression in patients with colorectal cancer receiving vitamin C versus placebo. (Reproduced, with permission, from Moertel CG et al: High-dose vitamin C versus placebo in the treatment of patients with advanced cancer who have had no prior chemotherapy: A randomized double-blind comparison. *N Engl J Med* 1985;**312**:137.)

randomized trial of high-dose vitamin C versus placebo in the treatment of advanced colorectal cancer in patients who had not had prior chemotherapy. In addition to analyzing survival as an outcome, the investigators used the Kaplan-Meier method and the logrank statistic to analyze progression of the disease. Progression was defined as any of the following: an increase of more than 50% in the product of the perpendicular diameters of any area of known malignancy, new area of malignancy, substantial worsening of symptoms or performance status, or weight loss of 10% or more. The results of their analysis are reproduced in Fig 11-13.

a. What conclusion can be drawn from the figure?

b. What is the median time to disease progression in each group?

c. Do you think the analysis of survival times found a statistically significant difference?

6. Determine the 90% confidence interval for the odds ratio calculated in Section 11.6.2. Are the results consistent with the results from the generalized Wilcoxon test? If not, explain why.

Statistical Methods for Multiple Variables

12

PRESENTING PROBLEMS

Presenting Problem 1. People who participate in health maintenance organizations (HMOs) have been shown to have reduced costs of medical care, but there has been some concern that these reductions have had adverse effects on health. In 1976, the RAND Corporation undertook an interventional study to investigate this issue; 1673 adults living in Seattle, Washington, were randomly assigned to either an HMO or a fee-for-service (FFS) plan for a period of 3 to 5 years (Ware et al, 1986, 1987). Within the FFS group, some patients were randomly assigned to receive free medical care and others shared in the costs.

The health status of the adults was evaluated at the beginning and again at the end of the study by using a battery of 11 measures in three categories: physiological health (blood pressure, vision); health habits (weight, cholesterol, smoking); and general health (physical, mental, social and role functioning, social contacts, perceptions of health). In addition, the number of days spent in bed because of poor health and the occurrence of serious symptoms were determined periodically throughout the study. These 13 measures, recorded at the beginning of the study—along with information on the participant's age, gender, income, and the system of health care to which he or she was assigned (HMO, free FFS or pay FFS)—were the independent variables used in the study. The dependent variables were the values of these same 13 measures at the end of the study.

The investigators wanted to know whether there are any significant relationships between independent (explanatory) variables and each dependent (response) measure. Since there are several explanatory variables, the simple regression methods introduced in Chapter 10 will not suffice, and a more advanced technique, called multiple regression, must be used. There are 13 outcomes of interest to the investigators, and multiple regression can be used to predict each outcome separately. The results from one of the regression analyses from the 1987 publication are given in Table 12–3.

Presenting Problem 2. Smoking has been ac-

cepted as a risk factor for coronary artery disease and myocardial infarction. However, Hartz et al (1984) also wanted to investigate the possibility that toxic effects of smoking are associated with ventricular wall motion abnormalities, independent of the association of smoking with coronary occlusion. They collected data on 4763 male patients between the ages of 35 and 74 who had diagnostic coronary angiography and ventriculography between 1972 and 1981. (Although data collection continued over a 10-year period, the study is actually a cross-sectional design, since all information about each patient was collected at one point in time). Men were classified as nonsmokers (those who had never smoked), mild smokers (less than 10 pack-year history of smoking) and moderate smokers (more than 10 pack-year).

Data on ventricular wall motion were recorded for each of six left ventricular segments, and patients were classified into three categories: normal, hypokinetic (decreased wall motion), and akinetic or dyskinetic (absent or paradoxical wall motion). Hypokinetic ventricles were further divided into four subgroups containing one, two, three to four, or five to six hypokinetic segments. As a measure of coronary occlusion, the degree of obstructive disease (percentage of effective stenosis from 0 to 100) in each of the three main coronary arteries was determined and then added to produce an occlusion score ranging from 0 to 300. Ventricular wall motion or hypokinesis and occlusion are positively associated; ie, men with higher occlusion scores have a greater degree of hypokinesis. Therefore, any relationship between smoking and hypokinesis may actually be a relationship between smoking and occlusion.

To determine whether there is a relationship between smoking and ventricular wall motion that is independent of occlusion, investigators must control for the possible confounding effect of coronary occlusion. Multiple regression techniques can be used to control for confounding, a technique commonly referred to as analysis of covariance.

Presenting Problem 3. Previous studies have shown that a high dietary intake of potassium is associated with lower blood pressure. Hypertension, of course, is a well-accepted risk factor for stroke. Studies in hypertensive rats have shown that a high potassium intake protects against strokes even though blood pressure is not affected. Khaw and Barrett-Connor (1987) designed a cohort study to examine the relationship between dietary intake of potassium and mortality due to stroke; 859 men and women were included in the study and were followed for 12 years. The investigators also collected data on other factors that might have contributed to mortality, including the subject's blood pressure, cholesterol, fasting plasma glucose, body-mass index, and history of smoking tobacco. The investigators wished to test their hypothesis that a high dietary intake of potassium is related to a lower risk of stroke.

This study also involves multiple explanatory variables. The response variable is duration of survival, not simply survival versus mortality. Although duration of survival is numerical, longitudinal studies typically include censored observations because of patients entering or leaving the study at different times. In Chapter 11, we saw that special methods must be used with censored observations; when the purpose of a study is to determine what factors are related to a censored outcome, such as survival, the Cox regression model, sometimes called the Cox proportional hazards model, is appropriate. Selected results from the published analysis are given in Table 12-4.

Presenting Problem 4. Childhood enuresis can indicate an underlying problem that is benign or one that is more serious, such as urinary tract obstruction. In a cross-sectional study of a general population of 5-year-old to 13-year-old children, Foxman, Valdez, and Brook (1986) tried to determine factors associated with enuresis. The investigators wanted to know whether a child wet his or her bed at least once during the previous 3 months, and whether the child wet at least once per week in the previous 3 months. Data on 1700 children were gathered as part of a health questionnaire completed by parents in families enrolled in a large insurance experiment described in the article. Independent (explanatory) variables included on the questionnaire were age, gender, and race of the child; education of the mother (in years); family income; family size; and psychological distress.

One feature of the regression techniques discussed in this chapter is that they may be used to examine or explain relationships as well as to predict. However, the dependent measure in this problem is a nominal variable with two values—either enuresis is present or it is not—and multiple regression is not the best method to use in this situation. A technique similar to multiple regression, called logistic regression, can be used when there

are two or more explanatory variables and when the response is dichotomous; the results from a logistic regression analysis are presented in Table 12-5.

Presenting Problem 5. Sports injuries arise from several different factors. Buckley (1988) investigated the risk that a player receives a concussion in college football by analyzing the games played by 49 college teams in the 8-year period 1975–1982. Although not explicitly stated, historical records were probably used in the study; however, the research question indicates that this study was cross-sectional. In the abstract, the investigator stated that the hypothesis tested in the study was that the variables of team (offense and defense), player position, situation (rushing and passing), and activity (blocking and tackling) had no effect on the occurrence of game-related concussions. The frequency of injury under each condition was determined by the investigator and the raw frequencies were then adjusted to reflect the facts that a player is involved in the activity of blocking more often than tackling and that rushing plays occur more often than passing plays. The adjusted frequencies of game-related concussions by team, situation, and activity are given in Table 12-6.

Each independent variable in this research problem is measured on a categorical or nominal scale (team, player position, situation, and activity), as is the dependent (response) variable (whether or not a concussion occurred). If there are only two variables, the chi-square method introduced in Chapter 9 can be used to determine whether a relationship exists between them; with three or more nominal or categorical variables, a statistical method called log-linear analysis is appropriate.

Presenting Problem 6. Many studies have reported the use of perioperative parenteral nutrition in patients having major surgery; some studies conclude this procedure is beneficial, but others conclude that there is no benefit. It is difficult for physicians to interpret the literature when studies report conflicting results. As you now know, studies frequently fail to find significance because of low power associated with small sample sizes. A traditional way to resolve the problem of conflicting results in medicine is by reviewing many studies and summarizing their strengths and weaknesses in what are commonly called *review articles*. Detsky et al (1987), however, used a relatively new method to combine the results of several studies in a statistical manner. They applied meta-analysis to 18 controlled trials conducted on perioperative total parenteral nutrition so that overall conclusions regarding safety, efficacy, effectiveness, cost, and cost effectiveness of the procedure could be drawn; Table 12-7 gives a summary of the results of the meta-analysis.

12.1 PURPOSE OF THE CHAPTER

The purpose of this chapter is to present a conceptual framework that applies to almost all statistical procedures discussed so far in this text. We also describe briefly, using a conceptual framework only, some of the more advanced techniques used in medicine.

12.1.1 A Conceptual Framework

The previous chapters illustrated statistical techniques appropriate when there are a limited number of observations on each subject in a study. For example, a t test is used when there are two groups of subjects and the measure of interest is a single numerical variable—such as in the study comparing the urinary 5-HIAA for patients with and without carcinoid heart disease to see whether this feature clinically distinguishes among patients with this condition (see Presenting Problem 3 in Chapter 7). When the outcome of interest is nominal, the chi-square test can be used—such as in the study of the occurrence of lethal shock in two groups of patients, those who did and did not receive a prophylactic vaccine (Presenting Problem 2 in Chapter 9). Regression analysis is used to predict one numerical measure from another—such as in the study predicting the value of C1q binding from IgG-containing complexes in patients with systemic lupus erythematosus (Presenting Problem 1 in Chapter 10).

Alternatively, each of these examples can be viewed conceptually as involving a set of subjects with two observations on each subject: (1) for the t test, one numerical variable, urinary 5-HIAA level, and one nominal (or group membership) variable, the presence or absence of carcinoid heart disease; (2) for the chi-square test, two nominal variables, occurrence or nonoccurrence of lethal shock and receipt or nonreceipt of vaccine; (3) for regression, two numerical variables, C1q binding and IgG-containing complex. The advantage of this perspective will become apparent as we discuss the more advanced techniques necessary when many variables are included in a study.

To practice viewing research questions from a conceptual perspective, let us reconsider the problem of determining whether there are differences in immune function in women who have low, moderate, or high levels of social adjustment. Here, again, the problem may been seen as involving a set of subjects with two observations per subject: one numerical variable, immune function (as measured by natural killer cell activity), and one ordinal (or group membership) variable with three values (treated as if nominal), social adjustment. If there were only two values for social adjustment, the t test would be used. However, with more than two groups, one-way analysis of variance (ANOVA) is appropriate (see Presenting Problem 1, Chapter 8).

Many problems in medicine have more than two observations per subject because of the complexity involved in studying disease in humans. In fact, many of the Presenting Problems used in this text have had multiple observations, although we have chosen to simplify the problems by examining only selected variables. However, one method involving more than two observations per subject has been discussed, two-way ANOVA. Recall that in Presenting Problem 2 of Chapter 8, acetylcholine synthesis was examined in younger and older patients with and without Alzheimer's disease. In this example, there are three observations per subject: one numerical variable, choline acetyltransferase level; a nominal (or group membership) variable, presence or absence of disease; and another nominal (or group membership) variable, age. (Although age is actually a numerical measure, the investigators divided age into two categories, less than and greater than 80 years, and treated it as though it were a nominal measure.)

If the term **independent variable** is used to designate the group membership variables (eg, presence or absence of carcinoid heart disease) or the X variable (eg, IgG-containing complex), and the term **dependent** is used to designate the variables whose means are compared (eg, urinary 5-HIAA) or the Y variable (eg, C1q binding), the above observations can be summarized as in Table 12–1. For the sake of simplicity, this summary omits ordinal variables. When independent variables are measured on an ordinal scale, they are generally treated as if they were nominal (eg, using "high" versus "low" Apgar scores to predict infant survival during first months of life); when dependent variables are ordinal (eg, predicting level of functional capacity in rheumatoid arthritis categorized as Class 1-Class 4 from age at onset), a method such as regression is used, treating the ordinal variable as though it were numerical.

Table 12-1. Summary of conceptual framework[1] for questions involving two variables.

Independent Variable	Dependent Variable	Method
Nominal	Nominal	Chi-square
Nominal (dichotomous)	Numerical	t-test[1]
Nominal (more than 2 values)	Numerical	One-way ANOVA[1]
Nominal	Numerical (censored)	Actuarial methods
Numerical	Numerical	Regression[2]

[1]Assuming the necessary assumptions (eg, normality, independence, etc) are met.
[2]Correlation is appropriate when neither variable is designated as independent or dependent.

12.1.2 Introduction to Methods for Multiple Variables

Statistical techniques involving multiple variables are used increasingly in medical research, and several of them are illustrated in this chapter. The multiple regression model, in which several independent variables are used to explain, or predict, the values of a single numerical response, is presented first, partly because it is a natural extension of the regression model for one independent variable illustrated in Chapter 10. More importantly, however, all the other advanced methods except meta-analysis can be viewed as modifications or extensions of the multiple regression model. All except meta-analysis involve more than two observations per subject and are concerned with explanation or prediction.

The goal in this chapter is to present the logic of the different methods listed in Table 12–2 and to illustrate how they are used and interpreted in medical research. Although the models are sometimes given in equation form, we will not perform any calculations. These methods are generally not mentioned in traditional introductory texts, and most people who take statistics courses do not learn about them until their third or fourth course. They are being used with increasing frequency in medicine, however, partly because of the increased involvement of statisticians in medical research and partly because of the availability of complex statistical computer programs. In truth, few of these methods would be used very much in any field were it not for computers because of the time-consuming and complicated computations involved. Therefore, to read the literature with confidence, especially studies designed to identify prognostic or risk factors, physicians must have a nodding acquaintance with the methods described in this chapter.

Before we examine the advanced methods, though, a comment on terminology is necessary. Some statisticians reserve the term **multivariate** to refer to situations that involve more than one dependent (or response) variable. If we use this definition, then the RAND study of HMOs vs FFS systems is a multivariate study requiring multivariate techniques of analysis, because there are 13 outcome measures in the study. The study by Khaw and Barrett-Connor investigating the risk of mortality from stroke would not be classified as multivariate according to this definition, however, because even though 15 risk factors were evaluated, the only response variable was mortality. By this strict definition, multiple regression and most of the other methods discussed in this chapter would not be classified as multivariate techniques. Other statisticians, ourselves included, use the term *multivariate* more freely to refer to methods that examine the simultaneous effect of multiple variables; by this definition, all the techniques discussed in this chapter (with the possible exception of meta-analysis) are classified as multivariate methods.

12.2 PREDICTING WITH MORE THAN ONE VARIABLE: MULTIPLE REGRESSION

12.2.1 Review of Regression

Simple linear regression (Chapter 10) is the model used when investigators wish to predict the value of a response (dependent) variable denoted Y, from an explanatory (independent) variable X. The regression model is

$$Y = a + bX$$

For simplicity of notation, note that in this chapter, we use Y to denote the dependent variable, even though it is Y', the predicted value, that is actually given by this equation. In this chapter, we also use a and b, the sample estimates, instead of the population parameters, β_0 and β_1.

With only one explanatory variable, the simple regression model has a geometric interpretation in which a is the intercept and b is the slope. More generally, however, b is called the **regression coefficient**, and the t test may be used to see whether there is a significant relationship between X and Y. For example, with dependent variable C1q binding and independent variable IgG-containing complex, the regression equation determined in Chapter 10, Exercise 1, part B, is

Table 12-2. Summary of conceptual framework[1] for questions involving two or more independent (explanatory) variables.

Independent Variable	Dependent Variable	Method
Nominal	Nominal	Log-linear
Nominal and numerical	Nominal (dichotomous)	Logistic regression
Nominal and numerical	Nominal (2 or more values)	Discriminant analysis[1]
Nominal	Numerical	ANOVA[1]
Numerical	Numerical	Multiple regression[1]
Numerical and nominal	Numerical (censored)	Cox regression
Nominal with confounding factors	Numerical	ANCOVA[1]
Numerical only	—	Factor analysis[1] and cluster analysis[1]

[1]Certain assumptions (eg, multivariate normality, independence, etc) are needed to use these methods.

$$Y = -2.03 + 0.94\ X$$

This regression line intersects the Y-axis at -2.03 and has a positive slope equal to 0.94, indicating that C1q binding increases by 0.94 with each 1.0 increase in IgG-containing complex.

If the regression coefficient is found to be significant (as it is, in this example), the equation can be used to predict a future patient's C1q binding from the IgG-containing complex. For example, if a patient's IgG-containing complex is 30, the C1q binding is predicted to be

$$Y = -2.03 + (0.94)(30) = 25.96$$

The actual value for a patient will probably differ; in fact, the three patients in the study with IgG-containing complex equal to 30 actually have C1q-binding levels of 25, 39, and 20 (see Table 10–1).

12.2.2 Multiple Regression

The extension of simple regression to two or more independent variables is straightforward. For example, if there are four independent variables, the **multiple regression** model is

$$Y = a + b_1X_1 + b_2X_2 + b_3X_3 + b_4X_4$$

where X_1 is the first independent variable and b_1 is the regression coefficient associated with it, X_2 is the second independent variable and b_2 is the regression coefficient associated with it, etc. Although the derivation of the formulas for a and b were not presented in Chapter 10, the formulas themselves were given. We do not give the formulas for the regression coefficients in multiple regression, however, because they become more and more complex as the number of independent variables increases; and no one calculates them by hand, in any case.

Any arithmetic equation in the form of the above equation is called a **linear combination;** thus, the response variable Y can be expressed as a (linear) combination of the explanatory variables. Note that a linear combination is really just a weighted average that gives a single number (or index) after the X's are multiplied by their associated b's and the bX products are added. Thus, a linear combination is an efficient way to summarize the value of several variables as one value.

As in simple regression, the dependent (or response) variable Y is a numerical measure. The traditional multiple regression model calls for the independent variables to be numerical measures as well; however, nominal independent variables may be used, as discussed in the next section, but nominal dependent variables may not be. Therefore, to summarize, the appropriate technique used with numerical independent variables and a numerical dependent variable is multiple regression analysis, as indicated in Table 12–2.

12.2.3 Interpreting the Multiple Regression Equation

The RAND investigators (Presenting Problem 1) found two statistically significant differences between patients who were enrolled in the HMO and those in the free FFS plan: A slightly higher rate of bed-days and a smaller percentage reporting serious symptoms occurred for the HMO systems. (The bed-day variable was obtained as follows: Once a year, each participant was asked to report the number of days spent in bed from poor health during the past 30 days; this number was then averaged over the number of years the participant was in the study.) Further analysis indicated that these differences were concentrated in low-income patients who began the experiment with health problems.

Let us examine the regression equation to predict number of bed-days in greater detail. Table 12–3 is reproduced as published by the investigators (Ware et al, 1987). The regression coefficients in Table 12–3 can be used to predict the number of bed-days by multiplying a given patient's value for each independent variable X by the corresponding regression coefficient b and then summing to obtain the predicted number of bed-days. The first variable is a nominal variable called "FFS freeplan"; a patient who is in the FFS freeplan is assigned a value of 1 for this variable, and all patients not in the FFS freeplan are assigned a value of 0. This procedure is called **dummy coding,** and it allows investigators to include nominal variables in a regression equation in a straightforward manner. Nominal variables with dummy coding are sometimes referred to as dummy variables. There is also a nominal variable "FFS payplan," and it is a dummy variable as well. A separate dummy variable for HMO is not needed since a patient in the HMO group is identified by having a 0 value for both the freeplan and the payplan variables. The dummy variables are interpreted as follows: A patient who is in the FFS freeplan has 0.017 fewer bed-days on the average (the 1 for being in the FFS freeplan is multiplied by -0.017) than a patient in HMO; a patient in the FFS payplan has 0.014 fewer bed-days than a patient in the HMO. (The determination of which value is assigned 1 and which is assigned 0 is an arbitrary decision made by the researcher but can be chosen to facilitate interpretations of interest to the researcher.)

Next, the patient's personal-functioning score (a numerical variable based on answers to two questionnaires administered at enrollment, with higher scores indicating greater activity) is multiplied by -0.0002. The patient's mental health score is multiplied by -0.00006; and his or her health perceptions score is multiplied by −0.002. All coefficients indicate that patients with higher scores on these indexes have fewer bed-days.

To illustrate the process of calculating the predicted number of bed-days for the variables considered thus far, let's suppose that a patient in the FFS payplan has a personal-functioning score of 95 out of 100, a mental health score of 70 out of 100, and a health perceptions score of 80 out of 100. The first six terms in the equation, including the intercept, are

$$Y = 0.613 - 0.017(1) - 0.014(0) - 0.0002(95)$$
$$- 0.00006(70) - 0.002(80)$$

The process just outlined continues until the value of each variable has been multiplied by the appropriate regression coefficient; then, the products are added to obtain the predicted number of bed-days.

Table 12–3. Regression coefficients and *t* test values for predicting bed-days in RAND study.[1]

Explanatory Variables and Other Measures (X)	Dependent-Variable Equation	
	Coefficient (b)	t Test
Intercept	0.613	22.36
FFS freeplan	−0.017	−2.17
FFS payplan	−0.014	−2.18
Personal functioning	−0.0002	−1.35
Mental health	−0.00006	0.25
Health perceptions	−0.002	−5.17
Age	−0.0001	−0.54
Male	−0.026	−4.58
Income	−0.021	−1.65
Three-year term	0.002	0.44
Took physical	−0.003	−0.56
Income*freeplan	0.021	0.86
Income*payplan	0.0002	0.01
Health*freeplan	0.0002	0.33
Health*payplan	0.0006	1.47
Health*income	0.001	1.88
Health*term3	0.0007	1.79
Health*income* freeplan	−0.0034	−2.13
Health*income* payplan	0.0018	1.42
Bed-day00	0.105	6.15
Sample size	1568	
R-squared	0.12	
Residual standard error	0.01	

[1]Reproduced, with permission, from Ware JE et al: *Health Outcomes for Adults in Prepaid and Fee-for-Service Systems of Care.* (R–3459–HHS. Santa Monica, Calif.: The RAND Corporation, 1987, p. 59.

Other variables in the equation include age, gender (a dummy variable), income, whether the patient was in the study for 3 years or 5 years, and whether the patient took the initial physical examination (both dummy variables), and the bed-days measured at enrollment. There are also some variables with asterisks: income*freeplan, income*payplan, etc. An asterisk indicates multiplication in computer symbols. Therefore, the value of income*payplan is obtained by multiplying the patient's income by a 1 if the patient is a member of the FFS paying group and by a 0 otherwise. These variables are included in the regression equation because the investigators thought there might be an interaction among the variables. Recall from the discussion in Chapter 8 that an interaction occurs if, say, the number of bed-days is large for persons with high income on the FFS freeplan but small for persons with low income on the FFS freeplan. Studies in medicine include interaction terms in the regression equation less often than studies in other fields, but interaction terms are seen on occasion in the medical literature.

As a related issue, multiple regression measures only the linear relationship between the independent variables and the dependent variable, just as in simple regression. For example, the regression equation in this example assumes that the increase in number of bed-days for people between the ages of 20 and 30 is the same as the increase in number of bed-days for people between the ages of 60 and 70. If investigators suspect that the relationship between a given independent variable and the dependent measure is not linear, they may include squared terms (such as age*age) or the logarithm of age [such as ln(age)] in the regression equation, depending on the nature of the suspected curvilinear relationship.

Regression coefficients are interpreted differently in multiple regression than in simple regression. In simple regression, the regression coefficient *b* indicates the amount the predicted value of *Y* changes each time *X* increases by one unit. In multiple regression, a given regression coefficient indicates how much the predicted value of *Y* changes each time *X* increases by one unit, *holding the values of all other variables in the regression equation constant*. For example, the number of predicted bed-days is reduced by 0.026 day if the patient is a male, assuming all other variables are held constant. This feature of multiple regression makes it an ideal method to control for baseline differences, as we discuss in Section 12.3.

12.2.4 Statistical Tests for the Regression Coefficient

Table 12–3 also gives the value for *t* tests associated with each regression coefficient. The *t* test can be used to determine whether or not each regression coefficient is different from zero, or the t distribution can be used to form confidence intervals for each regression coefficient. The investigators in the RAND study chose to do the t test. They also stated that a result was "termed significant if it was likely to occur by chance no more often than one time in twenty (two-tailed test)." Thus, the investigators used $\alpha = .05$ for each statistical test, and they tested for both a positive and a negative relationship between number of bed-days and each independent variable.

The critical value of the t test depends on the sample size, but with $n = 1568$ patients, the value of z may be used; and the two-tailed value for $\alpha = .05$ is ± 1.96. Therefore, regression coefficients with an associated t statistic less than -1.96 or greater than $+1.96$ are declared significant. Accordingly, the significant explanatory variables in predicting bed-days are FFS freeplan and FFS payplan (as discussed above), health perceptions (a positive perception is associated with fewer bed-days), male (another dummy variable with males coded 1 and females 0; males have fewer bed-days), health*income*freeplan interaction indicating differential effects on bed-days depending on the patient's score on this combination of factors), and bed-day00 (the number of bed-days for 30 days prior to enrollment is positively associated with the number of bed-days at the end of the study).

12.2.5 Standardized Regression Coefficients

Some authors present regression coefficients that can be used with individual subjects to obtain predicted Y values, as in the RAND study. But the size of the regression coefficients cannot be used to decide which independent variables are the most important, because their size is also related to the scale on which the variables are measured, just as in simple regression. For example, in the RAND study, the variable "male" was coded 1 if male and 0 if female, and the variable "age" was coded as the number of years of age at enrollment in the study. Then, if gender and age were equally important in predicting number of bed-days, the regression coefficient for male would be much larger than the regression coefficient for age so that the same amount would be added to the prediction for each variable. These regression coefficients are sometimes called *unstandardized;* the only conclusion that can be drawn from them is that their positive or negative sign describes the direction of the relationship or the correlation between them and the dependent variable Y. As another example, in the RAND study, personal functioning is negatively associated with bed-days, since people with higher levels of functioning spend fewer days in bed ($b = -0.0002$). However, the number of days a person spent in bed at the beginning of the study, bed-day00, is positively associated with the number of bed-days at the end of the study ($b = +0.105$).

One way to eliminate the effect of scale is to *standardize* the regression coefficients. Standardizing is done by subtracting the mean value of X and dividing by the standard deviation before analysis, so that all variables have a mean of 0 and a standard deviation of 1. Then, one can compare the magnitudes of the regression coefficients and draw conclusions about which explanatory variables play an important role. (One can also determine the standardized regression coefficients after the regression equation has been calculated for the original values.) The larger the standardized coefficient, the larger the value of the t statistic is. The major disadvantage of **standardized regression coefficients** is that they cannot be easily used to predict outcome values.

12.2.6 Multiple R

Multiple R is the multiple regression analogue of the Pearson product moment correlation coefficient r. It is also called the coefficient of multiple determination, but most authors use the shorter term. As an example, suppose the predicted number of bed-days is calculated for each patient in the RAND study; then, the correlation between the predicted number of bed-days and the actual number of bed-days is calculated. This correlation is the multiple R. If the multiple R is squared (R^2), it is a measure of how much of the variation in the actual number of bed-days is accounted for by knowing the information included in the regression equation. The term R^2 is interpreted in exactly the way r^2 is in simple correlation and regression, with 0 indicating no variance accounted for and 1.00 indicating 100% of the variance accounted for. (Recall that in simple regression, the correlation between the actual value Y of the dependent variable and the predicted value, denoted Y', is the same as the correlation between the dependent variable and the independent variable; ie, $r_{Y'Y} = r_{XY}$. Thus, R and R^2 in multiple regression play the same role as r and r^2 in simple regression. However, the statistical test for R and R^2 used the F distribution instead of the t distribution.)

The computations are time-consuming, but fortunately, computers do them for us. In the RAND study, $R^2 = .12$, indicating that a relatively small amount of the variability in bed-days is accounted for by knowing the patient's health care system, gender, age, etc. Therefore, factors other than those included in the study are needed to predict number of bed-days.

Other outcomes examined in the RAND study were easier to predict. For example, the R^2 values for some of the other regression equations determined in the RAND study were substantially higher: .37 in predicting health perceptions, .62

in predicting smoking, .30 in predicting diastolic blood pressure, .40 in predicting cholesterol level, and .85 in predicting weight.

12.2.7 Stepwise Multiple Regression

The primary purpose of the RAND study was explanation—ie, using multiple regression analysis to determine variables that help explain certain health outcomes. When the regression equation is to be used for prediction of future subjects as well as for explanation, investigators may want to include only the variables that add to prediction in a statistically significant way. This technique has the obvious advantage of requiring less data collection for future applications of the equation.

Reducing the number of variables can be accomplished in several ways. In one approach, all variables are introduced into the regression equation, as in the RAND study; then, the variables that do not have significant regression coefficients are eliminated from the equation. Thus, in a study whose purpose is to predict number of bed-days, all variables except FFS freeplan, FFS payplan, health perceptions, male, health*income*freeplan, and bed-day00 are eliminated with this approach. The regression equation is then recalculated by using only the variables retained, because the regression coefficients have different values when some variables are removed from the analysis.

Computer programs may also be used to select an optimal set of explanatory variables. One such procedure is called *forward selection*. Forward selection begins with one variable in the regression equation; then, additional variables are added one at a time until all statistically significant variables are included in the equation. The first variable in the regression equation is the X variable that has the highest correlation with the response variable Y. The next X variable considered for the regression equation is one that will increase the multiple R^2 by the largest amount. If the increase in R^2 is statistically significant by the F test, it is included in the regression equation. This step-by-step procedure continues until there are no remaining X variables that produce a significant increase in R^2. The values for the regression coefficients are calculated, and the regression equation resulting from this forward selection procedure can be used to predict outcomes for future subjects.

A similar *backward elimination* procedure is also available; in it, all variables are initially included in the regression equation, as in the RAND study. However, instead of the use of t tests to determine which variables are significant predictors, the single X variable that reduces R^2 by the smallest increment is removed from the equation. If the resulting decrease is not statistically significant, that variable is permanently removed from the equation. Next, the remaining X variables are examined to see which produces the next smallest decrease in R^2. This procedure continues until the removal of an X variable from the regression equation causes a significant reduction in R^2. That X variable is retained in the equation, and the regression coefficients are calculated.

When features of both the forward selection and the backward elimination procedures are used together, the method is called **stepwise regression** (stepwise selection). Stepwise selection begins in the same manner as forward selection. However, at each step after the addition of a new X variable to the equation, all previously entered X variables are checked to see whether they maintain their level of significance. Previously entered X variables are retained in the regression equation only if their removal would cause a significant reduction in R^2. There are subtle advantages in the forward vs backward vs stepwise procedures that cannot be covered in this text. They will not generally produce identical regression equations, but conceptually, all approaches determine a "parsimonious" equation using a subset of explanatory variables.

12.2.8 Polynomial Regression

Polynomial regression is a special case of multiple regression in which the independent variables are powers of X. Polynomial regression provides a way to fit a regression model to curvilinear relationships and is an alternative to transforming the data to a linear scale. For example, the following equation can be used to predict a quadratic relationship:

$$Y' = b_0 + b_1X + b_2X^2$$

If analysis of the residuals fails to indicate an adequate fit, a cubic term, a fourth-power term, etc, can also be included until an adequate fit is obtained.

12.2.9 Missing Observations

When studies involve several variables, some observations on some subjects may be missing. Controlling the problem of missing data is easier in studies in which information is collected prospectively; it is much more difficult when information is obtained from already existing records, such as patient charts. Two important factors are whether a small percentage or a large percentage of the observations are missing and whether missing observations are randomly missing or missing because of some causal factor.

For example, suppose a researcher designs a case-control study to examine the effect of leg length inequality on the incidence of loosening of the femoral component after total hip replacement. Cases are patients who have developed loosening of the femoral component, and controls are patients who have not. In the routine follow-up, leg length inequality was measured in some patients by using weight-bearing anterior-posterior (AP) hip and lower-extremity films. Other patients had measurements taken using non-weight-bearing films. However, the type of film ordered during follow-up may well be related to whether the patient turns out to be a case or a control; ie, patients with symptoms are more likely to have had the weight-bearing films, and patients without symptoms are more likely to have had the routine films. A researcher investigating this question must not base the leg length inequality measures on weight-bearing films only, because controls would be less likely than cases to have weight-bearing film measures in their records. In this situation, the missing leg length information occurred because of symptoms and not randomly.

The potential for missing observations increases in studies involving multiple variables. Depending on the cause of the missing observations, solutions include dropping from the study subjects who have missing observations, deleting from the study those variables that have missing values, or substituting some value for the missing data, such as the mean across all other subjects. Investigators in this situation should seek advice from a statistician on the best way to handle the problem.

12.2.10 Cross-Validation

The statistical procedures for all regression models are based on correlations among the variables, which, in turn, are related to the amount of variation in the variables included in the study. Some of the observed variation in any variable, however, occurs simply by chance; and the same degree of variation would not occur if another sample were selected and the study were replicated. The mathematical procedures for determining the regression equation cannot distinguish between real and chance variation. Therefore, if the equation is to be used to predict scores for future subjects, it should be validated on a second sample, a process called **cross-validation.** Cross validating the regression equation gives a realistic evaluation of the usefulness of the prediction it provides.

Alternatively, one can estimate the magnitude of R or R^2 in another sample without actually performing the cross-validation. This R^2 will be smaller than the R^2 for the original sample because the mathematical formula used to obtain the estimate removes the chance variation. For this reason, the formula is called a formula for *shrinkage.* Many computer programs provide both R^2 for the sample used in the analysis as well as R^2 adjusted for shrinkage. For example, see $R-sq(adj)$ in the output for MINITAB in Fig 10–17.

12.2.11 Sample Size Requirements

There are no formulas for determining how large a sample is needed in multiple regression or any multivariate technique. There are, however, some rules of thumb that may be used for guidance. A common recommendation by statisticians calls for ten times as many subjects as there are independent variables. Assumptions about normality are complicated, depending on whether the independent variables are viewed as fixed or random (as in fixed-effects model or random-effects model in ANOVA), and they are beyond the scope of this text. However, having a large ratio of subjects to variables decreases problems that may arise because assumptions are not met. To ensure that estimates of regression coefficients and multiple R and R^2 are accurate representatives of actual population values, investigators should never perform regression unless there are at least twice as many subjects as variables.

12.3 CONFOUNDING VARIABLES: ANALYSIS OF COVARIANCE

Analysis of covariance (ANCOVA) is the statistical technique used to control for the influence of a confounding variable. **Confounding variables** occur most often when subjects cannot be assigned at random to different groups, ie, when the groups of interest already exist. For example, Presenting Problem 2 analyzes ventricular wall motion abnormalities in smokers and nonsmokers. The investigators in this study cannot assign people at random to be smokers and nonsmokers; these groups already exist and must be analyzed as such. When randomization cannot be used in assigning subjects to groups, investigators should control statistically for possible confounding factors.

The regression coefficient for any given variable is interpreted as the change in the dependent variable, holding all other independent variables constant; therefore, the regression model is a perfect way to control for a confounding (nuisance) variable. In Presenting Problem 2, the investigators wanted to know whether smokers have more ventricular wall motion abnormalities than nonsmokers. They can use a t test to determine whether the mean numbers of wall motion abnormalities are different in these two groups. However, the researchers know that wall motion abnormalities are also related to the degree of coronary stenosis, and smokers generally have a greater degree of coronary stenosis. Thus, any difference observed in the mean number of wall abnormalities between smokers and nonsmokers may really be a difference in the amount of coronary stenosis between these two groups of patients.

This situation is illustrated in the graph of hypothetical data in Fig 12–1; in the figure, the relationship between occlusion scores and wall motion abnormalities appears to be the same for smokers and nonsmokers. However, nonsmokers have both lower occlusion scores and lower numbers of all motion abnormalities; smokers have higher occlusion scores and higher numbers of wall motion abnormalities. The question is whether the difference in wall motion abnormalities is due to smoking, to occlusion, or to both.

In this study, the investigators must control for the degree of coronary stenosis so that it does not confound (or confuse) the relationship between smoking and wall motion abnormalities. A very useful method to control for confounding variables is analysis of covariance (ANCOVA). Table 12–2 specifies ANCOVA whenever the independent measures are grouping variables on a nominal scale, such as smoking versus nonsmoking, and there are confounding variables, such as degree of coronary occlusion. [If the dependent measure is also nominal, such as whether or not a patient has survived to a given point in time, the Mantel-Haenzel chi-square procedure (briefly discussed in Chapter 11) can be used to control for the effect of a confounding (nuisance) variable.] ANCOVA can be performed by using the methods of ANOVA discussed in Chapter 8; however, most medical studies employing ANCOVA use the regression methods discussed in this chapter.

If ANCOVA is used in this example, the occlusion score is called the **covariate;** and the mean number of wall motion abnormalities in smokers and nonsmokers is said to be *adjusted for* the occlusion score (or degree of coronary stenosis). Put another way, ANCOVA simulates the Y outcome that would be observed if the value of X were held constant, ie, if all the patients had the same degree of coronary stenosis. This adjustment is achieved by calculating a regression equation to predict Y from the covariate and from a dummy variable coded 1 if the subject is a member of the group (ie, a smoker) and 0 otherwise. For example, the regression equation determined for the hypothetical observations in Fig 12–1 is

$Y = -0.1898 + 0.0113 \times$ Occlusion score $+ 1.2758$ if a smoker

The equation illustrates that smokers have a larger number of predicted wall motion abnormalities, because 1.2758 is added to the equation if the subject is a smoker. In addition, the equation can be used to obtain the mean number of wall motion abnormalities in each group, adjusted for degree of coronary stenosis.

If the relationship between coronary stenosis and ventricular motion is ignored, the mean number of wall motion abnormalities, calculated from the observations in Fig 12–1, is 3.33 for smokers and 1.00 for nonsmokers. If, however, ANCOVA is used to adjust for degree of coronary stenosis, the adjusted mean wall motion is 2.81 for smokers and 1.53 for nonsmokers, a difference of 1.28, which is represented by the regression coefficient for the dummy variable for smoking. In ANCOVA, the adjusted Y mean for a given group is obtained by (1) finding the difference between the group's mean on the independent variable X, denoted $\overline{X}_j$, and the grand mean $\overline{X}.$; (2) multiplying the difference by the regression coefficient; and (3) subtracting this product from the unadjusted Y mean. Thus, for group j, adjusted mean $\overline{Y}_j = \overline{Y}_j - b(\overline{X}_j - \overline{X}.)$. (See Exercise 1.)

In this example, ANCOVA results in a small difference in the mean number of wall motion abnormalities for smokers and nonsmokers adjusted for occlusion. This result is consistent with our knowledge that coronary stenosis alone has some effect on abnormality of wall motion; the unadjusted means contain this effect as well as any effect from smoking. Therefore, controlling for the effect of coronary stenosis results in a smaller difference in number of wall motion abnormalities, a difference related only to smoking.

Fig 12–2 illustrates schematically the way ANCOVA adjusts the mean of the independent variable. Using unadjusted means is analogous to using a separate regression line for each group. For example, the mean value of Y for group 1 is found by using the regression line drawn through the group 1 observations to project the mean value of X onto the Y-axis, denoted $\overline{Y}_1$ in Fig 12–2. Similarly, the mean of Group 2 is found at $\overline{Y}_2$ by using the regression line drawn through the observations in that group. Conceptually the Y means ad-

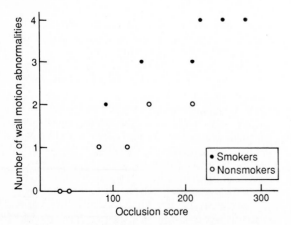

Figure 12–1. Relationship between degree of coronary stenosis and ventricular wall motion abnormalities in smokers and nonsmokers (hypothetical data).

justed for the covariate (stenosis) are analogous to the projections using a common regression line, ie, a regression equation calculated for the entire sample of observations that takes the covariate into account. The adjusted means for groups 1 and 2, Adj. $\overline{Y}_1$ and Adj. $\overline{Y}_2$, are illustrated by the projections from the regression line.

ANCOVA assumes that the relationship between the covariate (X variable) and the independent variable (Y) is the same in both groups, ie, that any relationship between coronary stenosis and wall motion abnormality is the same for smokers and nonsmokers. This assumption is equivalent to requiring that the regression slopes be the same in both groups; geometrically, ANCOVA asks whether there is a difference between the intercepts, assuming the slopes are equal.

There are many instances in medical research in which ANCOVA is appropriate. For example, age is a variable that affects almost anything studied in medicine; if preexisting groups in a study have different age distributions, investigator must adjust for age before comparing the groups on other variables. The methods illustrated in Chapter 4 to adjust mortality rates for characteristics such as age and birth weight are used when information is available on groups of individuals; when information is available on individuals themselves, ANCOVA is used.

Before leaving this section, we point out some important aspects of ANCOVA. First, although there were only two groups in the example we used, ANCOVA can be used to adjust for the ef-fect of a confounding variable in more than two groups. In addition, one can adjust for more than one confounding variable in the same study, and the confounding variable may be nominal as well as numerical. For instance, the RAND study comparing HMO and FFS health plans may be viewed as an example of ANCOVA. The investigators in this study were interested in whether the type of health care systems had an effect on the number of bed-days. They examined three groups of patients: (1) those in an HMO; (2) those in a fee-for-service plan in which subjects participated in paying for care (FFS payplan); and (3) those in a fee-for-service plan in which subjects were provided with free care (FFS freeplan). The investigators knew that many other factors besides type of plan may be related to number of bed-days, however, and they used regression to control for these variables. The regression coefficients for type of health plan were statistically significant in the analysis given in Table 12–3, indicating that patients in both FFS plans had fewer bed-days, controlling for the influence of all other variables considered in the study.

Finally, ANCOVA can be considered as a special case of the more general question of comparing two regression lines (discussed in Chapter 10, Section 10.5.) In ANCOVA, we assume that the slopes are equal, and attention is focused on the intercept. However, we can also perform the more global test of both slope and intercept by using multiple regression. In Chapter 10, Presenting Problem 1, interest was in comparing the regression lines predicting C1q binding from IgG complexes in patients with systemic lupus erythematosus (SLE) and in patients with steroid-responsive nephrotic syndrome (SRNS); we noted that the best procedure involved the use of dummy coding. If we let X be IgG complex, Y be C1q binding, and Z be a dummy variable—where $Z = 1$ if the patient is a member of the group with SLE and $Z = 0$ if the patient has SRNS—then the multiple regression model for testing whether the two regression lines are the same (coincident) is

$$Y = a + b_1X + b_2Z + b_3XZ$$

The regression lines have equal slopes and are parallel when b_3 is zero; ie, there is no interaction between the independent variable X and the group membership variable Z. The regression lines have equal intercepts and equal slopes (are coincident) if both b_2 and b_3 are zero; thus, the model becomes the simple regression equation $Y = a + bX$. The statistical test for b_2 and b_3 is the t test discussed in Section 12.2.4.

The comparison of two regression lines is relatively simple to do by using stepwise multiple regression, where the variables X, Z, and their product XZ are added to the regression equation one

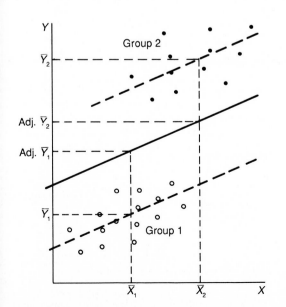

Figure 12–2. Illustration of means adjusted using analysis of covariance.

at a time. The point at which the addition of a new term does not add significantly to the prediction of Y determines which model applies.

12.4 PREDICTING A CENSORED OUTCOME: PROPORTIONAL HAZARDS MODEL

In Chapter 11, we found that special methods must be used when the outcome has not yet been observed for all subjects in the study sample. Studies of survival naturally fall into this category; investigators usually cannot wait until all patients in the study die before presenting information on survival. From an analysis perspective, the problem is one of **censored observations,** a situation in which subjects have been observed unequal lengths of time and the outcome is not yet known for all of them. We examined methods for comparing duration of survival of two or more groups of patients in which the groups are formed on the basis of one variable. But many times in clinical trials or cohort studies, investigators wish to look at the simultaneous effect of several variables on length of survival; ie, they would like to do an ANCOVA on censored data. For example, in the study described in Presenting Problem 3, Khaw and Barrett-Connor (1987) wanted to determine whether dietary potassium is related to mortality from stroke, independent of other known cardiovascular risk factors. They analyzed data from a cohort study of 859 people, but only 24 stroke-associated deaths had occurred in the 12 years the study had been under way; thus, the observations on survival were censored.

Table 12–2 indicates that the regression technique developed by Cox (1972) is appropriate when there are time-dependent censored observations. This technique is called the **Cox regression,** or **proportional hazards, model.** In essence, this model allows the covariates (independent variables) in the regression equation to vary with time. The dependent variable is the survival time of the ith patient, denoted Y_i. Both numerical and nominal independent variables may be used in the model.

The Cox regression coefficients can be used to determine the relative risk (introduced in Chapter 4) between each independent variable and the outcome variable, adjusted for the effect of all other variables in the equation, as we will see in the analysis of Presenting Problem 3. Thus, instead of giving adjusted means, as ANCOVA does in re-

gression, the Cox model gives adjusted relative risks. One may also use stepwise methods to select the independent variables that add significantly to the prediction of the outcome, as in multiple regression; however, a chi-square test (instead of the F test) is used to test for significance. Unfortunately, this model involves a complicated exponential equation that is beyond the scope of this text; we can only discuss its application by using Presenting Problem 3.

In the study described in Presenting Problem 3, investigators used the Cox model to examine the relationship of dietary potassium to stroke mortality, adjusted for age, systolic blood pressure, blood cholesterol level, obesity, fasting blood glucose level, cigarette smoking, and estrogen use in women. From the regression equation, they determined the adjusted relative risk for each of these variables—ie, the relative risk independent of the other independent variables. For example, the relative risk for stroke in women having a daily intake of 10mmol potassium is 0.56, indicating that potassium has a protective effect against stroke, independent of the patient's age, blood pressure, cholesterol, etc. (See Table 12–4.)

Note that the upper confidence interval limit in Table 12–4 for the relative risk associated with potassium intake is less than 1 in women, indicating

Table 12–4. Independent relative risks for stroke-associated mortality in men and women.[1]

Risk Factor	Relative Risk and 95% Confidence Intervals	
	Men	**Women**
Potassium (per 10 mmol)	0.65 (0.41–1.00)	0.56 (0.38–0.82)
Age (per 5 years)	1.88 (1.00–3.63)	1.59 (1.00–2.59)
Systolic blood pressure (per 20 mm Hg)	1.01 (0.49–2.05)	1.35 (0.89–2.04)
Cholesterol (per 40 mg/ dL)	0.62 (0.29–1.31)	0.89 (0.54–1.47)
Fasting plasma glucose (per 20 mg/ dL)	1.64 (1.13–2.30)	1.30 (0.90–1.88)
Body-mass index (per 0.5 kg/m²)	1.18 (0.56–2.50)	1.08 (0.68–2.91)
Smoking (yes or no)	3.44 (0.61–19.5)	1.66 (0.89–5.70)
Estrogen use (yes or no)		0.79 (0.21–3.00)

[1]Adapted and reproduced, with permission, from Table 4 in Khaw K, Barrett-Connor E: Dietary potassium and stroke-associated mortality. *N Engl J Med* 1987; **316:**235.

---

*Recall from Chapter 11 that a hazard function gives the probability that a patient dies during a specific time interval, given that the patient lived until the beginning of the interval. In the Cox model, the covariates have a multiplicative, or proportional, effect on the probability of dying—thus, the term *proportional hazards* model.

that potassium intake is inversely related to stroke, after adjustment for age, systolic blood pressure, blood cholesterol level, obesity, fasting blood glucose level, cigarette smoking, and estrogen use—ie, increased consumption of potassium is associated with lower mortality from stroke. The upper limit in men is equal to 1.00, indicating a probable inverse relationship in men as well. Other interesting associations with mortality from this study include the positive association with age (even when adjusted for the other risk factors) and, for men, the positive relationship with fasting plasma glucose. (No other factors are significant because the 95% confidence intervals include 1.)

12.5 PREDICTING NOMINAL OR CATEGORICAL OUTCOMES

The regression models discussed in the previous sections have one common feature: The dependent Y variable is numerical. This section describes three models that can be used when the dependent variable is nominal. From Table 12–2, the three methods are log-linear analysis, logistic regression, and discriminant analysis. The choice of which method to use depends on whether the Y value has more than two values and on the scale of measurement for the independent variables.

12.5.1 Logistic Regression

Logistic regression is a model appearing with increasing frequency in the medical literature; it is used when the independent variables include both numerical and nominal measures and the outcome variable is dichotomous, having only two values. Presenting Problem 4 illustrates the use of logistic regression to predict whether or not a child wets the bed, a yes-or-no outcome. The logistic model gives the probability that the outcome, such as bed-wetting, occurs as an exponential function of the independent variables. For example, with three independent variables, the model is

$$p_x = \frac{1}{1 + \exp[-(b_0 + b_1 x_1 + b_2 x_2 + b_3 x_3)]}$$

where b_0 is the intercept, b_1, b_2, and b_3 are the regression coefficients, and exp indicates that the base of the natural logarithm (2.718) is taken to the power shown in parentheses. The equation can be derived by using a stepwise method similar to the one for multiple regression; a chi-square test (instead of the F test) is used to determine whether a variable adds significantly to the prediction.

In the study described in Presenting Problem 4, the variables used by the investigators to predict enuresis included age of the child (in years), gender (a dummy variable with girls coded 1, boys

coded 0), race (1 for white, 0 for nonwhite), number of years of education of the mother, family income (transformed to a logarithm scale), family size (transformed to the square root scale), and psychologic distress (0–10, with 10 being most distressed). Only three of these variables were found to be significant in predicting the occurrence of enuresis. Note that some authors, such as the investigators reporting on dietary potassium and stroke (Presenting Problem 3), present results on all predictor variables, regardless of whether or not they are significant. Other authors give the results only for significant variables, as do the investigators in the study described in Presenting Problem 4. The results of the analysis are given in Table 12–5.

The three significant variables given in Table 12–5 are psychologic distress, gender, and age, with the last two being inversely related to occurrence of enuresis; ie, males and younger children are more likely to have enuresis. The logistic regression for predicting wetting the bed at least once in the past 3 months is

$$p = \frac{1}{1 + \exp\{-[(0.68) + (0.16 \times \textbf{Psychologic distress}) - (0.43 \text{ if a girl}) - (0.32 \times \textbf{age})]\}}$$

For example, a five-year-old girl with a psychologic distress score of 6 has a probability equal to .40 of wetting the bed at least once in the last three months:

$$p = \frac{1}{1 + \exp\{-[(0.68) + (0.16 \times 6) - (0.43) - (0.32 \times 5)]\}}$$

$$= \frac{1}{1 + \exp[-(-0.39)]}$$

Table 12–5. Logistic regression predicting occurrence and frequency of enuresis.[1]

Independent Variable	Coefficients of Dependent Variables	
	Wet at Least Once in Previous 3 Months ($n = 1704$)	Wet at Least 1 Time per Week ($n = 1702$)
Intercept	0.68[2]	−0.86[2]
Psychologic distress	0.16[3]	0.14[2]
Gender (girls)	−0.43[4]	−0.20
Age	−0.32[3]	−0.25[3]

[1]Reproduced, with permission, from Table 2 in Foxman B, Valdez B, Brook RH: Childhood enuresis: Prevalence, perceived impact, and prescribed treatments. *Pediatrics* 1986; **77**:482.
[2]$p < .05$.
[3]$p < .001$.
[4]$p < .01$.

$$= \frac{1}{1 + \exp(0.39)}$$

$$= \frac{1}{1 + 1.48}$$

$$= .40$$

A major advantage of using logistic regression is that it requires no assumptions about the distribution of the independent variables. Another advantage is that the regression coefficient can be interpreted in terms of relative risks in cohort studies or odds ratios in case-control studies. In other words, the relative risk of wetting the bed if the subject is a girl is $\exp(-0.43) = .65$; thus, gender is an inversely related factor for girls, since the relative risk is less than 1.

12.5.2 Discriminant Analysis

Logistic regression is used almost exclusively in the biological sciences. A related technique, **discriminant analysis,** although used with less frequency in medicine, is a common technique in the social sciences. It is similar to logistic regression in that it is used to predict a nominal or categorical outcome.

Discriminant analysis differs from logistic regression in two important ways: (1) It assumes that the independent variables follow a multivariate normal distribution, so it must be used with caution if some X variables are nominal; and (2) it can be used with a dependent variable that has more than two values. Therefore, if a researcher wants to determine factors that distinguish, say, between patients with microcytic, macrocytic, or normocytic anemia, discriminant analysis should be used, because the dependent variable, type of anemia, has more than two values.

The procedure involves determining several discriminant functions, which are simply linear combinations of the independent variables, that separate, or discriminate among the groups as much as possible. The number of discriminant functions needed is determined by a multivariate test statistic called Wilks' lambda. The discriminant functions' coefficients can be standardized and then interpreted in the same manner as in multiple regression to draw conclusions about which variables are important in discriminating among the groups.

In many research situations, either logistic regression or discriminant analysis can be used, depending on how the problem is defined. For example, discriminant analysis can be used in the enuresis research instead of logistic regression. Another way to define this problem is to discriminate between children who wet their beds and those who do not. The investigators in Presenting

Problem 4 chose to perform two logistic analyses, one to predict whether or not the child had wet the bed in the past 3 months and a second to predict whether the child wet the bed at least once per week during the previous 3 months. Alternatively, wetting the bed can be classified in three categories: (1) no bed-wetting in the last 3 months, (2) bed-wetting once weekly or less in the last 3 months, and (3) bed-wetting more than once weekly in the last 3 months. Discriminant analysis can then be used because there are more than two outcome values, and only one analysis is required. The choice between logistic regression and discriminant function analysis depends primarily on the way the investigator conceptualizes the research problem and, to some extent, on the traditional methods used in a discipline.

Although discriminant analysis is most often employed to explain or describe factors that discriminate among groups of interest, the procedure can also be used to classify future subjects. Classification involves determining a separate prediction equation corresponding to each group that gives the probability of belonging to that group. For classification of a future subject, a prediction is calculated for each group; and the individual is classified as belonging to the group he or she most closely resembles.

In the following chapter we illustrate methods for adjusting the probability of an event from information related to the occurrence of the event; specifically, the probability that a patient has a disease once the results of a diagnostic test are known. These same methods can be used in discriminant analysis to adjust the prediction that an individual belongs to any given group. For example, in the childhood enuresis study, 86% of the children in the study had not wet their bed in the past month. Therefore, in the prediction of whether any particular child has wet the bed during the past month, the correct prediction will be made 86% of the time if the prediction of a dry bed is made. This prior information can be taken into consideration when considering the other factors that may influence whether or not a child has wet the bed.

12.5.3 Log-Linear Analysis

Log-linear analysis may be interpreted as a regression model in which all the variables, both independent and dependent, are measured on a nominal scale. For example, all the variables in Presenting Problem 5 for predicting a game-related concussion are nominal variables: team (offense or defense), situation (rushing or passing), activity (blocking or tackling), and concussion (yes or no). With more than one independent variable, the data can be displayed in complex contingency tables, such as Table 12–6, in which the fre-

Table 12-6. Adjusted frequency of game-related concussions by team, situation, and activity for NAIRS-I/II college football.[1]

Team	Situation	Activity	
		Tackle	Block
Offense	Rushing	41	274
	Passing	17	39
Defense	Rushing	109	197
	Passing	18	30

[1]Reproduced, with permission, from Table 4 in Buckley WE: Concussions in college football. *Am J Sports Med* 1988; **16**:51.

quency of concussion is given for the offensive team in both rushing and passing situations and for both tackle and block activities, and then is repeated for the defensive team in each of the four possible situation-activity combinations. The situation can quickly become complex with a large number of variables. Recall that chi-square analysis is used when there are two variables (both nominal) and the question is whether or not there is a relationship between the two variables. If there are more than two variables (all nominal), log-linear analysis can determine whether there is a relationship among them. The technique is called log-linear because it involves using the logarithm of the observed frequencies in the contingency table.

The frequencies of game-related concussions from Presenting Problem 5 are given in Table 12-6. The variable "team" (offense vs defense) results in equal exposure to risk: 11 players on offense and 11 players on defense. However, the "situation" and "activity" variables do not result in equal exposure; eg, there are more running than passing situations and more blocking than tackling activities. Therefore, the author adjusted the frequencies of concussion prior to the analysis to reflect equal exposure by assuming a ratio of 6:1 blocks to tackles and 2.2 rushing plays for every passing play. These ratios were based on NCAA yearly football statistics. The adjusted frequencies were then analyzed by using log-linear analysis.

Recall that the chi-square statistic uses expected values calculated by assuming that row and column factors are independent; ie, expected cell frequency = (row frequency × column frequency) ÷ total frequency. Similarly, log-linear analysis is based on a multiplicative model. For example, in very simple terms, the log-linear model in the concussion example can be written as

$$F = \tau \times \tau_1 \times \tau_2 \times \tau_3$$

where F is the predicted frequency in each cell (or combination of factors) and τ is the frequency based on equal numbers in each cell, assuming there is no relationship between the independent variables and the frequencies of concussion.*

As an example, let's assume that there is no relationship between team and concussion, situation and concussion, or activity and concussion. Then, the best guess of the frequency of occurrence of each situation in Table 12-6 is the total number of concussions (725) divided by the number of cells in the table (8), or 90.6. The term τ_1 is the modification to the predicted cell frequencies F if the marginal frequencies (ie, row or column totals) of the first independent variable, whether the team is on offense or defense, are considered. For example, if it matters whether a team is on offense or defense, then the 371 concussions that occurred on offense are evenly divided among the 4 cells related to offense (with $F = 92.75$ in each cell), and the 354 concussions that occurred on defense are evenly divided among the 4 cells related to defense (with $F = 88.5$ in each cell). Similarly, τ_2 is the modification to F if the marginal frequencies of the second independent variable are considered, etc.

From the log-linear analysis of observations in Table 12-6, Buckley (1988) concluded that the highest risk of concussion is to an offensive player involved in a block on a rushing play. At the second highest level of risk are defensive players involved in a block during a rushing play. Furthermore, rushing plays result in the highest risk of injury, regardless of team or activity. The lowest risk occurs during passing plays for both offensive and defensive teams, regardless of whether they are blocking or tackling.

Log-linear analysis may also be used to analyze multidimensional contingency tables in situations in which there is no distinction between independent and dependent variables, ie, when investigators simply want to examine the relationship among a set of nominal measures. The fact that log-linear analysis does not require distinguishing between independent and dependent variables points to a major difference between it and other regression models—namely, that the regression coefficients are not interpreted in log-linear analysis.

---

*The actual log-linear model is generally transformed to an additive model by taking the logarithm of the multiplicative model; ie,

$$\ln F = \ln \tau + \ln \tau_1 + \ln \tau_2 + \ln \tau_3$$

In addition, $\ln \tau$ and $\ln \tau_i$ are generally written as λ and λ_i; so the model appears as

$$\ln F = \lambda + \lambda_1 + \lambda_2 + \lambda_3$$

12.6 COMBINING THE RESULTS FROM SEVERAL STUDIES: META-ANALYSIS

Meta-analysis is a way to combine results of several independent studies on a specific topic. Meta-analysis is different from the methods discussed in the preceding sections because its purpose is not to identify risk factors or to predict outcomes for individual patients; rather, it is a technique applicable to any research question. It is included in this chapter because it is an important but relatively advanced technique.

The idea of summarizing a set of studies in the medical literature is not new; review articles written by knowledgeable authors have long had an important role in helping practicing physicians keep up to date and make sense of the many studies on any given topic. Meta-analysis takes the review article a step further by using statistical procedures to combine the results from different studies. Glass (1977) developed the technique because many research projects are designed to answer similar questions but they do not always come to similiar conclusions. The problem for the practitioner is to determine which study to believe, a problem unfortunately too familiar to readers of medical research reports. A concise review of meta-analysis applications in medicine is given by L'Abbé, Detsky, and O'Rourke (1987).

Sacks et al (1987) reviewed meta-analyses of clinical trials and concluded that there are four purposes of meta-analysis: (1) to increase statistical power by increasing the sample size; (2) to resolve uncertainty when reports do not agree; (3) to improve estimates of effect size; and (4) to answer questions not posed at the beginning of the study. Purpose 3 requires some expansion since the concept of effect size is central to meta-analysis. Cohen (1977) developed this concept and defined **effect size** as the "degree to which the phenomenon is present in the population." An effect size may be thought of as an index of how much difference there is between two groups—generally, a treatment group and a control group. The effect size is based on means if the outcome is numerical, on proportions if the outcome is nominal, or on correlations if the outcome is an association. It is the effect sizes that are statistically combined in meta-analysis.

The investigators in Presenting Problem 6 used meta-analysis to evaluate the efficacy of perioperative total parenteral nutrition (TPN). They examined the literature, using manual and computerized searches, for articles on parenteral nutrition and its synonyms and found 18 trials written in English that reported endpoints of either complications from surgery or death. For each study, the authors developed a quality score based on the methods used to enroll patients, assign patients to treatment groups, and assess clinical outcomes. For example, studies were given higher quality scores if they were randomized, if strict criteria were used to assess complications, and if clinical outcomes were blindly assessed. Some basic information about the studies evaluated in this meta-analysis are given in Table 12–7.

The authors of the meta-analysis article calculated the difference between the complication rates observed in each study (controls minus TPN, so a positive difference indicates that the complication rate with TPN is lower) and the 95% confidence intervals for the differences. These differences and intervals are illustrated in Fig 12–3. You may wonder why the 95% confidence limits for the difference between proportions in Fig 12–3 are not symmetric about the difference—ie, why the X does not appear in the center of the confidence interval. The reason is that the authors used a different method to calculate confidence limits from the one we discussed in Chapter 9. The method used in the article is based on Fisher's exact test and is viewed by some statisticians (Yates, 1984) as being more appropriate with small samples in which the normal approximation may not apply.

From the data in Table 12–7 and Fig 12–3, it appears that only study i reported a statistically significant difference in complications rates in favor of TPN. However, several of the studies had

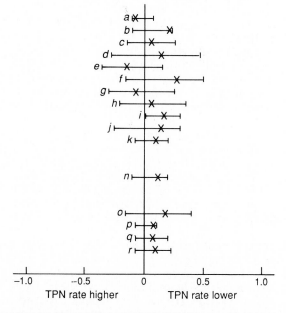

Figure 12-3. Differences in rates of complications. X is the difference between the control group and the TPN group; the bar gives 95% confidence limits for the differences. Letters a–r correspond to the studies; studies l and m did not report complication rates. (Adapted and reproduced, with permission, from Detsky AS et al: Perioperative parenteral nutrition: A meta-analysis. *Ann Intern Med* 1987;**107**:195.)

Table 12-7. Characteristics of studies reviewed on parenteral nutrition.[1]

Study[2]	Randomized	Quality Score	Number of Complications[3] Control	Number of Complications[3] Treatment	Positive Treatment Effect
a	Yes	0.28	0/24	2/20	No
b	Quasi[4]	0.10	1/5	0/10	No
c	Yes	0.20	5/26	4/30	No
d	Quasi	0.18	3/9	2/10	No
e	Quasi	0.54	4/23	8/24	No
f	Yes	0.22	4/10	1/10	No
g	Yes	0.16	1/9	2/12	No
h	Yes	0.36	18/32	15/30	No
i	Yes	0.22	19/59	19/112	Yes
j	Yes	0.10	2/10	1/10	No
k	Yes	0.22	9/55	5/58	No
l	Yes	[5]	[6]	[6]	No
m	Yes	[5]	[6]	[6]	No
n	Yes	[5]	4/27	1/25	No
o	No[7]	0.04	5/15	5/29	No
p	No	0.30	1/10	0/20	No
q	No	0.33	23/105	7/54	No
r	No	0.30	34/177	8/60	No

[1]Adapted and reproduced, with permission, from Tables 1 and 2 in Detsky AS et al: Perioperative parenteral nutrition: A meta-analysis. *Ann Intern Med* 1987; **107**:195.
[2]See the article by Detsky et al (1987) for article references corresponding to letters a–r.
[3]The number of complications is given in the numerator; the denominator contains the number of subjects in the group.
[4]Assignment was by birth date, chart number, or other nonrandom method.
[5]Abstracts that could not be evaluated in detail.
[6]Not given.
[7]Crossover permitted by physician after random assignment.

relatively small sample sizes, and the failure to find a significant difference may be due to low power. Using meta-analysis to combine the results from these studies can provide insight on this issue.

A meta-analysis does not simply add the means or proportions across studies to determine an "average" mean or proportion. Although there are several different methods for combining results, they all use the same principle of determining an effect size in each study and then combining the effect sizes in some manner. For example, the effect size in a study comparing two means is the difference between the means divided by the pooled, or "average," standard deviations from the two groups. Similarly, if two proportions or rates are the statistics used in a study, the effect size is the difference between the two proportions divided by the pooled standard deviations of the proportions. Effect sizes are amenable to interpretation themselves. For example, an effect size of 0.8 or more, defined by Cohen (1977) as a "large" effect, indicates that the means of the two groups

are separated by 0.8 standard deviation; effect sizes of 0.5 and 0.2, defined as "moderate" and "small," respectively, indicate that the means of the two groups are separated by 0.5 and 0.2 standard deviation. Fig 12-4 illustrates effect sizes of 0.8, 0.5, and 0.2 in normally distributed populations.

The methods for combining the effect of sizes include the z approximation for comparing two proportions (Chapter 9); the t test for comparing two means (Chapter 7); and the P-values for the comparisons. The values of z or t or the P-value corresponding to the effect size in each study are the numbers combined in the meta-analysis to provide a pooled (overall) P-value for the combined studies.

The authors in the meta-analysis study described in Presenting Problem 6 calculated the effect size for the difference in proportions, determined the z value for the difference in each study, and then pooled (averaged) the z values to obtain an overall measure of difference in complication rates. They presented their results separately for

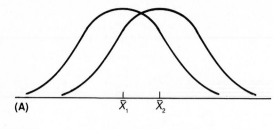

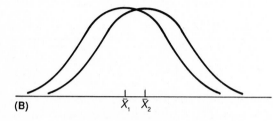

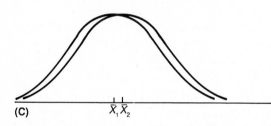

Figure 12-4. Illustration of effect sizes. (*A*) Effect size of 0.8; ie, means differ by 0.8 standard deviation. (*B*) Effect size of 0.5; ie, means differ by 0.5 standard deviation. (*C*) Effect size of 0.2; ie, means differ by 0.2 standard deviation.

their evaluation for the meta-analysis; and some studies had high rates of iatrogenic complications with TPN. In addition, many of the studies included patients who were not malnourished, and possibly TPN is not beneficial for this selected group of patients. From the meta-analysis, the authors conclude that routine use of TPN in unselected patients is not justified. They recommend that future studies be done in selected subgroups of patients at high risk to determine the efficacy of TPN in this group.

There is one potential bias in meta-analysis studies that the cautious reader must keep in mind: Most meta-analyses are based on the published literature, and some people believe that it is easier to publish studies with "significant" results than studies that show no difference. This potential bias is of less concern for Presenting Problem 6 because the conclusion was that TPN does not make a difference, at least in the unselected population of patients on which the studies were based. If the results indicated a preference for TPN, however, the question of publication bias would be appropriate. In this situation, one could estimate how many studies showing no difference would have to be done but not published to raise the pooled *P*-value above the .05 level so that the combined results would no longer be significant. The reader can have more confidence in the conclusions from a meta-analysis that finds a significant effect if a large number of unpublished negative studies would be required to repudiate the overall significance.

12.7 OTHER METHODS FOR MULTIPLE VARIABLES

To make this chapter as complete as possible, we briefly mention four other methods occasionally referred to in the medical literature: factor analysis, cluster analysis, multivariate analysis of variance (MANOVA), and canonical correlation. These methods tend to be more common in the psychiatric and behavioral sciences literature, but they are sometimes seen in the literature of other

randomized trials and nonrandomized trials; these results are given in Table 12-8. The pooled results of the 11 trials that were randomized or quasi-randomized (using birth dates, chart numbers, etc) indicate only weak evidence that total parenteral nutrition reduces the risk for complications from major surgery, with a pooled *P*-value of. 21. The 95% confidence interval includes zero, ranging from −0.023 to 0.128. However, there were many flaws in the designs of these studies, according to

Table 12-8. Pooled analysis of complication rates from parenteral nutrition.[1]

Complications from Surgery	Proportion with Complication		Difference	*P*-Value	95% Confidence Limits
	Control	Treatment			
Randomized papers[2]	0.251	0.18	0.052	.21	−0.023, 0.128
All randomized trials[3]	0.24	0.17	0.057	.13	−0.013, 0.127
All trials[3]	0.235	0.16	0.071	.014	0.021, 0.124

[1]Adapted and reproduced, with permission, from Table 3 in Detsky AS et al: Perioperative parenteral nutrition: A meta-analysis. *Ann Intern Med* 1987; **107**:195.
[2]Ignores abstracts.
[3]Includes abstracts.

specialties as well. These methods all involve multiple measurements on each subject, but they have different purposes—ie, they are used to answer different research questions.

12.7.1 Factor Analysis

Educators frequently wish to know how students score on a set of factors or dimensions, such as ability in reading comprehension, arithmetic computation, and spatial perception skills. Similarly, psychiatrists need to know how a patient scores on the Minnesota Multiphasic Personality Inventory (MMPI), a commonly used personality test, in order to decide how to counsel the patient. In both examples, tests with a large number of items have been developed; students or patients take the test, and scores on various items are combined to produce scores on the relevant factors. **Factor analysis** is the statistical method used to determine which items are combined to generate a given factor, and it is most commonly used in the psychiatric literature. In a research problem in which factor analysis is appropriate, a large number of people are measured on a set of variables. There is no designation of independent and dependent variables—or, if so, we say there are only independent variables.

Conceptually, a factor analysis works as follows: First, as we stated above, a large number of people are measured on a set of items; a rule of thumb calls for at least ten times as many subjects as items. The second step involves calculating correlations. To illustrate, suppose there are 500 arthritis patients who have answered 50 items about their level of physical activity. Factor analysis answers the question of whether or not some of the items group together in a logical way, say items that measure the same underlying component of physical activity. For example, some items may relate to use of the hands, other items may relate to walking, and still others may relate to general activity. If two items measure the same component, such as picking up a fork and using a comb, they can be expected to have higher correlations with each other than with other items, such as getting up from a chair. Therefore, in the second step, correlations are calculated among all 50 questions, producing a correlation matrix with 50 rows and 50 columns. (Each entry in the correlation matrix is the correlation between two items; the diagonal entries are equal to one since the correlation between a variable and itself will be perfect.)

In the third step, factor analysis manipulates the correlations among the items to produce linear combinations of the items, similar to a regression equation. The difference is that each linear combination, called a *factor,* is determined so that the first one accounts for the most variation among the items, the second factor accounts for the most

residual variation after the first factor is taken into consideration, etc. Although there are as many factors as there are variables, generally a much smaller number of factors will account for enough of the variation among subjects that this smaller number of factors may be used to draw inferences about a patient's level of activity. For example, it is much more convenient to refer to scores for use of the hands, walking, and general activity than to refer to scores on the original 50 items. Thus, the fourth step involves determining how many factors are needed and how they should be interpreted.

Investigators who use factor analysis usually have an idea of what the important factors are, and they design the items accordingly. There are many other issues in factor analysis, such as how to derive the linear combinations, how many factors to retain for interpretation, and how to interpret the factors. Using factor analysis, as well as the other multivariate techniques, requires considerable statistical skill.

12.7.2 Cluster Analysis

A statistical technique similar conceptually to factor analysis is **cluster analysis.** The difference is that cluster analysis attempts to find similarities among the *subjects* that were measured instead of among the measures that were made. The object in cluster analysis is to determine a classification or taxonomic scheme that accounts for variance among the subjects. Another way to conceptualize cluster analysis is to think of it as similar to discriminant analysis, except that the investigator does not know which group the subject belongs to. As in factor analysis, there is no dependent variable.

Cluster analysis is frequently used in archeology and paleontology to analyze objects to see whether they have similarities that imply they belong to the same taxon. Biologists use this technique to help determine classification keys, such as a key to classify leaves or flowers according to the appropriate species. Journalists and marketing analysts also use cluster analysis, referred to in these fields as Q-type factor analysis, as a way to classify readers and consumers into groups with common characteristics. Physicians have not used this technique to a great extent, except in areas of interdisciplinary research with one of these other fields.

12.7.3 Multivariate Analysis of Variance (MANOVA)

Multivariate analysis of variance and canonical correlation are similar to each other in that they both involve *multiple dependent* variables as well as multiple independent variables; thus, they are not listed in Table 12–2. **Multivariate analysis of** variance (**MANOVA**) conceptually (although not computationally) is a simple extension of the ANOVA designs discussed in Chapter 8 to situations in which there are two or more dependent variables. As with ANOVA, MANOVA is appropriate when the independent variables are nominal or categorical and the outcomes are numerical.

For example, in Presenting Problem 2 of Chapter 8, Francis and his colleagues (1985) examined neurochemical changes in brain biopsy specimens of patients with Alzheimer's disease and of controls. The patients and controls were also classified by age, and a two-way ANOVA was done to determine whether there were differences in choline acetyltransferase between patients and controls or between those with early and late onset. In actuality, choline acetyltransferase was only one of five neurochemical activities or substances analyzed; Table 8–4 shows that differences in serotonin, norepinephrine, 3-methoxy-4-hydroxyphenylglycol, and 5-hydroxyindoleacetic acid were also analyzed. Instead of performing five separate ANOVAs, the authors in this study could have first performed a MANOVA to ask the global question; Are there any differences according to disease status or to age? If the answer was yes—from a test of the null hypothesis of no differences, using the multivariate statistic called Wilks' lambda—follow-up ANOVAs could have been done to see which of the five substances were different.

The motivation for doing MANOVA prior to univariate ANOVA is similar to the reason for performing univariate ANOVA prior to t tests: to eliminate doing many significance tests and increasing the likelihood that a chance difference is declared significant. In addition, MANOVA permits the statistician to look at complex relationships among the dependent variable.

12.7.4 Canonical Correlation Analysis

Canonical correlation analysis also involves both multiple independent and multiple dependent variables. This method is appropriate when *both* the independent variables and the outcomes are numerical and the research question focuses on the relationship between the set of independent variables and the set of dependent variables. For example, suppose the researchers in the RAND study of different health care systems wished to examine the overall relationship between indicators of health outcome (independent variables: type of system, the patient's personal functioning, mental health, health perceptions, age, gender, income, etc) measured at the beginning of the study and the set of outcomes (dependent variables: physical functioning, role functioning, mental health, social contacts, smoking, serious symptoms, bed-days, cholesterol, weight, blood pressure, etc) measured at the end of the study. Ca-

nonical correlation analysis forms a linear combination of the independent variables to predict not just a single outcome measure but a linear combination of outcome measures. The two linear combinations of independent variables and dependent variables, each resulting in a single number (or index), are determined so that the correlation between them is as large as possible. The correlation between the pair of linear combinations (or numbers or indices) is called the canonical correlation. Then, as in factor analysis, a second pair of linear combinations is derived from the residual variation after the effect of the first pair is removed; etc. Generally, the first two or three pairs of linear combinations account for sufficient variation, and they can be interpreted to gain insights about related factors or dimensions.

Canonical correlation analysis is not yet used often in medicine, but it is a relatively new statistical method. There are many research situations in medicine that lend themselves to this type of analysis, and we may see more instances of its use in the future. For instance, for the RAND data, we might hypothesize three pairs of linear combinations: One pair of linear combinations might indicate that patients' personal functioning and mental health is related to their subsequent level of physical and role functioning; the second pair might relate the patient's age, gender, and health perceptions to their weight, cholesterol, and blood pressure levels; and the third pair might relate their health care system and income with their mental health and the number of bed-days.

12.8 SUMMARY OF ADVANCED METHODS

The advanced methods presented in this chapter are used in approximately 6–9% of the articles in medical and surgical journals (Emerson and Colditz, 1983; Resnick, Dawson-Saunders, and Folse, 1987). We think their use is growing for a number of reasons, including increasing statistical knowledge by investigators, increased collaboration with statisticians in medical studies, the growing concern by journal editors that poorly designed and analyzed studies not be published in their journals, and the widespread availability of statistical packages for computers. Unfortunately for the consumer of the medical literature, these methods are complex and are not easy to understand. As with other complex statistical techniques, investigators should consult with a statistician if one of these advanced methods is planned. This chapter attempts to present these methods in a conceptually oriented way so that medical students and physicians can understand their purpose and the situations where they are appropriate.

In summary, advanced methods should be used in studies whenever there are three or more characteristics (measurements) per subject. Flowchart

Table 12-9. Prediction of cerebral palsy.[1,2]

Stage[3]	Prevalence of Antecedent in NCPP (%)	Predicted Risk[4] (%)	95% Confidence Limits	% of CP Cases with Antecedent
Before pregnancy ($R^2 = 0.0121$)				
Maternal mental retardation[5]	0.4	2.3	0.9/5.5	2.7
Motor deficit in older sibling	1.2	1.4	0.7/2.9	4.4
Hyperthyroidism[5]	0.6	1.8	0.8/4.4	2.7
Maternal seizures[5]	0.4	1.6	0.6/4.4	2.7
Prior fetal deaths > 2	0.6	1.2	0.4/3.2	2.2
Pregnancy ($R^2 = 0.0102$)				
Severe proteinuria[5]	1.1	1.3	0.6/2.6	4.3
Third-trimester bleeding[5]	13.7	0.6	0.4/0.9	22.3
Thyroid and estrogen use	0.14	1.9	0.4/8.0	1.1
Asymptomatic heart disease	0.9	1.2	0.5/2.8	2.7
Incompetent cervix (rubella)[5]	0.3	1.5	0.5/5.0	1.6
Labor and delivery ($R^2 = 0.0635$)				
Gestational age ≤ 32 wk	3.3	1.4	0.9/2.0	21.0
Lowest fetal heart rate ≤ 60 beats/min[5]	1.0	1.4	0.7/2.7	5.7
Breech presentation[5]	2.6	0.8	0.5/1.3	11.1
Chorionitis	2.3	0.7	0.4/1.2	10.1
Placental weight ≤ 325 g	8.7	0.5	0.4/0.8	25.6
Placental complications	3.2	0.6	0.4/1.0	11.2
Immediately post partum ($R^2 = 0.0610$)				
Birth weight ≤ 2000 g	1.7	1.6	1.0/2.6	22.2
Time to cry ≥ 5 min[5]	1.1	1.0	0.6/1.9	12.0
Moro's reflex asymmetric	4.5	0.4	0.3/0.7	18.8
White race	45.8	0.3	0.2/0.5	55.0
Microcephaly[5] (male sex)[5]	2.0	0.6	0.3/1.2	4.9
Nursery period ($R^2 = 0.0876$)				
Neonatal seizures[5]	0.3	9.6	6.0/15.2	12.2
Major non-CNS malformation[5]	6.8	0.8	0.6/1.2	21.7
Antibiotics, no infection[5]	9.2	0.7	0.5/1.0	20.8
Infection	1.2	1.2	0.7/2.3	7.5

[1]Reproduced, with permission, from Table 1 in Nelson KB, Ellenberg JH: Antecedents of cerebral palsy. *N Engl J Med* 1986, **315**:81.

[2]NCPP denotes the Collaborative Perinatal Project of the National Institute of Neurological and Communicative Disorders and Stroke; CP cerebral palsy; R^2 the proportion of log-likelihood explained by the logistic regression, which is a measure of the predictive ability of the model; and CNS central nervous system.

[3]R^2 represents the value attained when all the listed factors were included in the model.

[4]The risk of cerebral palsy is estimated from the multiple logistic regression model for each stage, and for each antecedent assumes that only the ancedent of interest is present.

[5]Indicates the major predictors among children weighing 2500 g or more at birth. Factors in parentheses were important only among this group of children.

C-5 in Appendix C can be consulted for guidance on the appropriate multivariate methods for different research questions.

EXERCISES

1. Using the formula below, verify the adjusted mean number of ventricular wall motion ab-

normalities in smokers and nonsmokers from the hypothetical data in Section 12.3. That is, for group j,

$$\text{Adjusted mean } \overline{Y}_j = \overline{Y}_j - b(\overline{X}_j - \overline{X}.)$$

2. Explain why family income was transformed to a logarithmic scale in Presenting Problem 4.

3. Use the logistic regression equation given in

Section 12.5.1 to find the probability that a 5-year-old boy with a psychologic distress score of 6 has wet the bed in the past three months. Compare your answer with the result 0.40 for a 5-year-old girl.

4. Nelson and Ellenberg (1986) presented the results of a "multivariate analysis" of the risk factors for cerebral palsy. They studied 54,000 pregnancies over a 7-year period and collected data on more than 400 potential risk factors. Variables that related to cerebral palsy (CP) at the .05 level of significance using univariate (rather than multivariate) analyses or were of interest on the basis of earlier reports were examined with multivariate techniques. They used stepwise logistic regression at five different time periods beginning before pregnancy and extending through the child's first month of life. An excerpt from the Methods section of the paper states:

Each stage, from before pregnancy through the nursery period, was examined separately with use of the stepwise multiple logistic procedure. Then, in sequence, important factors from earlier stages were examined in conjunction with factors at the subsequent stage. At each stage, the inclusion of a variable required, in addition to statistical significance, that it contribute an increase in R^2, a measure of predictive ability in a logistic model, or at least 5 percent.

Table 12–9 reproduces a table from the article.

a. Why did the authors require a variable to increase R^2 by 5% before adding it to the regression equation?

b. From Table 12–9, which stage was best predicted?

c. Of all the variables, which is the most predictive of CP?

d. Why do the 95% confidence intervals for the relative risk for many predictor variables contain 1 when the predictor variable was supposed to be significantly related to CP in order to be used in the regression equation?

5. Bale, Bradbury, and Colley (1986) performed a study to consider the physique and anthropometric variables of athletes in relation to their type and amount of training and to examine these variables as potential predictors of distance running performance. Sixty runners were divided into three groups: (1) elite runners with 10-kilometer (km) runs in less than 30 min; (2) good runners with 10-km times between 30 and 35 min, and (3) average runners with 10-km times between 35 and 45 min. Anthropometric data included body density, percentage fat, percentage absolute fat, lean body mass, ponderal index, biceps and calf circumferences, humerus and femur widths, and various skinfold measures. The authors wanted to determine whether the anthropometric variables were able to differentiate between the groups of runners. What is the best method to use for this research question?

6. Use the regression equation found in the RAND study (Table 12–3) to predict the number of bed-days during a 30-day period for a 70-year-old woman in the FFS payplan who has the following scores for independent variables (asterisks designate dummy variables given a value of 1 if yes and 0 if no):

Personal functioning	80
Mental health	80
Health perceptions	75
Age	70
Income	10 (from a formula used in the RAND study)
Three-year term*	Yes
Took physical*	Yes
Bed-day00	14

7. Exercise 4 in Chapter 3 referred to a study by Nathan et al (1984) to evaluate the clinical information value of the glycosylated hemoglobin assay in patients with diabetes. A multiple regression equation to predict physicians' estimates of blood glucose was reported; the equation included urine test results, insulin use, nocturia, and polydipsia, and it had a multiple correlation coefficient of $R = 0.39$. When the result of random blood glucose testing was added to the regression equation, R increased to 0.58. What is the increase in percentage of variation accounted for in physicians' estimates with information on glucose testing, and how should this finding be interpreted?

8. Foltin (1987) determined the degree of osteoporosis in 358 patients with tibial condyle fractures according to a four-point scale. Five types of fractures were also distinguished. These two variables, along with age of patient (five categories), were analyzed to evaluate the relationships among them. What is the preferred statistical method for this analysis?

Evaluating Diagnostic Procedures

13

PRESENTING PROBLEMS

Presenting Problem 1. A 70-year-old woman came to the physician's office because of fatigue, pain in her hands and knees, and intermittent, sharp pains in her chest. Physical examination revealed an otherwise healthy female—cardiopulmonary exam was normal and there was no swelling of her joints. A possible diagnosis in this case is systemic lupus erythematosus (SLE). The question is whether or not to order an ANA (antinuclear antibody) test and, if so, how to interpret the results of a positive or negative test.

Presenting Problem 2. Many published studies have raised questions about the **reliability (sensitivity** and **specificity)** of urine tests performed by reference laboratories to detect the presence of commonly abused drugs, such as cocaine, methadone, and amphetamines. Investigators from the Centers for Disease Control (CDC) conducted a cross-sectional study to (1) determine the error rates of 13 drug abuse screening laboratories that serve a large number of drug abuse treatment facilities and (2) classify each laboratory's performance as acceptable or unacceptable (Hansen, Caudill, and Boone, 1985). Urine samples were mailed to the selected laboratories using the *blind test method* of proficiency testing; ie, the urine specimens were submitted without the laboratory's knowledge that they were test samples. Each laboratory in the study received approximately 100 urine samples, of which 30–40% contained concentrations of a given drug (or its metabolite) above that considered to be a minimum reporting level. In addition, each laboratory received urine samples that did not contain drugs. Data summarizing the laboratories' performance on both positive samples and negative samples are given in Table 13–1. A physician ordering a urine test for cocaine needs to know how to interpret the results. What are the sensitivity and specificity of the tests for cocaine?

Presenting Problem 3. Steinberg et al (1986) of the Office of Medical Practice Evaluation at the Johns Hopkins Hospital in Baltimore point out that the efficacy of a given diagnostic test in medical practice is most often reported in terms of sensitivity and specificity, but it is the clinician's perception of the efficacy of a test that determines when the test is used and how results are interpreted. Therefore, they studied physicians' perceptions of the value of the liver spleen scan (LSS) in the detection of liver metastases.

The investigators interviewed 42 physicians who had ordered 62 LSSs to evaluate the possible presence of a liver metastases. The physicians' estimates of the sensitivity and specificity of the LSS in detection of metastases as well as estimates of **likelihood ratios** (ratios of **true-positives** to **false-positives**) for various scan results were calculated from interviews in which each physician was asked to estimate the probability that an LSS would be normal or abnormal in a series of patients with known pathologic conditions. This cross-sectional study found wide variability in the physicians' perceptions of the sensitivity and specificity of the LSS in detection of liver metastases. The likelihood ratios for scans showing a focal defect suggested that some physicians believed that a focal defect was unequivocally diagnostic of a liver metastases; ie, there are no false-positive results.

In clinical medicine, treatment decisions are profoundly affected by the physician's interpretation of a test result. Physicians' understanding of a test's sensitivity and specificity will undoubtedly reduce the magnitude of errors of inference that may adversely affect patient management.

13.1 PURPOSE OF THE CHAPTER

Decision making is a term that applies to actions people take many times each day. Many decisions—such as what time to get up in the morning, where and what to eat for lunch, and where to park the car—are often made with little thought or planning. Others—such as how to prepare for an examination, whether or not to purchase a new car and, if so, what make and model—require some planning and may even include a conscious outlining of the steps involved. This chapter and the following chapter address the second type of decision making as applied to problems within the context of medicine. These problems include evaluating the accuracy of a diagnostic procedure, interpreting the results of a positive or negative procedure in a specific patient, **modeling** complex patient problems, and selecting the most appropriate approach to the problem. They might be best described as the application of probabilistic and statistical principles to *individual patients.* The topics presented may be broadly defined as methods in **medical decision making** or **analysis,** and they are not generally covered in introductory biostatistics textbooks.

We decided to include this material for several reasons. The reviews of journal articles previously cited indicate that 2% of articles in psychiatry (Hokanson et al, 1986) and 7% of articles in surgery (Reznick, Dawson-Saunders, and Folse, 1987) and family practice (Fromm and Snyder, 1986) use the methods discussed in this chapter and the next. More importantly, medical decision making is becoming an increasingly important area of research in medicine, for a variety of reasons that will be touched upon in the following discussion. Thus, more and more journal articles deal with topics such as evaluating new diagnostic procedures, determining the most cost-effective approach for dealing with certain diseases or conditions, and evaluating different options available for management of a specific patient. These methods also form the basis for cost-benefit analysis, which we expect to be an area of continued concern in medicine.

Finally, correct application of the principles of medical decision making help a health care provider make better diagnostic and management decisions. Physicians and other health care providers who read the medical literature and wish to evaluate new procedures and recommended programs in medicine need to understand the basic principles of medical decision making.

For these reasons, this chapter departs from previous chapters in which the emphasis was on understanding the logic of and basis for statistical procedures, leaving the details of calculations to computer programs. People who make decisions about patients must be able to perform the calculations outlined in this chapter. For further details, you may also want to consult articles discussing the basic concepts involved in medical decision making (eg, Doubilet, 1988; Eraker et al, 1986; Griner et al, 1981; McNeil, Keeler, and Adelstein, 1975; Metz, 1978; Murphy et al, 1987; Shewchuk and Francis, 1988; Sox, 1986; Sox and Liang, 1986).

We begin the presentation with a discussion of the threshold model of decision making, which provides a unified way to look at the decision of whether or not to perform a diagnostic procedure; we use Presenting Problem 1 to illustrate the model. Next, the concepts of sensitivity and specificity are defined and illustrated by using Presenting Problem 2 on urine tests for the presence of various drugs. The way sensitivity and specificity are used to make decisions about individual patients is also illustrated, using Presenting Problem

1 again. Four different methods that lead to equivalent results are presented. Then, an extension of the diagnostic-testing problem in which the test results are numbers, not simply positive or negative, is given. Presenting Problem 3 is used to illustrate how physicians can revise probabilities more accurately.

13.2 EVALUATING DIAGNOSTIC PROCEDURES WITH THE THRESHOLD MODEL

Consider the patient described in Presenting Problem 1, the 70-year-old woman who has fatigue, joint pain, and intermittent, sharp pains in her chest. Before deciding whether or not to order a diagnostic test for systemic lupus erythematosus (SLE), the physician must first consider the possibility that the woman has lupus; ie, the physician must believe that there is some chance, or probability, that the patient has lupus before he or she can make any decisions about how to investigate this possibility. This probability may simply be the **prevalence** of the disease if a screening test is being considered. If a history and physical examination have been performed, the prevalence is adjusted, upward or downward, according to the patient's characteristics (such as age, gender, and race) and symptoms and signs. Physicians sometimes use the term *index of suspicion* for the prevalence (probability) of a given disease prior to their performing a diagnostic procedure; it is also called the **prior probability,** using the terminology of decision analysis. It may also be thought of in the context of a threshold model (Pauker and Kassirer, 1980).

The **threshold model** is illustrated in Fig 13–1A. The physician's estimate that the patient has the disease, from information available without using the diagnostic test, is called the probability of disease. It helps to think of the probability of disease as a line that extends from 0 to 1. According to Pauker and Kassirer, the **testing threshold,** T_t, is the point on the probability line at which there is no difference between the value of not treating the patient and performing the test. Similarly, the **treatment threshold,** T_{rx}, is the point on the probability line at which there is no difference between the value of performing the test and treating the patient without doing a test. The points at which the thresholds occur depend on several factors: how risky the diagnostic test is, how beneficial the treatment is to patients who have the disease, how risky the treatment is to patients with and without the disease, and how accurate the test is.

Fig 13–1B illustrates the situation in which the test is quite accurate and has very little risk to the patient. In this situation, the physician is likely to test at a lower probability of disease as well as at

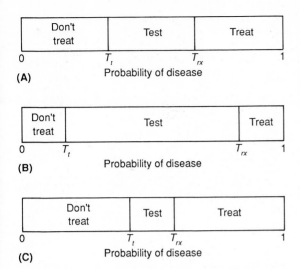

Figure 13-1. Threshold model of decision making. (*A*) Threshold model. (*B*) Accurate, or low-risk, test. (*C*) Inaccurate, or high-risk, test. (Adapted and reproduced, with permission, from Pauker SG, Kassirer JP: The threshold approach to clinical decision making. *N Engl J Med* 1980;**302**:1109.)

a high probability of disease. Fig 13-1C illustrates the opposite situation, in which the test has low accuracy or is risky to the patient. In this case, the test is less likely to be performed. Pauker and Kassirer further show that the test and treatment thresholds can be determined for a diagnostic procedure if the risk of the test, the risk and the benefit of the treatment, and the accuracy of the test are known. While that analysis is beyond the scope of this text, the statistical method to determine the accuracy of the test can and will be presented. After completing this chapter, you may wish to consult the original article by Pauker and Kassirer (1980).

13.3 MEASURING THE ACCURACY OF DIAGNOSTIC PROCEDURES

There are two aspects to the accuracy of a diagnostic test or procedure. The first aspect is how good the test is in detecting the condition it is testing for and thus being positive in patients who actually have the condition. This aspect is called the **sensitivity** of the test. If a test has high sensitivity, it has a low **false-negative** rate; ie, the test does not falsely give a negative result in many patients who have the disease. There many equivalent ways to define sensitivity: the probability of a positive test result in patients who have the condition; the proportion of patients with the condition who test positive; the *true-positive* rate. Some people

use aids such as *positivity in disease* or *sensitive to disease* to help them remember the definition of sensitivity.

The second aspect of accuracy is how well the test correctly identifies those patients who do not have the condition; this aspect is called the **specificity** of the test. If the specificity of a test is high, the test has a low **false-positive** rate; ie, the test does not falsely give a positive result in many patients without the disease. There are also many equivalent ways to define specificity: the probability of a negative test result in patients who do not have the condition; the proportion of patients without the condition who test negative; 1 minus the false-positive rate. The phrases for remembering the definition of specificity are *negative in health* or *specific to health*.

Sensitivity and specificity of a diagnostic procedure are commonly determined by selecting a group of patients known to have the disease (or condition) and another group known not to have the disease (or condition) and administering the test to both groups. The sensitivity is then calculated as the proportion (or percentage) of patients known to have the disease who have a positive test; specificity is the proportion (or percentage) of patients known to be free of the disease who have a negative test. Of course, we do not always have a **gold standard** immediately available or one totally free from error. Sometimes, we must wait for autopsy results for definitive classification of the patient's condition, as with Alzheimer's disease.

In Presenting Problem 2, the CDC wanted to evaluate the accuracy of 13 laboratories in testing urine for a variety of drugs. Each laboratory received 100 samples of urine; a known concentration of the drug was added to 30–40% of the samples. An average of 32 positive challenges were presented to each of the 13 laboratories, and 36% of them were correctly identified. Also, an average of 61 negative challenges were presented to each laboratory, and 99% of them were identified correctly. (See Table 13-1.)

Let's use the information associated with detection of cocaine to develop a 2 × 2 table from which we can calculate sensitivity and specificity of the drug-testing procedure. Table 13-2 illustrates the basic setup for the 2 × 2 table method. Traditionally, the columns represent the disease (or condition), using D^+ and D^- to denote the presence and absence of disease (the drug cocaine, in this example). The rows represent the tests, using T^+ and T^- for positive and negative test result. **True-positive** (TP) results go in the upper left cell, the T^+D^+ cell. False-positives (FP) occur when the test is positive but no drug is present, the upper right T^+D^- cell. Similarly, **true-negatives** (TN) occur when the test is negative in samples that do not contain the drug, the T^-D^- cell in the lower right; and false-negatives (FN) are in the

Table 13-1. Comparisons of laboratory performance from blind study.[1]

Drug or Drug Class	Positive Samples		Negative Samples	
	No. of Challenges	Number Correct	No. of Challenges	Number Correct
Barbiturates	455	187	689	689
Amphetamines	572	177	637	618
Methadone	533	469	663	583
Cocaine	416	150	793	785
Codeine	481	216	715	708
Morphine	468	178	728	713

[1]Adapted and reproduced, with permission, from Tables 7 and 8 in Hansen JH, Caudill SP, Boone J: Crisis in drug testing. *JAMA* 1985; **253**:2382.

lower left T^-D^+ cell corresponding to a negative test in samples with the drug present.

In the CDC drug study, 416 samples contained cocaine; therefore, 416 goes at the bottom of the first column headed by D^+. There were 793 samples without the drug, and this is the total of the second (D^-) column. Since there were 150 positive tests among the 416 samples with cocaine, 150 goes in the T^+D^+ (true-positive) cell of the table, leaving 266 of the 416 samples as false-negatives. The 8 false-positive samples represent the 1% of samples in which the test was positive but the drug was not present, the T^+D^- cell of the table. Thus, there are 785 true-negatives. Table 13-3 shows the completed table.

Using Table 13.3, we can calculate sensitivity and specificity of the test for presence of cocaine in urine. Try it before reading further. (The sensitivity of the drug test is the proportion of the samples with cocaine that test positive, 150 out of 416, or 36%. The specificity is the proportion of the samples without cocaine that test negative, 785 out of 793, or 99%.)

13.4 USING SENSITIVITY AND SPECIFICITY TO REVISE PROBABILITIES

Unfortunately, the values of sensitivity and specificity cannot be used alone to determine the value of a diagnostic test in a specific patient; instead, they are combined with a physician's index of sus-

picion (or the prior probability) that the patient has the disease in order to determine the probability of disease (or nondisease) given the knowledge of the test result. Note that a physician's index of suspicion is not always based on probabilities determined by experiments or observations; sometimes, it must simply be a "best guess," which is simply an estimate lying somewhere between the prevalence of the disease being investigated in this particular patient population and certainty. A physician's best guess generally begins with baseline prevalence and then is revised upward (or downward) based on clinical signs and symptoms. Some vagueness is acceptable in the initial estimate of the index of suspicion; in the next chapter, we will discuss how to evaluate the effect of the initial estimate on the final decision.

Some physicians learn how to manipulate the index of suspicion (or prior probability) quite well; others have more difficulty with this concept. We present four different methods that can be used to manipulate the prior probability, because we have found that some people prefer one method to another. You can select the method that makes the most sense to you or is the easiest to remember and apply.

13.4.1 The 2 × 2 Method

In Presenting Problem 1, a decision must be reached on whether or not to order an ANA test

Table 13-2. Basic setup for 2 × 2 table.

Test	Disease	
	Positive D+	Negative D-
Positive, T+	TP (true-positive)	FP (false-positive)
Negative, T-	FN (false-negative)	TN (true-negative)

Table 13-3. 2 × 2 table for evaluating sensitivity and specificity of test for cocaine in urine.

Lab Finding	Cocaine in Sample	
	Present	Absent
Positive for cocaine	(TP) 150	(FP) 8
Negative for cocaine	(FN) 266	(TN) 785
	416	793

for SLE for the 70-year-old woman. This decision depends on three pieces of information: How likely SLE is in the patient prior to performing any tests; how accurate the ANA test is for SLE in the diseased population (sensitivity); and how frequently the ANA test is negative in the nondiseased population (specificity).

What is your index of suspicion of SLE in this patient? Consider that SLE is rare in patients of this age. However, the patient is female and is complaining of fatigue and joint pain. In addition, she has had intermittent, sharp chest pain, all making SLE more likely. A reasonable prior probability of SLE in a patient like this one is 1 or 2 in 100.

How will this probability change if an ANA test is positive? If it is negative? Answers to these questions help us determine the testing threshold and indicate whether the test should be performed. Obviously, we must know how sensitive and specific the ANA test is for lupus in order to answer these questions. From the literature (Tan et al, 1982), the ANA test is very sensitive to SLE, being positive 95–99% of the time when the disease is present. It is not, however, very specific: Positive results are also obtained with connective tissue diseases other than SLE, and the occurrence of a positive ANA in the normal healthy population also increases with age. The estimate of specificity of the ANA test for SLE is approximately 50% when used in a population of people with connective tissue diseases other than SLE. Therefore, in a patient with a baseline 2% chance of SLE, how will the results of an ANA test that is 95% sensitive and 50% specific for SLE change the probabilities of disease and nondisease? These new probabilities are called the **predictive values of a positive test** or the **posterior probabilities.**

In the 2 × 2 method for determining predictive values of a diagnostic test, the first step is to include the information reflecting the physician's best guess that the patient has the disease before the test is done, which is the index of suspicion (or prior probability) of disease. In the 2 × 2 method, many people find it easier to work with whole numbers rather than percentages. So another way of saying that there is a 2% chance that this patient has SLE is to say that 20 out of 1000 patients like this one would have the disease. In Table 13–4, this number (20) is written at the bottom of the D^+ column. Similarly, 980 patients out of 1000 would not have lupus, and this number is written at the bottom of the D^-column.

The second step is to fill in the cells of the table by using the information on the test's sensitivity and specificity. Table 13–4 shows that the true-positive rate, or sensitivity, corresponds to the T^+D^+ cell (labeled TP). If the ANA test is 95% sensitive for SLE, then 95% of the 20 patients with lupus, or 19 patients, will be true-positives, and 1 will be a false-negative (see Table 13–5). From the same reasoning, a test that is 50% specific will result in 490 true-negatives in the 980 patients without SLE, and there will be 980 − 490 = 490 false-positives.

The third step is to add across the rows. For row 1, there are 19 + 490 = 509 people like this patient who will have a positive result on an ANA test (see Table 13–6). Similarly, there are 491 people who will have a negative test.

The fourth, and final, step involves the calculations for predictive values. From Table 13–6, of the 509 people with a positive test, 19 actually have the disease, giving a result of 3.7%. Similarly, 490 of the 491 people with a negative test do not have SLE, giving a result of 99.8%. The percentage 3.7% is called the **predictive value of a positive test** and gives the percentage of patients with a positive ANA test result who actually have SLE (or the probability of SLE, given a positive ANA test). The percentage 99.8% is called the **predictive value of a negative test** and gives the probability that the patient does not have SLE when the ANA test is negative. Two other probabilities can be estimated from this table as well, although they do not have specific names: 490/509 = .963 is the probability the patient does not have SLE, even though the ANA test is positive; and 1/490 = .002 is the probability that the patient does have SLE, even though the ANA test is negative.

To summarize this example, the ANA test is very sensitive but not very specific for SLE. When it is used for a patient who has a low probability of lupus, the ANA test provides considerable information if it is negative, virtually ruling out the disease. Thus, in general, tests that have high sen-

Table 13-4. Step one: Adding the prior probabilities to the 2 × 2 table.

Test	Disease	
	D^+	D^-
T^+	(TP)	(FP)
T^-	(FN)	(TN)
	20	980

Table 13-5. Step 2: Using sensitivity and specificity to determine number of true-positives, false-negatives, true-negatives, and false-positives in 2 × 2 table.

Test	Disease	
	D^+	D^-
T^+	(TP) 19	(FP) 490
T^-	(FN) 1	(TN) 490
	20	980

Table 13-6. Step 3: Completed 2 × 2 table for calculating predictive values.

A. Completed Table

Test	Disease D⁺	Disease D⁻	
T^+	(TP) 19	(FP) 490	509
T^-	(FN) 1	(TN) 490	491
	20	980	1000

B. Step 4

Predictive value of a positive test	$PV^+ = TP/(TP + FP) = 19/509$ $= .037$
Predictive value of a negative test	$PV^- = TN/(TN + FN) = 490/491$ $= .998$

sitivity are useful for ruling out a disease in patients for whom the prior probability is low. A positive test does not tell us much, however, partly because of its low specificity. Even though a positive test increases the probability of SLE from 2% prior to the test to 3.7% after the test, most physicians would require more evidence before instituting treatment for SLE, because a variety of conditions other than lupus can also cause a positive ANA. The other reason a positive test is not very instructive is that the prior probability of SLE is so low. We return to this problem after illustrating the three other methods for determining predictive values.

13.4.2 The Decision Tree Method

Using Presenting Problem 1 again, we illustrate the **decision tree** method for revising the initial probability of disease, a 2% chance of lupus in this example. Trees are useful for diagraming what can occur when a test is ordered. Fig 13-2 illustrates that prior to ordering a test, the patient can be in one of two conditions: with the disease (or condition) or without the disease (or condition). This situation is represented by the branches, with one labeled D^+ indicating a disease present and the other labeled D^- representing no disease.

In the second step, the prior probabilities are included on the tree. Fig 13-3 shows the 2% best guess that this patient has SLE, written above the D^+ branch. The chance that the patient does not have SLE, 98% (found by subtracting 2% from 100%), is written above the D^- line.

Fig 13-4 illustrates the third step in determining the predictive values for a diagnostic test. Because the ANA test is not 100% accurate, the test can be either positive or negative, regardless of the patient's true condition. These situations are denoted by T^+ for a positive test and T^- for a negative test and are illustrated in the decision tree by the two branches connected to both the D^+ and D^- branches.

In the fourth and fifth steps, information on sensitivity and specificity of the test is added to the tree. In the fourth step, we concentrate on the 2% of the time SLE is present, the D^+ branch; an ANA test will be positive in approximately 95% of these patients. Fig 13-5 shows the 95% sensitivity of the test written on the T^+ line. Therefore, in 5% of the cases (100% − 95%), the test will be negative, written on the T^- line. This information is then combined to obtain the numbers at the end of the lines: The result for 95% of the 2% of the patients with SLE, or 2% × 95% = 1.9%, is written at the end of the D^+T^+ branch; the result 2% × 5% = 0.1% for patients with SLE who have a negative test is written at the end of the D^+T^- branch.

In the fifth step, we concentrate on the 98% of the patients who do not have SLE. (See Fig 13-6.) If the test is 50% specific in the patient population of interest, we expect 50% of those without lupus to have a negative test, written on the T^- line. The remaining 50% are false-positives, written on the T^+ line. Multiplying the percentages gives 49% at the end of the D^-T^+ branch and 49% at the end of the D^-T^- branch.

The tree for the entire situation at this point is shown in Fig 13-7. Note that the percentages at the ends of the four branches add to 100%.

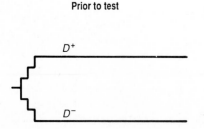

Figure 13-2. Step 1: Decision Prior to ordering diagnostic test.

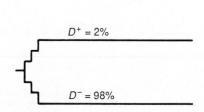

Figure 13-3. Step 2: Decision tree with prior probabilities on branches.

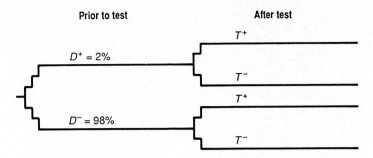

Figure 13-4. Step 3: Decision tree with test result branches.

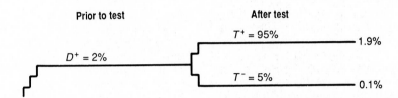

Figure 13-5. Step 4: Decision tree with sensitivity information included on D^+ branch.

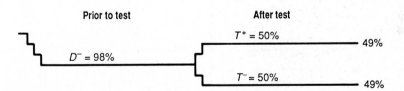

Figure 13-6. Step 5: Decision tree with specificity information included on D^- branch.

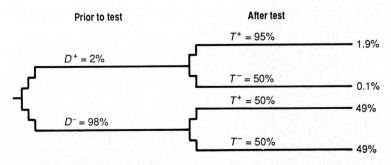

Figure 13-7. End of step 5: Decision tree with test result information.

The next step is to reverse the tree so that it more accurately describes the situation confronting the physician: The physician does not know whether or not the woman has lupus; the result of the ANA test is all the information that is "given." Fig 13–8 shows that the tree is reversed by putting the *test* first and then drawing a conclusion about whether the disease is likely from the test results.

To complete the reversed tree, we transfer the number from the ends of the branches on the first tree to the ends of the branches on the reversed tree. (See Fig 13–9.) Thus, the D^+T^+ 1.9% goes to the T^+D^+ line; the D^+T^- 0.1% goes to the T^-D^+ line; the D^-T^+ 49% goes to the T^+D^- line; and the D^-T^- 49% goes to the T^-D^- line.

Finally, working backward in the *reversed tree*, we add the numbers at the ends of the branches to obtain the percentages of patients in whom positive and negative tests are found. Fig 13–10 illustrates that in this clinical problem, a positive test is expected in 50.9% of the patients (1.9% + 49%). Similarly, a negative test is expected in 49.1% of the patients (49% + 0.1%).

The answers to the key questions (or the predictive values) can now be determined. If this patient has a positive ANA test, the revised probability that she has SLE is found by dividing 1.9% by 50.9%, giving a 3.7% chance. If the test is negative, the chance that the patient does not have lupus is given by 49.0% divided by 49.1%, or a 99.8% chance. The decision tree can also be used to determine the probability that a patient with this symptom complex who has a positive test will not have SLE, which is 96.3% (49% divided by 50.9%). Similarly, only 0.2% (0.1% divided by 49%) of those with a negative test will actually have SLE. These conclusions are, of course, exactly the same as the conclusions reached by using the 2 × 2 table method in the previous section.

13.4.3 Bayes' Theorem

Another method for calculating the predictive value of a positive test involves the use of a mathematical formula. The formula gives the predictive value of a positive test, or the chance that a patient with a positive test has the disease. The symbol P stands for the probability that an event will happen (see Chapter 5), and $P(D^+ \mid T^+)$ is the probability that the disease is present, given that the test is positive. As we discussed in Chapter 5, this probability is a **conditional probability** in which the event of the disease being present is dependent, or conditional, on having a positive test result. The formula, known as **Bayes' theorem,** can be rewritten from the form we used in Chapter 5, as follows:

$$P(D^+ \mid T^+) = \frac{P(T^+ \mid D^+)P(D^+)}{P(T^+ \mid D^+)P(D^+) + P(T^+ \mid D^-)P(D^-)}$$

This formula specifies the probability of disease, given that a positive test occurs. The two probabilities in the numerator are (1) the probability that a test is positive, given that the disease is present (or the sensitivity of the test) and (2) the best guess (or prior probability) that the patient has the disease to begin with. The denominator is simply the probability that a positive test occurs at all, $P(T^+)$, which can occur in one of two ways: a positive test when the disease is present, and a positive test when the disease is not present. The first product in the denominator is the same as the numerator. The second product in the denominator is the probability of a positive test, given that the disease is not present, multiplied (or weighted) by the probability that the disease is not present. The first quantity in this term is simply the false-positive rate, and the second can be thought of as 1 minus the probability that the disease is present.

Rewriting Bayes' theorem in terms of sensitivity and specificity, we obtain

$$P(D^+ \mid T^+) = \frac{\text{Sens.} \times \text{Prior prob.}}{(\text{Sens.} \times \text{Prior prob.} + [\text{False-pos.rate} \times (1 - \text{Prior prob.})]}$$

Presenting Problem 1 is again employed to illustrate the use of Bayes' formula. Recall that the prior probability of SLE is .02, and sensitivity and

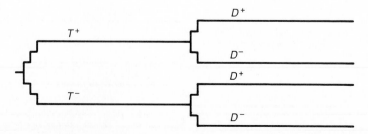

Figure 13–8. Step 6: Reversing tree to correspond with situation facing physicians.

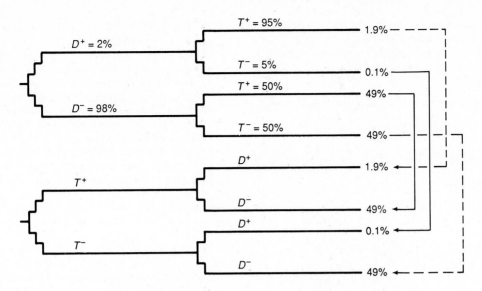

Figure 13-9. Step 7: Transferring numbers to reversed tree.

specificity of the ANA test for SLE are 95% and 50%, respectively. For Bayes' theorem, most people find it easier to work with decimals rather than percentages. In the numerator, the sensitivity times the probability of disease is .95 × .02. In the denominator, that quantity is repeated and added to the false-positive rate, .50, times 1 minus the probability of SLE, .98. Thus, we have

$$P(D^+|T^+) = \frac{.95 \times .02}{(.95 \times .02) + (.50 \times .98)}$$

$$= \frac{.019}{.019 + .49}$$

$$= .037$$

This result, of course, is exactly the same as the result obtained with the 2 × 2 table and the decision tree methods.

A similar formula may be derived for the predictive value of a negative test:

$$P(D^-|T^-) = \frac{P(T^-|D^-)P(D^-)}{P(T^-|D^-)P(D^-) + P(T^-|D^+)P(D^+)}$$

Written in terms of sensitivity, specificity, and prior probability, the formula is

$$P(D^-|T^-) = \frac{\text{Specificity} \times (1 - \text{Prior prob.})}{[\text{Specificity} \times (1 - \text{Prior prob.})] + (\text{False-neg. rate} \times \text{prior prob.})}$$

Calculation of the predictive value using Bayes' theorem for a negative test in the SLE example is left as an exercise.

13.4.4 The Likelihood Ratio

The final method for revising prior probabilities uses a quantity called the likelihood ratio and works with prior **odds** rather than prior probabilities. The likelihood ratio is being used with increasing frequency in the medical literature. Therefore, even if you decide not to use this particular approach to revising probabilities, you will find it useful to know about.

The **likelihood ratio** expresses the odds that the test result occurs in patients with the disease versus those without the disease. Thus, there is one likelihood ratio for a positive test and another for

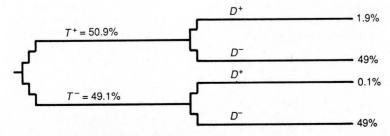

Figure 13-10. Final reversed decision tree.

a negative test. For a positive test, the likelihood ratio is the odds of a positive test in patients with the disease, or the sensitivity, versus a positive test in patients without the disease, or the false-positive rate. Thus,

$$LR^+ = \frac{\text{Sensitivity}}{\text{False-positive rate}}$$

The likelihood ratio is multiplied by the prior, or **pretest, odds** to obtain the **posttest odds** of a positive test. Thus,

Pretest odds × Likelihood ratio = Posttest odds

In Presenting Problem 1, the sensitivity of the ANA test for SLE is assumed to be 95%; the specificity is 50%, so the false-positive rate is 100% − 50% = 50%. Therefore, the likelihood ratio for a positive test is

$$LR^+ = \frac{.95}{.50} = 1.90$$

Before using the likelihood ratio, we must convert the prior probability into prior odds. The prior probability of lupus before the ANA test is .02 and the odds are found by dividing the prior probability by one minus the prior probability. We obtain

$$\text{Pretest odds} = \frac{\text{Prior probability}}{1 - \text{Prior probability}}$$

$$= \frac{.02}{1 - .02} = \frac{.02}{.98} = .0204$$

Combining the pretest odds with the likelihood ratio, we obtain the posttest odds:

$$.0204 \times 1.90 = .0388$$

Since the posttest odds are really .0388 to one, although the "to 1" part does not appear in the above formula, these odds can be used as odds themselves. Or they can be reconverted to a probability (the predictive value of a positive test) by dividing the odds by one plus the odds. That is,

Posterior probability = Predictive value

$$= \frac{\text{Posttest odds}}{1 + \text{Posttest odds}}$$

$$= \frac{.0388}{1 + .0388}$$

$$= .037$$

This probability is, of course, the same result we found with previous methods.

We emphasize that there is nothing wrong with thinking in terms of odds instead of probabilities. And with odds, the problem of converting back and forth to probabilities does not exist.

A major advantage of the likelihood ratio method is the need to remember only one number, the ratio, instead of two numbers, sensitivity and specificity. Sackett, Haynes, and Tugwell (1985) indicate that likelihood ratios are much more stable (or robust) for changes in prevalence than are sensitivity and specificity; these authors also give the likelihood ratios for some common symptoms, signs, and diagnostic tests (Chap. 4 and Table 5–25).

We have demonstrated that the calculations in the four methods for determining predictive values of diagnostic tests produce the same results. Therefore, the choice is one of preference and convenience. While the likelihood ratio method may seem complicated, a nomogram published by Fagan (1975) makes this approach somewhat simpler to use. In this nomogram, reproduced in Fig 13–11, the pretest and posttest odds have been converted to prior and posterior probabilities, eliminating the need to perform this extra calculation.

To use the nomogram, we place a straightedge at the point of the prior probability, denoted *P(D)*, on the right side of the graph and the likelihood ratio in the center of the graph; the revised probability, or predictive value *P(D|T)*, is then read from the left-hand side of the graph. In our example, the prior probability of .02 and the likelihood ratio of 1.90 result in a revised probability of between .02 and .05 from the nomogram, consistent with the calculations above.

13.4.5 Summary of Methods

We have illustrated four equivalent methods for revising the probability that a patient has a disease from the results of a diagnostic test. This process assumes that the prior probability of the disease, the accuracy and risks associated with the test, and the risks and benefits of treatment interact in such a way to position the problem between the testing threshold and the treatment threshold. Illustrating all four methods gives you an opportunity to choose the approach that makes the most sense to you and is easiest to understand and remember. We personally find the 2 × 2 table approach the easiest, but the other three methods also appear in the literature. We will use the decision tree approach again in the next chapter to model more complex problems.

Before leaving this section, we include some final comments. The ANA test for SLE exemplifies a fairly typical situation in medicine in which a

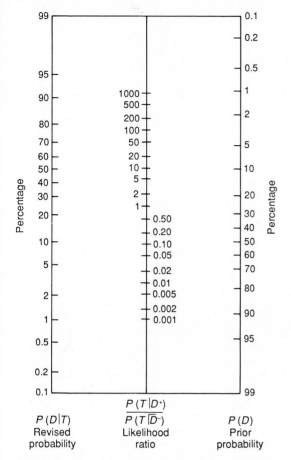

Figure 13-11. Nomogram for using Bayes' theorem. (Adapted and reproduced with permission, from Fagan TJ: Nomogram for Bayes theorem. (Letter.) *N Engl J Med* 1975;**293**:257.)

very sensitive test (95–99% of SLE patients are ANA-positive) is used to detect the presence of a disease with low prevalence (or prior probability); ie, the test is used in a screening capacity. By itself, the ANA has little diagnostic meaning. When it is used indiscriminately to screen for a disease that has a low prevalence (annual incidence of SLE is approximately 6–7 per 100,000 and prevalence is approximately 1 in 2000), there will be a high rate of false positivity. In the estimate of prior probability in this example, the clinician would also take into account that only 15% of patients with SLE develop the presenting symptoms after age 55.

The ANA become useful when used in conjunction with clinical findings that suggest the possibility of an underlying connective tissue disease such as SLE. Even if specificity were better, such as .95, a patient like the woman in Presenting Problem 1 would have lupus only 28% of the time with a positive ANA. When the prior probability

is very low—as it is in this case—even a very sensitive and specific test will increase the probability only to a moderate level. For this reason, a positive result based on a very sensitive test is often followed by a specific test, such as the anti-DNA antibody test for the SLE example.

As another example of a test with high sensitivity for a particular disease, consider serum calcium. It is a good screening test because it is almost always elevated in patients with primary hyperparathyroidism; ie, it rarely "misses" a person with primary hyperparathyroidism. However, serum calcium is not specific for this disease because other conditions, such as malignancy, sarcoidosis, multiple myeloma, or vitamin D intoxication, may also be associated with an elevated serum calcium. Therefore, a more specific test, such as radioimmunoassay for parathyroid hormone, may be ordered after finding an elevated serum calcium. The posterior probability calculated by using the serum calcium test becomes the new index of suspicion (prior probability) for analyzing the effect of the radioimmunoassay.

The diagnosis of AIDS in low-risk populations provides an example of the important role played by prior probability. Some states in the United States have considered or passed a law requiring premarital testing for the HIV antibody in couples applying for a marriage license. The ELISA test is highly sensitive and specific; some estimates range as high as 99% for each. However, the prevalence of HIV antibody in a low-risk population, such as people getting married in a midwestern community, is very low; estimates range from 1 in 1000 to 1 in 10,000. How useful is a positive test in such situations? For the higher estimate of 1 in 1000 for the prevalence and 99% sensitivity and specificity, 99% of the people with the antibody will test positive (99% × 1 = 0.99 person), as will 1% of the 999 people without the antibody (9.99 people). Therefore, among those with a positive ELISA test (0.99 + 9.99 = 10.98 people), less than 1 person is truly positive (the positive predictive value is actually about 9% for these numbers).

The above examples illustrate three important points:

1. To rule out a disease, physicians want a negative result to be reliable; therefore, there must be few false-negatives. A sensitive test is the best choice if factors such as cost and risk are similar.

2. To find evidence of a disease, physicians want a positive result to indicate a high probability that the patient has the disease. Therefore, there must be few false-positives, and a highly specific test is the choice.

3. To make accurate diagnoses, physicians must understand the role of prior probability of disease. If the prior probability of disease is extremely small, a positive result will not mean very

much and will need to be followed by a test that is highly specific. The usefulness of a negative result depends on the sensitivity of the test.

13.5 ROC CURVES

The above procedures for revising the prior (pretest) probability of a disease on the basis of information from a diagnostic test are applicable if the diagnostic procedure is either positive or negative. Many tests, however, have values measured on a **numerical scale.** When test values are measured on a continuum, sensitivity and specificity rates depend on where the cutoff between positive and negative is set. This situation can be illustrated by two normal (gaussian) distributions of laboratory test values, one distribution for people who have the disease and one for people who do not have the disease. Fig 13–12 presents two hypothetical distributions corresponding to this situation in which the mean value for people with the disease is 75 and for those without the disease is 45. If the cutoff point is placed at 60, about 10% of the people without the disease will be incorrectly classified as "abnormal" because their test value is greater than 60, and about 10% of the people with the disease will be incorrectly classified as "normal" because their test value is less than 60. In other words, this test has sensitivity of 90% and specificity of 90%.

Suppose a physician decides he or she wants to use a test with greater sensitivity; ie, the physician prefers to have more false-positives than to miss people who really have the disease. Fig 13–13 illustrates what happens if the sensitivity is increased by lowering the cutoff point to 55 for a normal test. The sensitivity is increased, but at the cost of a lower specificity.

A more efficient way to display the relationship between sensitivity and specificity for tests that

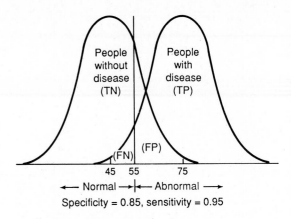

Figure 13-13. Two hypothetical distributions with cutoff at 55.

have continuous outcomes is with **receiver operating characteristic,** or **ROC, curves.** ROC curves were developed in the communications field as a way to display signal-to-noise ratios. If we think of true-positives as being the "correct signal" from a diagnostic test and false-positives as being "noise," we can see how this concept applies. The ROC curve is a plot of the sensitivity (or true-positive rate) to the false-positive rate. The dotted line in the middle of the graph in Fig 13–14 corresponds to a test that is positive or negative just by chance. The closer an ROC curve is to the upper left-hand corner of the graph, the more accurate it is, because the true-positive rate is 1 and the false-positive rate is 0. As the criterion for a positive test becomes more stringent, the point on the curve corresponding to sensitivity and specificity (point A) moves down and to the left (lower sensitivity, higher specificity); if less evidence is re-

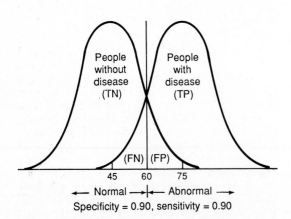

Figure 13-12. Two hypothetical distributions with cutoff at 60.

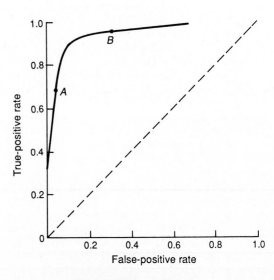

Figure 13-14. ROC curve.

quired for a positive test, the point on the curve corresponding to sensitivity and specificity (point *B)* moves up and to the right (higher sensitivity, lower specificity).

ROC curves are useful graphic means for comparing two diagnostic tests. For example, both radionuclide (RN) scanning and computerized tomography (CT) have been used to detect brain tumors. Griner et al (1981) illustrated ROC curves for both procedures; their findings are reproduced in Fig 13–15. In a comparison of the two procedures in terms of accuracy, CT scanning was found always to be superior to RN scanning. At a sensitivity of 90%, the CT scan gives 5% false-positives, but the RN scan gives 50% false-positives. Similarly, for a low false-positive rate, such as 5%, the sensitivity of the CT scan is higher than that of the RN scan, 90% versus 65%.

A statistical test can be performed to determine whether two ROC curves are significantly different. The procedure involves determining the area under each ROC curve and uses a modification of the Wilcoxon rank-sum procedure to compare them (Hanley and McNeil, 1983).

13.6 ILLUSTRATION OF PHYSICIANS' ABILITIES TO REVISE PROBABILITIES

The methods for revising probabilities described in this chapter are equivalent, and you should feel free to use the method that is easiest to understand and remember. All the methods are based on two assumptions: (1) The diseases or diagnoses being considered are mutually exclusive and include the actual diagnosis; and (2) the results of each diagnostic test are independent from the results of all other tests.

The first assumption is easy to meet if the diagnostic hypotheses are stated in terms of the probability of disease, $P(D^+)$, versus the probability of no disease, $P(D^-)$, as long as D^+ refers to a specific disease, such as SLE. Also, a more general form of Bayes' theorem can be used to consider the probability of several different diseases. For example, Sonnenberg, Kassirer, and Kopelman (1986) analyze the case of a 75-year-old woman who had suspected renal cell carcinoma and consider three diagnoses: renal cell carcinoma, transitional cell carcinoma, and benign lesion. In this situation, the authors assume that only these three conditions are possible and that only one of them is present.

The second assumption of mutually independent diagnostic tests is more difficult to meet. Two tests T_1 and T_2 for a given disease are independent if the result of T_1 does not influence the chances associated with the result of T_2. When applied to individual patients, independence means that if T_1 is positive in Patient A, T_2 is no more likely to be positive in Patient A than in any other patient in the population that Patient A represents. Even though the second assumption is violated in many medical applications of decision analysis, the methods described in this chapter appear to be fairly robust, as we will see in the next chapter.

A reasonable question is whether physicians really need to learn how to revise pretest probabilities or whether the correct interpretation does not become intuitive after some clinical experience. While some physicians do learn intuitively how to revise the chance that a patient has a disease from the results of a diagnostic test, many apparently do not. One the of the problems many people have, including physicians, is the base-rate fallacy. Let us look at an example from Kahneman and Tversky (1972):

> Two cab companies operate in a given city, the Blue and the Green (according to the color of the cab they run). Eighty-five percent of the cabs in the city are Blue, and the remaining 15% are Green. A cab was involved in a hit-and-run accident at night. A witness later identified the cab as a Green cab. The court tested the witness' ability to distinguish between Blue and Green cabs under nighttime visibility conditions. It found the witness was able to identify each color correctly about 80% of the time, but confused it with the other color about 20% of the time. What do you think are the chances that the errant cab was indeed Green, as the witness claimed? (Kahneman and Tversky, 1972, 13)

The answer is .41 (the problem is left as an exercise). However, many people believe it should be .80. They ignore the fact that only 15% of the cabs are Green, or they do not correctly use the information about the accuracy of the witness. We hope that you see the analogy to diagnostics

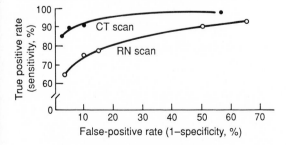

Receiver operator characteristic of computerized tomography (CT) and radionuclide scanning (RN) for detection of brain tumor.

Figure 13–15. Illustration of using 2 ROC curves to compare performance of 2 diagnostic procedures. (Reproduced, with permission, from Griner PF et al: Selection and interpretation of diagnostic tests and procedures: Principles and applications. *Ann Intern Med* 1981;**94**:557.)

tests. Another researcher determined that physicians as well as the general public fall prey to the base-rate fallacy—and, in fact, that this fallacy is also found in the recommendations for diagnostic testing in some medical textbooks (Bar-Hillel, 1980).

To illustrate how physicians are subject to the base-rate fallacy, let us examine Presenting Problem 3 in more detail. Although the diagnostic modality evaluated in the study is quickly becoming outdated with more advanced technologies, the methods the investigators used are applicable in evaluating any diagnostic procedure. In this study, physicians were asked to designate why they were ordering a liver spleen scan (Steinberg et al, 1986). If liver metastases was given as a reason, they were further asked to state (1) the pretest probability of liver metastases—how likely they thought liver metastases were in the patient; and (2) the probability that the liver spleen scan would be normal, would show a focal defect, or would show a nonhomogeneous distribution of tracer without focal defect.

From the answers to these questions, the investigators worked backward to infer each physician's estimate of sensitivity and specificity of liver spleen scans for liver metastases. The authors then compared the physicians' answers to the second question with what they should have been according to the pretest probability given in question 1 and the inferred sensitivity and specificity. And what did they find? First of all, the investigators found that the physicians' estimates of sensitivity and specificity varied over a broad range, regardless of the definition of a positive scan. For example, when a positive scan was defined as showing a focal defect, estimates of sensitivity of the liver spleen scan ranged from .10 to 1.00 in patients with underlying parenchymal disease, and from .20 to 1.00 in patients without underlying parenchymal disease. Similarly, when a positive scan was defined as either focal defect or nonhomogeneous distribution of tracer without focal defect, the estimates of sensitivity ranged from .75 to 1.00 and from .50 to 1.00, respectively.

The investigators also found that a physician's stated posttest probability often did not agree with what the posttest probability should have been based on the physician's inferred estimates of sensitivity and specificity. For example, 69% of the physician's estimates differed by .01–.24, and 19% differed by .25 or more.

As a result of their study, these investigators recommended that hospitals make available to physicians the performance capabilities (sensitivity and specificity) of liver spleen scans and other diagnostic procedures. Furthermore, they recommended that physicians be given microcomputer assistance in estimating the impact of a diagnostic test on the probability of disease. While these aims are laudable, it is unlikely that many hospitals will have such aids available to physicians for all diagnostic procedures. Therefore, as in the case of reading and evaluating the medical literature, physicians must be able to perform these calculations themselves. The techniques presented in this chapter and the next will help medical students and physicians do just that.

13.7 SUMMARY

Topics in this chapter and the following chapter are departures from the topics considered in traditional introductory texts in biostatistics. However, the increase in journal articles illustrating methods in decision making indicates that practitioners should be familiar with the concepts. Equally important, the methods for calculating the probability of disease discussed in this chapter are methods that every clinician needs to be able to use each day. These methods allow clinicians to integrate the results of published studies into their own practice of medicine. There is ample evidence that many physicians need more insight into how these methods work.

We presented four equivalent methods for determining how likely a disease (or condition) is in a given patient from the results of a diagnostic procedure. Three pieces of information are needed: (1) How likely the disease or condition is prior to any procedure—ie, the base rate (or prevalence); (2) how accurately the procedure identifies the condition when it is present (sensitivity); and (3) how accurately the procedure identifies the absence of the condition when it is indeed absent (specificity). We can make an analogy with hypothesis testing: A false-positive is analogous to a Type I error, falsely declaring a significant difference; and sensitivity is like power, correctly concluding a difference when it is present.

The logic discussed in this chapter is applicable in many situations other than diagnostic testing. For example, the answer to each history question or the result of each component of the physical examination may be interpreted in a similar manner. When the outcome from a procedure or inquiry is expressed as a numerical value, rather than as simply "positive" or "negative," receiver operating characteristic (ROC) curves can be used to evaluate the ramifications of decisions.

Unfortunately, articles in the literature purporting to evaluate diagnostic procedures do not always contain the information needed for readers to interpret them correctly, and many contain misleading information. A common error is investigators' presenting information related only to sensitivity and ignoring the performance of the procedure in patients without the condition. Misleading information results when investigators calculate predictive values by using the same subjects they used to determine sensitivity and specificity. As readers, we can only assume they do not recognize the crucial role that prior probability (or prevalence) plays in interpreting the results of both a positive and a negative procedure.

Not many introductory texts discuss topics in medical decision making, partly because it is a relatively new topic and partly because much of the research in this field is interdisciplinary. However, a survey of the biostatistics curriculum in medical schools (Dawson-Saunders et al, 1987) found that these topics are now being taught at 87% of the medical schools. If you are interested in learning more about this growing field, consult the books by Weinstein et al (1980) and by Sox et al (1988), or the articles by Albert (1978), Griner et al (1981), Sox (1986), and Pauker and Kassirer (1987). The text on clinical epidemiology by Sackett, Haynes, and Tugwell (1985) covers these topics in Chapters 3 and 4.

EXERCISES

1. Suppose in Presenting Problem 1 that the woman had swelling of the joints in addition to her other symptoms of fatigue, joint pain, and intermittent, sharp chest pain. In this case, the probability of lupus is higher, perhaps 20%.
   a. What is the probability of lupus with a positive test?
   b. What are the chances of lupus with a negative test?
2. Use Bayes' theorem and the likelihood ratio method to calculate the probability of no lupus when the ANA test is negative, using a pretest probability of lupus of 2%.
3. Given the information on Blue and Green cabs from Section 13.6, calculate the probability the errant cab was indeed Green.
4. Refer to Presenting Problem 4 in Chapter 9. Suppose venography is the gold standard, indicating presence or absence of deep venous thrombosis. Evaluate sensitivity and specificity of thermography, using the data in Table 9–10.
5. A 43-year-old white male comes to your office for an insurance physical exam. Routine urinalysis reveals glucosuria. You have recently learned of a newly developed test that produced positive results in 138 of 150 known diabetics and in 24 of 150 persons known not to have diabetes.
   a. What is the sensitivity of the new test?
   b. What is the specificity of the new test?
   c. What is the false-positive rate of the new test?
   d. Suppose a fasting blood sugar is obtained with known sensitivity and specificity of 0.80 and 0.96, respectively. If this test is ap-

plied to the same group that the new test used (150 persons with diabetes and 150 persons without diabetes), what is the predictive validity of a positive test?
   e. For the current patient, after the positive urinalysis, you think the chance that he has diabetes is about 90%. If the fasting blood sugar test is positive, what is the revised probability of disease?
6. Consider a 22-year-old woman who comes to your office with palpitations. Physical examination shows a healthy female with no detectable heart murmurs. In this situation, your guess is that this patient has a 25–30% chance of having mitral valve prolapse, based on prevalence of the disease and physical findings for this particular patient. Echocardiograms are fairly sensitive for detecting mitral valve prolapse in patients who have it—approximately 90% sensitive. Echocardiograms are also quite specific, showing only about 5% false-positives; in other words, a negative result will be correctly obtained in 95% of people who do not have mitral valve prolapse.
   a. If the echocardiogram is positive for this woman, how does that change your opinion of the 30% chance of mitral valve prolapse? That is, what is your best guess of mitral valve prolapse with a positive test?
   b. If the echocardiogram is negative, how sure can you be that this patient does not have mitral valve prolapse?
7. Borowitz and Glascoe (1986) reported a study to determine whether the language portion of the Denver Developmental Screening Test (DDST) is a sensitive screen of speech and language development in preschool-aged children with suspected developmental problems. Seventy-one children were given the DDST plus

Table 13-7. Evaluation of Denver Developmental Screening Test.[1]

| Speech/ Language Deficit | No. (%) of Deficits Detected by[2] | | Z Value |
	Speech/ Language Screening	DDST Language Sector	
Any	65 (92)	30 (42)	−5.16[3]
Articulation	60 (84)	28 (39)	−4.49[3]
Expressive	56 (79)	30 (42)	−4.56[3]
Receptive	36 (51)	25 (34)	−1.24

[1]Reproduced, with permission, from Table 1 in Borowitz KC, Glascoe P: Sensitivity of the Denver Developmental Screening Test in speech and language screening. *Pediatrics* 1986; **78**:1075.
[2]N = 71.
[3]P < .001.

the Preschool Language Scale (PLS), a well-accepted measure of language development, and the latter was used as the gold standard. On each test, children were assigned either a pass or a fail measurement. They were evaluated on three subtests measuring articulation and expressive and receptive abilities. The authors used the information given in Table 13–7.

a. Comment on the information presented. Is this information adequate for evaluating the accuracy of the DDST? What are the sensitivity and specificity for the DDST?

b. What statistical procedure was used to compare the results of the PLS and DDST? Was this test appropriate?

8. Assume a diagnostic test is 90% sensitive and 95% specific. Complete the following chart. What conclusions should be drawn from this chart?

Prevalence	Predictive Value of a Positive Result	Predictive Value of a Negative Result
0.1%		
1%		
2%		
5%		
10%		
20%		
50%		
80%		

Clinical Decision Making

14

PRESENTING PROBLEMS

Presenting Problem 1. There is no consensus regarding the management of incidental intracranial saccular aneurysms. Some experts advocate surgery, because of their own good results; others point out that prognosis without surgery is relatively benign, especially for aneurysms smaller than 10 mm. The decision regarding surgery depends on several factors and is complicated because rupture of an incidental aneurysm is a long-term risk, spread out over many years, but surgery represents an immediate risk. Some patients may prefer to avoid surgery, even at the cost of later excess risk; others may not.

Decision analysis was used by van Crevel, Habbema, and Braakman (1986) to evaluate the dilemma of whether surgery should be performed in patients with incidental intracranial saccular aneurysms. To approach the problem, they considered a fictitious 45-year-old woman with migraine (but otherwise healthy) who had been having attacks for the past 2 years. Her attacks were invariably right-sided and did not respond to medication. There was no family history of migraine. Because of her clinical symptoms, the neurologist suspected an arteriovenous malformation and ordered four-vessel angiography, which showed an aneurysm of 7 mm on the left middle cerebral artery. Should the neurologist advise the patient to have preventive surgery? The decision is diagramed in Fig 14-1.

Presenting Problem 2. Decision analysis was used by Phillips et al (1987) to estimate the clinical and economic implications of testing for cervical infection caused by *Chlamydia trachomatis* in women during routine gynecologic visits. Infections caused by *C trachomatis* are among the most common sexually transmitted diseases in the United States. Severe consequences of *C trachomatis* include morbidity associated with urethritis, mucopurulent cervicitis, and salpingitis. Salpingitis can lead to infertility, ectopic pregnancy, and chronic pelvic pain, and *C trachomatis* has been implicated as the etiologic agent in 20–40% of patients with this condition. The prevalence of chlamydial infection is reported to be 4–9% in primary-care practices and even higher in other settings, 6–23% in family planning clinics, and 20–30% in clinics

for patients with sexually transmitted diseases. The investigators examined the consequences of routine culture, of test by direct immunofluorescence or enzyme immunoassay (rapid tests), and of no test. They also calculated the costs associated with each strategy to determine a threshold prevalence of infection at which medical costs would be reduced by testing routinely for *C trachomatis*. The strategy is diagramed in Fig 14-4.

Presenting Problem 3. The fecal occult blood test, such as the Hemoccult test, is a simple, inexpensive screening test for colorectal cancer in asymptomatic patients. Major drawbacks with this procedure are the number of cancers missed (false-negatives) and the number of false-positive results. Several procedures have been recommended in the literature as appropriate for follow-up to a positive Hemoccult. Many of the recommended procedures call for sigmoidoscopy, barium enema, repeat Hemoccult, and colonoscopy, but the recommended order of these procedures varies widely. Brandeau and Eddy (1987) evaluated 22 protocols for working up a typical asymptomatic patient who has a positive fecal occult blood test by using information on the prevalences of cancers, adenomas, and other conditions in such patients; the natural history of colorectal cancer; the effectiveness of screening tests; risks; and costs. They determined the number of cancers and polyps each protocol would be expected to identify and the cost involved. They also found the marginal costs and effectiveness of the best protocols from both the patient's and a societal perspective. Selected results from their analysis are given in Figs 14-6 and 14-7 and in Table 14-1.

14.1 THE DECISION PROCESS

This chapter extends the topic of decision making introduced in the previous chapter to more complex problems. The challenge in many clinical situations is to compare two options for solving a problem. For example, in Presenting Problem 1, the problem is whether or not to operate on a patient with an incidental aneurysm. Surgery is termed the *active option*. The active option often carries with it an increased benefit if it is successful; however, there are also increased risks and costs associated with surgery. The *passive option* is not to perform surgery.

Both active and passive options can have a positive or a negative outcome. Thus, the surgery option is positive if successful but negative if death or disability results; the passive option is positive if there is no rupture but negative if there is a rupture, especially if the rupture is followed by disability or death. The physician's role is to decide which outcome is more likely and then, balancing this probability with the risks and the benefits of the options, discuss the options with the patient and determine an approach. The increasing emphasis on controlling the costs associated with medical care and concerns regarding litigation have caused the profession to examine this process in more detail.

14.1.1 Purpose of the Chapter

The methods discussed in this chapter are applicable to decisions about individual patients, such as whether or not to perform surgery, and to health policy decisions, such as whether or not to recommend screening of the general population. We use the problem of deciding whether or not to perform surgery on a patient with an incidental aneurysm, described in Presenting Problem 1, to illustrate some of the components of medical decision making. Presenting Problem 2 illustrates the use of decision analysis in screening. Finally, we illustrate the application of decision analysis as a method for evaluating a set of suggested proposals for follow-up to a positive Hemoccult in an asymptomatic patient (Presenting Problem 3). Our discussion provides only a brief introduction to this rapidly growing field, and the calculations

we perform illustrate specific aspects of the process; as usual, we assume computer programs are used with actual applications. For further details, consult the text by Wienstein et al (1980).

14.1.2 Components of Making a Decision

14.1.2.a Defining the Problem, Alternative Actions, and Possible Outcomes: Decision analysis can be applied to any problem in which there is a choice among different alternative actions. Recall from Chapter 13 that Pauker and Kassirer (1980) developed the threshold model of decision making as one way to view the problem of whether to treat a patient, perform a diagnostic procedure, or do nothing. There are two major components of any decision, and *specifying the alternative actions* is the first component. The second is *determining the possible outcomes* from each alternative action. Fig 14–1 illustrates both components. The alternative actions are to operate or not to operate; and if no operation is performed, to determine whether or not the aneurysm subsequently ruptures, causing subarachnoid hemorrhage. The outcomes are death, disability, or recovery after a rupture; no rupture; and death, disability, or success following surgery. These components represent the branches on the decision tree.

The point at which a branch occurs is called a *node;* note that in Fig 14–1, nodes are identified by squares and circles. The squares denote decision nodes, ie, points at which the decision is under the control of the decision maker; thus, whether or not to operate is a decision over which the physician and the patient have control. The circles denote chance nodes, ie, points at which the results occur by chance; thus, whether or not the aneurysm ruptures without surgery is a chance outcome of the decision not to operate.

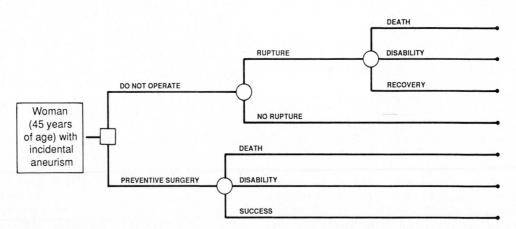

Figure 14–1. Decision tree for management and outcome in incidental aneurysms. (Reproduced, with permission, from van Crevel H, Habbema JD, Braakman R: Decision analysis of the management of incidental intracranial saccular aneurysms. *Neurology* 1986;**36**:1335.)

14.1.2.b Determining Probabilities: The next step in developing a decision tree is to assign a *probability* for each branch leading from each chance node. For example, what is the probability of aneurysm rupture if no surgery is performed? What is the probability of death with surgery? Of disability with surgery? Of success with surgery? If the aneurysm does rupture, what is the probability of a good outcome (recovery) versus a poor outcome (disability or death)?

14.1.2.c Deciding on the Value of the Outcomes: The final step in defining a decision problem requires *assigning a value, or utility,* to each outcome. With some decision problems, the outcome is cost, and dollar amounts can be used as the utility of each outcome. For example, in Presenting Problem 1, some way must be found to quantify the outcomes of death, disability, etc. The actual analysis of the decision tree involves combining the probabilities of each action with the utility of each outcome so that the "best" decision can be made at the decision nodes under the control of the decision maker. Each step and the analysis of the aneurysm problem are illustrated in the next section.

14.2 MAKING A DECISION FOR AN INDIVIDUAL PATIENT

To decide whether or not to perform surgery on a 45-year-old woman with a 7-mm aneurysm on the left middle cerebral artery, the investigators in Presenting Problem 1 defined the decision alternatives and possible outcomes as illustrated in Fig 14–1. This step is the "art" of any decision analysis. The authors define *recovery* as meaning that the patient functions at her present, presurgery level, although minor symptoms or signs may be present; *disability* means that she cannot function at her present level.

14.2.1 Determining the Probability of Each Branch

To determine the probabilities for the decision tree, the investigators surveyed the medical literature and found they needed to take the following information into consideration.

1. The risk of rupture depends on the life expectancy or age of the patient, and the size of the aneurysm influences the risk of rupture, with smaller aneurysms carrying a lower risk. The risk of rupture over a patient's lifetime is approximately $1 - (1 - R)^L$ where R is the annual risk and L is the life expectancy. For a 45-year-old woman with a 7-mm aneurysm, the annual risk *(R)* of rupture without surgery is about 1%, and the life expectancy *(L)* is approximately 35 years; therefore, the lifetime risk is approximately .29.

2. The mortality from subarachnoid hemorrhage is high, about 55%.

3. Serious morbidity after subarachnoid hemorrhage (hemiparesis, dysphasia, or mental deterioration) is also high, estimated at 15%.

4. Surgical mortality and morbidity for incidental aneurysm are much lower than for ruptured aneurysms, estimated at 2% and 6%, respectively.

5. The attitude of the patient toward short-term and long-term risks must be considered. Operation carries an immediate risk, while a rupture, if it occurs, appears on the average after about one-half of the patient's life expectancy has elapsed.

In Fig 14–2, the above probabilities have been added to the branches of the decision tree. (The expected utilities listed in the figure are discussed in Section 14.2.3.) For example, since the probability of rupture without surgery is estimated as .29, the probability of no rupture is $1 - .29 = .71$. Similarly, the probability of success following surgery is found by subtracting from 1 the probabilities of death and disability, ie, $1 - .02 - .06 = .92$.

14.2.2 Determining the Utility of Each Outcome

14.2.2.a Objective (Quantitative) Outcomes Versus Subjective (Qualitative) Outcomes: When outcomes for a decision are cost, number of years of life, or some other variable that has an inherent numerical value, these values may be used as the utilities. When outcomes do not have a numerical value, as in our example, investigators must find a way to give them a value. This process is known as assigning **utilities.**

The scale used for utilities is arbitrary; the investigators in Presenting Problem 1 developed a scale for the utility of each outcome ranging from 0 for death to 100 for perfect health. They decided that disability following surgery should be valued at 75. Although this decision is not completely arbitrary, other individuals might assign different values for disability. For example, some people feel that having a serious disability is a terrible outcome, almost as bad as dying, and they might give this outcome a utility of 10 or 20 rather than 75. For this reason, the term *subjective utilities* is sometimes used to describe the utilities determined in this kind of situation.

14.2.2.b An Example of Determining Subjective Utilities: Subjective utilities can be obtained informally, as just described, or by a more rigorous process called a lottery technique. This technique involves a process called *game theory*.

To illustrate, suppose you are asked to play a game in which you can choose a prize of $50 or you can play the game with a 50–50 chance of winning $100 (and gaining $0 if you lose). Here, the *expected value* of playing the game is .50 × $100

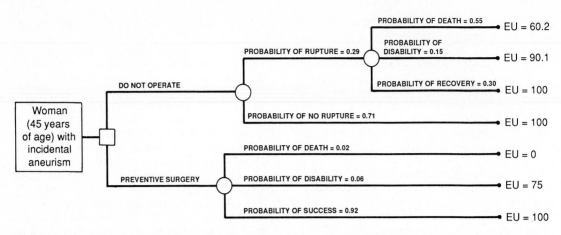

Figure 14-2. Decision tree for aneurysms with probabilities and utilities included. EU stands for expected utility. (Adapted and reproduced, with permission, from van Crevel H, Habbema JD, Braakman R: Decision analysis of the management of incidental intracranial saccular aneurysms. *Neurology* 1986;**36**:1335.)

= $50, the same as the prize. Would you take the sure $50 or play the game? If you choose not to gamble and take $50 instead, the next question asks whether you would play the game if the chance of winning was increased from 50% to 60%, resulting in an expected value of $60, $10 more than the prize. If you would still take $50, the chance would be upped to 70%, etc, until they reached a point at which you could not decide whether to play the game or take the prize, called the *point of indifference;* this is the value you attached to playing this game. You would be described as risk-adverse, since you refused to gamble even when the odds were in your favor, ie, when the expected value of the game was more than the prize.

Suppose now that a colleague would play the game rather than take the $50 prize if the chance of winning $100 was 50-50. Then, the next question would be whether the colleague would still play if the chance of winning $100 was only 40%, etc, until the point of indifference was reached. Your colleague would be described as risk-seeking because he or she was willing to gamble even when the odds were not in his or her favor and the expected value of the game was less than the prize.

14.2.2.c The Subjective Utilities in the Aneurysm Example: A similar process can be used to determine the values of survival, death, and disability. The patient or decision maker is given a set of alternative scenarios and asked to choose between them. In the aneurysm example, a patient might be asked to choose between the option of living 10 years with mental deterioration and the option of undergoing a procedure resulting in disability-free survival for 10 years 50% of the time and in immediate death 50% of the time. This set of scenarios would be systematically varied until the patient's point of indifference was reached,

and this point would be used to determine how much the patient values life with disability. Although the authors do not specify how they determined the value of 75 for disability, this is the value used in the decision analysis.

The determination of utilities for the outcomes following a rupture is more difficult. The same values could be used (ie, 0 for death, 75 for disability, and 100 for recovery); however, this technique ignores the number of years the patient lives prior to the rupture of the aneurysm. As an alternative, a lottery procedure similar to the one described above could be used to determine how much the patient values death and disability after a period of survival. The authors assumed the patient was somewhat risk-adverse and used a procedure called *discounting* to determine the utility of each outcome at some later date, assuming no operation is performed. A utility of 100 is still used for no rupture and for recovery following a rupture. However, disability following a rupture at some time in the future is considered more positive than disability following immediate surgery and is given a utility of 90.1. Similarly, death following a future rupture is preferred to death following immediate surgery and is given a utility of 60.2. These utilities have been appended to the ends of the branches of the decision tree in Fig 14-2.

14.2.3 Analyzing the Decision Tree

The decision tree is analyzed by a process known as calculating the *expected utilities,* or "folding back the tree." Folding back the tree begins with the outcomes and works backward through the tree to the point where a decision must be made. In our example, the first step is to determine the expected utility (EU) of the outcomes related to a ruptured aneurysm without an operation, obtained by multiplying the probability of each outcome by the utility for that outcome and summing the relevant products, ie, the probability of death times the utility of death, plus the probability of disability times the utility of disability, plus the probability of recovery times the utility of recovery:

EU (rupture
no operation) = (.55 × 60.2) + (.15 × 90.1)
                   + (.30 × 100)
                 = 33.1 + 13.5 + 30
                 = 76.6

The result 76.6 indicates the "average value" over all outcomes from the decision not to perform surgery in patients who subsequently have a rupture. Since the utility scale in this example is arbitrary, the value of 76.6 must be compared with values of 100 for no rupture and 0 for surgery with perioperative death.

This process is repeated for each chance node, one step at a time, back through the tree. Continuing with the example, the expected utility of not operating is the probability of rupture times the utility associated with this outcome, just found to be 76.6, plus the probability of no rupture times the utility; ie,

EU (no operation) = (.29 × 76.6) + (.71 × 100)
                      = 93.3

Similarly, the expected value of preventive surgery is

EU (surgery) =(.02 × 0) + (.06 × 75) + (.92 × 100)
               = 96.5

The expected utilities have been added to the decision tree in Fig 14–3. The expected utility of

surgery is 96.5 (based on 100 for a woman who lives and dies according to life table chances), compared with 93.3 for nonsurgical management. Therefore, surgery reduces the loss of utility caused by an incidental aneurysm from 6.7 (100 − 93.3) to 3.5 (100 − 96.5); with a life expectancy of 35 years, this reduction translates into 2.3 fewer years (6.7% × 35 years) with conservative management and 1.2 fewer years (3.5% × 35 years) with surgical management, on the average.

The optimal decision (performing surgery, in this example) is the one with the largest expected value, and the decision maker's choice, in terms of increased survival with surgery of 1.2 years, appears relatively easy to make. In situations in which the expected utility of two decisions is very close, the situation is called a "toss-up," and considerations such as how good the estimates used in the analysis are become more important. In fact, some of you may have been uncomfortable with the numbers used in the decision analysis and may have speculated on the effect different numbers would have on the decision. Therefore, the final step in the analysis is to determine how the decision would change if some of the numbers in the analysis were changed. This step will give an indication of how robust the decision is, ie, how applicable it is in situations other than the specific one included in the analysis.

14.2.4 Evaluating the Decision: Sensitivity Analysis

Accurate probabilities for each branch in a decision tree are frequently difficult to obtain from the literature. Investigators often have to use estimates made for related situations. For example, in Presenting Problem 1, the authors state that annual risk of rupture of an incidental aneurysm is

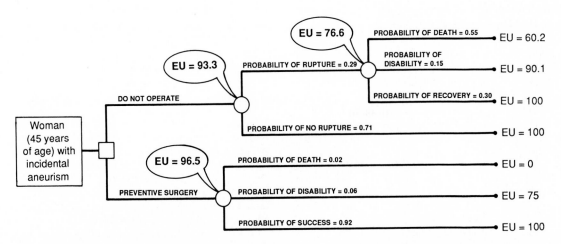

Figure 14-3. Decision tree for aneurysms with completed analysis. (Adapted and reproduced, with permission, from van Crevel H, Habbema JD, Braakman R: Decision analysis of the management of incidental intracranial saccular aneurysms. *Neurology* 1986;**36**:1335.)

not precisely known. Some studies have shown that a first rupture occurs at a rate of about 1% a year, and this is the value used in the decision analysis. However, the investigators would like to know whether the decision would be the same if the rate was either more or less than 1%. Similarly, the decision depends on the probability of mortality from subarachnoid hemorrhage, estimated at .55 in the above analysis; and again, the investigators would like to know whether the decision would change if that probability changed. The procedure for evaluating the way the decision changes as a function of changing probabilities and utilities is called **sensitivity analysis.**

To illustrate the logic involved in performing a sensitivity analysis, we determine the risk of aneurysm rupture that would change the optimal choice of treatment from surgical to conservative management. From Fig 14–3, the expected utility when the probability of rupture is .29 is 93.3; we want to know what the probability of rupture would need to be in order for the expected utility to be the same as for preventive surgery, ie, for it to be 96.5. Letting X stand for the probability of rupture, we must solve the following equation for X:

$$76.6X + 100(1 - X) = 96.5$$

Solving for X gives .15. Thus, when the probability of rupture is .15 instead of .29, the decision will be a toss-up, because the expected utilities of the two options, conservative treatment and preventive surgery, will be the same. Working backward in the formula for lifetime risk equal to .15—ie, solving $1 - (1 - R)^{35} = .15$ for R—gives an annual risk equal to .00463, or slightly less than 0.5%. Therefore, the decision to perform preventive surgery remains the same until the annual risk for someone with a 35-year life expectancy decreases to half the value used in the decision analysis.

Although our illustration of sensitivity analysis varied only one component, risk of rupture, one can also perform an analysis to determine the sensitivity of the final decision to other assumptions used in the decision. In addition, one can determine the sensitivity of two or more assumptions simultaneously. Most statisticians and researchers in decision analysis recommend that all published reports of a decision analysis include a sensitivity analysis.

The authors of the decision analysis on management of incidental aneurysms performed a sensitivity analysis to determine whether or not the decision was stable. They found that the decision will not change as long as the probabilities are within the following ranges: annual risk of rupture, 0.5–2%; mortality from rupture, 50–60%; disability after rupture, 10–20%; surgical mortality, 1–4%; surgical morbidity, 4–10%; and dis-

ability utility, 62.5–87.5. They also gave information on the decrease in benefits from surgery with older patients. For example, with surgical mortality of 2% and morbidity of 4%, the break-even value of surgical management is a life expectancy of 12 years; ie, surgery is beneficial only to patients who have 12 or more years of life-expectancy.

This example illustrates the use of decision trees for making a decision regarding a particular patient, a 45-year-old woman with a 7-mm aneurysm. The same tree could be used with another patient; but the lifetime probability of rupture would have to be adjusted for the patient's age, and different utilities would be needed to reflect the patient's values. Trees for decisions that do not depend on the specific features of any given patient are also possible, as we will see in the next section.

14.3 MAKING A DECISION ON HEALTH POLICY

The investigators in Presenting Problem 2 used decision analysis to calculate the economic implications of strategies for managing women seeking routine gynecologic care. The strategies they analyzed included routine culture for *C trachomatis,* a routine test with direct immunofluorescence or enzyme immunoassay (rapid tests), and no test. The decision tree for this problem is given in Fig 14–4, and it illustrates the use of subtrees, a helpful technique when trees become complex with repetitive branches.

14.3.1 Designing the Decision Tree

The basic tree considers three options: rapid test, culture, or no test. In each situation, the patient either does or does not have a chlamydial infection. The bracket in Fig 14–4 for the rapid test and culture branches indicates that all subsequent branches are the same and only one repetition is illustrated. In the no-test branch with infection, as in all other branches leading to untreated or uncured infection, the reader is directed to subtree A (illustrated in the bottom portion of Fig 14–4), which contains the ramifications of chlamydial infection of the cervix: cervicitis, acute salpingitis, and no complications. With acute salpingitis, the patient may be hospitalized or treated as an outpatient, and the reader is referred to subtree B, which contains the complications of this condition: infertility, chronic pelvic pain, ectopic pregnancy, or none. Note that the decision tree in Fig 14–4 shows many possible outcomes.

The decision model was designed for women without symptoms or signs of acute salpingitis, urethritis, or mucopurulent cervicitis and for women without gonorrhea, since these patients

Decision tree

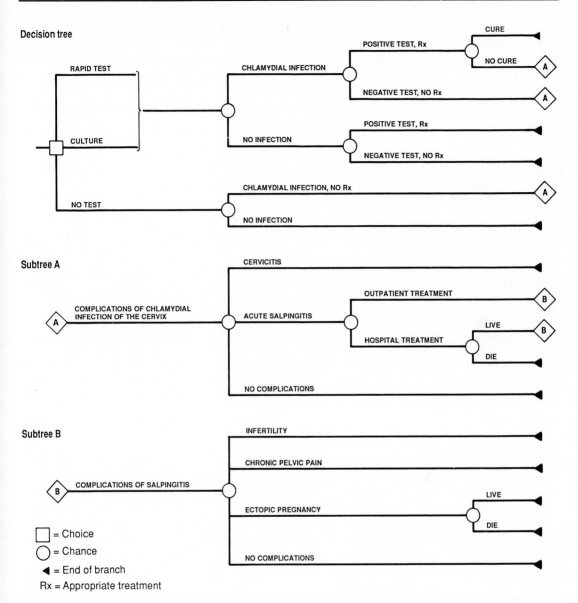

Figure 14–4. Decision tree for screening strategies for *C trachomatis* infection. (Adapted and reproduced, with permission, from Phillips RS et al: Should tests for *Chlamydia trachomatis* cervical infection be done during routine gynecologic visits? An analysis of the costs of alternative strategies. *Ann Intern Med* 1987;**107**:188).

should receive therapy with antibiotics active against chlamydial infection. Pregnant women were also excluded because complications from the infection differ from those in nonpregnant women.

This particular decision problem provides a good illustration of how judgments are involved in the process of determining the branches for the tree. For example, an additional strategy in this example is to treat the patients without first performing a culture. This strategy is illustrated in a letter discussing this decision tree from Ingelfinger (1988), who suggested the strategy of giving all patients a prescription for nystatin to be filled if they

develop vaginitis. Another possible branch in this decision problem is one for complications from treating women who do not have a chlamydial infection. In this study, the risk of complications from treating the false-positive patients is quite small; but in other situations, it would be an important branch to include on the tree.

14.3.2 Determining the Probabilities

Because the decision tree is so complex, the probabilities and other assumptions used in the analysis are listed by the authors rather than being shown graphically. They are as follows:

1. Sensitivity and specificity of cultures were 0.75 and 1.00, respectively; for the two rapid tests (direct immunofluorescence test and enzyme immunoassay), sensitivity = 0.60 and specificity = 0.98.

2. Patients with a positive test would be treated as recommended by the Centers for Disease Control (CDC), with a cure rate of 90%. A complication of vaginitis occurs in 15% of patients with a positive test.

3. For each patients with a positive test, one partner would also be treated.

4. Patients with acute salpingitis would be treated with a regimen recommended by the CDC, with two follow-up visits and a follow-up culture; 21% are expected to be hospitalized for treatment.

5. An estimated 32% of women with undetected *C trachomatis* cervical infections seek medical attention for related problems within 1 year; the estimate is 15% for acute salpingitis and 17% for symptomatic cervicitis.

6. Estimates of risks for adverse sequelae from acute salpingitis within 10 years are infertility in 18%, chronic pelvic pain in 15%, and ectopic pregnancy in 5%.

7. Risk of death is .0025 for patients hospitalized with salpingitis and .00009 for patients with ectopic pregnancy.

8. An estimated 50% of women would attempt pregnancy after an episode of salpingitis and be at risk for ectopic pregnancy; 50% who experience infertility would seek medical evaluation. All women with chronic pelvic pain seek medical care.

9. Direct costs included culture, $40; rapid tests, $15. Estimates of outpatient care hospitalization were obtained from insurance records.

10. Indirect costs included lost wages, lost household management, and lost lifetime earnings in the event of death.

14.3.3 Stating Results of the Decision Analysis

Using the above estimates for probabilities and utilities of the appropriate branches, the authors found that using a rapid test and subsequently treating women who have positive results would be cost-efficient if the prevalence of infection is 7% or greater. Routine cultures would result in lower costs than no test if the prevalence of infection is 14% or greater.

The authors performed a sensitivity analysis to see how the three different strategies compare, varying the factors listed above over a "reasonable range." They found that the rapid test strategy resulted in costs lower than the culture strategy unless the prevalence of infection exceeds 42%. The factors most important in influencing the decision for rapid tests were the estimate of acute salpingitis resulting from untreated chla-

mydial infection, the costs of the tests, and the probability of adverse sequelae occurring after treatment of cervical infection. For example, if the risk for salpingitis is increased from 15% to 20%, the threshold prevalence of infection before treatment is recommended drops from 7% to 5%. A lower cost of the rapid tests also resulted in a lower threshold of infection required for cost-effectiveness. Similarly, if the culture costs only $20 instead of $40, it would be the test of choice with a prevalence of 7%. The estimates of sensitivity and specificity had a minimal impact on the decision: If the sensitivity decreases from 80% to 60%, the threshold increases to 9%; if the specificity decreases from 98% to 95%, the threshold increases to 8%. The results of the sensitivity analysis related to the costs of the tests are presented graphically in Fig 14–5. The graph clearly illustrates the 7% and 14% thresholds and shows that no testing is cost-effective only when the prevalence is less than 7%.

From their analysis, the investigators recommend a rapid test for routine testing of women seeking gynecologic care, because the prevalence of infection reported among women seen in office practices exceeds the 7% threshold. At this level of prevalence, only about 69% of the patients with a positive test have the infection, and about one-third of the patients with positive tests have false-positive results. They do not recommend follow-up of a positive rapid test with a culture because of the degree of insensitivity of the culture procedure owing to the use of the cervical swab. Any decision to implement such a plan in another setting would, of course, depend on the preva-

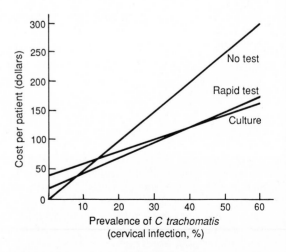

Figure 14–5. Sensitivity analysis of screening strategy to prevalence of *C trachomatis*. (Adapted and reproduced, with permission, from Phillips RS et al: Should tests for *Chlamydia trachomatis* cervical infection be done during routine gynecologic visits? An analysis of the costs of alternative strategies. *Ann Intern Med* 1987;**107**:188.)

lence of infection and the cost of rapid tests in that setting. For example, if the prevalence of chlamydial infection is less than 7%, the cost of rapid tests would need to be correspondingly less than $15 in order to justify screening of all women.

14.4 USING DECISION ANALYSIS TO EVALUATE SEVERAL PROTOCOLS

Diagnostic tools available for working up an asymptomatic patient with a positive Hemoccult test (Presenting Problem 3) include rigid and flexible sigmoidoscopy, colonoscopy, and the barium enema. An ideal diagnostic protocol would combine these tests so that all cancers could be found without undue costs, risks, or discomfort to the patient. The ideal does not exist, however, because none of the tests are perfect; and as more tests are done, the costs and risks increase accordingly. Some organizations recommend that all persons over age 50 be screened annually, and an appropriate way to work up positive patients must be found in order to keep these costs and risks in hand.

Several protocols have been recommended in the literature, and the range of procedures used in actual practice varies widely. Therefore, Brandeau and Eddy (1987) designed a decision analysis study to determine the protocol that is the most cost effective. They examined 22 protocols: 7 recommended in the literature and 15 that could be justified on a logical basis. We discuss 7 of the 22 protocols in this section; they are reproduced in Fig 14-6. Four protocols are from the literature (protocols 1, 2, 3, 4) and three approaches were generated by the authors (protocols A, B, C). If you are interested in more details on these protocols or the other protocols evaluated in the study, consult the article by Brandeau and Eddy (1987).

14.4.1 Protocols Evaluated

In addition to evaluating several protocols recommended in the literature, the investigators were interested in some specific research questions. One question of interest was whether a negative barium enema should be followed by a colonoscopy; this question considers the difference between Barium Enema & Colonoscopy (protocol A) and Barium Enema (protocol B). A second question was whether there is any value to preceding colonoscopy with a barium enema if colonoscopy will eventually be used in all cases; thus, protocol C (Colonoscopy) was also evaluated.

A summary of the seven protocols follows.

1. Sigmoidoscopy: Sigmoidoscopy, if negative, is followed by barium enema. If it is positive, it is followed by colonoscopy.

2. Rigid Sigmoidoscopy: Rigid sigmoidoscopy, if negative, is followed by repeat Hemoccult. If the repeat Hemoccult is positive, both a barium enema and colonoscopy are done, in that order.

3. Repeat Hemoccult: Repeat Hemoccult, if positive, is followed by both sigmoidoscopy (type not indicated) and barium enema, in that order.

4. Flexible Sigmoidoscopy: Flexible sigmoidoscopy, if positive, is followed by colonoscopy. If it is negative, it is followed by barium enema, which is followed by colonoscopy if positive. If barium enema is negative, it is followed by repeat Hemoccult. If repeat Hemoccult is positive, it is followed by colonoscopy. If colonoscopy is negative, it is followed by upper GI series.

A. Barium Enema (& Colonoscopy): Barium enema is always followed by colonoscopy. If colonoscopy is negative, it is followed by repeat barium enema. If repeat barium enema is positive, it is followed by second colonoscopy.

B. Barium Enema: Barium enema is followed by colonoscopy only if positive. If colonoscopy is negative, it is followed by repeat barium enema. If repeat barium enema is positive, it is followed by second colonoscopy.

C. Colonoscopy: Colonoscopy is followed by barium enema only if negative. If barium enema is positive, it is followed by second colonoscopy.

14.4.2 Assumptions Made in the Analysis

14.4.2.a Estimating Patient Conditions: A decision problem of this scope requires the investigators to estimate the proportion of Hemoccult-positive patients who have cancer, polyps, and miscellaneous bleeding and who are true false-positives having no bleeding. From information in the medical literature, they assumed that 8% of the patients with a positive Hemoccult have cancer, 40% have polyps, 36% have miscellaneous bleeding, and 16% have no bleeding.

14.4.2.b Determining the Characteristics of the Procedures: Several characteristics of the procedures must be taken into consideration in the design of a decision strategy. Rigid sigmoidoscopy, flexible sigmoidoscopy, barium enema study, and colonoscopy can search different regions of the bowel, and they have different accuracies for detecting cancer and polyps; these accuracies can also depend on the results of the tests that precede them. For example, colonoscopy and flexible sigmoidoscopy are thought to be more accurate if they follow a barium enema, because the barium enema results can guide the endoscopist's search for abnormal lesions. These tests also have different diagnostic ability, since all but the barium enema permit a biopsy at the time of the examination. The tests also cause different amounts of discomfort, have different complication rates,

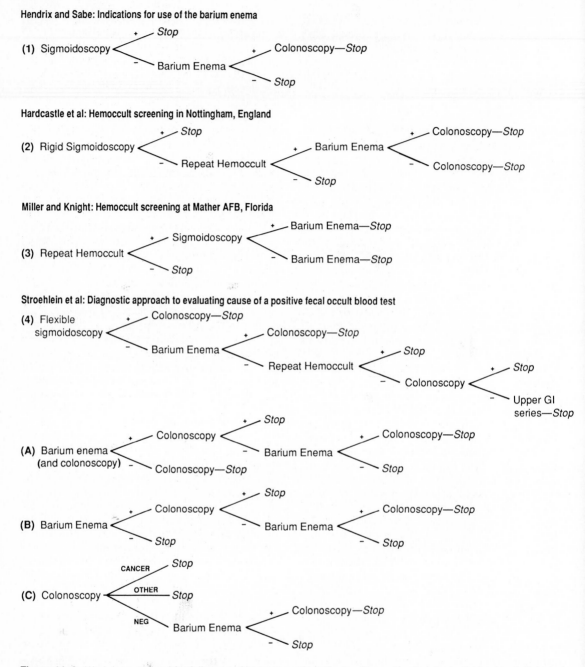

Figure 14-6. Workups proposed for follow-up of asymptomatic patients with positive fecal occult blood test. (Adapted and reproduced, with permission, from Brandeau ML, Eddy DM: The workup of the asymptomatic patient with a positive fecal occult blood test. *Med Decis Making* 1987;**7**:32.)

and are associated with different costs. Again, the literature was consulted to obtain information regarding these characteristics, which is summarized next.

1. Rigid sigmoidoscopy: Approximately 40–45% of colonic cancers occur within the reach (18–20 cm) of a rigid sigmoidoscope. Within this range, the sensitivity for detecting cancer and pol-

yps is estimated to be 40–60% and 35–50%, respectively. Many patients experience discomfort, and colonic perforation occurs 1 in 10,000 procedures. The cost of this procedure varies from $25 to $150; an estimate of $49 was used in the analysis.

2. Flexible sigmoidoscopy: Within a range of 50–55 cm, the sensitivity of flexible sigmoidos-

copy is 60–90%. Patients tolerate this procedure better than rigid sigmoidoscopy, but the complication rate is about the same. Cost is approximately $105.

3. Colonoscopy: A colonoscopy can examine the entire colon. Its sensitivity is estimated as 80–90%, increasing to 90–98% if preceded by a barium enema. Complication rates are higher, with bleeding in 18 of 10,000 cases and, when combined with barium enema, perforation in 3 of 1000 cases. It is also more expensive, costing $400–$800; a cost of $680 was used in the analysis.

4. Barium enema: A barium enema has a sensitivity for cancer and polyps of 40–60%; this value can be increased with careful patient preparation to 95% for cancer and 92% for polyps larger than 5 mm. False-positive results are about 5%. Colonic perforation occurs in 2 of 10,000 patients, and there is also an unmeasured risk from radiation. The cost, with air contrast, is about $175.

5. Repeat Hemoccult: A repeat Hemoccult has a sensitivity for both cancer and polyps of about 55%. Its cost is minimal, about $8.

6. Upper GI (gastrointestinal) series: This procedure has about 5% sensitivity for detecting miscellaneous bleeding and zero sensitivity for cancer and polyps. Cost is about $150.

14.4.3 The Decision Tree

A decision tree was developed to evaluate each of the protocols. Fig 14–7 shows the tree for protocol 1 (Sigmoidoscopy), in which a rigid sigmoidoscopy is followed by an air contrast barium enema if negative, which in turn is followed by a colonoscopy if positive.

The probabilities of cancer, polyps, bleeding, and no bleeding are discussed above. The other probabilities in Fig 14–7 were estimated from a review of the literature by the authors. For exam-

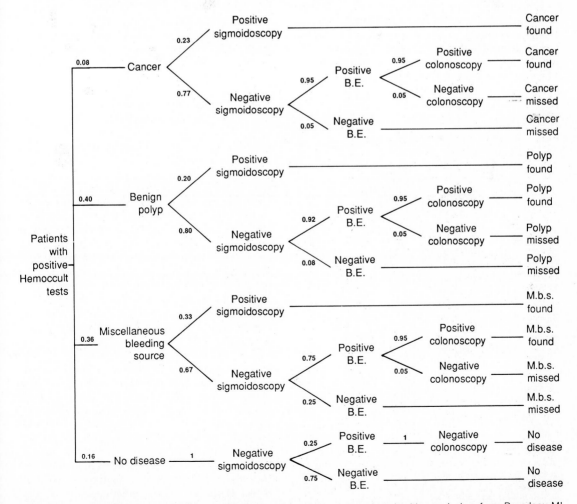

Figure 14–7. Example of decision tree for protocol 1. (Redrawn and reproduced, with permission, from Brandeau ML, Eddy DM: The workup of the asymptomatic patient with a positive fecal occult blood test. *Med Decis Making* 1987;**7**:32.)

ple, the probability is .23 that the sigmoidoscopy will find a cancer; ie, sensitivity is 50% for detecting the 45% of cancers in the range of a sigmoidoscope. Similarly, if the sigmoidoscopy fails to detect cancer in a patient with cancer, an air contrast barium enema will be positive 95% of the time, and a positive barium enema will lead to a positive colonoscopy 95% of the time. Thus, the probability that cancer will be found by using this protocol, if cancer is present, is .23 (the probability of a positive sigmoidoscopy) plus .77 × .95 × .95 (the probability of negative sigmoidoscopy followed by positive barium enema and positive colonoscopy), or .23 + .69 = .92, or 92%. Therefore, this protocol will miss a cancer in 8% of the patients. Similarly, the probability of finding polyps when present is .20 + (.80 × .92 × .95) = .90, or 90%; and this process is continued for the remainder of the branches of the tree.

14.4.4 Results of the Decision Analysis

The investigators developed a similar decision tree for each of the protocols evaluated in their study. Using a computer program, they evaluated each tree to determine the effectiveness of the protocol in detecting cancer and polyps. The fraction of cancers found, cancers and polyps found, and the average cost per patient are given in Table 14–1. In general, there is a positive relationship between the cost of a protocol and its effectiveness in de-

tecting lesions. The procedure that detects the most cancers and polyps is protocol A, Barium Enema & Colonoscopy; however, it puts patients through a great deal of discomfort and is the most costly ($876). Protocol C, Colonoscopy alone, does almost as well, and it costs $110 less.

The authors could not designate one protocol as the very best. However, they could determine which strategies were better in a relative sense. The authors prepared a graph comparing the fraction of cancers and precancerous lesions found for the average cost per patient in all 22 protocols they evaluated. The protocols that appear along the line connecting protocols 3 (Repeat Hemoccult) and A (Barium Enema & Colonoscopy) are said to dominate over protocols below the line because they detect more lesions for the same cost. The graph for all 22 protocols is reproduced in Fig 14–8; the seven we discussed are identified on the graph. Four of the seven strategies are on the line and are relatively better than the other three: protocols three (Repeat Hemoccult), A (Barium Enema & Colonoscopy), B (Barium Enema), and C (Colonoscopy). Fig 14–8 may be used to make recommendations about the protocols. For example, protocol 2 (Rigid Sigmoidoscopy) should not be used, since it detects only 2% more lesions than protocol 3 (Repeat Hemoccult) but costs about $250 more.

14.4.5 Conclusions From the Decision Analysis

From the results of the decision analysis of the 22 different protocols they analyzed, the investigators drew several conclusions. Performing a barium enema prior to sigmoidoscopy or colonoscopy increases the overall sensitivity for cancers and precancerous lesions by about only 0.3%, with an average cost increase of about $45. Furthermore, using a barium enema as a "screening" procedure and following it with a colonoscopy only if it is positive (protocol B) produces a $165 reduction in cost with only a 2% reduction in the number of lesions found. These conclusions are generally the same even if the sensitivity of colonoscopy without a preceding barium enema is reduced from 90% to 80%. However, for physicians who do not have access to colonoscopy, the most effective workup is to perform a barium enema and follow it with flexible sigmoidoscopy in all patients. Overall, the effectiveness of workups with sigmoidoscopy ranged from 60% to 93%, compared with 91% to 97% for workups with colonoscopy.

Four of the seven protocols in the literature and the American Cancer Society recommend repeat Hemoccults. However, the analysis shows that a repeat Hemoccult is not advantageous. This test has a false-negative rate of 45% for cancers and polyps. Therefore, although a repeat Hemoccult

Table 14-1. Selected results from fecal occult blood study.[1]

Protocol[2]	Fraction Found		Average Cost per Patient
	Cancers	Cancers & Polyps	
1. Sigmoidoscopy	0.886	0.860	$522
2. Rigid Sigmoidoscopy	0.627	0.598	430
3. Repeat Hemoccult	0.607	0.580	149
5. Flexible Sigmoidoscopy	0.896	0.870	797
A. Barium Enema & Colonoscopy	0.988	0.970	876
B. Barium Enema	0.967	0.941	687
C. Colonoscopy	0.981	0.960	766

[1]Adapted and reproduced, with permission, from Table 3 in Brandeau ML, Eddy DM: The workup of the asymptomatic patient with a positive fecal occult blood test. *Med Decision Making* 1987; 7:32.
[2]See Fig 14–6 for a complete description of each protocol; this table lists the first step only.

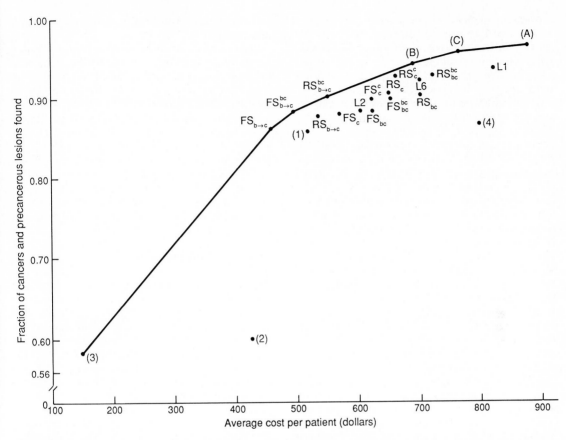

Figure 14-8. Information on cost and effectiveness of each protocol. (Adapted and reproduced, with permission, from Brandeau ML, Eddy DM: The workup of the asymptomatic patient with a positive fecal occult blood test. *Med Decis Making* 1987;**7**:32.)

may provide a cost-effective way of eliminating some patients with initial false-positive Hemoccults, too many cancers are missed (false-negative results) with these protocols.

14.5 EXTENSIONS OF DECISION THEORY

Medical decision making is a rapidly evolving field, with results of new decision analyses being published with increasing frequency. There is also a significant growth in the number of new methods being developed for this area. Three will be discussed here: multiple-testing strategies, Markov models, and artificial intelligence.

14.5.1 Multiple-Testing Strategies

For many clinical problems, protocols specify the combination of two diagnostic tests for reaching a treatment decision. In some situations, the two tests are administered at the same time, as in testing a patient with severe chest pain for a myocar-

dial infarction (MI) by performing serial electrocardiograms (ECG) and cardiac enzyme measurements. Here, the physician may decide against a diagnosis of MI if either the ECG or the cardiac enzymes are normal. However, if both are abnormal in the characteristic pattern, MI is the most likely diagnosis.

In other clinical situations, two tests are administered in series; ie, the result of the first test is used to determine whether or not the second test should be done, as in patients with a possible diagnosis of primary hyperparathyroidism. Generally, the first test is a measurement of the patient's serum calcium level. If it is elevated, the second step is to measure the patient's level of parathyroid hormone (PTH), which will be inappropriately high in patients with primary hyperparathyroidism. If serum calcium is normal, there is no need for additional testing.

Hershey, Cebul, and Williams (1987) evaluated situations for which two tests are recommended, and they were interested in whether there are any instances in which either parallel or series testing is always preferable to a single test. For example, two types of clinical decisions involve parallel testing: Use test 1 and test 2 at the same time and treat only if *both* are positive, as with the MI example above; or use test 1 and test 2 at the same time and treat if *either* is positive. Similarly, two decision approaches involve series testing: Perform test 1 and, if positive, perform test 2, as in the hyperparathyroid example; or perform test 1 and, if negative, perform test 2.

The authors compared each of these decisions with the decisions made using test 1 only and using test 2 only. They determined that two tests, whether in parallel or series, are never preferable to a single test over all ranges of prevalence (pretest probability). In other words, there are always some values of prevalence for which single testing is the approach of choice—generally, in the midrange of disease probabilities. Above this range, the either-test-positive criterion is better; and below this range, the both-tests-positive criterion is better. In the range of intermediate pretest probabilities, however, the best strategy is to perform only one test and act on the basis of the results of that test.

14.5.2 Markov Models

One of the criticisms of decision analysis is that many decisions must be simplified if we are to use the methods. For example, consider Presenting Problem 1, where the question is how to manage patients who have incidental aneurysms. A patient with an aneurysm may develop a subarachnoid hemorrhage in any given year, or the aneurysm may remain without complications. If a hemorrhage occurs, the patient may be left with serious disability, or the patient may recover. The risks of both a rupture and a disability following rupture depend on the age of the patient. The investigators in Presenting Problem 1 took these issues into consideration by including the patient's life expectancy as part of the calculation of risk.

Another way to analyze this problem is to use a technique called a Markov process, which permits patients to move back and forth from one "state of health" to another. To illustrate the logic involved in a Markov process, let's suppose that a million hypothetical patients, each with an incidental aneurysm, were followed through time. During any given year, a patient in the "aneurysm state" could move to the "aneurysm rupture state" and then on to the "disability state," the "recovery state," or the "death state." After recovering, a patient could decide to have surgery or not to have surgery and could move to the "aneurysm state" again in the future. The chance that a patient will move from one health state to another is called the transition probability for that state. The Markov model involves computing the transition probabilities so that they can be used in a decision analysis; they allow investigators to

determine the expected survival for any of the million patients.

For example, for the 45-year-old woman with a 7-mm aneurysm, the investigators can determine the probability that she will stay in the "aneurysm state" during the next year as well as the probabilities that she will move to the "disability state" or the "death state." If she remains in the "aneurysm state" for this next year, then they can calculate the probabilities for the year following; etc. From these probabilities, the physician and patient can make an informed decision regarding therapy. When computer programs are used for the calculations, the Markov model is a useful method for analyzing complex problems that evolve and change over time.

14.5.3 Artificial Intelligence

Artificial intelligence (AI) is a field of computer science concerned with designing computer programs that understand language, reason, and solve problems—ie, computerized decision trees. Most AI programs in medicine determine diagnoses and make recommendations for therapy (Clancey and Shortliffe, 1984). Some of the programs are known as consultation or expert programs because they provide expert advice on how to handle a patient problem. These systems use clinical algorithms (protocols) in making decisions about how to manage a patient with a particular problem. Usually, the protocols are in the form of a decision tree that follows simple branching logic illustrated in the Presenting Problems. As examples, MYCIN provides therapy advice for certain types of infections; INTERNIST is an experimental program for computer-assisted diagnosis in general internal medicine; PUFF interprets measurements from respiratory tests administered to patients in a pulmonary function laboratory and makes the diagnosis; and AIRHEUM provides physicians who are not rheumatologists with diagnostic advice in that specialty area.

Many AI systems were originally designed by expert physicians for use by paramedical personnel; recently, they have also been used to supplement and extend medical education. Some of the more complex programs are linked to videolaser disks so that sound and visual information are included in the program. It remains to be seen how large a role AI systems will play in medicine in the future, but some believe that the primary challenges to tomorrow's physician will be in the area of management, with AI computer programs being used to aid in most diagnostic decisions. If you are interested in this topic, you may wish to consult texts, each a collection of articles related to decision making in medicine: *Readings in Medical Artificial Intelligence* edited by Clancey and Shortliffe (1984) and *Clinical Decisions and Laboratory Use* edited by Connelly et al (1982).

sidered in a sensitivity analysis of the decision to maintain or to change the schedule?

b. The investigators calculated the expected numbers of different outcomes under the current vaccination schedule and the proposed schedule. They presented the results in terms of a "base case" scenario in which pertussis incidence in the 7-to-12 month age group is assumed to be 7.6 per 100,000 in children under 2 months of age, 17.7 in children aged 2–6 months, 6.2 in children aged 7–12 months, and 9.7 in children aged 13–47 months. The results are given in the Table 14–2. From this analysis, which schedule is better for minimizing total pertussis, the current schedule or the proposed schedule?

c. Which schedule is better from the perspective of adverse effects from the vaccine?

15

Reading the Medical Literature

15.1 PURPOSE OF THE CHAPTER

This final chapter has several purposes. Most importantly, it ties together the concepts and skills presented in previous chapters and applies these concepts very specifically to reading medical journal articles. Throughout the text, we have attempted to illustrate strengths and weaknesses of some of the studies discussed, but this chapter focuses particularly on the attributes of studies that indicate to readers that they can use the results with confidence. The chapter begins with a brief summary of major types of medical studies. Next, the anatomy of a typical journal article is examined in considerable detail, and the contents of each component—Abstract or Summary, Introduction, Methods, Results, Discussion and Conclusions—are discussed. In this examination, we discuss common shortcomings or threats to the validity of different studies.

There are, of course, different reasons for reading the medical literature. Some articles are of interest because the practitioner wants only to maintain an awareness of advances in a field. In these instances, the reader may decide to skim the article without considering how well the study was designed and carried out; the reader may be able to depend on experts in the field who write review articles providing the relatively superficial level of information the practitioner requires. On other occasions, however, the reader wants to know whether the conclusions of the study are valid, perhaps so that they can be used to determine patient care or to plan a research project. In these cases, the reader will read and evaluate the article with a critical eye in order to detect poorly done studies with unwarranted conclusions.

To assist readers in their critical reviews, we present a checklist for evaluating the validity of a journal article. The checklist notes some of the characteristics that a well-designed and well-written article should have. It combines the results from our medical students and physician colleagues who read and critiqued medical articles. It also reflects the opinions expressed in an article describing how journal editors and statisticians can interact to improve the quality of published medical research (Marks et al, 1988). As we indicated in Chapter 1, Williamson, Goldschmidt, and Colton (1986) found that only about 20% of the studies reviewed in 28 articles assessing their validity met the criteria the authors defined for scientific adequacy. The checklist should assist you as a reader in using your time most effectively, ie, in continuing to read the valid articles and in identifying poorly done studies so that you can move on to more productive reports.

15.2 REVIEW OF MAJOR STUDY DESIGNS

Chapter 2 introduced the major types of study designs used in medical research, broadly divided into **experimental studies,** or clinical trials, and **observational studies:** cohort, case-control, cross-sectional, and case-series studies. Each design has certain advantages over the others as well as some specific disadvantages; they are briefly summarized below. (A more detailed discussion is presented in Chapter 2.)

Clinical trials provide the strongest evidence for causation because they are experiments and, as such, are subject to the least number of problems or biases. They are the study of choice when the objective is to evaluate the effectiveness of a treatment or a procedure. Clinical trials that use **randomization** are the strongest design of all. Drawbacks to using clinical trials include their expense and the generally long period of time needed to complete them.

Cohort studies are the best observational study design for investigating the causes of a condition, the course of a disease, or the **risk factors.** Causation cannot be proved with cohort studies, though, because they do not involve interventions. However, because they are **longitudinal studies,** they incorporate the correct time sequence to provide strong evidence for possible causes and effects. In addition, in cohort studies that are **prospective studies,** rather than **historical cohort studies,** investigators can control many sources of bias. Cohort studies have disadvantages, of course. If they take a long time to complete, they are frequently weakened by patient attrition. They are also expensive to carry out if the disease or outcome of interest is rare or requires a long to time develop.

Case-control studies are the most efficient way to study rare diseases, examine conditions that take a long time to develop, or investigate a preliminary hypothesis. They are the quickest and generally the least expensive studies to design and carry out. However, case-control studies also are the most vulnerable to possible biases, and they depend entirely on high-quality existing records. A major issue in case-control studies is the selection of an appropriate control group; and some statisticians have recommended the use of two control groups: one similar in some ways to the cases (such as having been hospitalized or treated during the same period of time) and another made up of healthy subjects.

Cross-sectional studies are best for determining the status of a disease or condition at a particular point in time; they are similar to case-control studies in being relatively quick and inexpensive to complete. Because cross-sectional studies provide only a "snapshot in time," they may lead to misleading conclusions if interest focuses on a disease or other time-dependent process.

Case-series studies are the weakest study design and represent a description of generally unplanned observations; in fact, many would not call them studies at all. Their primary use is to provide insights for research questions to be addressed by subsequent, planned studies.

15.3 THE ABSTRACT & THE INTRODUCTION SECTION OF A RESEARCH REPORT

Most journal articles include an abstract or summary of the article prior to the body of the article itself. Most of us are guilty of reading *only* the abstract on occasion, perhaps because we are in a great hurry or have only a cursory, tangential interest in the topic. However, this practice is unwise when one needs to know whether the conclusions stated in the article are justified and can be used to make decisions. This section discusses the Abstract and Introduction portions of a research report and outlines the information they should contain.

15.3.1 The Abstract

The major purposes of the abstract are (1) to tell readers enough about the article so that they can decide whether to read it in its entirety and (2) to identify the focus of the study. The International Committee of Medical Journal Editors (1988) has recommended that the abstract "state the purposes of the study or investigation, basic procedures (selection of study subjects or experimental animals; observational and analytic methods),

main findings (specific data and their statistical significance, if possible) and the principal conclusions."

If the topic is of interest to readers, we suggest they then ask two questions to decide whether or not to read the article: (1) If the study has been properly designed and analyzed, would the results be important and worth knowing about? (2) With a sufficiently large sample size, a small difference between two groups will be statistically significant even though it may not be clinically significant; if the results are statistically significant, does the magnitude of the change or effect also have clinical significance or importance? If the answers to these questions are affirmative, then readers will probably want to continue to read the report.

15.3.2 The Introduction

15.3.2.a Reason the Study Is Needed: The Introduction section of a research report is usually fairly short. Generally, the authors briefly review the findings from previous studies that indicate the need for the present study. In some cases, the study described in the report is a natural outgrowth or the next logical step of previous studies. In other cases, the previous studies have been inadequate in one way or another. The overall purpose of this information is twofold: to provide the necessary background information to place the present study in its proper context, and to provide reasons why the present study needed to be done. In some journals, justification for doing the study is given in the Discussion section of the article instead of in the Introduction section.

15.3.2.b Purpose of the Study: Regardless of where background information on the study is given, the Introduction section is the place where the investigators communicate to the reader the purpose of their study. The purpose of the study is frequently presented in the last paragraph or last sentences at the end of the Introduction. The purpose should be stated clearly and succinctly, in a manner analogous to a "15-second summary" of a patient case. For example, in the study described in Presenting Problem 1 of Chapter 7, Lisboa et al (1985) do this very well; they stated: "The aim of the present study was to evaluate inspiratory muscle function both through mouth and transdiaphragmatic pressures and correlate it with ventilatory disability and respiratory failure in adult patients with severe kyphoscoliosis." Readers should also be able to determine whether the purpose for the study was conceived prior to data collection or whether it evolved after the authors viewed their data; the latter situation is much more likely to capitalize on chance findings. The lack of a clearly stated research question is the most common reason medical manuscripts are rejected by journal editors (Marks et al, 1988).

15.3.2.c Population Included in the Study: In addition to stating the purpose of the study, the Introduction section sometimes contains information on where the study was done, the length of time covered by the study, and who the subjects in the study were. Alternatively, this information may be contained in the Method section. The reader should ask whether the location of the study and the subjects included in the study are applicable in the reader's practice environment.

The time period covered by a study gives important clues regarding the validity of the results. If the study on a particular treatment modality covers too long a period of time, patients at the beginning of the study may differ in important ways from patients at the end of the study. For example, major changes may have occurred in the way the disease in question is diagnosed, and patients near the end of the study may have had their disease diagnosed at an earlier stage than did patients who entered the study early (see Detection Bias, below). On the other hand, if the purpose of the study is to examine sequelae of a condition or procedure, then the period of time covered by the study must be sufficiently long to detect consequences.

15.4 THE METHOD SECTION OF A RESEARCH REPORT

The Method section contains information about how the study was done, including type of study design used. Simply knowing the study design gives the reader a great deal of information. In addition, the Method section contains information regarding subjects who participated in the study or, in animal or inanimate studies, information on the animals or materials. The procedures used should be described in sufficient detail so that the reader knows how measurements were made. The study outcomes should be specified along with the criteria used to assess them. The Method section also should include information on the sample size for the study and on the statistical methods used to analyze the data; this information is often placed at the end of the Method section. Each of these topics is discussed in detail in this section.

Of utmost importance, however, is how well the study has been designed. The most critical statistical errors, according to the statistical consultant to *The New England Journal of Medicine,* involve improper research design: "An investigator can always re-analyze the data but cannot re-design the study after data collection has occurred" (Marks et al, 1988).

15.4.1 Subjects in the Study

15.4.1.a How Subjects Were Chosen: There are several critical pieces of information that authors of journal articles should provide about subjects included in their study so that readers can judge the applicability of the study results. Of foremost importance is how the patients were selected for the study and, if appropriate, how treatment assignments were made.

Randomized selection or assignment greatly enhances the generalizability of the results and avoids biases that otherwise may occur in patient selection (see Section 15.4.2). Some authors believe it is sufficient merely to state that "subjects were randomly selected" or "treatments were randomly assigned," but some statisticians (DerSimonian et al, 1982) recommend that the type of randomization process be specified as well. Authors who report the methods used to randomize provide some assurance to the readers that randomization actually occurred, because some investigators have a faulty view of what constitutes randomization. For example, an investigator may believe that assigning alternate patients to the treatment and the control makes the assignment random. However, as we emphasized in Chapter 5, randomization involves one of the precise methods that ensure that each subject (or treatment) has a known probability of being selected.

15.4.1.b Eligibility Criteria: The authors should present information to illustrate that major selection biases (discussed in the next section) have been avoided, an aspect especially important in **nonrandomized trials.** The issue of which patients serve as controls was discussed in Chapter 2 related to case-control studies. In addition, the eligibility criteria for both inclusion and exclusion of subjects in the study must be specified in detail. The reader should be able to state, given any hypothetical subject, whether or not this person would be included in or excluded from the study.

15.4.1.c Patient Follow-Up: For similar reasons, sufficient information must be given regarding procedures the investigators used to follow up patients, and they should state the numbers lost to follow-up. Some articles include this information under the Results section. These aspects are illustrated by Presenting Problem 1 of Chapter 9, a study designed by Graham et al (1985) to evaluate the use of cimetidine (300 mg four times daily) compared with **placebo** in the treatment of benign gastric ulcer. The first three paragraphs reproduced below describe the inclusion and exclusion criteria. The final paragraph gives information on follow-up and dropouts.

Patients were accepted into the study if they had benign gastric ulcer disease. The diagnosis of gastric ulcer was confirmed in each patient by endoscopic examination. Malignancy was excluded by cytologic findings and histologic examination of a minimum of four biopsy samples. The ulcer was measured in longest dimension during the endoscopic examination at entry; only those patients with ulcers from 0.5 to 2.5 cm in diameter were eligible. If several gastric ulcers were present, each ulcer was evaluated and a separate biopsy was done; the largest ulcer was termed the "index" ulcer. Patients with only pyloric ulcer were not entered, but patients with a gastric ulcer coexisting with another ulcer in the pyloric channel or in the duodenum were not excluded.

Patients who had previously been treated with cimetidine, antacids, or both, could be admitted provided such treatment had lasted no more than 5 days and had been discontinued before the patients received coded medication. The "wash-out" period was 72 hours for cimetidine and 24 hours for antacids.

In addition to exclusions for malignancy and ulcer size, patients were excluded if they had a history of gastric surgery or vagotomy, except simple plication of an ulcer more than 6 months before entry; pyloric obstruction or recent gastrointestinal bleeding; concomitant disease related to ulcer such as Zolinger-Ellison syndrome; and a history of recent ingestion of ulcerogenic drugs such as aspirin or other nonsteroidal anti-inflammatory agents. Chronic use of such drugs was prohibited and patients were questioned about drug use, including over-the-counter preparations, at each visit.

One hundred seventy-two patients participated in the study; 87 were randomly assigned to receive cimetidine and 85 to receive placebo. Twenty-two patients were excluded from analysis: 13 because five investigators failed to contribute a required minimum of 5 patients; 2 because their index ulcers were either larger or smaller than the entry criterion; 2 because they received an initial supply of coded drug but did not return for follow-up; 1 for failure to take the coded medicine; 1 (a patient receiving placebo) who developed hepatitis 2 weeks after starting therapy; 1 patient whose ulcer was located in the pylorus; 1 who had chronic renal failure; and 1 who had a malignant ulcer. Twelve of the excluded patients had been randomly assigned to receive cimetidine and 10 to receive placebo.[*]

The description of follow-up and dropouts should be sufficiently detailed to permit the reader to draw a diagram of the information. Occasionally, an article will present such a diagram.

15.4.2 Bias Related to Subject Selection

Bias in studies is something that should not happen; it is an error related to selecting subjects or procedures or to measuring a characteristic. Biases are sometimes called **measurement errors** or

systematic errors to distinguish them from **random error (random variation)** that occurs any time a sample is selected from a population. This section discusses selection bias, a type of bias common in medical research.

Selection biases can occur in any study, but they are easier to control in clinical trials and cohort designs. The reader should be made aware of selection biases, even though it is impossible to predict exactly how their presence affects the conclusions. Sackett (1979) enumerates 35 different biases. We discuss some of the major ones that seem especially important to the clinician. If you are interested in a more detailed discussion, consult the article by Sackett and the text by Feinstein (1985), which devotes several chapters to the discussion of bias (especially Chap 4, Section 2, and Chaps 15–17).

15.4.2.a Prevalence or Incidence (Neyman) Bias:
Prevalence bias occurs when a condition is characterized by early fatalities (some subjects die before they are diagnosed) or "silent" cases (cases in which the evidence of exposure disappears when the disease begins). Prevalence bias can result whenever there is a time gap between exposure and selection of study subjects and the "worst" cases have died. A cohort study begun prior to the onset of disease is able to detect occurrences properly. However, a case-control study that begins at a later date consists only of the residual—the people who did not die. This bias can be prevented in cohort studies and avoided in case-control studies by limiting eligibility for the study to newly diagnosed or incident cases. The practice of limiting eligibility is common in population-based case-control studies in cancer epidemiology.

To illustrate prevalence or incidence bias, let's suppose that there are two groups of people, those with risk factor for a given disease (eg, hypertension as a risk factor for stroke) and those without the risk factor. Suppose 1000 people with hypertension and 1000 people without hypertension have been followed for 10 years. At this point, we might have the following situation:

Patients	No. of Patients in 10-Year Cohort Study		
	Alive With Cerebro-vascular Disease	Dead from Stroke	Alive with No Cerebro-vascular Disease
With hypertension	50	250	700
Without hypertension	80	20	900

---

[*]Reprinted, with permission, from Graham KY et al: Healing of benign gastric ulcer: Comparison of cimetidine and placebo in the U.S. *Ann Intern Med* 1985; **102**: 573.

A cohort study begun 10 years ago would conclude correctly that those patients with hypertension are more likely to develop cerebrovascular disease than patients without hypertension (300 to 100) and far more likely to die from it (250 to 20).

However, suppose a case-control study is undertaken at the end of the 10-year period without limiting eligibility to newly diagnosed cases of cerebrovascular disease. Then the following situation will occur:

Patients	No. of Patients in Case-Control Study at End of 10 Years	
	With Cerebrovascular Disease	Without Cerebrovascular Disease
With hypertension	50	700
Without hypertension	80	900

The **odds ratio** is calculated as $(50 \times 900)/(80 \times 700) = 0.80$, and it appears that the risk factor "hypertension" is actually a protective factor for the disease. Therefore, the bias introduced in an improperly designed case-control study of a disease that kills off one group faster than the other can lead to a conclusion exactly opposite from the correct one that would be obtained from a well-designed case-control study or a cohort study.

15.4.2.b Admission Rate Bias (Berkson's Fallacy):
Admission rate bias occurs when there are differential study admission rates; it can cause major distortions in risk ratios. As an example, admission rate bias can occur when patients with the disease (cases) who have the risk factor are admitted to the hospital more frequently than either the cases without the risk factor or the controls with the risk factor; ie, because the admissions rates are different for the cases and the controls.

This fallacy was first pointed out by Berkson (1946) in evaluating an earlier study that had concluded that tuberculosis might have a protective effect on cancer. This conclusion was reached after a case-control study found a negative association between tuberculosis and cancer: The frequency of tuberculosis among hospitalized cancer patients was less than the frequency of tuberculosis among the hospitalized control patients who did not have cancer. These counterintuitive results occurred because a smaller proportion of patients with both cancer and tuberculosis were hospitalized and thus available for selection as cases in the study; chances are that patients with both diseases were more likely to die than patients with cancer or tuberculosis alone.

Admission rate bias is an important problem to be aware of, because many case-control studies reported in the medical literature use hospitalized patients as sources for both cases and controls. The only way to control for this bias is to include an unbiased control group, best accomplished by choosing controls from a wide variety of disease categories or from a population of healthy subjects.

15.4.2.c Nonresponse Bias or the Volunteer Effect:
The volunteer effect is a common problem in social science survey research and can be avoided through carefully designed survey procedures. However, it is also a problem in medicine when patients either volunteer or refuse to participate in studies.

For example, the nationwide Salk polio vaccine trials in 1954 used two different study designs to evaluate the effectiveness of the vaccine (Meier, 1978). In some communities, children were randomly assigned to receive either the vaccine or a placebo injection. Some communities, however, refused to participate in a randomized trial; they agreed, instead, that second graders could be offered the vaccination and first and third graders could constitute the controls. In analysis of the data, researchers found that families who volunteered their children for participation in the nonrandomized study tended to be better-educated and to have a higher income than families who refused to participate. They also tended to be absent from school with a higher frequency than nonparticipants.

Although in this example we might guess how absence from school could bias results, it is not always easy to determine how selection bias will affect the outcome of the study; it may cause the experimental treatment to appear either better or worse than it should. Therefore, investigators should reduce the potential for nonresponse bias as much as possible by using all possible means to increase the response rate and obtain the participation of most eligible patients.

15.4.2.d Membership Bias:
Membership bias is essentially a problem of preexisting groups. It also arises because one or more of the same characteristics that cause people to belong to the groups are related to the outcome of interest. For example, investigators have not been able to perform a clinical trial to examine the effects of smoking; some researchers have claimed it is not smoking itself that causes cancer but some other factor, one that simply happens to be more common in smokers. Membership bias is a very important bias for readers of the medical literature to be aware of because it cannot be prevented and it makes the study of the effect of potential risk factors related to life-style very difficult.

A problem similar to membership bias is called the healthy worker effect; it was recognized in epidemiology when workers in a hazardous environment were unexpectedly found to have a higher survival rate than the general public. After further investigation, the cause of this incongruous finding was determined: Good health is a prerequisite in persons who are hired for work, but being healthy enough to work is not a requirement for persons in the general public.

15.4.2.e Procedure Selection Bias: Procedure selection bias occurs when treatment assignments are made on the basis of certain characteristics of the patients with the result that the treatment groups are not really similar. This bias frequently occurs in studies that are not randomized and is especially a problem in studies using historical controls. A good example is the comparison of a surgical versus a medical approach to a problem, such as coronary artery disease. In early studies comparing surgical and medical treatment, patients were not randomized, and there was evidence that patients who received surgery were healthier than those treated medically; ie, only healthier patients were subjected to the risks associated with the surgery. The CASS study (1983) discussed in Presenting Problem 4 in Chapter 5 was undertaken in part to resolve these questions. Procedure selection bias is an important bias to be aware of because there are many published studies that describe a series of patients, some treated one way and some treated another way, and then proceed to make comparisons and draw inappropriate conclusions as a result.

15.4.3 Procedures Used in the Study

15.4.3.a Terms and Measurements: The procedures used in the study are also described in the Method section. It is here that the authors provide definitions of any unusual measures used in the study, such as operational definitions developed by the investigators. If unusual instruments or methods are used, the authors should provide a reference and a brief description. For example, in Presenting Problem 1 in Chapter 3, examining the incidence of primary cardiac arrest during vigorous exercise, Siscovick et al (1984) needed a way to quantify exercise. They stated, "We used the Minnesota Leisure-Time Activity Questionnaire to quantify the average amount of energy expended in such activity over the previous year. This instrument takes into account the intensity, duration, and frequency of specific activities and yields estimates in kilocalories per day of the energy expended and in minutes per day of the time spent in high-intensity activity. High-intensity activity was activity that required approximately 6 kcal or more per minute—eg, jogging, swimming, playing singles tennis, and chopping wood" (Siscovick et al, 1984, 875). In addition, the authors provided a reference for the Minnesota Leisure-Time Activity

Questionnaire for readers who might wish more information.

15.4.3.b Data Quality: The Method section is also the place where the reliability of methods used should be described, unless the methods have been previously established as reliable. For example, recall that in Presenting Problem 1 in Chapter 4, the relationship between diet, serum lipoprotein, and progression of coronary atherosclerotic lesions was evaluated (Arntzenius et al, 1985). The investigators used the mean change in diameter of a patients's coronary arteries based on angiography as a measure of progression of lesion. They wanted to provide some information to indicate the reliability of measurements of vessel diameter. In the Method section of the article, they stated, "In order to evaluate the variability of the measurements, end-diastolic cine frames of 13 routine coronary angiograms were analyzed twice by one technician, with a median interval of 28 days between analyses. The average difference between duplicate measurements was found to be 0.00 mm; the variability, defined as the standard deviation of the differences between repeated measurements, was found to be 0.10 mm." This statement, of course, refers to the issue of **intrarater reliability.** It is also useful to have another researcher replicate the readings to provide information on **interrater reliability.** This method is commonly used in studies involving interpretation of radiographic or histologic sections to provide information on reliability of the readings.

15.4.4 Common Procedural Biases

Several biases may occur in the measurement of various patient characteristics and in the procedures used or evaluated in the study. Some of the more common biases are described below.

15.4.4.a Procedure Bias: Procedure bias is discussed by Feinstein (1985); it occurs when groups of subjects are not treated in the same manner. For example, the procedures used in an investigation may lead to detection of other problems in patients in the treatment group and make these problems appear to be more prevalent in this group. As another example, the patients in the treatment group may receive more attention and be followed up more vigorously than those in another group, thus stimulating greater compliance with the treatment regimen. The way to avoid this bias is by (1) carrying out all maneuvers except the experimental factor in the same way in all groups and (2) examining all outcomes by using similar procedures and criteria.

15.4.4.b Recall Bias: Recall bias is a problem that may occur when patients are asked to recall certain events and subjects in one group are more likely to remember the event than those in the other group. For example, people take aspirin commonly and for many reasons. But patients diagnosed as having peptic ulcer disease may recall the ingestion of aspirin with greater frequency than those without gastrointestinal problems.

15.4.4.c Insensitive-Measure Bias:
Measuring instruments may not be able to detect the characteristic of interest or may not be properly calibrated. For example, using routine x-rays to detect osteoporosis is an insensitive method because approximately 30% bone loss must occur before it can be detected. Newer densitometry techniques are more sensitive and thus avoid insensitive-measure bias.

15.4.4.d Detection Bias:
Detection bias can occur with the introduction of a new diagnostic technique capable of detecting the condition of interest at an earlier stage. Survival for patients diagnosed with the new procedure will inappropriately appear to be longer, merely because the condition was diagnosed earlier.

A spin-off of detection bias, called the Will Rogers phenomenon (because of his attention to human phenomena), was described by Feinstein, Sosin, and Wells (1985). They found that a cohort of subjects with lung cancer first treated in 1953–1954 had lower 6-month survival rates for patients with each of the three main stages (tumor, nodes, and metastases) as well as for the total group than did a 1977 cohort treated at the same institutions. Newer imagining procedures were used with the later group; however, according to the "old" diagnostic classification, this group had a prognostically favorable "zero-time shift" in that they were diagnosed at an earlier stage of disease.

In addition, by demonstrating metastases in the 1977 group that were missed in the earlier group, the new technological approaches resulted in "stage migration"; ie, members of the 1977 cohort were diagnosed as being in a more advanced, or "bad," stage, but they would have been diagnosed as in a "good" stage in 1953–1954. The individuals who "migrated" from the good-stage group to the bad-stage group tended to be from the bottom of the distribution in the good-stage group, ie, from among the patients with the worst prognosis; so removing them resulted in an increase in survival rates in the good group. At the same time, these individuals, now assigned to the bad group, were better off than most other patients in the bad-stage group; so their addition to the bad-stage group resulted in an increased survival in this group as well. The authors stated that the 1953–1954 and 1977 cohorts actually had similar survival rates when patients in the 1977 group were classified according to the symptom stages

that would have been in effect had there been no advances in diagnostic techniques.

15.4.4.e Compliance Bias:
Compliance bias occurs when patients find it easier or more pleasant to comply with one treatment than another. For example, in the treatment of hypertension, a comparison of alpha-methyldopa versus hydrochlorothiazide may demonstrate better results because some patients do not take alpha-methyldopa owing to its side effects, such as drowsiness, fatigue, or impotence in male patients.

15.4.5 Assessing Study Outcomes

15.4.5.a Reliability of Assessment:
The outcomes in the study should be clearly stated so that readers can evaluate the results. Some outcomes are relatively simple to assess, such as mortality. For example, in the study on total versus segmental mastectomy in Presenting Problem 3 of Chapter 11, Fisher et al (1985) examined overall survival in women who had various procedures, an outcome that is easy to assess. Another outcome in this study, however, was disease-free survival. In discussing the protocol, the authors stated:

> All patients assigned to the three groups were followed with respect to disease-free survival, distant disease-free survival, and overall survival. Failure times were computed from the time of the initial operation. Recurrences of tumor in the chest wall and operative scar, but not in the ipsilateral breast, were classified as local treatment failures. Tumors in the internal mammary, supraclavicular, or ipsilateral axillary nodes were classified as regional treatment failures. Patients classified as having any distant disease included those with a distant metastasis as a first treatment failure, a distant metastasis after a local or regional recurrence, or a second cancer (including tumor in the second breast). Overall survival refers to survival with or without recurrent disease.*

From this description, can we determine the methods the investigators used to identify disease-free status in patients? Unfortunately, from this report, we cannot determine the **reliability** of the judgments regarding disease-free status. To determine reliability, we need more information regarding the follow-up of patients and the methods used to ensure that those patients classified as disease-free actually were disease-free.

15.4.5.b Blindedness:
Another aspect of assessing the outcome is related to ways to increase ob-

---

*Reprinted, with permission, from Fisher B et al: Five-year results of a randomized clinical trial comparing total mastectomy and segmental mastectomy with or without radiation in the treatment of breast cancer. *N Eng J Med* 1985; **312**:666.

items in the checklist. However, practitioners do not have time to read all the articles published, and they must make some choices about which ones are most important and best presented. Frequent readers of clinical trials may wish to consult two articles dealing specifically with published results of clinical trials (Pocock, Hughes, & Lee, 1987; Zelen, 1983).

The checklist below is fairly exhaustive, and most readers will not want to use all of it unless they are reviewing an article for their own use or for a report. The items on the checklist are included in part as a reminder to the reader to look for these characteristics. We have placed an asterisk (*) beside those items we believe are the most critical; these items are the ones readers should use when a less exhaustive checklist is desired.

15.7.1 General Articles

I. In the Abstract section
  *A. Is the study, if properly designed and analyzed, important and worth knowing about?
  *B. If the results are statistically significant, do they also have clinical significance?
II. In the Introduction section
  A. Why is the study needed?
  *B. What is the purpose of the study?
  C. What has been done before, and how does this study differ? (This information may be in the Discussion section.)
  *D. What is the population to which the study findings apply? Does the location of the study have relevance? Is the time period covered by the study appropriate? (This information may be in the Method section.)
III. In the Method section
  *A. What design is used in the study (clinical trial, cohort, case-control, cross-sectional, case-series)?
  *B. What are the criteria for inclusion and exclusion of subjects? What limitations result?
  *C. How are subjects chosen or recruited? Randomly? If not, are they representative of the population? (How are subjects assigned to a trial?)
  *D. Is there a control group? If so, how is it chosen?
  E. How are patients followed up? Who are the dropouts, and how many are there?
  F. How is data quality ensured? Response rates? Reliability (second-party review)? Compliance?
  G. What are the independent variables (or factors) studied, and what are the dependent variables (outcome or response variables)? Are they clearly defined?

*H. Do the authors explain or give a reference to any unusual method used in the study?
  *I. Are the statistical methods used in the study specified in sufficient detail?
  *J. Is there a statement about the sample size or power? Statements on power are especially critical in a negative study.
  K. In a multicenter trial, what is the number of centers involved, and what methods of quality assurance are used?
IV. In the Results section
  *A. Do the results relate to research questions and the purpose of the study?
  B. Do the statistical tests answer the research questions? Are all relevant outcomes reported?
  *C. Are actual values reported (eg, means, standard deviations, proportions), not just the results of statistical tests? In paired designs, is the magnitude and range of the differences reported?
  D. Are many comparisons made or many statistical tests performed?
  E. Are groups similar on baseline measures? If not, were appropriate analyses done to take differences into consideration?
  F. Are appropriate graphics used to present results clearly?
V. In Conclusion and Discussion section(s)
  *A. Are the questions posed in the study adequately addressed?
  *B. Are the conclusions justified from the data? Do the authors extrapolate beyond the data?
  C. Are shortcomings of the study addressed and constructive suggestions given for future research?

15.7.2 Clinical Trials

A. Is the number of therapies considered appropriate?
  *B. Is the choice of control appropriate?
  *C. Were the patients randomized, and if so, how? If not:
    1. How were patients selected for the study to avoid selection biases?
    2. If historical controls were used, were methods and criteria the same for the experimental group, and were cases and controls compared on prognostic factors?
  *D. Was a power analysis done to determine sample size?
  E. Was study protocol adequately described?
  *F. Were there multiple endpoints? Were subgroup analyses performed and, if so, reported appropriately? Were repeated measures made over time, and if so, how were they analyzed?

*G. Were there censored observations, and if so, how were they analyzed?

15.7.3 Cohort Studies

*A. Are the patients representative of the population to which the findings are to be applied?
B. Is there evidence of volunteer bias?
*C. Was there adequate follow-up time?
D. Were there few dropouts or patients lost to follow-up?

15.7.4 Case-Control Studies

A. Were records of cases and controls reviewed blindly?
*B. How were possible selection biases controlled?
C. How were possible measurement biases controlled?

15.7.5 Decision Analysis Studies

*A. Is the outcome clearly defined with criteria for assessment?
B. Is there any evidence of circular reasoning (using findings as both predictors and criteria)?
C. Was the outcome assessed blindly?
*D. Was information on false-positive rate or specificity given in addition to information on sensitivity?
E. If predictive values were given, was the dependence on prevalence emphasized?

15.7.6 Meta-Analysis or Review Studies

*A. Do the authors specify how the literature search was conducted, and was an effort made to reduce publication bias?
B. If significant findings were concluded, did the authors specify the number of additional negative studies that would be needed to eliminate the observed significance?

EXERCISES

For Questions 1–39, choose the single best answer.

Questions 1–3: In a sample of 400 healthy men, the mean serum globulin level is found to be 30 g/L, with a standard deviation of 3 g/L.
1. The coefficient of variation of serum globulin in this sample of men is:
   A. 0.10.
   B. 0.15.
   C. 3.
   D. 10.
   E. 15.
2. If it is reasonable to assume that serum globulin levels follow a normal distribution, then approximately 50% of the men will have a value:
   A. between 27 and 33 g/L.
   B. between 24 and 36 g/L.
   C. below 27 or above 33 g/L.
   D. below 30 g/L.
   E. above 33 g/L.
3. Again, assuming that the distribution of serum globulin levels is normal, a healthy man would have a serum globulin level higher than 36 g/L:
   A. 1% of the time.
   B. 2.5% of the time.
   C. 5% of the time.
   D. 10% of the time.
   E. 16.5% of the time.
4. If the correlation between two measures of mental status is .70, we can conclude that:
   A. the value of one measure increases by .70 when the other measure increases by 1.
   B. 49% of the observations fall on the regression line.
   C. 70% of the observations fall on the regression line.
   D. 70% of the variation in one measure is accounted for by the other.
   E. 49% of the variation in one measure is accounted for by the other.

Questions 5–6: An evaluation of an antibiotic in the treatment of possible occult bacteremia was undertaken. Five hundred children with fever but no focal infection were randomly assigned to the antibiotic or to a placebo medication. A blood sample for culture was obtained prior to beginning therapy, and all patients were reevaluated after 48 hours.
5. The design used in this study is best described as a:
   A. cohort study.
   B. crossover study.
   C. clinical trial.
   D. randomized clinical trial.
   E. placebo-controlled trial.
6. The authors reported the proportion of children with major infectious morbidity among those with bacteremia was 13% in the placebo group and 10% in the antibiotic group. The 95% confidence interval for the difference in proportions was −2.6% to +8.6%. Thus, the most important conclusion is that:
   A. there is a statistically significant difference in major infectious morbidity when placebo is compared with antibiotic.
   B. the proportion of children with major infectious morbidity is the same with placebo and antibiotic.

C. there is no statistically significant difference in the proportions who received placebo and antibiotic.

D. the study has low power to detect a difference owing to the small sample size, and no conclusions should be drawn until a larger study is done.

E. using a chi-square test to determine significance is preferable to determining a confidence interval for the difference.

7. If the relationship between two measures is linear and the coefficient of determination has a value near 1, a scatterplot of the observations:

A. is a horizontal straight line.

B. is a vertical straight line.

C. is a straight line that is neither horizontal nor vertical.

D. is a random scatter of points about the regression line.

E. has a positive slope.

Questions 8–10: A study was undertaken to evaluate the use of computerized tomography (CT) in the diagnosis of lumbar disk herniation. Eighty patients with lumbar disk herniation confirmed by surgery were evaluated with CT, as were 50 patients without herniation. The CT results were positive in 56 of the patients with herniation and in 10 of the patients without herniation.

8. The sensitivity of CT for lumbar disk herniation in this study is:

A. 10/50, or 20%.

B. 24/80, or 30%.

C. 56/80, or 70%.

D. 40/50, or 80%.

E. 56/66, or 85%.

9. The false-positive rate in this study is:

A. 10/50, or 20%.

B. 24/80, or 30%.

C. 56/80, or 70%.

D. 40/50, or 80%.

E. 56/66, or 85%.

10. Computerized tomography is used in a patient who has a 50–50 chance of having a herniated lumbar disk, according to the patient's history and physical examination. What are the chances of herniation if the CT is positive?

A. 35/100, or 35%.

B. 50/100, or 50%.

C. 35/50, or 70%.

D. 40/55, or 73%.

E. 35/45, or 78%.

11. In a placebo-controlled trial of the use of oral aspirin-dipyridamole to prevent arterial restenosis after coronary angioplasty, 38% of patients receiving the drug had restenosis, and 39% of patients receiving placebo had restenosis. In reporting this finding, the authors stated "$P > .05$," which means that:

A. chances are greater than 1 in 20 that a difference would again be found if the study were repeated.

B. the probability is less than 1 in 20 that a difference this large could occur by chance alone.

C. the probability is greater than 1 in 20 that a difference this large could occur by chance alone.

D. treated patients were 1/20 less likely to have restenosis.

E. there is a 95% chance that the study is correct.

12. A study of the relationship between the concentration of lead in the blood and hemoglobin resulted in the following prediction equation: $Y = 15 - 0.1(X)$, where Y is the predicted hemoglobin and X is the concentration of lead in the blood. From the equation, the predicted hemoglobin for a person with blood lead concentration of 20 mg/dL is:

A. 13.

B. 14.8.

C. 14.9.

D. 15.

E. 20.

Questions 13–14: The graph in Fig 15–1 summarizes the gender-specific distribution of values on a laboratory test.

13. From Fig 15–1, we can conclude that:

A. values on the laboratory test are lower in men than in women.

B. the distribution of laboratory values in women is bimodal.

C. laboratory values were reported more often for women than for men in this study.

D. half of the men have laboratory values between 30 and 43.

E. the standard deviation of laboratory values is equal in men and women.

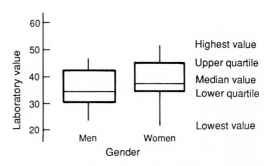

Figure 15-1. Gender-specific distribution of values on laboratory test.

14. The most appropriate statistical test to compare the distribution of laboratory values in men with that of women is:
    A. chi-square.
    B. the paired *t* test.
    C. the independent-groups *t* test.
    D. correlation.
    E. regression.

15. A study was undertaken to evaluate any increased risk of breast cancer among women who use birth control pills. The relative risk was calculated. A type I error in this study consists of concluding:
    A. a significant increase in the relative risk when the relative risk is actually 1.
    B. a significant increase in the relative risk when the relative risk is actually greater than 1.
    C. no significant increase in the relative risk when the relative risk is actually 1.
    D. no significant increase in the relative risk when the relative risk is actually greater than 1.
    E. no significant increase in the relative risk when the relative risk is actually less than 1.

16. The scale used in measuring serum creatinine (mg/dL) is:
    A. nominal.
    B. ordinal.
    C. interval.
    D. discrete.
    E. qualitative.

17. The scale used in measuring presence or absence of a risk factor is:
    A. nominal.
    B. ordinal.
    C. interval.
    D. continuous.
    E. quantitative.

18. Which of the following sources is most likely to provide an accurate estimate of the prevalence of muscular dystrophy in a community?
    A. A survey of practicing physicians asking how many patients they are currently treating.
    B. Information from hospital discharge summaries.
    C. Data from autopsy reports.
    D. A telephone survey to a sample of randomly selected homes in the community asking how many people living in the home have the disease.
    E. Examination of the medical records of a representative sample of people living in the community.

Questions 19–20: In an epidemiologic study of carbon-black workers, 500 workers with respiratory disease and 200 workers without respiratory disease were selected for study. The investigators obtained a history of exposure to carbon-black dust in both groups of workers. Among workers with respiratory disease, 250 gave a history of exposure to carbon-black dust; among the 200 workers without respiratory disease, 50 gave a history of exposure.

19. The odds ratio is:
    A. 1.0.
    B. 1.5.
    C. 2.0.
    D. 3.0.
    E. not determinable from the above information.

20. This study is best described as a:
    A. case-control study.
    B. cohort study.
    C. cross-sectional study.
    D. controlled experiment.
    E. randomized clinical trial.

21. A physician wishes to study whether a particular risk factor is associated with some disease. If, in reality, the presence of the risk factor leads to a relative risk of disease of 4.0, the physician wants to have a 95% chance of detecting an effect this large in the planned study. This statement is an illustration of specifying:
    A. a null hypothesis.
    B. a Type I, or alpha, error.
    C. a Type II, or beta, error.
    D. statistical power.
    E. an odds ratio.

22. The most likely explanation for a lower crude annual mortality rate in a developing country than in a developed country is that the developing country has:
    A. an incomplete record of deaths.
    B. a younger age distribution.
    C. an inaccurate census of the population.
    D. a less stressful life-style.
    E. lower exposure to environmental hazards.

Questions 23–25: Fig 15–2 is reproduced from the study of total mastectomy versus segmental mastectomy with and without radiation in the treatment of breast cancer (Fisher et al, 1985).

23. A major shortcoming of Fig 15–2 is:
    A. that the use of three graphs instead of one makes it difficult to compare survival.
    B. that it is inappropriate to present *P*-values on graphs.
    C. suppression of zero on the *Y*-axis.
    D. that the survival figures are adjusted for the number of nodes.
    E. that survival should be expressed as a proportion.

24. The most appropriate method to adjust survival for the number of positive nodes is:
    A. multiple regression with a dummy variable.
    B. logistic regression.

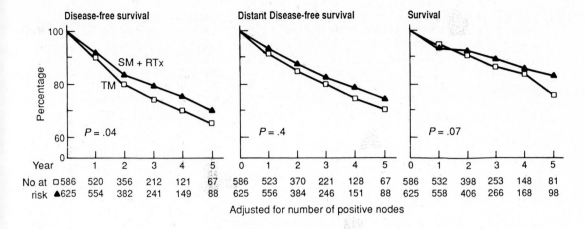

Figure 15-2. Survival curves for patients with breast cancer. (Reproduced, with permission, from Fisher B et al: Five-year results of a randomized clinical trial comparing total mastectomy and segmental mastectomy with or without radiation in the treatment of breast cancer. *N Engl J Med* 1985;**312**:665.)

C. discriminant analysis.

D. analysis of variance.

E. Kaplan-Meier product limit curves.

25. From Fig 15-2, we can conclude that:

A. disease-free survival is most likely with $P = .04$.

B. most patients have been in the study for 3 years or more.

C. simple survival is the easiest outcome to predict.

D. there is no significant difference in survival rates.

E. the findings are inconclusive until all patients have been in the study the entire 5 years.

26. Birth weights of a population of infants at 40 weeks gestational age are approximately normally distributed, with a mean of 3000 g. Roughly 68% of such infants weigh between 2500 and 3500 g at birth. If a sample of 100 infants were studied, the standard error would be:

A. 50.

B. 100.

C. 250.

D. 500.

E. none of the above.

27. A significant positive correlation has been observed between alcohol consumption and the level of systolic blood pressure in men. From this correlation, we may conclude that:

A. there is no association between alcohol consumption and systolic pressure.

B. men who consume less alcohol are at lower risk for increased systolic pressure.

C. men who consume less alcohol are at higher risk for increased systolic pressure.

D. high alcohol consumption can cause increased systolic pressure.

E. low alcohol consumption can cause increased systolic pressure.

28. In a randomized trial of patients who received a cadaver renal transplant, 100 were treated with cyclosporin and 50 were treated with conventional immunosuppression therapy. The difference in treatments was not statistically significant at the 5% level. Therefore:

A. this study has proven cyclosporin is not effective.

B. cyclosporin could be significant at the 1% level.

C. cyclosporin could be significant at the 10% level.

D. the groups have been shown to be the same.

E. the treatments should not be compared because of the differences in the sample sizes.

29. The statistical method used to develop guidelines for diagnostic related groups (DRGs) was:

A. survival analysis.

B. *t* tests for independent groups.

C. multiple regression.

D. analysis of covariance.

E. logistic regression.

30. Suppose the confidence limits for the mean of a variable are 8.55 and 8.65. These limits are:

A. less precises but have a higher confidence than 8.20 and 9.00.

B. more precise but have a lower confidence than 8.20 and 9.00.

C. less precise but have a lower confidence than 8.20 and 9.00.

D. more precise but have a higher confidence than 8.20 and 9.00.

E. indeterminate since the level of confidence is not specified.

31. A senior medical student wants to plan her elective schedule. The probability of getting an elective on maternal-fetal care is .6, and the probability of getting an elective on sports medicine is .5. The probability of getting both electives in the same semester is .3. What is the probability of getting into maternal-fetal care or sports medicine or both?
   A. .3.
   B. .5.
   C. .6.
   D. .8.
   E. 1.1.

32. A clinic wants to survey patients to determine the proportion of patients who have an immediate family member who is also a patient at the clinic. Every 50th patient chart is selected, and the patient (or the mother, if a child) is contacted. This procedure is best described as:
   A. random sampling.
   B. stratified sampling.
   C. quota sampling.
   D. representative sampling.
   E. systematic sampling.

33. There is a .4 probability that a medical student will receive his first choice of residency programs. Four senior medical students want to know the probability that they all will obtain their first choice. The solution to this problem is best found by using:
   A. the binomial distribution.
   B. the normal distribution.
   C. the chi-square distribution.
   D. the z test.
   E. correlation.

34. The graph in Fig 15-3 gives the 5-year survival rates for patients with cancer at various sites. From this figure, we can conclude:
   A. that breast and uterine corpus cancer are increasing at higher rates than other cancers.
   B. that lung cancer is the slowest-growing cancer.
   C. that few patients are diagnosed with distant metastases.
   D. that survival rates for patients with regional involvement are similar, regardless of the primary site of disease.
   E. none of the above.

35. A study was undertaken to compare treatment options in black and white patients who are diagnosed as having breast cancer. The 95% confidence interval for the odds ratio for blacks being more likely to be untreated than whites was 1.1 to 2.5. The statement that most accurately describes the meaning of these limits is that:
   A. 95% of the time blacks are more likely than whites to be untreated.

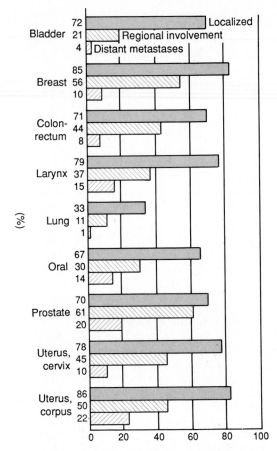

Figure 15-3. Five-year survival rates for patients with cancer. (Adapted and reproduced, with permission, from Rubin P [editor]: *Clinical Oncology: A Multidisciplinary Approach,* 6th ed. American Cancer Society, 1983.)

B. 95% of the odds ratios fall within these limits.
   C. the probability is 95% that odds ratios in similar studies would fall within these limits.
   D. since the observed odds ratio falls in the center of these limits, the probability is 95% that it is the correct value.
   E. blacks are 95 times more likely than whites to receive no treatment for breast cancer.

Questions 36-38: Physicians wish to determine whether the emergency room (ER) at the local hospital is being overused by patients with minor health problems. A random sample of 5000 patients was selected and categorized by age and degree of severity of the problem that brought them to the ER; severity was measured on a scale of 1 to 3, with 3 being most severe; the results are given in the table below:

Numbers of Patients

Age	Severity of Problem			Total
	Low	Medium	High	
< 5 years	1100	300	200	1600
5–14 years	500	900	300	1700
> 14 years	600	500	600	1700
Total	2200	1700	1100	5000

36. The joint probability that a patient 3 years old with a problem of high severity is selected for review is:
   A. 200/5000 = .04.
   B. 200/1600 = .125.
   C. 1600/5000 = .32.
   D. 1100/5000 = .22.
   E. 1600/5000 + 1100/5000 = .54.

37. If a patient comes to the ER with a problem of low severity, how likely is it that the patient is older than 14 years of age?
   A. 1700/5000 × 2200/5000 = .15.
   B. 600/2200 = .27.
   C. 1700/5000 = .34.
   D. 600/1700 = .35.
   E. 2200/5000 = .44.

38. The physicians performed a chi-square analysis to see whether there was an association between age of patient and severity of problem. In this analysis, the number of degrees of freedom for the chi-square test is:
   A. 3.
   B. 4.
   C. 6.
   D. 9.
   E. 4998.

39. Fig 15–4 was presented by Kremer et al (1987) to illustrate the way they conducted their study. From the figure, the study design is best described as a:

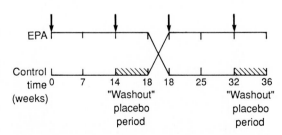

Figure 15-4. Illustration of study design. (Adapted and reproduced, with permission, from Kremer JM et al: Fish-oil fatty acid supplementation in active rheumatoid arthritis. *Ann Intern Med* 1987:**106**:497.)

   A. cohort study.
   B. crossover study.
   C. clinical trial.
   D. randomized clinical trial.
   E. placebo-controlled trial.

Questions 40–44: These questions comprise a set of one-best-choice items. For each of the situations outlined below, list the most appropriate statistical method to use in analyzing the data from the choices A–E that follow. Each choice may be used more than once and not all choices need be used.

   A. Independent-groups *t* test.
   B. Chi-square test.
   C. Wilcoxon signed-ranks test.
   D. Pearson correlation.
   E. Analysis of variance.

40. Determining the relationship between increased levels of dietary intake of fiber and decreased levels of fat consumption.

41. Comparing average body weight before and after a smoking cessation program.

42. Comparing gender of the head of household in families of patients whose medical costs are covered by insurance, Medicaid, or self.

43. Investigating a possible association between exposure to an environmental pollutant and miscarriage.

44. Comparing blood cholesterol levels in patients who follow either a diet low or moderate in fat and who take either a drug to lower cholesterol or placebo.

Questions 45–50: For these multiple true-false questions, proceed as follows:

Select A if 1, 2, and 3 only are correct.
Select B if 1 and 3 only are correct.
Select C if 2 and 4 only are correct.
Select D if 4 only is correct.
Select E if all are correct.

45. Purposes of a single-blind study include the reduction of:
   1. systematic error.
   2. observer bias.
   3. sampling error.
   4. subject bias.

46. The incidence of cancer was compared in two groups of women; one group had taken a certain drug in pregnancy and the other had not. The 95% confidence interval for the relative risk of developing cancer was 0.8 to 3.0. We can conclude that:
   1. in this study, the observed death rate from cancer is greater in the group not exposed to the drug.
   2. the relative risk of developing cancer is

statistically significantly higher in patients who are exposed to the drug.

3. there is a 5% chance that the relative risk of cancer incidence is less than 1.

4. exposure to the drug during pregnancy has no statistically significant effect on incidence of cancer.

47. A statistician is consulted to determine the sample size needed to define a significant difference between two forms of treatment. Minimal information required includes:

1. the level of significance desired.

2. the amount of difference the investigator wants to detect.

3. an estimate of the standard deviation.

4. the manner in which the treatments differ.

48. The Internal Revenue Service wants to compare the distributions for taxes paid by plastic surgeons and taxes paid by family practitioners. Appropriate methods for analyzing this question include the:

1. independent-groups *t* test.

2. paired *t* test.

3. Wilcoxon rank-sum test.

4. Wilcoxon signed-ranks test.

49. Fig 15-5 was used by Willett et al (1987) to illustrate the age-standardized rates of coronary heart disease among women according to history of hypercholesterolemia and use of cigarettes. Correct conclusions based on this graph include:

1. history of high cholesterol is associated with coronary heart disease.

2. use of cigarettes is associated with coronary heart disease.

3. the more a woman smokes cigarettes, the more likely she is to develop coronary heart disease.

4. history of high cholesterol and cigarette use are positively associated.

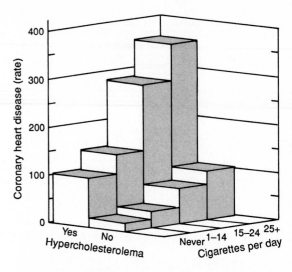

Figure 15-5. Age-standardized rates of coronary heart disease. (Adapted and reproduced, with permission, from Willett WC et al: Relative and absolute excess risks of coronary heart disease among women who smoke cigarettes. *N Engl J Med* 1987;**317**:1303.)

50. Creatinine levels in a group of 100 healthy adults were approximately normally distributed, with mean 1.0 mg/100 mL and standard deviation 0.2. From these data, the following conclusions are correct:

1. 95% of healthy adults may be expected to have creatine levels between 0.6 and 1.4 mg/100 mL.

2. 95% confidence limits for the mean are 1.0 ± 0.04 mg/100 mL.

3. 50% of healthy adults have creatine levels of 1.0 mg/100 mL or higher.

4. The coefficient of variation is 20%.

Glossary

absolute value   the positive value of a number, regardless of whether the number is positive or negative. The absolute value of *a* is symbolized |*a*|.

actuarial analysis   see *life table analysis*.

adjusted rate   a rate adjusted so that it is independent of the distribution of a possible confounding variable. For example, age-adjusted rates are independent of the age distribution in the population to which they apply.

age-specific mortality rate   the mortality rate in a specific age group.

alpha error   see *type I error*.

alpha value   the level of alpha (α) selected in a hypothesis test.

alternative hypothesis   the opposite of the null hypothesis, which is the conclusion when the null hypothesis is rejected.

analysis of covariance (ANCOVA)   a special type of analysis of variance or regression used to control for the effect of a possible confounding factor.

analysis of residuals   in regression, an analysis of the differences between *Y* and *Y'* to evaluate assumptions.

analysis of variance (ANOVA)   a statistical procedure that determines whether or not there are any differences among two or more groups of subjects on one or more factors. The *F* test is used in ANOVA.

bar chart or bar graph   a chart or graph used with nominal characteristics to display the numbers or percentages of observations with the characteristic of interest.

Bayes' theorem   a formula for calculating the conditional probability of one event, *P*(*A*|*B*), from the conditional probability of the other event, *P*(*B*|*A*).

bell-shaped distribution   a term used to describe the shape of the normal (gaussian) distribution.

beta error   see *type II error*.

bias   the error related to the ways targeted and sampled populations differ; also called measurement error.

binomial distribution   the probability distribution that describes the number of successes *X* observed in *n* independent trials, each with the same probability of occurrence.

bivariate plot   a two-dimensional plot or scatterplot of the values of two characteristics measured on the same set of subjects.

blind study   an experimental study in which subjects do not know the treatment patients are receiving; investigators may also be blind to the treatment patients are receiving.

block design   in analysis of variance, a design in which subjects within each block (or stratum) are assigned to a different treatment.

Bonferroni *t* procedure   a method for comparing means in analysis of variance; also called the Dunn multiple-comparison procedure.

box and whisker plot   a graph that displays both the frequencies and the distribution of observation. It is useful for comparing two distributions.

box plot   the same as box and whisker plot.

canonical correlation analysis   an advanced statistical method for examining the relationships between two sets of numerical measurements made on the same set of subjects.

case-control study   an observational study that begins with patient cases who have the outcome or disease being investigated and control subjects who do not have the outcome or disease; it then looks backward to identify possible precursors or risk factors.

case-series study   a simple descriptive account of interesting or intriguing characteristics observed in a group of subjects.

categorical observation   a variable whose values are categories (an example is type of anemia). See also *nominal scale*.

cause-specific mortality rate   the mortality rate from a specific disease.

cell   a category of counts or values in a contingency table.

censored observation   an observation whose value is unknown, generally because the subject has not been in the study long enough for the outcome of interest, such as death, to occur.

central limit theorem   a theorem that states that the distribution of means is approximately normal if the sample size is large enough ($n \geq 30$), regardless of the underlying distribution of the original measurements.

chance agreement   a measure of the proportion of times two or more raters would agree in their measurement or assessment of a phenomenon.

chi-square distribution   the distribution used to analyze counts in frequency tables.

chi-square test   the statistical test used to test the null hypothesis that proportions are equal or, equivalently, that nominal factors or characteristics are independent or not associated.

classes or **class limits**   the subdivisions of a numerical characteristic (or the widths of the classes) when it is displayed in a frequency table or graph (an example is ages by decades).

clinical trial   an experimental study of a drug or procedure in which the subjects are humans.

cluster analysis   an advanced statistical method that determines a classification or taxonomy from multiple measures of a set of objects or subjects.

cluster random sample   a two-stage sampling process

in which the population is divided into clusters, a random sample of clusters is chosen, and then random samples of subjects within the clusters are selected.

coefficient of determination the square of the correlation coefficient. It is interpreted as the amount of variance in one variable that is accounted for by knowing the second variable.

coefficient of variation the standard deviation divided by the mean (generally multiplied by 100). It is used to obtain a measure of relative variation.

cohort a group of subjects who remain together in the same study over a period of time.

cohort study an observational study that begins with a set of subjects who have a risk factor or have been exposed to an agent and a second set of subjects who do not have the risk factor or exposure. Both sets are followed prospectively through time to determine the outcome or consequence.

complementary event an event opposite to the event being investigated.

computer package a set of statistical computer programs for analyzing data.

concurrent controls control subjects assigned to a placebo or control condition during the same period of time that an experimental treatment or procedure is being evaluated.

conditional probability the probability of an event, such as *A*, given that another event, such as *B*, has occurred, denoted $P(A|B)$.

confidence bands dashed lines on each side of a regression line or curve that have a given probability of containing the line or curve in the population.

confidence interval the interval computed from sample data that has a given probability that the unknown parameter is contained within the interval. Common confidence intervals are 90%, 95%, and 99%.

confidence limits the limits of a confidence interval. These limits are computed from sample data and have a given probability that the unknown parameter is located between them.

confounded a term used to describe a study or observation that has one or more nuisance variables present that may lead to incorrect interpretations.

confounding variable a variable more likely to be present in one group of subjects than another that is related to the outcome of interest and thus potentially confuses, or "confounds," the results.

conservative a term used to describe a statistical test if it reduces the chances of a type I error.

contingency table a table used to display counts or frequencies for two or more nominal or quantitative variables.

continuity correction an adaptation to a test statistic when a continuous probability distribution is used to estimate a discrete probability distribution; eg, using the chi-square distribution for analyzing contingency tables.

continuous scale a scale used to measure a numerical characteristic with values that occur on a continuum (an example is age).

controlled for a term used to describe a confounding variable that is taken into consideration in the design or the analysis of the study.

controlled trial a trial in which subjects are assigned to a control condition as well as to an experimental condition.

control subjects in a clinical trial, subjects assigned to the placebo or control condition; in a case-control study, subjects without the disease or outcome.

corrected chi-square test a chi-square test for a 2 × 2 table that uses Yates' correction, making it more conservative.

correlation coefficient (Pearson product moment) a measure of the linear relationship between two numerical measurements made on the same set of subjects. It ranges from −1 to +1, with 0 indicating no relationship.

covariate the potentially confounding variable controlled for in analysis of covariance (or in logistic regression).

Cox regression a regression method used when the outcome is censored. The regression coefficients are interpreted as adjusted relative risk or odds ratios.

criterion variable the outcome (or dependent variable) that is predicted in a regression problem.

critical ratio the term for the *z* or t score used in statistical tests.

critical region the region (or set of values) in which a test statistic must occur in order for the null hypothesis to be rejected.

critical value the value that a test statistic must exceed (in an absolute value sense) in order for the null hypothesis to be rejected.

crossover study a clinical trial in which each group of subjects receives two or more treatments but in different sequences.

cross-product ratio see *relative risk*.

cross-sectional study an observational study that examines a characteristic (or set of characteristics) in a set of subjects at one point in time; a "snapshot" of a characteristic or condition of interest; called survey or poll in social science research.

cross-validation a procedure for applying the results of an analysis from one sample of subjects to a new sample of subjects to evaluate how well they generalize. It is frequently used in regression.

crude rate a rate for the entire population that is not specific or adjusted for any given subset of the population.

cumulative frequency or **percentage** in a frequency table, the frequency (or percentage) of observations that occur at a given value plus all lower values.

curvilinear relationship (between *X* and *Y*) a relationship that indicates that *X* and *Y* covary, but not in constant increments.

decision analysis a formal model for describing and analyzing a decision; also called medical decision making.

decision tree a diagram of a set of possible actions, with their probabilities and the values of the outcomes listed. It is used to analyze a decision process.

degree of freedom a parameter in some commonly used probability distributions, eg, the *t* distribution and the chi-square distribution.

dependent groups or samples samples in which the values in one group can be predicted from the values in the other group.

dependent-groups *t* test see *paired* t *test*.

dependent variable the variable whose values are the outcomes in a study; also called response or criterion variable.

descriptive statistics statistics such as the mean, the standard deviation, the proportion, and the rate used to describe attributes of a set of data.

dichotomous variable a nominal measure that has only two outcomes (examples are gender: male or female; survival: yes or no).

directional test see *one-tailed test*.

discrete scale a scale used to measure a numerical characteristic that has integer values (an example is number of pregnancies).

discriminant analysis a regression technique for predicting a nominal outcome that has more than two values; a method used to classify subjects or objects into groups; also called discriminant function analysis.

distribution (population) the values of a characteristic or variable along with the frequency of their occurrence. Distributions may be based on empirical observations or may be theoretical probability distributions (eg, normal, binomial, chi-square).

dot plot a graphical method for displaying the frequency distribution of numerical observations for one or more groups.

double-blind trial a clinical trial in which neither the subjects nor the investigator(s) know which treatment subjects have received.

dummy coding a procedure in which a code of 0 or 1 is assigned to a nominal predictor variable used in regression analysis.

Dunnett's procedure a multiple-comparison method for comparing multiple treatment groups with a single control group following a significant F test in analysis of variance.

effect or effect size the magnitude of a difference or relationship. It is the basis for statistical methods used in meta-analysis.

error mean square the mean square in the denominator of F in ANOVA.

estimation the process of using information from a sample to draw conclusions about the values of parameters in a population.

event a single outcome (or set of outcomes) from an experiment.

expected frequencies in contingency tables, the frequencies observed if the null hypothesis is true.

experiment (in probability) a planned process of data collection.

experimental study a comparative study involving an intervention or manipulation. It is called a trial when human subjects are involved.

explanatory variable see *independent variable*.

factor a characteristic that is the focus of inquiry in a study; used in analysis of variance.

factor analysis an advanced statistical method for analyzing the relationships among a set of items or indicators to determine the factors or dimensions that underlie them.

factorial design in ANOVA, a design in which each subject (or object) receives one level of each factor.

false-negative a test result that is negative in a person who has the disease.

false-positive a test result that is positive in a person who does not have the disease.

F distribution the probability distribution used to test the equality of two estimates of the variance (by comparing the ratio of the variances with 1). It is the distribution used with the F test in ANOVA.

first quartile the 25th percentile.

Fisher's exact test an exact test for 2×2 contingency tables. It is used when the sample size is too small for use of the chi-square test.

Fisher's z transformation a transformation applied to the correlation coefficient so that it is normally distributed.

frequency the number of times a given value of an observation occurs. It is also called counts.

frequency distribution in a set of numerical observations, the list of values that occur along with the frequency of their occurrence. It may be set up as a frequency table.

frequency polygon a line graph connecting the midpoints of the tops of the columns of a histogram. It is useful in comparing two frequency distributions.

frequency table a table showing the number or percentage of observations occurring at different values (or ranges of values) of a characteristic or variable.

F test the statistical test for comparing two variances. It is used in ANOVA.

gaussian distribution see *normal distribution*.

Gehan's test a statistical test of the equality of two survival curves.

generalized Wilcoxon test see *Gehan's test*.

geometric mean the nth root of the product of n observations, symbolized GM or G. It is used with logarithms or skewed distributions.

gold standard in diagnostic testing, a test that always identifies the true condition—diseased or disease-free—of a patient.

hazard function the probability that a person dies in a certain time interval, given that the person has lived until the beginning of the interval. Its reciprocal is mean survival time.

histogram a graphical display of a frequency distribution of numerical observations.

historical cohort study a cohort study that uses existing records or historical data to determine the effect of a risk factor or exposure on a group of patients.

historical controls in clinical trials, previously collected observations on patients used as the control values against which the treatment is compared.

homogeneity the situation in which the standard deviation of the dependent (Y) variable is the same regardless of the value of the independent (X) variable; an assumption in ANOVA and regression.

homoscedasticity see *homogeneity*.

hypothesis test an approach to statistical inference resulting in a decision to reject or not to reject the null hypothesis.

incidence a rate giving the proportion of people who develop a given disease or condition within a specified period of time.

independent events events whose occurrence or outcome has no effect on the probability of each other or of another event.

independent groups or samples samples for which the values in one group cannot be predicted from the values in the other group.

independent-groups t test see *two-sample t test*.

independent observations observations determined

at different times or by different individuals without the knowledge of the value of the first observation.

independent variable the explanatory or predictor variable in a study. It is sometimes called a factor in ANOVA.

inference (statistical) the process of drawing conclusions about a population of observations from a sample of observations.

interaction a relationship between two independent variables such that they have a different effect on the dependent variable; ie, the effect of one level of a factor *A* depends on the level of factor *B*.

intercept in a regression equation, the predicted value of *Y* when *X* is equal to zero.

interquartile range the difference between the 25th percentile and the 75th percentile.

interrater reliability the reliability between measurements made by two different persons (or raters).

intervention the maneuver used in an experimental study. It may be a drug or a procedure.

intrarater reliability the reliability between measurements made by the same person (or rater) at two different points in time.

joint probability the probability of two events' both occurring.

Kaplan-Meier product limit method a method for analyzing survival for censored observations. It uses exact survival times in the calculations.

kappa κ a statistic used to measure interrater or intrarater agreement for nominal measures.

level of significance the probability of incorrectly rejecting the null hypothesis in a test of hypothesis. Also see *alpha value* and *P-value*.

life table analysis a method for analyzing survival times for censored observations that have been grouped into intervals.

likelihood ratio in diagnostic testing, the ratio of true-positives to false-positives.

linear combination a weighted average of a set of variables or measures. For example, the prediction equation in multiple regression is a linear combination of the predictor variables.

linear regression (of *Y* on *X*) the process of determining a regression or prediction equation to predict *Y* from *X*.

linear relationship (between *X* and *Y*) a relationship indicating that *X* and *Y* covary according to constant increments.

logarithm (natural) the exponent indicating the power to which *e* (2.718) is raised to obtain a given number.

logistic regression the regression technique used when the outcome (dependent variable) is dichotomous.

log-linear analysis a statistical method for analyzing the relationships among three or more nominal variables. It may be used as a regression method to predict a nominal outcome from nominal independent variables.

logrank test a statistical method for comparing two survival curves when there are censored observations.

longitudinal study a study that takes place over an extended period of time.

Mann-Whitney-Wilcoxon test: see *Wilcoxon rank-sum test*.

Mantel-Haenzel chi-square test a statistical test of

two or more 2 × 2 tables. It is used to combine tables or to control for confounding factors.

marginal frequencies (probabilities) the row and column frequencies (or probabilities) in a contingency table—ie, the frequencies listed on the margins of the table.

matched-groups *t* test see *paired* t *test*.

matching (or matched groups) the process of making two groups homogeneous on possible confounding factors. It is sometimes done prior to randomization in clinical trials.

McNemar test the chi-square test for comparing proportions from two dependent or paired groups.

mean the most common measure of central tendency, denoted by μ in the population and by $\overline{X}$ in the sample. In a sample, the mean is the sum of the *X* values divided by the number *n* in the sample ($\Sigma X/n$).

mean square among groups an estimate of the variation in analysis of variance. It is used in the numerator of the *F* statistic.

mean square within groups an estimate of the variation in analysis of variance. It is used in the denominator of the *F* statistic.

measurement error the amount by which a measurement is incorrect because of problems inherent in the measuring process; also called bias.

measures of central tendency index or summary numbers that describe the middle of a distribution. See *mean; median; mode*.

measures of dispersion index or summary numbers that describe the spread of observations about the mean. See *range; standard deviation*.

median a measure of central tendency. It is the middle observation; ie, the one that divides the distribution of values into two halves. It is also equal to the 50th percentile.

medical decision making or analysis the application of probabilities to the decision process in medicine. It is the basis for cost-benefit analysis.

meta-analysis a method for combining the results from several independent studies of the same outcome so that an overall *P*-value may be determined.

model class the interval (generally from a frequency table or histogram) that contains the highest frequency of observations.

mode the value of a numerical variable that occurs the most frequently.

model or modeling a statistical statement of the relationship among variables.

morbidity rate the number of patients in a defined population who develop a morbid condition over a specified period of time.

mortality rate the number of deaths in a defined population over a specified period of time. It is the number of people who die during a given period of time divided by the number of people at risk during the period.

multiple-comparison procedure a method for comparing several means.

multiple comparisons comparisons resulting from many statistical tests performed for the same observations. It is not appropriate unless the relevant omnibus test is done.

multiple *R* in multiple regression, the correlation between actual and predicted values of *Y* (ie, r_{YY}).

multiple regression a multivariate method for determining a regression or prediction equation to predict Y from a set of variables, $X_1, X_2, \ldots, X_p$.

multivariate a term that refers to a study or analysis involving a multiple independent or dependent variables.

multivariate analysis of variance (MANOVA) an advanced statistical method that provides a global test when there are multiple dependent variables and the independent variables are nominal. It is analogous to analysis of variance with multiple outcome measures.

mutually exclusive events two or more events for which the occurrence of one event precludes the occurrence of the others.

Newman-Keuls procedure a multiple-comparison method for making pairwise comparisons between means following a significant F test in analysis of variance.

nominal scale the simplest scale of measurement. It is used for characteristics that have no numerical values (examples are race and gender). It is also called a categorical or qualitative scale.

nondirectional test see *two-tailed test*.

non-mutually exclusive events two or more events for which the occurrence of one event does not preclude the occurrence of the others.

nonparametric method a statistical test that makes no assumptions regarding the distribution of the observations.

nonprobability sample a sample selected in such a way that the probability that a subject is selected is unknown.

nonrandomized trial a clinical trial in which subjects are assigned to treatments on other than a randomized basis. It is subject to several biases.

normal distribution a symmetric, bell-shaped probability distribution with mean μ and standard deviation σ. If observations follow a normal distribution, the interval ($\mu \pm 2\sigma$) contains 95% of the observations. It is also called the gaussian distribution.

null hypothesis the hypothesis being tested about a population. *Null* generally means "no difference" and thus refers to a situation in which there is no difference (eg, between the means in a treatment group and a control group).

numerical scale the highest level of measurement. It is used for characteristics that can be given numerical values; the differences between numbers have meaning (examples are height, weight, blood pressure level). It is also called an interval or ratio scale.

objective probability an estimate of probability from observable events or phenomena.

observational study a study that does not involve an intervention or manipulation. It is called case-control, cross-sectional, or cohort, depending on the design of the study.

observed frequencies the frequencies that occur in a study. They are generally arranged in a contingency table.

odds the probability that an event will occur divided by the probability that the event will not occur; ie, odds = $P/(1 - P)$, where P is the probability.

odds ratio an estimate of the relative risk calculated in case-control studies. It is the odds that a patient was exposed to a given risk factor divided by the odds that a control was exposed to the risk factor.

one-tailed test a test in which the alternative hypothesis specifies a deviation from the null hypothesis in one direction only. The critical region is located in one end of the distribution of the test statistic. It is also called a directional test.

ordinal scale used for characteristics that have an underlying order to their values; the numbers used are arbitrary (an example is Apgar scores).

outcome (in an experiment) the result of an experiment or trial.

outcome variable the dependent or criterion variable in a study.

paired t test the statistical method for comparing the difference (or change) in a numerical variable observed for two paired (or matched) groups. It also applies to before and after measurements made on the same group of subjects.

parameter the population value of a characteristic of a distribution (eg, the mean μ).

percentage a proportion multiplied by 100.

percentile a number that indicates the percentage of a distribution that is less than or equal to that number.

placebo a sham treatment or procedure. It is used to reduce bias in clinical studies.

point estimate a general term for any statistic (eg, mean, standard deviation, proportion).

Poisson distribution a probability distribution used to model the number of times a rare event occurs.

pooled standard deviation the standard deviation used in the independent-groups t test when the standard deviations in the two groups are equal.

population the entire collection of observations or subjects that have something in common and to which conclusions are inferred.

posterior probability the conditional probability calculated by using Bayes' theorem. It is the predictive value of a positive test (true-positives divided by all positives) or a negative test (true-negatives divided by all negatives).

post hoc comparisons methods for comparing means following analysis of variance.

posttest odds in diagnostic testing, the odds that a patient has a given disease or condition after a diagnostic procedure is performed and interpreted. They are similar to the predictive value of a diagnostic test.

power the ability of a test statistic to detect a specified alternative hypothesis or difference of a specified size when the alternative hypothesis is true. More loosely, it is the ability of a study to detect an actual effect or difference.

predictive value of a negative test the proportion of time that a patient with a negative diagnostic test result does not have the disease being investigated.

predictive value of a positive test the proportion of time that a patient with a positive diagnostic test result has the disease being investigated.

pretest odds in diagnostic testing, the odds that a patient has a given disease or condition before a diagnostic procedure is performed and interpreted. They are similar to prior probabilities.

prevalence the proportion of people who have a given disease or condition at a specified point in time. It is not truly a rate, although it is often incorrectly called prevalence rate.

prior probability   the unconditional probability used in the numerator of Bayes' theorem. It is the prevalence of a disease prior to performing a diagnostic procedure.

probability   the number of times an outcome occurs in the total number of trials. If *A* is the outcome, the probability of *A* is denoted *P(A)*.

probability distribution   a frequency distribution of a random variable, which may be empirical or theoretical (eg, normal, binomial).

product limit method   see *Kaplan-Meier product limit method.*

proportion   the number of observations with the characteristic of interest divided by the total number of observations. It is used to summarize counts.

proportional hazards model   see *Cox regression.*

prospective study   a study designed before data are collected.

P-value   the probability of observing a result as extreme as or more extreme than the one actually observed based on chance alone (ie, if the null hypothesis is true).

qualitative observations   characteristics measured on a nominal scale.

quantitative observations   characteristics measured on a numerical scale; the resulting numbers have inherent meaning.

quartile   the 25th percentile or the 75th percentile, called the first and third quartiles, respectively.

random assignment   the use of random methods to assign different treatments to patients or vice versa.

random error or **variation**   the variation in a sample that can be expected to occur by chance.

randomization   the process of assigning subjects to different treatments (or vice versa) by using random numbers.

randomized clinical trial   an experimental study in which subjects are randomly assigned to treatment groups.

random sample   a sample of *n* subjects (or objects) selected from a population so that each has known chance of being in the sample.

random variable   a variable in a study in which subjects are randomly selected or randomly assigned to treatments.

range   the difference between the largest and the smallest observation.

rank-order scale   a scale for observations arranged according to their size, from lowest to highest or vice versa.

ranks   a set of observations arranged according to their size, from lowest to highest or vice versa.

rate   a proportion associated with a multiplier, called the base, (eg, 1000, 10,000, 100,000) and computed over a specific period of time.

ratio   a part divided by another part. It is the number of observations with the characteristic of interest divided by the number without the characteristic.

regression (of *Y* on *X*)   the process of determining a prediction equation for predicting *Y* from *X*.

regression coefficient   the *b* in the simple regression equation $Y = a + bX$. It is sometimes interpreted as the slope of the regression line. In multiple regression, the *b*'s are weights applied to the predictor variables.

relative risk/risk ratio the ratio of the incidence of a given disease in exposed or at risk persons to the incidence of the disease in unexposed persons. It is calculated in cohort or prospective studies.

reliability   a measure of the reproducibility of a measurement. It is measured by kappa for nominal measures and by correlation for numerical measures.

repeated-measures design   a study design in which subjects are measured at more than one point in time. It is also called a split-plot design in ANOVA.

representative sample   a sample that is similar in important ways to the population to which the findings of a study are generalized.

residual   the difference between the predicted value and the actual value of the outcome (dependent) variable in regression.

response variable   see *dependent variable.*

retrospective cohort study   see *historical cohort study.*

retrospective study   a study undertaken in a post hoc manner, ie, after the observations have been made.

risk factor   a term used to designate a characteristic that is more prevalent among subjects who develop a given disease or outcome than among subjects who do not. It is generally considered to be causal.

robust   a term used to describe a statistical method if its value is not affected to a large extent by a violation of the assumptions of the test.

ROC (receiver operating characteristic) curve   in diagnostic testing, a plot of the true-positives on the *Y*-axis versus the false-positives on the *X*-axis.

sample   a subset of the population.

sampled population   the population from which the sample is actually selected.

sampling distribution (of a statistic)   the frequency distribution of the statistic for many samples. It is used to make inferences about the statistic from a single sample.

scale of measurement   the degree of precision with which a characteristic is measured. It is generally categorized into nominal (or categorical), ordinal, and numerical (or interval and ratio) scales.

scatterplot   a two-dimensional graph displaying the relationship between two characteristics or variables.

Scheffe's procedure   a multiple-comparison method for comparing means following a significant *F* test in analysis of variance. It can be used to make any comparisons among means, not simply pairwise. It is the most conservative multiple-comparison method.

self-controlled study   a study in which the subjects serve as their own controls, achieved by measuring the characteristic of interest before and after an intervention.

sensitivity   the proportion of time a diagnostic test is positive in patients who have the disease or condition. A sensitive test has a low false-negative rate.

sensitivity analysis   in decision analysis, a method for determining the way the decision changes as a function of probabilities and utilities used in the analysis.

simple random sample   a random sample in which each of the *n* subjects (or objects) in the sample has an equal chance of being selected.

skewed distribution   a distribution in which there are a relatively small number of outlying observations in one direction only. If the observations are small, the distribution is skewed to the left, or negatively skewed; if the observations are large, the distribution is skewed to the right, or positively skewed.

slope (of the regression line) the amount Y changes for each unit that X changes. It is designated by b in the sample.

Spearman's rank correlation (rho) a nonparametric correlation that measures the tendency for two measurements to vary together. It is also the Pearson correlation between the rank orderings of two measurements.

specificity the proportion of time that a diagnostic test is negative in patients who do not have the disease or condition. A specific test has a low false-positive rate.

specific rate a rate that pertains to a specific group or segment of the observations (examples are age-specific mortality rate and cause-specific mortality rate).

standard deviation the most common measure of dispersion or spread. It can be used with the mean to describe the distribution of observations. It is the square root of the average of the squared deviations of the observations from their mean.

standard error the standard deviation of the sampling distribution of a statistic.

standard error of the estimate a measure of the variation in a regression line. It is based on the differences between the predicted values and the actual values of the dependent variable Y.

standard error of the mean the standard deviation of the mean in a large number of samples.

standardized regression coefficient a regression coefficient that has the effect of the measurement scale removed so that the size of the coefficient can be interpreted.

standard normal distribution the normal distribution with mean 0 and standard deviation 1, also called z distribution.

statistic a summary number for a sample (eg, the mean), often used as an estimate of a parameter in the population.

statistical significance generally interpreted as a result that would occur by chance less than 1 time in 20, with a P-value less than or equal .05. It occurs when the null hypothesis is rejected.

statistical test the procedure used to test a null hypothesis (eg, t test, chi-square test).

stem and leaf plot a graphical display for numerical data. It is similar to both a frequency table and a histogram.

stepwise regression in multiple regression, a sequential method of selecting the variables to be included in the prediction equation.

stratified random sample a sample consisting of random samples from each subpopulation (or stratum) in a population. It is used so the investigator can be sure that each subpopulation is appropriately represented in the sample.

subjective probability an estimate of probability that reflects a person's opinion or best guess from previous experience.

sums of squares a quantity calculated in analysis of variance and used to obtain the mean squares for the F test.

suppression of zero a term used to describe a misleading graph that does not have a break (a jagged line) in the Y-axis to indicate that part of the scale is missing.

survey an observational study that generally has a cross-sectional design.

survival analysis the statistical method for analyzing survival data when there are censored observations.

symmetric distribution a distribution that has the same shape on both sides of the mean. The mean and median are equal for this distribution. It is the opposite of a skewed distribution.

systematic error a measurement error that is the same (or constant) over all observations. See also bias.

systematic random sample a random sample obtained by selecting each kth subject or object.

target population the population to which the investigator wishes to generalize.

t distribution a symmetric distribution with mean 0 and a standard deviation larger than 1 for small sample sizes. As n increases, the t distribution approaches the normal distribution.

test statistic the specific statistic used to test the null hypothesis (eg, the t statistic or chi-square statistic).

testing threshold in diagnostic testing, the point at which the optimal decision is to perform a diagnostic test.

third quartile the 75th percentile.

threshold model a model for deciding when a diagnostic test should be ordered, as opposed to doing nothing or treating the patient without performing the test.

transformation a change in the scale for the values of a variable.

treatment threshold in diagnostic testing, the point at which the optimal decision is to treat the patient without first performing a diagnostic test.

trial an experiment involving humans, commonly called a clinical trial. It is also a replication (repetition) of an experiment.

true-negative a test result that is negative in a person who does not have the disease.

true-positive a test result that is positive in a person who has the disease.

t test the statistical test for comparing a mean with a norm or for comparing two means with small sample sizes ($n < 30$). It is also used for testing whether a correlation coefficient or a regression coefficient is 0.

Tukey's HSD (honestly significant difference) test a post hoc test for making multiple pairwise comparisons between means following a significant F test in analysis of variance. It is the method most highly recommended by statisticians.

two-sample t test the statistical test used to test the null hypothesis that two independent (or nonrelated) groups have different means.

two-tailed test a test in which the alternative hypothesis specifies a deviation from the null hypothesis in either direction. The critical region is located in both ends of the distribution of the test statistic. It is also called a directional test.

two-way analysis of variance ANOVA with two independent variables.

type I error the error that results when one rejects the null hypothesis when it is true or when one concludes that there is a difference when there is none.

type II error the error that results when one does not reject the null hypothesis when it is false or when one does not detect a difference when there is a difference.

unbiasedness (of a statistic) a term used to describe a statistic whose mean from a large number of samples is equal to the population parameter.

uncontrolled study an experimental study that has no control subjects.

utility the value of the different outcomes in a decision tree.

validity the property of a measurement that measures the characteristic it purports to measure.

variable a characteristic of interest in a study that has different values for different subjects or objects.

variance the square of the standard deviation. It is a measure of dispersion in a distribution of observations in a population or a sample.

variation the variability in measurements of the same object or subject. It may occur naturally or may represent an error in measurement.

vital statistics mortality and morbidity rates used in epidemiology and public health.

weighted average a number formed by multiplying each number in a set of numbers by a value called a weight and then adding the resulting products.

Wilcoxon rank-sum test a nonparametric test for comparing two independent samples with ordinal data or with numerical observations that are not normally distributed.

Wilcoxon signed-ranks test a nonparametric test for comparing two dependent samples with ordinal data or with numerical observations that are not normally distributed.

Yates' correction the process of subtracting 0.5 from the numerator at each term in the chi-square statistic for 2×2 tables prior to squaring the term.

z approximation (to the binomial) the z-test used to test the equality of two independent proportions.

z distribution the normal distribution with mean 0 and standard deviation 1. It is also called the standard normal distribution.

z ratio the test statistic used in the z-test. It is formed by subtracting the hypothesized mean from the observed mean and dividing by the standard error of the mean.

z score the deviation of X from the mean divided by the standard deviation.

z test the statistical test for comparing a mean with a norm or comparing two means for large samples ($n > 30$).

z transformation a transformation that changes a normally distributed variable with mean $\overline{X}$ and standard deviation s to the z distribution with mean 0 and standard deviation 1.

References

The Use of Statistics in the Medical Literature

Altman DG: Statistics: necessary and important. (Commentary.) *Br J Obstet Gynaecol* 1986; **93**:1-5.

Avram M et al: Statistical methods in anesthesia articles: An evaluation of two American journals during two six-month periods. *Anesth Analg* 1985; **64**:604-611.

Bartko JJ: Rationale for reporting standard deviations rather than standard errors of the mean. *Am J Psychiatry* 1985; **142**:1060.

DerSimonian R et al: Reporting on methods in clinical trials. *N Engl J Med* 1982; **306**:1332-1337.

Emerson JD Colditz GA: Use of statistical analysis in *The New England Journal of Medicine. N Engl J Med* 1983; **309**:709-713.

Fromm BS, Snyder VL: Research design and statistical procedures used in *The Journal of Family Practice. J Fam Pract* 1986; **23**:564-566.

Garfunkle JM: Analysis of statistical analysis. (Editor's column.) *J Pediatr* 1986; **109**:827.

Hokanson JA, Luttman DJ, Weiss GB: Frequency and diversity of use of statistical techniques in oncology journals. *Cancer Treat Rep* 1986; **70**:589-594.

Hokanson JA et al: Spectrum and frequency of use of statistical techniques in psychiatric journals. *Am J Psychiatry* 1986; **143**:1118-1125.

Hokanson JA, et al: Statistical techniques reported in pathology journals during 1982-1985. *Arch Pathol Lab Med* 1987; **111**:202-207.

Hokanson JA et al: The reporting of statistical techniques in otolaryngology journals. *Arch Otolaryngol Head Neck Surg* 1987; **113**:45-50.

International Committee of Medical Journal Editors: Uniform requirements for manuscripts submitted to biomedical journals. *Can Med Assoc J* 1988; **138**:321-328.

Marks RG et al: Interactions between statisticians and biomedical journal editors. *Stat Med* 1988; **7**:1003-1011.

Pocock SJ, Hughes MD, Lee RJ: Statistical problems in the reporting of clinical trials. *N Engl J Med* 1987; **317**:426-432.

Reznick RK, Dawson-Saunders E, Folse JR: A rationale for the teaching of statistics to surgical residents. *Surgery* 1987; **101**:611-617.

Williamson JW, Goldschmidt PG, Colton T: The quality of medical literature: An analysis of validation assessments. In: *Medical Uses of Statistics.* Bailar JC, Mosteller F (editors). Massachusetts Medical Society, 1986.

Zelen M: Guidelines for publishing papers on cancer clinical trials; Responsibilities of editors and authors. *J Clin Oncol* 1983; **1**:164-169.

Articles Published in the Medical Literature

Anasetti C et al: Marrow transplantation for severe aplastic anemia. *Ann Intern Med* 1986; **104**:461-466.

Arntzenius AC et al: Diet, lipoproteins, and the progression of coronary atherosclerosis. *N Engl J Med* 1985; **312**:805-811.

Bale P, Bradbury D, Colley E: Anthropometric and training variables related to 10 km running performance. *Br J Sports Med* 1986; **20**:170-173.

Bartle WR, Gupta AK, Lazor J: Nonsteroidal anti-inflammatory drugs and gastrointestinal bleeding. *Arch Intern Med* 1986; **146**:2365-2367.

Baumgartner JD et al: Prevention of gram-negative shock and death in surgical patients by antibody to endotoxin core glycolipid. *Lancet* 1985; **2**:59-63.

Berry AJ et al: The prevalence of hepatitis B viral markers in anesthesia personnel. *Anesthesiology* 1984; **60**:6-9.

Borowitz KC, Glascoe FP: Sensitivity of the Denver Developmental Screening Test in speech and language. *Pediatrics* 1986; **78**:1075-1078.

Brandeau ML, Eddy DM: The workup of the asymptomatic patient with a positive fecal occult blood test. *Med Decision Making* 1987; **7**:32-46.

Buckley WE: Concussions in college football. *Am J Sports Med* 1988; **16**:51-56.

Camitta BM et al: A prospective study of androgens and bone marrow transplantation for treatment of severe aplastic anemia. *Blood* 1979; **53**:504-514.

CASS Principal Investigators and Associates: Coronary Artery Surgery Study (CASS): A randomized trial of coronary bypass surgery. *Circulation* 1983; **68**:951-960.

Colditz GA et al: A prospective study of parental history of myocardial infarction and coronary heart disease in women. *N Engl J Med* 1987; **316**:1105-1110.

Dawson-Saunders B, Paiva REA, Doolen DR: Using ACT scores and grade-point averages to predict students' MCAT scores. *J Med Educ* 1986; **61**:681-683.

de Champlain J et al: Circulating catecholamine levels in human and experimental hypertension. *Circ Res* 1976; **38**:109-114.

Detsky AS et al: Perioperative parenteral nutrition: A meta-analysis. *Ann Intern Med* 1987; **107**:195-203.

DiMaio et al: Screening for fetal Down's syndrome in pregnancy by measuring maternal serum alpha-fetoprotein levels. *N Engl J Med* 1987; **317**:342-346.

Doll R, Hill AB: Smoking and carcinoma of the lung. *Br Med J* 1950; **2**:739-748.

Doll R, Peto R: Mortality in relation to smoking: 20 years' observations on male British doctors. *Br Med J* 1976; **2**(6051):1525-1536.

Einarsson K et al: Influence of age on secretion of cholesterol and synthesis of bile acids by the liver. *N Engl J Med* 1985; 313:277–282.

Feldman W et al: The use of dietary fiber in the management of simple, childhood, idiopathic, recurrent, abdominal pain. *Am J Dis Child* 1985; 139:1216–1218.

Fisher B et al: Five-year results of a randomized clinical trial comparing total mastectomy and segmental mastectomy with or without radiation in the treatment of breast cancer. *N Engl J Med* 1985; 312:665–673.

Fletcher EC et al: Undiagnosed sleep apnea in patients with essential hypertension. *Ann Intern Med* 1985; 103:190–195.

Foltin E: Bone loss and forms of tibial condylar fracture. *Arch Orthop Trauma Surg* 1987; 106:341–348.

Foxman B, Valdez B, Brook RH: Childhood enuresis; Prevalence, perceived impact, and prescribed treatments. *Pediatrics* 1986; 77:482–487.

Francis PT et al: Neurochemical studies of early-onset Alzheimer's disease. *N Engl J Med* 1985; 313:7–11.

Funkhouser AW et al: Estimated effects of a delay in the recommended vaccination schedule for diphtheria and tetanus toxoids and pertussis vaccine. *JAMA* 1987; 247:1341–1346.

Gage TP: Managing the cancer risk in chronic ulcerative colitis. *J Clin Gastroenterol* 1986; 8:50–57.

Ginsberg JM et al: Use of single voided urine samples to estimate quantitative proteinuria. *N Engl J Med* 1983; 309:1543–1546.

Goldsmith AM et al: Sequential clinical and immunologic abnormalities in hemophiliacs. *Arch Intern Med* 1985; 145:431–434.

Gordon T, Kannel WB: The Framingham, Massachusetts, Study twenty years later. In: *The Community as an Epidemiologic Laboratory.* Kessler IJ, Levin ML (editors). Johns Hopkins Press, 1970.

Graham DY et al: Healing of benign gastric ulcer: Comparison of cimetidine and placebo in the United States. *Ann Intern Med* 1985; 102:573–576.

Greenwood M: The natural duration of cancer. *Rep Public Health Med Subjects* 1926; 33:1–26.

Hamill PVV et al: Physical growth: National Center for Health Statistics percentiles. *Am J Clin Nutr* 1979; 32:607–629.

Han T et al: Prognostic importance of cytogenetic abnormalities in patients with chronic lymphocytic leukemia. *N Engl J Med* 1984; 310:288–292.

Hansen HJ, Caudill SP, Boone J: Crisis in drug testing. *JAMA* 1985; 253:2382–2387.

Hartz AJ et al: The association of smoking with cardiomyopathy. *N Engl J Med* 1984; 311:1201–1206.

Helmrich SP et al. Venous thromboembolism in relation to oral contraceptive use. *Obstet Gynecol* 1987; 69:91–95.

Hooton TM, Running K, Stamm WE: Single-dose therapy for cystitis in women. *JAMA* 1985; 253:387–390.

Hornig CR, Dorndorf W, Agnoli AL: Hemorrhagic cerebral infarction—a prospective study. *Stroke* 1986; 17:179–184.

Horvath EP et al: Effects of formaldehyde on the mucous membranes and lungs. *JAMA* 1988; 259:701–707.

Ingelfinger JA: Routine testing for chlamydial cervical infections. (Letter to the Editor.) *Ann Intern Med* 1988; 108:153.

Irwin M et al: Life events, depressive symptoms, and immune function. *Am J Psychiatry* 1987; 144:437–441.

Jacobson L, Westrom L: Objectivized diagnosis of acute pelvic inflammatory disease. *Am J Obstet Gynecol* 1969; 105:1088–1098.

Kalman CM, Laskin OL: Herpes zoster and zosteriform herpes simplex virus infections in immunocompetent adults. *Am J Med* 1986; 81:775–778.

Khaw K, Barrett-Connor E: Dietary potassium and stroke-associated mortality: A 12-year prospective population study. *N Engl J Med* 1987; 316:235–240.

Kilbourne EM et al: Clinical epidemiology of toxic-oil syndrome. *N Engl J Med* 1983; 309:1408–1414.

Knutson RA et al: Use of sugar and providone-iodine to enhance wound healing: Five years' experience. *South Med J* 1981; 74:1329–1335.

Kremer JM et al: Fish-oil fatty acid supplementation in active rheumatoid arthritis. *Ann Intern Med* 1987; 106:497–503.

Leveno KJ et al: A prospective comparison of selective and universal electronic fetal monitoring in 34,995 pregnancies. *N Engl J Med* 1986; 315:615–619.

Levinsky RJ et al: Circulating immune complexes in steroid-responsive nephrotic syndrome. *N Engl J Med* 1978; 298:126–129.

Lisboa C et al: Inspiratory muscle function in patients with severe kyphoscoliosis. *Am Rev Respir Dis* 1985; 132:48–52.

MacMahon B, Yen S, Trichopoulos D: Coffee and cancer of the pancreas. *N Engl J Med* 1981; 304:630–633.

Mantzouranis EC, Rosen FS, Colten HR: Reticuloendothelial clearance in cystic fibrosis and other inflammatory lung diseases. *N Engl J Med* 1988; 319:338–343.

Maxwell MH et al: Error in blood pressure measurement due to incorrect cuff size in obese patients. *Lancet* 1982; 33–36.

McAuliffe WE et al: Psychoactive drug use among practicing physicians and medical students. *N Engl J Med* 1986; 315:805–810.

Meier P: The biggest public health experiment ever. Pages 3–15 in: *Statistics: A Guide to the Unknown,* 2nd ed. Tanur JM et al: (editors). Holden-Day, 1978.

Melbye M et al: Long-term seropositivity for human T-lymphotropic virus type III in homosexual men without the acquired immunodeficiency syndrome: Development of immunologic and clinical abnormalities. *Ann Intern Med* 1986; 104:496–500.

Moertel CG et al: High-dose vitamin C versus placebo in the treatment of patients with advanced cancer who have had no prior chemotherapy. *N Engl J Med* 1985; 312:137–141.

Morris PJ et al: Cyclosporin conversion versus conventional immunosuppression: Long-term follow-up and histological evaluation. *Lancet* 1987; 586–591.

Multiple Risk Factor Intervention Trial Research Group: Multiple risk factor intervention trial. *JAMA* 1982; 248:1465–1477.

Murphy JM et al: Performance of screening and diagnostic tests. *Arch Gen Psychiatry* 1987; 44:550–555.

Nathan DM et al: The clinical information value of the glycosylated hemoglobin assay. *N Engl J Med* 1984; 310:341–346.

National Center for Health Statistics: *Vital Statistics of the United States, 1983.* Vol 2: *Mortality,* Part A. US

Government Printing Office, DHHS Publ. No. (PHS) 87-1122, 1987.

Nelson KB, Ellenberg JH: Antecedents of cerebral palsy. *N Engl J Med* 1986; **315**:81-86.

Norwegian Multicenter Study Group: Timolol-induced reduction in mortality and reinfarction in patients surviving acute myocardial infarction. *N Engl J Med* 1981; **304**:801-807.

O'Malley MS, Fletcher SW: Screening for breast cancer with breast self examination. *JAMA* 1987; **257**:2196-2203.

Oren J, Kelly D, Shannon DC: Identification of a high-risk group for sudden infant death syndrome among infants who were resuscitated for sleep apnea. *Pediatrics* 1986; **77**:495-499.

Orloff MJ et al: Protacaval shunt as emergency procedure in unselected patients with alcoholic cirrhosis. *Surg Gynecol Obstet* 1975; **141**:59-68.

Phillips RS et al: Should tests for *Chlamydia trachomatis* cervical infection be done during routine gynecologic visits? *Ann Intern Med* 1987; **107**:188-194.

Preminger GM et al: Percutaneous nephrostolithotomy vs open surgery for renal calculi. *JAMA* 1985; **254**:1054-1058.

Robin ED: The cult of the Swan-Ganz catheter: Overuse and abuse of pulmonary flow catheters. *Ann Intern Med* 1985; **103**:445-449.

Ross EM, Roberts WC: The carcinoid syndrome: Comparison of 21 necropsy subjects with carcinoid heart disease to 15 necropsy patients without carcinoid heart disease. *Am J Med* 1985; **79**:339-354.

Rubin P (editor): *Clinical Oncology for Medical Students and Physicians,* 6th ed. The American Cancer Society, 1983.

Scully RE, McNeely BU, Mark EJ: Normal reference laboratory values. *N Engl J Med* 1987; **314**:39-49.

Siscovick DS et al: The incidence of primary cardiac arrest during vigorous exercise. *N Engl J Med* 1984; **311**:874-877.

Society of Actuaries and Association of Life Insurance Medical Directors of America: *Blood Pressure Study 1979.* Recording & Statistical Corporation, 1980.

Steere AC et al: Successful parenteral penicillin therapy of established Lyme arthritis. *N Engl J Med* 1985; **312**:869-874.

Steinberg EP et al: A case study of physicians' use of liver-spleen scans. *Arch Intern Med* 1986; **146**:253-258.

Stone HH, Mullins RJ, Scovill WA: Vagotomy plus Belroth II gastrectomy for the prevention of recurrent alcohol-induced pancreatitis. *Ann Surg* 1985; **201**:684-688.

Tan EM et al: The 1982 revised criteria for the classification of systemic lupus erythematosus. *Arthritis Rheum* 1982; **25**:1271-1277.

Tirlapur VG, Mir MA: Effect of low calorie intake on abnormal pulmonary physiology in patients with chronic hypercapneic respiratory failure. *Am J Med* 1984; **77**:987-994.

van Crevel H, Habbema JDF, Braakman R: Decision analysis of the management of incidental intracranial saccular aneurysms. *Neurology* 1986; **36**:1335-1339.

Ware JE et al: Comparison of health outcomes at a health maintenance organization with those of fee-for-service care. *Lancet* 1986; 1017-1022.

Ware JE et al: *Health Outcomes for Adults in Prepaid and Fee-for-Service Systems of Care.* The RAND Corporation, 1987.

Watz R, Ek I, Bygdeman S: Noninvasive diagnosis of acute deep vein thrombosis. *Acta Med Scand* 1979; **206**:463-466.

Weisman MH et al: Measures of bone loss in rheumatoid arthritis. *Arch Intern Med* 1986; **146**:701-704.

Willett WC et al: Relative and absolute excess risks of coronary heart disease among women who smoke cigarettes. *N Engl J Med* 1987; **317**:1303-1309.

Medical Statistics and Epidemiologic Articles

Anderson S et al: *Statistical Methods for Comparative Studies.* Wiley, 1980.

Armitage P: *Statistical Methods in Medical Research.* Blackwell Scientific, 1971.

Bailar JC et al: Studies without internal controls. *N Engl J Med* 1984; **311**:156-162.

Berkson J: Limitations of the application of four-fold table analyses to hospital data. *Biometrics Bull* 1946: **2**:47-53.

Brown CG et al: The beta error and sample size determination in clinical trials in emergency medicine. *Ann Emerg Med* 1987; **183**:79-83.

Chalmers TC: A challenge to clinical investigators. *Gastroenterology* 1969; **57**:631-635.

Colton T: *Statistics in Medicine.* Little, Brown, 1974.

Cox DR: Regression models and life tables. *J R Stat Soc Series B* 1972; **34**:187-220.

Cutler S, Ederer F: Maximum utilization of the lifetable method in analyzing survival. *J Chronic Dis* 1958; **8**:699-712.

Daniel WW: *Biostatistics: A Foundation for Analysis in the Health Sciences.* Wiley, 1974.

Elveback LR, Guillier CL, Keating FR: Health normality, and the ghost of Gauss. *JAMA* 1970; **211**:69-75.

Feinstein AR: *Clinical Epidemiology: The Architecture of Research.* Saunders, 1985.

Fisher B et al: The Will Rogers phenomenon. *N Engl J Med* 1985; **312**:1604-1608.

Fleiss JL: *Statistical Methods for Rates and Proportions,* 2nd ed. Wiley, 1981.

Fletcher RH, Fletcher SW, Wagner EH: *Clinical Epidemiology,* 2nd ed. Williams and Wilkins, 1988.

Frieman JA et al: The importance of beta, the Type II error and sample size in the design and interpretation of the randomized control trial. *N Engl J Med* 1978; **299**:690-694.

Gardner MJ, Altman DG: Confidence intervals rather than *P* values: Estimation rather than hypothesis testing. *Br Med J* 1986; **292**:746-750.

Gordon T et al: Some methodologic problems in the long term study of cardiovascular disease: Observations on the Framingham Study. *J Chronic Dis* 1959; **10**:186-206.

Greenberg RS: Prospective studies. Pages 315-319 in: *Encyclopedia of Statistical Sciences.* Vol 7. Kotz S, Johnson NL (editors). Wiley, 1986.

Greenberg RS: Retrospective studies. Pages 120-124 in: *Encyclopedia of Statistical Sciences.* Vol 8. Kotz S, Johnson NL (editors). Wiley, 1988.

Kalbfleisch JD, Prentice RL: *The Statistical Analysis of Failure Time Data.* Wiley, 1980.

Kleinbaum DG, Kupper LL, Morgenstern H: *Epidemiologic Research.* Lifetime Learning, 1982.

L'Abbé KA, Detsky AS, O'Rourke K: Meta-analysis in clinical research. *Ann Intern Med* 1987; **107**:224–233.

Lee ET: *Statistical Methods for Survival Data Analysis.* Lifetime Learning, 1980.

McMaster University Health Sciences Centre, Department of Clinical Epidemiology and Biostatistics: Clinical disagreement, I: How it occurs and why. *Can Med Assoc J* 1980; **123**:499–504.

McMaster University Health Sciences Centre, Department of Clinical Epidemiology and Biostatistics: Clinical Disagreement, II: How to avoid it and learn from it. *Can Med Assoc J* 1980; **123**:613–617.

Moses LE, Emerson JD, Hosseini H: Analyzing data from ordered categories. *N Engl J Med* 1984; **311**:442–448.

Murphy EA: *Biostatistics in Medicine.* Johns Hopkins Press, 1982.

Sackett DL: Bias in analytic research. *J Chronic Dis* 1978; **32**:51–63.

Sackett DL, Haynes RB, Tugwell P: *Clinical Epidemiology.* Little, Brown, 1985.

Sacks H, Chalmers TC, Smith H: Randomized versus historical controls for clinical trials. *Am J Med* 1982; **72**:233–240.

Sacks HS et al: Meta-analysis of randomized controlled trials. *N Engl J Med* 1987; **316**:450–455.

Siegel S: *Nonparametric Statistics.* McGraw-Hill, 1956.

Sokal RR, Rohlf FJ: *Biometry,* 2nd ed. Freeman, 1981.

derived from the same cases. *Radiology* 1983; **148**:839–843.

Hershey JC, Cebul RD, Williams SV: The importance of considering single testing when two tests are available. *Med Decision Making* 1987; **7**:212–219.

Kahneman D, Tversky A: On prediction and judgment. *Oregon Res Inst Bull* 1972; **12**(4): 1-30.

Kassirer JP et al: Decision analysis: A progress report. *Ann Intern Med* 1987; **106**:270–291.

McNeil BJ, Keeler E, Adelstein SJ: Primer on certain elements of medical decision making. *N Engl J Med* 1975; **293**:211–215.

Metz CE: Basic principles of ROC analysis. *Semin Nucl Med* 1978; **8**:283–298.

Pauker SG, Kassirer JP: Decision analysis. *N Engl J Med* 1987; **316**:250–258.

Pauker SG, Kassirer JP: The threshold approach to clinical decision making. *N Engl J Med* 1980; **302**:1109–1117.

Shewchuk RM, Francis KT: Principles of clinical decision making—an introduction to decision analysis. *Phys Ther* 1988; **68**:357–359.

Sonnenberg FA, Kassirer JP, Kopelman RI: An autopsy of the clinical reasoning process. *Hosp Pract* 1986; 45–56.

Sox HC: Probability theory in the use of diagnostic tests. *Ann Intern Med* 1986; **104**:60–66.

Sox HC, Liang MH: The erythrocyte sedimentation rate. *Ann Intern Med* 1986; **146**:253–258.

Sox HC et al: *Med Decision Making.* Butterworths, 1985.

Weinstein MC et al: *Clinical Decision Analysis.* Saunders, 1980.

Medical Decision Making

Albert DA: Decision theory in medicine: A review and critique. *Milbank Mem Fund Q* 1978; **56**:362–401.

Bar-Hillel M: The base-rate fallacy in probability judgments. *Acta Psychol* 1980; **44**:211–233.

Clancy WJ, Shortliffe EH: Introduction: Medical artificial intelligence programs. Pages 1–17 in: *Readings in Medical Artificial Intelligence.* Addison-Wesley, 1984.

Connelly DP et al (editors): *Clinical Decisions and Laboratory Use.* University of Minnesota Press, 1982.

Detsky AS, Redelmeier D, Abrams HB: What's wrong with decision analysis? Can the left brain influence the right? *J Chronic Dis* 1987; **40**:831–836.

Doubilet PM: Statistical techniques for medical decision making: Applications to diagnostic radiology. *Am J Radiol* 1988; **150**:745–750.

Eraker SA et al: To test or not to test—to treat or not to treat. *J Gen Intern Med* 1986; **1**:177–182.

Fagan TJ: Nomogram for Bayes' theorem. *N Engl J Med* 1975; **293**:257.

Feinstein AR: The "chagrin factor" and qualitative decision analysis. *Arch Intern Med* 1985; **145**:1257–1259.

Griner PF et al: Selection and interpretation of diagnostic tests and procedures. *Ann Intern Med* 1981; **94**:553–600.

Hanley JA, McNeil BJ: A method of comparing the areas under receiver operator characteristic curves

General Statistical and Miscellaneous Readings

Box GEP: Non-normality and tests on variance. *Biometrika* 1953; **40**:318–335.

Box GEP: Some theorems on quadratic forms applied in the study of the analysis of variance. *Ann Math Stat* 1954; **25**:290–302, 484–498.

Browne RH: On visual assessment of the significance of a mean difference. *Biometrics* 1979; **35**:657–665.

Cohen J: *Statistical Power Analysis for the Behavioral Sciences,* rev. ed. Academic Press, 1977.

Conover WJ, Iman RL: Rank transformations as a bridge between parametric and nonparametric statistics. *Am Stat* 1981; **35**:124–129.

Dawson-Saunders B et al: The instruction of biostatistics in medical school. *Am Stat* 1987; **41**:263–266.

Dunn OJ, Clark VA: *Applied Statistics: Analysis of Variance and Regression.* Wiley, 1974.

Gehan EA: A generalized Wilcoxon test for comparing arbitrarily singly-censored samples. *Biometrika* 1965; **52**:15–21.

Glass GV: Integrating findings: The meta-analysis of research. Pages 351–379 in: *Review of Research in Education,* Shulman LS (editor). Peacock, 1977.

Grizzle JE: Continuity correction in the X^2-test for 2 × 2 tables. *Am Stat* 1967; **21**:28–32.

Hays WL: *Statistics for the Social Sciences,* 2nd ed. Holt, Rinehart and Winston, 1973.

Iman RL: Graphs for use with the Lilliefors test for normal and exponential distributions. *Am Stat* 1982; **36**:109–112.

Iman RL: Use of a *t*-statistic as an approximation to the exact distribution of the Wilcoxon signed ranks test statistic. *Comm Stat* 1974; **3**:795–806.

Kirk RE: *Experimental Design: Procedures for the Behavioral Sciences,* 2nd ed. Brooks/Cole, 1982.

Kleinbaum DG, Kupper LL: *Applied Regression Analysis and Other Multiveriate Methods.* Duxbury Press, 1978.

Mendenhall W: *Introduction to Probability and Statistics,* 7th ed. P.W.S. Publishing, 1987.

Miller I, Freund JD: *Probability and Statistics for Engineers,* 3rd ed. Prentice-Hall, 1985.

Pedhazur EJ: *Multiple Regression in Behavioral Research,* 2nd ed. Holt, Rinehart and Winston, 1982.

Snedecor GW, Cochran WG: *Statistical Methods,* 7th ed. Iowa State University Press, 1980.

Stoline MR: The status of multiple comparisons; Simultaneous estimation of all pairwise comparisons in one-way ANOVA designs. *Am Stat* 1981; **35**:134-141.

Tarone RE, Ware J: On distribution-free tests for equality of survival distributions. *Biometrika* 1977; **64**:156–160.

Tukey J: *Exploratory Data Analysis.* Addison-Wesley, 1977.

Wainer H: *How to Display Data Badly.* Educational Testing Service Technical Report No. 82–83. Educational Testing Service, 1982.

Walker H: *Studies in the History of Statistical Method.* Williams and Wilkins, 1931.

Winer BJ: *Statistical Principles in Experimental Design,* 2nd ed. McGraw-Hill, 1971.

Yates F: Tests of significance for 2 × 2 contingency tables. *J R Stat Soc Series A* 1984; **147**:426–463.

Appendix A:
Tables

Table A-1. Random Numbers.

927415	956121	168117	169280	326569	266541
926937	515107	014658	159944	821115	317592
867169	388342	832261	993050	639410	698969
867169	542747	032683	131188	926198	371071
512500	843384	085361	398488	774767	383837
062454	423050	670884	840940	845839	979662
806702	881309	772977	367506	729850	457758
837815	163631	622143	938278	231305	219737
926839	453853	767825	284716	916182	467113
854813	731620	978100	589512	147694	389180
851595	452454	262448	688990	461777	647487
449353	556695	806050	123754	722070	935916
169116	586865	756231	469281	258737	989450
139470	358095	528858	660128	342072	681203
433775	761861	107191	515960	759056	150336
221922	232624	398839	495004	881970	792001
740207	078048	854928	875559	246288	000144
525873	755998	866034	444933	785944	018016
734185	499711	254256	616625	243045	251938
773112	463857	781983	078184	380752	492215
638951	982155	747821	773030	594005	526828
868888	769341	477611	628714	250645	853454
611034	167642	701316	589251	330456	681722
379290	955292	664549	656401	320855	215201
411257	411484	068629	050150	106933	900095
407167	435509	578642	268724	366564	511815
895893	438644	330273	590506	820439	976891
986683	830515	284065	813310	554920	111395
335421	814351	508062	663801	365001	924418
927660	793888	507773	975109	625175	552278
957559	236000	471608	888683	146821	034687
694904	499959	950969	085327	352611	335924
863016	494926	871064	665892	076333	990558
876958	865769	882966	236535	541645	819783
619813	221175	370697	566925	705564	472934
476626	646911	337167	865652	195448	116729
578292	863854	145858	206557	430943	591126
286553	981699	232269	819656	867825	890737
819064	712344	033613	457019	478176	342104
383035	043025	201591	127424	771948	762990
879392	378486	198814	928028	493486	373709
924020	273258	851781	003514	685749	713570
502523	157212	472643	439301	718562	196269
815316	651530	080430	912635	820240	533626
914984	444954	053723	079387	530020	703312
312248	619263	715357	923412	252522	913950
030964	407872	419563	426527	565215	243717
870561	984049	445361	315827	651925	464440
820157	006091	670091	478357	490641	082559
519649	761345	761354	794613	330132	319843

Table A-2. Areas under the standard normal curve.[1]

z	Area Between −z & +z	Area in Two Tails (< −z & > +z)	Area in One Tail (< −z or > +z)
0.00	0.000	1.000	0.500
0.05	0.040	0.960	0.480
0.10	0.080	0.920	0.460
0.15	0.119	0.881	0.440
0.20	0.159	0.841	0.421
0.25	0.197	0.803	0.401
0.30	0.236	0.764	0.382
0.35	0.274	0.726	0.363
0.40	0.311	0.689	0.345
0.45	0.347	0.653	0.326
0.50	0.383	0.617	0.309
0.55	0.418	0.582	0.291
0.60	0.451	0.549	0.274
0.65	0.484	0.516	0.258
0.70	0.516	0.484	0.242
0.75	0.547	0.453	0.227
0.80	0.576	0.424	0.212
0.85	0.605	0.395	0.198
0.90	0.632	0.368	0.184
0.95	0.658	0.342	0.171
1.00[2]	0.683	0.317	0.159
1.05	0.706	0.294	0.147
1.10	0.729	0.271	0.136
1.15	0.750	0.250	0.125
1.20	0.770	0.230	0.115
1.25	0.789	0.211	0.106
1.28[2]	0.800	0.260	0.100
1.30	0.806	0.194	0.097
1.35	0.823	0.177	0.089
1.40	0.838	0.162	0.081
1.45	0.853	0.147	0.074
1.50	0.866	0.134	0.067
1.55	0.879	0.121	0.061
1.60	0.890	0.110	0.055
1.645[2]	0.900	0.100	0.050
1.65	0.901	0.099	0.049
1.70	0.911	0.089	0.045
1.75	0.920	0.080	0.040
1.80	0.928	0.072	0.036
1.85	0.936	0.064	0.032
1.90	0.943	0.057	0.029
1.95	0.949	0.051	0.026
1.96[2]	0.950	0.050	0.025
2.00	0.954	0.046	0.023
2.05	0.960	0.040	0.020
2.10	0.964	0.036	0.018
2.15	0.968	0.032	0.016
2.20	0.972	0.028	0.014
2.25	0.976	0.024	0.012
2.30	0.979	0.021	0.011
2.326[2]	0.980	0.020	0.010
2.35	0.981	0.019	0.009
2.40	0.984	0.016	0.008
2.45	0.986	0.014	0.007
2.50	0.988	0.012	0.006
2.55	0.989	0.011	0.005
2.575[2]	0.990	0.010	0.005
2.60	0.991	0.009	0.005
2.65	0.992	0.008	0.004
2.70	0.993	0.007	0.003
2.75	0.994	0.006	0.003

Table A-2. (cont.)

z	Area Between −z & +z	Area in Two Tails (< −z & > +z)	Area in One Tail (< −z or > +z)
2.80	0.995	0.005	0.003
2.85	0.996	0.004	0.002
2.90	0.996	0.004	0.002
2.95	0.997	0.003	0.002
3.00	0.997	0.003	0.001

[1]Adapted and reproduced, with permission from Table 1 in Pearson ES, Hartley HO (editors): *Biometrika Tables for Statisticians,* 3rd ed. Vol 1. Cambridge University Press, 1966. Used with the kind permission of the Biometrika trustees.
[2]Commonly used values are underlined.

Table A-3. Percentage points or critical values for the *t* distribution corresponding to commonly used areas under the curve.[1]

Degrees of Freedom	Area in 1 Tail				
	0.05	0.025	0.01	0.005	0.0005
	Area in 2 Tails				
	0.10	0.05	0.02	0.01	0.001
1	6.314	12.706	31.821	63.657	636.62
2	2.920	4.303	6.965	9.925	31.598
3	2.353	3.182	4.541	5.841	12.924
4	2.132	2.776	3.747	4.604	8.610
5	2.015	2.571	3.365	4.032	6.869
6	1.943	2.447	3.143	3.707	5.959
7	1.895	2.365	2.998	3.499	5.408
8	1.860	2.306	2.896	3.355	5.041
9	1.833	2.262	2.821	3.250	4.781
10	1.812	2.228	2.764	3.169	4.587
11	1.796	2.201	2.718	3.106	4.437
12	1.782	2.179	2.681	3.055	4.318
13	1.771	2.160	2.650	3.012	4.221
14	1.761	2.145	2.624	2.977	4.140
15	1.753	2.131	2.602	2.947	4.073
16	1.746	2.120	2.583	2.921	4.015
17	1.740	2.110	2.567	2.898	3.965
18	1.734	2.101	2.552	2.878	3.922
19	1.729	2.093	2.539	2.861	3.883
20	1.725	2.086	2.528	2.845	3.850
21	1.721	2.080	2.518	2.831	3.819
22	1.717	2.074	2.508	2.819	3.792
23	1.714	2.069	2.500	2.807	3.767
24	1.711	2.064	2.492	2.797	3.745
25	1.708	2.060	2.485	2.787	3.725
26	1.706	2.056	2.479	2.779	3.707
27	1.703	2.052	2.473	2.771	3.690
28	1.701	2.048	2.467	2.763	3.674
29	1.699	2.045	2.462	2.756	3.659
30	1.697	2.042	2.457	2.750	3.646
40	1.684	2.021	2.423	2.704	3.551
60	1.671	2.000	2.390	2.660	3.460
120	1.658	1.980	2.358	2.617	3.373
∞	1.645	1.960	2.326	2.576	3.291

[1]Adapted and reproduced, with permission, from Table 12 in Pearson ES, Hartley HO (editors): *Biometrika Tables for Statisticians,* 3rd ed. Vol 1. Cambridge University Press, 1966. Used with the kind permission of the Biometrika Trustees.

Table A-4. Percentage points or critical values for the F distribution corresponding to areas of 0.05 and 0.01 under the upper tail of the distribution.[1]

Degrees of Freedom, Denominator	Area	Degrees of Freedom, Numerator									
		1	2	3	4	5	6	7	8	9	10
1	0.05	161.4	199.5	215.7	224.6	230.2	234.0	236.8	238.9	240.5	241.9
	0.01	4052	4999.5	5403	5625	5764	5859	5928	5981	6022	6056
2	0.05	18.51	19.00	19.16	19.25	19.30	19.33	19.35	19.37	19.38	19.40
	0.01	98.50	99.00	99.17	99.25	99.30	99.33	99.36	99.37	99.39	99.40
3	0.05	10.13	9.55	9.28	9.12	9.01	8.94	8.89	8.85	8.81	8.79
	0.01	34.12	30.82	29.46	28.71	28.24	27.91	27.67	27.49	27.35	27.23
4	0.05	7.71	6.94	6.59	6.39	6.26	6.16	6.09	6.04	6.00	5.96
	0.01	21.20	18.00	16.69	15.98	15.52	15.21	14.98	14.80	14.66	14.55
5	0.05	6.61	5.79	5.41	5.19	5.05	4.95	4.88	4.82	4.77	4.74
	0.01	16.26	13.27	12.06	11.39	10.97	10.67	10.46	10.29	10.16	10.05
6	0.05	5.99	5.14	4.76	4.53	4.39	4.28	4.21	4.15	4.10	4.06
	0.01	13.75	10.92	9.78	9.15	8.75	8.47	8.26	8.10	7.98	7.87
7	0.05	5.59	4.74	4.35	4.12	3.97	3.87	3.79	3.73	3.68	3.64
	0.01	12.25	9.55	8.45	7.85	7.46	7.19	6.99	6.84	6.72	6.62
8	0.05	5.32	4.46	4.07	3.84	3.69	3.58	3.50	3.44	3.39	3.35
	0.01	11.26	8.65	7.59	7.01	6.63	6.37	6.18	6.03	5.91	5.81
9	0.05	5.12	4.26	3.86	3.63	3.48	3.37	3.29	3.23	3.18	3.14
	0.01	10.56	8.02	6.99	6.42	6.06	5.80	5.61	5.47	5.35	5.26
10	0.05	4.96	4.10	3.71	3.48	3.33	3.22	3.14	3.07	3.02	2.98
	0.01	10.04	7.56	6.55	5.99	5.64	5.39	5.20	5.06	4.94	4.85
12	0.05	4.75	3.89	3.49	3.26	3.11	3.00	2.91	2.85	2.80	2.75
	0.01	9.33	6.93	5.95	5.41	5.06	4.82	4.64	4.50	4.39	4.30
15	0.05	4.54	3.68	3.29	3.06	2.90	2.79	2.71	2.64	2.59	2.54
	0.01	8.68	6.36	5.42	4.89	4.56	4.32	4.14	4.00	3.89	3.80
20	0.05	4.35	3.49	3.10	2.87	2.71	2.60	2.51	2.45	2.39	2.35
	0.01	8.10	5.85	4.94	4.43	4.10	3.87	3.70	3.56	3.46	3.37
24	0.05	4.26	3.40	3.01	2.78	2.62	2.51	2.42	2.36	2.30	2.25
	0.01	7.82	5.61	4.72	4.22	3.90	3.67	3.50	3.36	3.26	3.17
30	0.05	4.17	3.32	2.92	2.69	2.53	2.42	2.33	2.27	2.21	2.16
	0.01	7.56	5.39	4.51	4.02	3.70	3.47	3.30	3.17	3.07	2.98
40	0.05	4.08	3.23	2.84	2.61	2.45	2.34	2.25	2.18	2.12	2.08
	0.01	7.31	5.18	4.31	3.83	3.51	3.29	3.12	2.99	2.89	2.80
60	0.05	4.00	3.15	2.76	2.53	2.37	2.25	2.17	2.10	2.04	1.99
	0.01	7.08	4.98	4.13	3.65	3.34	3.12	2.95	2.82	2.72	2.63
120	0.05	3.92	3.07	2.68	2.45	2.29	2.17	2.09	2.02	1.96	1.91
	0.01	6.85	4.79	3.95	3.48	3.17	2.96	2.79	2.66	2.56	2.47
∞	0.05	3.84	3.00	2.60	2.37	2.21	2.10	2.01	1.94	1.88	1.83
	0.01	6.63	4.61	3.78	3.32	3.02	2.80	2.64	2.51	2.41	2.32

Table A-4. (cont.)

Degrees of Freedom Denominator	Area	Degrees of Freedom, Numerator								
		12	15	20	24	30	40	60	120	∞
1	0.05	243.9	245.9	248.0	249.1	250.1	251.1	252.2	253.3	254.3
	0.01	6106	6157	6209	6235	6261	6287	6313	6339	6366
2	0.05	19.41	19.43	19.45	19.45	19.46	19.47	19.48	19.49	19.50
	0.01	99.42	99.43	99.45	99.46	99.47	99.47	99.48	99.49	99.50
3	0.05	8.74	8.70	8.66	8.64	8.62	8.59	8.57	8.55	8.53
	0.01	27.05	26.87	26.69	26.60	26.50	26.41	26.32	26.22	26.13
4	0.05	5.91	5.86	5.80	5.77	5.75	5.72	5.69	5.66	5.63
	0.01	14.37	14.20	14.02	13.93	13.84	13.75	13.65	13.56	13.46
5	0.05	4.68	4.62	4.56	4.53	4.50	4.46	4.43	4.40	4.36
	0.01	9.89	9.72	9.55	9.47	9.38	9.29	9.20	9.11	9.02
6	0.05	4.00	3.94	3.87	3.84	3.81	3.77	3.74	3.70	3.67
	0.01	7.72	7.56	7.40	7.31	7.23	7.14	7.06	6.97	6.88
7	0.05	3.57	3.51	3.44	3.41	3.38	3.34	3.30	3.27	3.23
	0.01	6.47	6.31	6.16	6.07	5.99	5.91	5.82	5.74	5.65
8	0.05	3.28	3.22	3.15	3.12	3.08	3.04	3.01	2.97	2.93
	0.01	5.67	5.52	5.36	5.28	5.20	5.12	5.03	4.95	4.86
9	0.05	3.07	3.01	2.94	2.90	2.86	2.83	2.79	2.75	2.71
	0.01	5.11	4.96	4.81	4.73	4.65	4.57	4.48	4.40	4.31
10	0.05	2.91	2.85	2.77	2.74	2.70	2.66	2.62	2.58	2.54
	0.01	4.71	4.56	4.41	4.33	4.25	4.17	4.08	4.00	3.91
12	0.05	2.69	2.62	2.54	2.51	2.47	2.43	2.38	2.34	2.30
	0.01	4.16	4.01	3.86	3.78	3.70	3.62	3.54	3.45	3.36
15	0.05	2.48	2.40	2.33	2.29	2.25	2.20	2.16	2.11	2.07
	0.01	3.67	3.52	3.37	3.29	3.21	3.13	3.05	2.96	2.87
20	0.05	2.28	2.20	2.12	2.08	2.04	1.99	1.95	1.90	1.84
	0.01	3.23	3.09	2.94	2.86	2.78	2.69	2.61	2.52	2.42
24	0.05	2.18	2.11	2.03	1.98	1.94	1.89	1.84	1.79	1.73
	0.01	3.03	2.89	2.74	2.66	2.58	2.49	2.40	2.31	2.21
30	0.05	2.09	2.01	1.93	1.89	1.84	1.79	1.74	1.68	1.62
	0.01	2.84	2.70	2.55	2.47	2.39	2.30	2.21	2.11	2.01
40	0.05	2.00	1.92	1.84	1.79	1.74	1.69	1.64	1.58	1.51
	0.01	2.66	2.52	2.37	2.29	2.20	2.11	2.02	1.92	1.80
60	0.05	1.92	1.84	1.75	1.70	1.65	1.59	1.53	1.47	139
	0.01	2.50	2.35	2.20	2.12	2.03	1.94	1.84	1.73	1.60
120	0.05	1.83	1.75	1.66	1.61	1.55	1.50	1.43	1.35	1.25
	0.01	2.34	2.19	2.03	1.95	1.86	1.76	1.66	1.53	1.38
∞	0.05	1.75	1.67	1.57	1.52	1.46	1.39	1.32	1.22	1.00
	0.01	2.18	2.04	1.88	1.79	1.70	1.59	1.47	1.32	1.00

[1]Adapted and reproduced, with permission, from Table 18 in Pearson ES, Hartley HO (editors): *Biometrika Tables for Statisticians,* 3rd ed. Vol 1. Cambridge University Press, 1966. Used with the kind permission of the Biometrika Trustees.

Table A-5. Percentage points or critical values for the X^2 distribution corresponding to commonly used areas under the curve.[1]

Degrees of Freedom	Area in Upper Tail			
	0.10	0.05	0.01	0.001
1	2.706	3.841	6.635	10.828
2	4.605	5.991	9.210	13.816
3	6.251	7.815	11.345	16.266
4	7.779	9.488	13.277	18.467
5	9.236	11.071	15.086	20.515
6	10.645	12.592	16.812	22.458
7	12.017	14.067	18.475	24.322
8	13.362	15.507	20.090	26.125
9	14.684	16.919	21.666	27.877
10	15.987	18.307	23.209	29.588
11	17.275	19.675	24.725	31.264
12	18.549	21.026	26.217	32.909
13	19.812	22.362	27.688	34.528
14	21.064	23.685	29.141	36.123
15	22.307	24.996	30.578	37.697
16	23.542	26.296	32.000	39.252
17	24.769	27.587	33.409	40.790
18	25.989	28.869	34.805	42.312
19	27.204	30.144	36.191	43.820
20	28.412	31.410	37.566	45.315
21	29.615	32.671	38.932	46.797
22	30.813	33.924	40.289	48.268
23	32.007	35.173	41.638	49.728
24	33.196	36.415	42.980	51.179
25	34.382	37.653	44.314	52.620
26	35.563	38.885	45.642	54.052
27	36.741	40.113	46.963	55.476
28	37.916	41.337	48.278	56.892
29	39.088	42.557	49.588	58.302
30	40.256	43.773	50.892	59.703
40	51.805	55.759	63.691	73.402
50	63.167	67.505	76.154	86.661
60	74.397	79.082	88.379	99.607
70	85.527	90.531	100.425	112.317
80	96.578	101.879	112.329	124.839
90	107.565	113.145	124.116	137.208
100	118.498	124.342	135.807	149.449

[1]Adapted and reproduced, with permission, from Table 8 in Pearson ES, Hartley HO (editors): *Biometrika Tables for Statisticians*, 3rd ed. Vol 1. Cambridge University Press, 1966. Used with the kind permission of the Biometrika Trustees.

Table A-6. z transformation[1] of the correlation coefficient.[2]

r	z	r	z
0.00	0.000	0.50	0.549
0.01	0.010	0.51	0.563
0.02	0.020	0.52	0.576
0.03	0.030	0.53	0.590
0.04	0.040	0.54	0.604
0.05	0.050	0.55	0.618
0.06	0.060	0.56	0.633
0.07	0.070	0.57	0.648
0.08	0.080	0.58	0.663
0.09	0.090	0.59	0.678
0.10	0.100	0.60	0.693
0.11	0.110	0.61	0.709
0.12	0.121	0.62	0.725
0.13	0.131	0.63	0.741
0.14	0.141	0.64	0.758
0.15	0.151	0.65	0.775
0.16	0.161	0.66	0.793
0.17	0.172	0.67	0.811
0.18	0.182	0.68	0.829
0.19	0.192	0.69	0.848
0.20	0.203	0.70	0.867
0.21	0.213	0.71	0.887
0.22	0.224	0.72	0.908
0.23	0.234	0.73	0.929
0.24	0.245	0.74	0.951
0.25	0.255	0.75	0.973
0.26	0.266	0.76	0.996
0.27	0.277	0.77	1.020
0.28	0.288	0.78	1.045
0.29	0.299	0.79	1.071
0.30	0.310	0.80	1.099
0.31	0.321	0.81	1.127
0.32	0.332	0.82	1.157
0.33	0.343	0.83	1.188
0.34	0.354	0.84	1.221
0.35	0.365	0.85	1.256
0.36	0.377	0.86	1.293
0.37	0.388	0.87	1.333
0.38	0.400	0.88	1.376
0.39	0.412	0.89	1.422
0.40	0.424	0.90	1.472
0.41	0.436	0.91	1.528
0.42	0.448	0.92	1.589
0.43	0.460	0.93	1.658
0.44	0.472	0.94	1.738
0.45	0.485	0.95	1.832
0.46	0.497	0.96	1.946
0.47	0.510	0.97	2.092
0.48	0.523	0.98	2.298
0.49	0.536	0.99	2.647

[1] $z = 1/2 \{\ln[(1 + r)/(1 - r)]\}$.

[2] Adapted and reproduced, with permission, from Table 14 in Pearson ES, Hartley HO (editors): *Biometrika Tables for Statisticians,* 3rd ed. Vol 1. Cambridge University Press, 1966. Used with the kind permission of the Biometrika Trustees.

Appendix B:
Answers to Exercises

CHAPTER 2

1. This is a classic example of a crossover clinical trial in which patients receive one treatment and then cross over to the second treatment. See Fig B-1 for an illustration of how the authors described the study decision.
2. This article examines cross-sectional information collected at the same point in time; we would classify it as a cross-sectional study for the purpose of diagnosis and staging.
3. The study is observational and evaluates the history of patients with the symptoms and persons without the symptoms to identify possible exposure or risk factors; therefore, it is a classic case-control study.
4. Since a treatment, sugar and providone-iodine, was used, the study qualifies as a clinical trial. The controls were concurrent; ie, the standard therapy and the treatment with sugar and providone-iodine were evaluated during the same period of time. However, there is no indication the patients were randomly assigned to treatment and standard therapy; thus, the study is a nonrandomized clinical trial.
5. The group of subjects was identified and initial data collection occurred in 1976; the same subjects were followed up in future years. Therefore, the study design is best described as a cohort or prospective study to identify risk factors.
6. This study begins with patients with acute nonvariceal upper gastrointestinal tract bleeding and control subjects and examines their histories for information on nonsteroidal anti-inflammatory drug use—a typical case-control study.

CHAPTER 3

1. Both measures are numerical, and the desire is to exhibit the relationship between them; therefore, a scatterplot with neutrophil leukotriene B_4 production (ng/10^7) on the X-axis and number of tender joints on the Y-axis would be appropriate.

2. A reasonable approach in a problem like this one is to illustrate the distribution of cases across time with a histogram for each subset of patients. In this study, the authors drew three histograms: for patients with self-limited illness, patients with subsequent neuromuscular illness, and patients with intermediate illness. Fig B-2 reproduces the authors' figure. Note that time is displayed along the X-axis and the number of patients along the Y-axis.
3. The table presented in Bartle, Gupta, and Lazor (1986) is reproduced here in Table B-1. This table clearly communicates the information.
4. a. A scatterplot is best. See Fig B-3, which reproduces the figure in the article.
   b. A histogram or a frequency polygon is best for displaying percentages. An exam-

Study design for a double-blinded, placebo-controlled, crossover trial. Patients were begun on either fish-oil (EPA, for eicosapentaenoic acid) or placebo treatment for the initial 14 weeks. During weeks 14 through 18, both groups ingested placebo (washout period). At weeks 18 patients crossed over to receive the opposite supplement during weeks 18 through 32. Weeks 32 to 36 represent another washout period. Leukotrienes from stimulated neutrophils were measured in vitro at baseline and at weeks 14, 18, and 32 (arrows). Evaluations and procedures done at baseline and on weeks 7, 14, 18, 25, 32, and 36 included clinical examination, complete blood count, erythrocyte sedimentation rate, and platelet count, a visit with a dietician, pill counts and dispensing of new pills, and gas chromatograph analysis of plasma lipids.

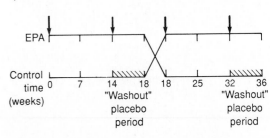

Figure B-1. Illustration of study design. (Adapted and reproduced, with permission, from Kremer JM et al: Fish-oil fatty acid supplementation in active rheumatoid arthritis: A double-blinded, controlled, crossover study. *Ann Intern Med* 1987;**106**:497.)

35	65	76	84	90	116
40	65	76	86	91	118
47	66	77	86	98	120
52	66	78	86	106	123
52	67	79	87	106	123
55	69	80	87	107	127
56	73	80	88	110	128
57	73	80	88	110	137
58	74	82	88	111	142
58	75	84	89	112	146

With 60 numbers, each value accounts for $100/60 = 1.67$ percentiles. The lower 2.5 percentile is found at the value halfway between the lowest and the second-lowest observations $(2.5/1.67 = 1.5)$ and is 37.5. The 97.5 percentile is located halfway between the 59th and 60th observations; it is 144. Therefore, the norms for percentage saturation of bile for this group of 60 healthy subjects are estimated to be 37.5 and 144. Of course, using a larger number of observations to establish norms for clinical use is desirable, if possible.

9. **a.** Approximately 26.25 lb (or 11.9 kg).
   b. Approximately 30.67 in (or 78 cm).
   c. Approximately 17 lb (or 7.8 kg.).
   d. This child is at the 10th percentile in height for her age and between the 90th and the 95th percentiles in weight; therefore, the child may be somewhat overweight for her length.
10. $0.29/0.12 = 2.42$, or 242%, indicating a great deal of variability in change in vessel diameter relative to the mean change.

CHAPTER 5

1. **a.** To show that gender and blood type are independent, we must show that $P(A)P(B) = P(A$ and $B)$ for each cell in the table:
   $$.42 \times .50 = .21$$
   $$.43 \times .50 = .215$$
   $$.11 \times .50 = .055$$
   $$.04 \times .50 = .02$$
   b. $P(\text{male and type O}) = P(\text{male}|\text{type O}) \times P(\text{type O}) = (.21|.42) \times (.42|1.00) = (.50) \times (.42) = .21$. This demonstrates that $P(\text{male}|\text{type O}) = P(\text{male})$ when these are independent.
2. **a.** $P(\text{chronic}) = (7 + 8 + 2)/47 = .36$.
   b. $P(\text{acute}) = (6 + 2 + 2)/47 = .21$.
   c. $P(\text{acute}|\text{seroconvert}) = 2/18 = .11$.
   d. $P(\text{seropositive}|\text{died}) = 2/8 = .25$.
   e. $P(\text{seronegative}) = 17/47 = .36$. Using the binomial distribution, we get

$$P(4 \text{ out of } 8) = \frac{8!}{4! \ 4!} (.36)^4(.64)^4$$

$$= \frac{40,320}{(24)(24)} (.0168)(.1678) = .1973, \text{ or } .20$$

3. Use the binomial distribution.
   a. $P(\text{infection}) = .30$

$$P(1 \text{ infection in 8 patients}) = \frac{8!}{1! \ 7!} (.30)^1(.70)^7$$

$$= 8(.30)(.0824)$$
$$= .1977, \text{ or } .20$$

   b. $P(\text{survival}) = .80$

$$P(7 \text{ survivals in 8 patients}) = \frac{8!}{7! \ 1!} (.80)^7(.20)^1$$

$$= 8(.2097)(.20)$$
$$= .3355, \text{ or } .34$$

4. $\lambda = 950/390 = 2.44$. A graph of the distribution is shown in Fig B-8, and we see that the distribution remains positively skewed, as in Fig 5-3. The probability of exactly 2 hospitalizations is $P(X = 2) = (2.44^2)(e^{-2.44})/2! = (5.954) (0.087)/2 = .259$

5. **a.** The probability that a normal healthy adult has a serum sodium above 147 meq/L is $P(z > 2) = .023$.
   b. $P[z < (130 - 141)/3] = P(z < -3.67) < .001$.
   c. $P[(132 - 141)/3 < z < (150 - 141)/3] = P(-3 < z < +3) = .997$.
   d. The top 1% of the standard normal distribution is found at $z = 2.326$. Therefore, $2.326 = (X - 141)/3$, or $X = 147.98$; so a serum sodium level of approximately 148 meq/L puts a patient in the upper 1% of the distribution.
   e. The bottom 10% of the standard normal distribution is found at $z = -1.28$. Therefore, $-1.28 = (X - 141)/3$, or $X = 137.16$; so a serum sodium level of approximately 137 meq/L puts a patient in the lower 10% of the distribution.

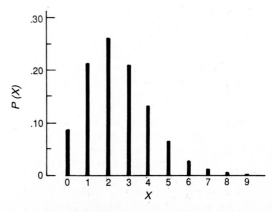

Figure B-8. Poisson distribution for $\lambda = 2.44$.

6. The distributions follow.

Probability	$\pi = 0.1$	$\pi = 0.3$	$\pi = 0.5$
$P(X = 0)$	.531	.118	.016
$P(X = 1)$	.354	.303	.094
$P(X = 2)$	.098	.324	.234
$P(X = 3)$	.015	.185	.313
$P(X = 4)$	.001	.060	.234
$P(X = 5)$	<.001	.010	.094
$P(X = 6)$	<.001	.001	.016

Thus, $P(X = 4$ when $\pi = 0.3) = (6!/4!2!)$ $(.3)^4(.7)^2 = [720/(24)(2)](.008)(.49) = .060$. Graphs of the above distributions illustrate that the binomial distribution is quite skewed when the proportion is 0.1 (as well as when the proportion is close to 1.0, such as 0.9 and 0.8). When the proportion is near 0.5, the distribution is nearly symmetric—perfectly so at 0.5.

CHAPTER 6

1. a. Mean number of months $= (12 + 13 + 14 + 15 + 16)/5 = 70/5 = 14.0$; the standard deviation is $\sqrt{[(12 - 14)^2 + (13 - 14)^2 + (14 - 14)^2 + (15 - 14)^2 + (16 - 14)^2]/5}$ $= 1.41$. (Note that the standard deviation uses 5 in the denominator instead of 4 because we are assuming that these 5 observations make up the entire population.)

   b. The mean of the mean number of months is $(12 + 12.5 + \cdots + 15.5 + 16)/25 = 350/25 = 14.0$; the standard deviation of the mean (or the standard error of the mean, SEM) is $\sqrt{[(12 - 14)^2 + (12.5 - 14)^2 + \cdots + (15.5 - 14)^2 + (16 - 14)^2]/25}$ $= 1.0$. Note that the mean of the means found in part b is the same as the mean found in part a, and the SEM is the same as $\sigma/\sqrt{n} = 1.41/\sqrt{2} = 1.0$

2. a. This question refers to individuals and is equivalent to asking what proportion of the area under the curve is greater than $(103 - 100)/3 = +1.00$ and less than $(97 - 100)/3 = -1.00$, using the z distribution; the area is 0.317, or 31.7% from Table A–1.

   b. This question concerns means. The standard error with $n = 36$ is $3/6 = 0.5$. The critical ratio for a mean equal to 99 is $(99 - 100)/0.5 = -2.00$ and for 101 is $+2.00$. The area below -2.00 and above $+2.00$ is 0.046; therefore, 4.6% of the *means* will be outside the limits of 99 and 101.

   c. 95% confidence limits are $101 \pm (1.96)(3/\sqrt{36}) = 101 \pm 0.98$, or 100.02 and 101.98.

   d. 95% confidence limits for $n = 144$ are $101 \pm (1.96)(3/\sqrt{144}) = 101 \pm 0.49$, or 100.51 and 101.49. The sample size was quadrupled from 36 to 144, which has the effect of reducing the standard error by half (because of the square root) and thereby decreasing the width of the confidence interval by half.

3. a. 90% confidence limits are $13.5 \pm (1.645)(1.0/\sqrt{49}) = 13.5 \pm 0.235$, or 13.265 and 13.735. 95% confidence limits are $13.5 \pm (1.96)(1.0/\sqrt{49}) = 13.5 \pm 0.28$, or 13.22 and 13.78. The 95% confidence interval is wider because of the increased confidence reflected by a larger value for z in the formula.

   b. Because the population mean, 14.0 g/dL, is outside the 95% confidence limits (ie, 14.0 > 13.78), we know that there is a difference at $P < .05$. To obtain a more precise P-value, we must perform a test of the hypothesis.

4. a. 99% confidence limits are $33.8 \pm (2.575)(23.6/\sqrt{43}) = 33.8 \pm 9.27$, or 24.53 and 43.07 pg/mL. The interpretation is that we have 99% confidence that the interval 24.53 to 43.07 pg/mL contains the true mean HCT in men with rheumatoid arthritis.

   b. Step 1. $H_0: \mu \geq 47.9$
   $H_1: \mu < 47.9$

   Step 2. The population standard deviation is known; therefore, use the z distribution.

   Step 3. Use $\alpha = .01$ as suggested.

   Step 4. The critical value from Table A–2 is -2.326, so we will reject the null hypothesis if the observed value for z is less than -2.326.

   Step 5. $z = (33.8 - 47.9)/(23.6/\sqrt{43}) = -14.10/3.6 = -3.92$.

   Step 6. Since -3.92 is less than -2.326, we conclude that the true mean HCT in men with rheumatoid arthritis is less than 47.9 pg/mL; ie, we reject the null hypothesis.

5. a. The bars indicate (mean ± standard error of the mean).

   b. We cannot tell from Fig 3–24. Each SEM must be multiplied by the square root of the number of patients in order to obtain the standard deviations. From information in the text of the article, the standard deviations are as follows:

Normal	$2.35 \times \sqrt{10} = 7.43$
CF	$2.43 \times \sqrt{15} = 9.41$

COPD	$2.63 \times \sqrt{6} = 6.44$
Immune deficiency	$2.12 \times \sqrt{7} = 5.61$
SLE	$6.32 \times \sqrt{5} = 14.13$
Splenectomy	$13.4 \times \sqrt{4} = 26.80$

These numbers indicate that patients with immune deficiency have the smallest standard deviation, followed by patients with COPD, normal patients, patients with CF, patients with SLE, and patients with splenectomy. Note that Fig 3-24 indicates that patients immune deficiency and those with CF have similar SEMs; however, the standard deviations are much greater in patients with CF. Using the SEM is frequently misleading to readers unless sample sizes are equal.

CHAPTER 7

1. The mean Pimax is $(54.8 + 62.0 + \cdots + 26.6)/9 = 371.4/9 = 41.27$ cm H_2O. The standard deviation is $\sqrt{[(54.8 - 41.27)^2 + \ldots + (26.6 - 41.27)^2]/8} = 16.23$.

2. A t statistic for testing whether mean Pimax is less than 60 is $t = (41.27 - 60)/(16.23/\sqrt{9}) = -3.46$. From Table A-3, the critical value for 8 degrees of freedom corresponding to .005 in one tail is ± 3.355 and corresponding to .0005 is ± 5.041. Since -3.46 is between these two values (when they are negative), we report $.0005 < P < .005$.

3. **a.** The 95% confidence interval for Pimax, using the z distribution, is $41.27 \pm (1.96)(16.23/\sqrt{9}) = 41 \pm 10.60$, or 30.67 to 51.87 cm H_2O. The 95% confidence interval using the t distribution (and 2.306 in place of 1.96) is 28.79 to 53.75 cm H_2O. Thus, inappropriately using the z distribution instead of the t distribution with a small sample produces confidence intervals that are too narrow or too conservative.

   b. The confidence intervals are narrower for 36 patients compared with 9 patients because the standard error of the mean (SEM) reduces from 5.41 to 2.71; in fact, quadrupling the sample size reduces the width of confidence interval by one-half.

   c. Kyphoscoliosis is not a common problem, and it is possible that only nine patients were available to the investigators. With uncommon problems, investigators sometimes opt for collaborating with colleagues at other institutions to increase the sample size available for study.

4. The first step is deciding whether or not the assumption of normality is justified—ie, whether arterial carbon dioxide tension is normally distributed in patients with chronic hypercapneic respiratory failure. If this assumption is reasonable, the t distribution may be used. The observations are made on the same patients before and after a diet; therefore, the paired t test is appropriate. The null hypothesis is that there is no change; ie, the mean change is zero. Calculations for the numbers in Table 7-2 are $t = 6.3/(6.5/\sqrt{8}) = 2.74$. The critical value of t from Table A-3 at .05, with 7 degrees of freedom, is 2.365 for a two-tailed test. Since 2.74 is greater than 2.365, we reject the null hypothesis of no difference and conclude that there is a difference in arterial CO_2 tension following diet.

5. We assume that heart rate is normally distributed in hypertensive patients, and we note that the standard deviations are of similar magnitude. Therefore, the two-group t test can be performed. The pooled standard deviation (from Section 7.4.2) is 12.21. The calculation for the 90% confidence interval for the difference in heart rates, using Table A-3 is $(90.7 - 77.8) \pm (1.725)(12.21)\sqrt{(1/13 + (1/9)} = 12.90 \pm 9.13$, or 3.77 to 22.03 beats per minute. Therefore, we conclude with 90% confidence that the interval from 3.77 to 22.03 beats per minute contains the true mean difference in heart rates among hyperadrenergic patients. This conclusion is consistent with the conclusion from the one-tailed t test in Section 7.4.2 in which the null hypothesis was rejected.

6. The "eyeball" test is illustrated in Fig B-9 and indicates that there is a statistically significant difference in serum catecholamine levels between hyperadrenergic and normoadrenergic patients. The conclusion from the graphs is that the groups differ significantly at $P < .05$, because the 95% confidence intervals for the means of the two respective groups do not overlap.

7. 95% confidence interval, using Table A-3 with $n - 1 = 7$ degrees of freedom, is $13.5 \pm (2.365)(8.2/\sqrt{8}) = 13.5 \pm 6.86$, or 6.64 to 20.36 mm Hg. 99% confidence interval is $13.5 \pm (3.499)(8.2/\sqrt{8}) = 13.5 \pm 10.14$, or 3.36 to 23.64 mm Hg. Comparing these intervals with the 90% confidence interval, we observe that the intervals are wider as the confidence level increases; this observation is consistent with the guideline that for us to have greater confidence that an interval contains the true mean, the intervals must encompass a larger range of possible values.

8. For $\alpha = .05$ and power of .90, $z_\alpha = 1.96$ and $z_\beta = -1.28$. Therefore, $n = \{[1.96 - (-1.28)](16.23)/10\}^2 = 27.65$, or 28, individuals.

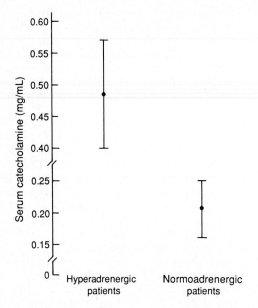

Figure B-9. 95% confidence limits for mean serum catecholamine in hyperadrenergic and normoadrenergic patients. (Adapted, with permission, from de Champlain J et al: Circulating catecholamine levels in human and experimental hypertension. *Circ Res* 1976;**38**:109.)

9. First, multiply each equation by $\sqrt{n}/\sqrt{n}$ to remove $1/\sqrt{n}$ from the denominator:

$$z_\alpha = \frac{(\overline{X} - \mu_0)\sqrt{n}}{\sigma}$$

and

$$z_\beta = \frac{(\overline{X} - \mu_1)\sqrt{n}}{\sigma}$$

Second, multiply each equation by σ to remove σ from the denominator:

$$z_\alpha\sigma = (\overline{X} - \mu_0)\sqrt{n}$$

$$z_\beta\sigma = (\overline{X} - \mu_1)\sqrt{n}$$

Third, subtract each side of the second equation from the respective sides of the first equation:

$$z_\alpha\sigma - z_\beta\sigma = (\overline{X} - \mu_0)\sqrt{n} - (\overline{X} - \mu_1)\sqrt{n}$$

Simplifying yields

$$(z_\alpha - z_\beta)\sigma = (\overline{X} - \mu_0 - \overline{X} + \mu_1)\sqrt{n}$$

$$= (\mu_1 - \mu_0)\sqrt{n}$$

Finally, solving for n gives

$$\frac{(z_\alpha - z_\beta)\sigma}{(\mu_0 - \mu_1)} = \sqrt{n}$$

and

$$n = \left[\frac{(z_\alpha - z_\beta)\sigma}{\mu_1 - \mu_0} \right]^2$$

10. The rule of thumb for α equal to .05 and power equal to .90 has $z_\alpha = 1.96$ and $z_\beta = -1.28$ for all computations; ie, only σ and $(\mu_1 - \mu_2)$ change.
Therefore,

$$n = 2\left[\frac{(z_\alpha - z_\beta)\sigma}{\mu_1 - \mu_2} \right]^2 = 2\left\{ \frac{[1.96 - (-1.28)]\sigma}{\mu_1 - \mu_2} \right\}^2$$

$$= 2\left[\frac{(3.24)\sigma}{\mu_1 - \mu_2} \right]^2 = 2\,(10.50)\left(\frac{\sigma}{\mu_1 - \mu_2} \right)^2$$

CHAPTER 8

1. The ANOVA table follows:

Source of Variation	Sum of Squares	df	Mean Square	F	P
Among groups	584.93	2	292.47	6.83	.01
Within groups	1456.24	34	42.83		
Total	2041.18				

The observed value for the F ratio is 6.83, with 2 and 34 degrees of freedom. The critical value from Table A–4 with 2 and 30 degrees of freedom (the closest value) is 5.39 at $P = .01$; therefore, there is sufficient evidence to conclude that the mean scores on the Hamilton Depression Rating Scale are not the same for all three groups of women.

2. The critical value for the Tukey HSD test is HSD = multiplier from Table 8–7 $\times \sqrt{MS_E/n}$ = 4.42 $\times \sqrt{42.83/12}$ = 8.35, assuming 12 patients in each group. The differences between the means are (1) moderate − low = 11.98 − 5.34 = 6.64; (2) high − low = 14.71 − 5.34 = 9.37; and (3) high − moderate = 14.71 − 11.98 = 2.84. Only the high and low groups are significantly different since their difference is greater than 8.35.

The critical value for the Scheffe procedure, using the value for the contrast found in Section 8.4.2.b, is $S = \sqrt{(k - 1)} \times$ multiplier $\times \sqrt{MS_E} \times$ contrast $= \sqrt{2} \times 5.31 \sqrt{42.83} \times 0.167$ = 3.259 × 2.674 = 8.72. Again, only the difference between the high and low groups surpasses 8.72, so they are the only two groups that differ significantly on the Hamilton Depression Scale, according to Scheffe's procedure. The conclusions in this example are the same using the Tukey and the Scheffe procedures.

3. Two comparisons are independent if they use nonoverlapping information.
   a. Independent because each comparison uses different data.
   b. Dependent because data on physicians is used in each comparison.
   c. Independent because each comparison uses different data.
   d. None of these three comparisons are independent from the other two because they all use data on medical students.
4. The grand mean is $[(34)(3.9) + (32)(4.7) + (14)(16.4)]/80 = 6.4$. Therefore, the estimate of the mean square among groups is $34(3.9 - 6.4)^2 + 32(4.7 - 6.4)^2 + 14(16.4 - 6.4)^2 = 1705$, divided by the number of groups minus 1, or 852.5. The estimate of the mean square error from the pooled variances is $[33(4.7)^2 + 31(3.9)^2 + 13(10.1)^2]/(33 + 31 + 13) = 32.8$. Thus, the F ratio is $852.5/32.8 = 25.99$, significant at $P < .01$ when compared with the critical value of 4.98 (with 2 and 60 degrees of freedom in Table A–4). The conclusion is that there is a significant difference in the mean total number of minutes of disordered breathing among controls, hypertensive patients with apnea, and hypertensive patients without apnea. However, the assumption of equal (homogeneous) variances is violated, a problem when n's are not equal. A possible way to deal with this problem is to use a logarithmic transformation of total number of minutes.
5. a. This is a one-way or one-factor ANOVA because there is only one non-error term, the among-groups term.
   b. Total variation is 2000.
   c. There were 4 groups of patients because the degrees of freedom are 3. There were 40 patients because $N - 4 = 36$.
   d. The F ratio is $(800/3)/33.3 = 8.01$.
   e. The critical value with 3 and 30 degrees of freedom is 4.51; with 3 and 40 degrees of freedom is 4.31; interpolating for 3 and 36 degrees of freedom gives 4.39.
   f. Reject the null hypothesis of no difference and conclude that mean blood pressure differs in groups that consume different amounts of alcohol. A post hoc comparison would be necessary to determine specifically which of the four groups differ.

CHAPTER 9

1. 90% confidence limits are consistent with a two-tailed test using $\alpha = .10$ and a one-tailed test using $\alpha = .05$; the 90% interval is $-0.062 \pm (1.645)(0.034) = -0.062 \pm 0.056$, or -0.118 to -0.006. This interval does not include zero and therefore is consistent with the results of the hypothesis test in Section 9.3.2.

2. a. The 2×2 table, with expected values in parentheses, follows.

	J5	Placebo	Total
Infection	16 (18.76)	23 (20.24)	39
No infection	110 (107.24)	113 (115.76)	223
Total	126	136	262

$X^2 = (16 - 18.76)^2/18.76 + (23 - 20.24)^2/20.24 + (110 - 107.24)^2/107.24 + (113 - 115.76)^2/115.76 = 0.40 + 0.38 + 0.07 + 0.07 = 0.92$. Table A–5 indicates that 2.706 is the critical value at $\alpha = .10$ with 1 degree of freedom; thus, the data in Table 9–2 do not indicate that there is a difference in infection rates for patients receiving J5 vaccine and patients receiving placebo.

   b. The relative risk of infection for patients receiving placebo versus those receiving the J5 vaccine is $(23/136)/(16/126) = 1.33$. The conclusion is that those receiving placebo appear to be at a slightly higher risk of infection; however, we do not know whether the relative risk of 1.33 is statistically greater than 1, ie, whether 1.33 could occur by chance. The method for forming confidence intervals for the risk ratio is discussed in the next chapter.

3. $X^2(2) = 14.53$, significant at $P < .01$; therefore, the study with 35 patients in three groups has low power to detect a difference if there is one.

4. The z test, confidence limits for the difference in two proportions, or the chi-square test may be used for each question. We illustrate the use of each.
   a. The cesarean rate is $1777/17,409 = 0.1021$ with selective monitoring and $1933/17,586 = 0.1099$ with universal monitoring. For a z test, the pooled proportion is $p = (1777 + 1933)/(17,409 + 17,586) = 0.1060$. Then, $z = (0.1021 - 0.1099)/\sqrt{0.106(1 - 0.106)(1/17,409 + 1/17,586)} = -0.0078/0.0033 = -2.36$, significant $(P < .02)$ from Table A–2. Therefore there is a statistically significant difference in cesarean rates with selective versus universal electronic fetal monitoring.
   b. The proportions are $196/7330 = 0.0267$ and $551/7288 = 0.0756$, respectively. The pooled proportion is 0.0511. A 99% confidence interval for the difference in proportions is $(0.0267 - 0.0756) \pm (2.575)\sqrt{0.0511(1 - 0.0511)(1/7330 + 1/7288)} = -0.0489 \pm 0.0094$, or -0.058 and -0.040. Since zero is not contained in this

interval, we have 99% confidence that among low-risk pregnancies there is a difference between proportions who had abnormal fetal heart rates with selective versus universal monitoring, and the difference is between 4.0% and 5.8%.

c. A chi-square test is illustrated here; expected values are in parentheses in the table.

	Selective	Universal	Total
Deaths	5 (4.51)	4 (4.49)	9
Live births	7325 (7325.49)	7284 (7283.51)	14,609
Total	7330	7288	14,618

$X^2 = (5 - 4.51)^2/4.51 + (4 - 4.49)^2/4.49 + (7325 - 7325.49)^2/7325.49 + (7284 - 7283.51)^2/7283.51 = 0.107$, which is not significant. Therefore, the difference in death rates is not statistically significant.

d. Although universal monitoring has no effect on the neonatal death rate among low-risk pregnancies, it is associated with an increased cesarean rate and with a higher proportion of infants with abnormal heart rates. Therefore, the authors' conclusions seem to be justified.

5. a. The numbers are too small for use of the z approximation; $n(1 - p) = 2$ for responders. The chi-square test can be used, but Fisher's exact test is the optimal method.

b. The same reasoning as in part A applies, except the problem occurs with the nonresponders.

c. If the number of months follows a normal distribution, the independent two-group t test may be used; if not, the Wilcoxon rank-sum test is appropriate with sample sizes under 30.

d. Same reasoning as in part c.

6. The 3×2 table, with expected values in parentheses, follows.

	Recurred	Did Not Recur	Total
Declined	49 (36.6)	12 (24.4)	61
Encouraged to abstain	24 (19.8)	9 (13.2)	33
Surgery	2 (18.6)	29 (12.4)	31
Total	75	50	125

$X^2 = 49.77$. With 2 degrees of freedom, the critical value is 13.816 at $\alpha = .001$. Therefore, the proportion of patients who had recurrent pancreatitis significantly differs depending on the treatments the patients received.

7. Using the shortcut formula for X^2 with 1 degree of freedom gives $X^2 = 133[(43)(37) - (23)(30)]^2/(73)(60)(66)(67) = 5.57$. This study shows that the proportion of patients taking cimetidine who have healed ulcers is greater than the proportion taking placebo ($P < .05$); therefore, there is some evidence that prescribing cimetidine for benign gastric ulcer is appropriate.

8. $n = \{[1.96 \sqrt{2 \times 0.10 \times 0.90} - (-0.525) \sqrt{(0.03 \times 0.97)} + (0.10 \times 0.90)]/(0.03 - 0.10)^2\} = (1.01/0.07)^2 = 14.47^2 = 209.31$, or 210, patients in each group.

CHAPTER 10

1. a. The mean C1q is $(35 + \cdots + 15)/29 = 36.83$, and the variance is 817.11 (the standard deviation is 28.59).

b. The standard deviation of IgG is $\sqrt{21,374/28} = 27.63$ and of C1q is $\sqrt{22,879/28} = 28.59$; so $b = r (s_y/s_x) = (.91)(28.59)/27.63 = 0.94$; $a = \overline{Y} - b\overline{X} = 36.83 - (0.94)(41.34) = -2.03$. Therefore, the regression equation is predicted C1q $= 0.94 \times$ IgG -2.03. (*Note:* If you use a computer to calculate the regression equation for the original observations, you may obtain C1q $= 0.943 - 2.179$; the differences are due to round-off error in using the above formulas and only two decimal places.)

c and d. The predicted and residual values, using $0.94 \times$ IgG $- 2.03$, are given in the table that follows.

Observation	C1q	Predicted C1q	Residual
1	35	29.93	5.07
2	16	12.07	3.93
3	98	87.27	10.73
4	99	84.45	14.55
5	1	1.73	−0.73
6	31	52.49	−21.49
7	37	31.81	5.19
8	38	62.83	−24.83
9	72	71.29	0.71
10	11	6.43	4.57
11	25	26.17	−1.17
12	75	55.31	19.69
13	2	−1.09	3.09
14	17	5.49	11.51
15	2	8.31	−6.31

Obser-vation	C1q	Predicted C1q	Residual
16	20	30.87	−10.87
17	72	82.57	−10.57
18	5	2.67	2.33
19	39	26.17	12.83
20	20	26.17	−6.17
21	63	43.09	19.91
22	59	54.37	4.63
23	27	31.81	−4.81
24	5	13.01	−8.01
25	28	27.11	0.89
26	53	53.43	−0.43
27	72	60.01	11.99
28	31	44.97	−13.97
29	15	37.45	−22.45

e. The mean predicted C1q is $(29.93 + \cdots + 37.45)/29 = 36.83$, the same as the mean C1q. The predicted mean $\overline{Y}'$ is always equal to the actual mean $\overline{Y}$.

f. The mean residual is $[5.07 + \cdots + (-24.83)]/29 = -0.01$; the value should be zero, and -0.01 represents round-off error. The variance of the residuals is 137.60.

g. The coefficient of determination is $r^2 = (817.11 - 137.60)/817.11 = .83$, which is equal to $(.91)^2 = .83$.

2. a. $X^2 = (8 - 4.07)^2/4.07 + (11 - 14.93)^2/14.93 + (10 - 13.93)^2/13.93 + (55 - 51.07)^2/51.07 = 6.24$. With 1 degree of freedom, this value is large enough to convince us that there is a significant relationship between years of work and the presence or absence of hepatitis B marker in workers in anesthesia.

b. No, the expected value for the first row and fourth column is $(5 \times 18)/84 = 1.07$, which is less than the recommended lower limit of 2.

3.

$$r \frac{s_Y}{s_X} = \frac{\Sigma(X - \overline{X})(Y - \overline{Y})}{\sqrt{\Sigma(X - \overline{X})^2}\sqrt{\Sigma(Y - \overline{Y})^2}} \times \frac{s_Y}{s_X}$$

$$= \frac{\Sigma(X - \overline{X})(Y - \overline{Y})}{\sqrt{\Sigma(\overline{X} - X)^2}\sqrt{\Sigma(Y - \overline{Y})^2}} \times \frac{\sqrt{\Sigma(Y - \overline{Y})^2/(n - 1)}}{\sqrt{\Sigma(X - \overline{X})^2/(n - 1)}}$$

$$= \frac{\Sigma(X - \overline{X})(Y - \overline{Y})}{\sqrt{\Sigma(X - \overline{X})^2}\sqrt{\Sigma(Y - \overline{Y})^2}} \times \frac{\sqrt{\Sigma(Y - \overline{Y})^2}}{\sqrt{\Sigma(X - \overline{X})^2}}$$

$$= \frac{\Sigma(X - \overline{X})(Y - \overline{Y})}{\sqrt{\Sigma(X - \overline{X})^2}} \times \frac{1}{\sqrt{\Sigma(X - \overline{X})^2}}$$

$$= \frac{\Sigma(X - \overline{X})(Y - \overline{Y})}{\Sigma(X - \overline{X})^2} = b$$

4. $r = b(s_X/s_Y) = (0.406)(3.448)/(2.275) = .615$, or .62.

5. A positive correlation indicates that the values of X and Y vary together; ie, large values of X are associated with large values of Y, and small values of X are associated with small values of Y. A positive slope of the regression line indicates that each time X increases by 1, Y increases by the amount of the slope, thereby pairing small values of X with small values of Y; a similar statement holds for large values.

6. Analysis of variance followed by post hoc comparisons is appropriate if the observations are normally distributed; t tests for independent groups are also acceptable if prior planned comparisons are made with the α value modified as in the Bonferroni procedure. If the observations are not normally distributed, the appropriate procedure is to transform them or to use nonparametric procedures for ranks, such as the Kruskal-Wallis analysis of variance procedure with Wilcoxon rank-sum tests as follow-up.

7. Probably to make comparisons with renal patients, using the independent-groups t test, although no comparisons were reported in the journal article.

8. The authors want to demonstrate that the slope of the regression line predicting the protein/creatinine ratio from 24-hour protein excretion varied by time of day. The graph demonstrates suppression of zero on the Y-axis.

9. The regression line goes through the point $(\overline{X}, \overline{Y})$, therefore, the point at which $\overline{X}$ intersects the regression line, when projected onto the Y-axis, is $\overline{Y}' = \overline{Y}$.

10. a. Both the ratio of OKT4 to OKT8 cells and the lifetime concentrate use have skewed distributions, and the logarithmic transformation makes them more closely resemble the normal distribution.

b. $r = -.453$ indicates a moderate-sized inverse (or negative) relationship; r^2 indicates that approximately 21% of the variation in one measure is accounted for by knowing the other.

c. The 95% confidence bands are specified as being related to single observations; therefore, 95% of predicted log(OKT4/OKT8) will fall within these lines. Note also the appropriate curve of the bands.

11. a. Case-control.

b. The odds ratio is $(20 \times 1157)/(41 \times 121) = 4.66$; 95% confidence limits are the antilogarithms of $\ln(4.66) \pm 1.96\sqrt{(1/20) + (1/41) + (1/121) + (1/1157)}$, or the antilogs of 0.97 and 2.11, which are 2.65 and 8.21, respectively.

c. The age-adjusted odds ratio is an estimate of the value of the odds ratio if the cases and controls had identical age distributions. Since the age adjustment increases the odds ratio from 4.66 to 8.1, the cases represented a younger group of women than the controls, and the age adjustment compensates for this difference. Thus, controlling for age, we have 95% confidence that the interval from 3.7 to 18 contains the true increase in risk of deep vein thrombosis (pulmonary embolism) with the use of oral contraceptives.

12. 99% confidence limits are the antilogarithms of $\ln(4.00) \pm 2.57 \sqrt{(1/8) + (1/11) + (1/10) + (1/55)}$ or the antilogs of 1.39 ± 1.49, which are 0.90 and 17.81. This confidence interval contains 1, indicating that there is not a significant increased risk when 99% confidence limits are used.

13. **a.** Bile acid synthesis, because the absolute value of the correlation is the highest.
   b. Bile acid synthesis, because it has the highest correlation.
   c. The relationships between age and bile acid synthesis appears to be relatively similar for men and women; however, the relationship between cholesterol secretion and age appears to be more positive in women than in men, and the relationship between age and pool-size cholic acid appears to be more negative in women than men (because the slopes are steeper in women).

CHAPTER 11

1. See Table 11-7A for the arrangements of the observations according to the length of time the patients were in the study. The calculations are summarized in Table 11-7B, and the graphs of the survival curves are presented in Fig 11-5.
2. Table 11-8 A contains the calculations for patients transplanted in 1978; calculations for patients from 1984 are in Table 11-8B. Survival curves are in Fig 11-6.
3. Patients may have had an abnormal karyotype for many months or years prior to the karyotype analysis; therefore, the time of karyotype analysis is an artificial starting point from which to analyze survival.
4. **a.** There do not appear to be any differences in survival among those receiving PO, IM, or no androgen. (In fact, this conclusion is correct, from statistical tests summarized by the authors.) Furthermore, the greatest mortality rates occur early in the study.
   b. It appears that the transplanted group has higher survival rates than the nontransplanted group; however, no statistical results are given in the figure.
   c. The median survival in transplanted patients cannot be determined because more than half are still alive. In nontransplanted

patients, median survival appears to be approximately 4 months; ie, 50% of these patients survive 4 months or less.
   d. (1) Recall from Chapter 7 that if the 95% confidence limits do not overlap, the two groups differ significantly ($P < .05$) at 6 months. (2) A more likely explanation is that the number of patients in the study beyond 6 months is not large enough to result in statistical significance—ie, there is a problem of low power.
   e. Although the authors do not state so, the small dots represent survival times of the 33 transplanted and 22 nontransplanted patients who survived at least 6 months.
5. **a.** There appears to be no difference in the lengths of time until disease progression in the vitamin C and placebo groups. In fact, the curve for vitamin C is lower than the curve for placebo at all points, indicating shorter times prior to disease progression in the vitamin C group.
   b. Median time to disease progression was approximately 3 months in vitamin C group and 4.5 months in the placebo group.
   c. As you probably suspect, the authors found no significant differences in survival between patients receiving vitamin C and those receiving placebo.
   The 90% confidence limits for the odds ratio equal to 2.71 are the antilogarithms of $\ln (2.71) \pm (1.645) \sqrt{(1/6 + (1/12) + (1/3) + (1/18)}$, or the antilogs of -0.318 and 2.312, which are 0.73 and 10.09. Since this interval contains 1, we conclude that the odds ratio is not significantly greater than 1. This conclusion is consistent with the results found earlier: P-values of .089 for the Gehan test and greater than .10 for the logrank test. (*Note:* If the odds ratio is estimated from a 2 × 2 table instead of from the expected and observed numbers of rejections found in Table 11-11, the value is 3.00 instead of 2.71.)

CHAPTER 12

1. For smokers, the adjusted mean is the mean in smokers, 3.33, minus the product of the regression coefficient, 0.0113, and the difference between the occlusion score in smokers and the occlusion score in the entire sample (estimated from Fig 12-1); ie, $3.33 - (0.0113)(198.33 - 151.67) = 2.80$. Similarly, the adjusted mean in nonsmokers is $1.00 - (0.0113)(105.00 - 151.67) = 1.53$. These are the same values we found (within round-off error) in Section 12.3.
2. Family income in the general population has a distribution with a positive skew; ie, many

families have low and moderate incomes, and a few families have very high incomes. Transforming the family income values to a logarithm scale makes the distribution more normal; statistical methods requiring normality can then be used.

3. The probability is $1/\{1 + \exp -[(0.68) + (0.16 \times 6) - (0.32 \times 5)]\}$, or $1/[1 + \exp(-0.04)]$, or 0.51, indicating that 5-year-old boys are more likely to have wet the bed than 5-year-old girls, after controlling for degree of psychological stress.

4. **a.** With 400 variables, a substantial number can be expected to be significant merely by chance. The requirement for R^2 to increase 5% assures us that variables in the equation have clinical as well as statistical significance.

   b. The nursery period, because it has the highest R^2; however, none of the values for R^2 are very large, indicating that cerebral palsy is difficult to predict, at least from the variables considered in this study.

   c. Neonatal seizures, because it has a predicted risk of 9.6%.

   d. A good question, and the answer is not obvious. It could be that these variables were included because previous reports had indicated them to be of interest. It is also possible that the univariate a priori tests were one-tailed with $\alpha = .05$, but the 95% confidence limits in Table 12–8 are equivalent to a one-tailed test for $\alpha = .025$ instead of .05.

5. If the investigators want to distinguish between three groups of runners, using the numerical anthropometric measures, discriminant analysis should be used. However, multiple regression could be used if the actual running time of each runner were used instead of dividing the runners into three groups; in this situation, the outcome measure is numerical.

6. The regression equation is $0.613 - (0.0002)(80) - (0.00006)(80) - (0.002)(75) - (0.0001)(70) - (0.021)(10) + (0.002)(1) - (0.003)(1) + (0.105)(14)$, which gives 1.694 predicted bed-days during a 30-day period.

7. R^2 with the blood glucose test results is $.58^2$, or .336; without the blood glucose testing, it is $.39^2$, or .152; therefore, an additional $.336 - .152 = .184$, or approximately 18%, of the variation in physicians' estimates is accounted for with this information. This finding implies (perhaps not surprisingly) that physicians depend more on blood glucose test information than the other variables in estimating a patient's blood glucose.

8. Log-linear analysis is best, since all the variables except age are nominal or ordinal, and the author used five categories of age to change it into an ordinal variable as well. Using log-linear methods, the author found that the influence of bone loss on fracture type was independent of age.

CHAPTER 13

1. **a.** With 95% sensitivity, $0.95 \times 20 = 19$ true-positives; with 50% sensitivity, there is a 50% false-positive rate and $0.50 \times 80 = 40$ false-positives. The probability of lupus with a positive test is $TP/(TP + FP) = 19/59 = 32.2\%$

   b. With 19 of 20 true-positives, there is 1 false-negative. Similarly, with 50% specificity, there are 40 true-negatives. Therefore, the chances of lupus with a negative test is $FN(FN + TN) = 1/41 = 2.4\%$.

2. Using Bayes' theorem with L = lupus and T = test, we have

$$P(D^-|T^-)\ P(T^-|D^-)P(D^-)/$$
$$= [P(T^-|D^-)P(D^-) + P(T^-|D^+)P(D^+)]$$
$$= (.50 \times .98)/[(.50 \times .98) + (.05 \times .02)]$$
$$= .49/(.49 + .001) = .998.$$

Using the likelihood ratio method requires us to "redefine" the pretest odds as the odds of *no* disease, ie, $.98/(1 - .98) = 49$. Similarly, the likelihood ratio for a negative test is the specificity divided by the false-negative rate (ie, the likelihood of a negative test for persons without the disease versus persons with the disease); therefore, the likelihood ratio is $.50/.05 = 10$. Multiplying, we get $49 \times 10 = 490$, the posttest odds. Reconverting to the posttest probability, or the predictive value of a negative test, gives $490/(1 + 490) = .998$, the same result as above.

3.

Witness	Actual Color	
	Green	Blue
Says green	12	17
Says blue	3	68

Therefore, when the witness says green, he or she is correct 12 out of 12 + 17 times, or 41%.

4. The sensitivity is 19/20, or 95%; specificity is 27/35, or 77%.

5. **a.** 138 positive results in 150 known diabetics $= 138/150 = 92\%$ sensitivity.

   b. $150 - 24 = 126$ negative results in 150 persons without diabetes gives $126/150 = 84\%$ specificity.

c. The false-positive rate is 24/150, or 100% − specificity = 16%

d. 80% sensitivity in 150 persons with diabetes gives 120 true-positives. 4% false-positives in 150 persons without diabetes is 6 persons. The chances of diabetes with a positive fasting blood sugar is thus 120/126 = 95.2%.

e. 80% sensitivity in 90 (out of 100) patients with diabetes = 72 true-positives; 4% false-positive rate × 10 patients without diabetes = 0.4 false-positive. Therefore, 72/72.4 = 0.9945, or 99.45%, of patients like this man who have a positive fasting blood sugar actually have diabetes.

6. a. Using Bayes' theorem with prior probability of .30; we have P(mitral valve prolapse, given a positive echocardiogram) = (.90)(.30)/[(.90)(.30) + (.05)(.70)] = .27/ (.27 + .035) = .885, ie, an 88.5% chance.

b. P(no mitral valve prolapse, given a negative echocardiogram) = (.95)(.70)/[(.95)(.70) + (.10)(.30)] = .665/(.665 + .03) = .957, or a 95.7% chance.

7. a. This is not the information we need. To use Table 13–7, we must assume that the children identified as having language deficit by the PLS were the same ones identified by the DDST; even then, only sensitivity can be evaluated. For example, on *articulation,* the DDST correctly identified 28 of 60, or 47%, of those identified by the PLS. We do not know, however, whether the DDST produces any false-positives; ie, did any of the 11 children who passed the PLS fail the DDST? Therefore, we cannot evaluate the specificity of the DDST. In fairness to the investigators, the title of the article referred only to the sensitivity of the DDST; however, as we have learned in this chapter, sensitivity alone is not sufficient to evaluate a diagnostic procedure.

b. The authors stated that the "Wilcoxon Matched Pairs Ranked Signs Test" was used; probably meaning the Wilcoxon matched-pairs signed-ranks test. Of more importance, however, is that the children either passed or failed each test, according to the article; and the results in the table are given as proportions. Pass or fail is, of course, a nominal scale. Although the investigators correctly recognized the need for a matched (or paired) nonparametric test, they used the test for ordinal observations or numerical observations that are not normally distributed. The appropriate test is the McNemar chi-square test for matched (or paired) proportions.

8. With low prevalence, a positive result on a diagnostic test does not provide a great deal of information, even when the test is very accurate. On the other hand, a negative test in a low-prevalence situation is quite useful, especially if the test is accurate. As the prevalence (or "index of suspicion") increases, however, the results of either a positive or a negative test that is accurate are useful.

Prevalence	Predictive Value of a Positive Result	Predictive Value of a Negative Result
0.1%	1.8%	99.99%
1%	15.4%	99.89%
2%	26.9%	99.8%
5%	48.7%	99.5%
10%	66.7%	98.8%
20%	81.8%	97.4%
50%	94.7%	90.5%
80%	98.6%	70.4%

CHAPTER 14

1. A patient's preference may change depending on the value of the outcome; ie, a patient who chooses surgical over medical management to avoid almost certain death without the surgery may opt for medical management if the outcome without surgery is mild to moderate disability instead of death. Another factor affecting a patient's preference is the risk of the procedure or treatment, eg, weighing the risk of morbidity and mortality associated with carotid endarterectomy versus the risk of stroke if the procedure is not performed. A patient's preference can also be affected by the timing of the outcome; an elderly patient may make one decision to avoid immediate major disability and a different decision if the disability is more likely to occur in 5–10 years.

2. At approximately 42%, where the lines for culture and rapid test cross.

3. For a sensitivity of .60 and a specificity of .98 from Section 14.3.2, with 7% prevalence, there are 4.2 true-positives and .02 × 93 = 1.86 false-positives. Therefore, PV+ = 4.2/ (4.2 + 1.86) = .693, or 69%.

4. a. .20, seen on the top branch.

b. No, this is not obvious. Although the author provides the proportion of patients who have a positive exam, .26 (and negative exam, .74), these numbers include false-positives (and false-negatives) as well as true-positives (and true-negatives). The author also gives the predictive value of a positive test, .78 (and of a negative test, .997); and by using quite a bit of algebra, we can work backward to obtain the estimates of .989 for sensitivity and .928 for specificity. However, readers of the article should not be expected to do these manipulations; authors should give precise values used in any analysis.

c. For no test or therapy: $(.20)(48) + (.80)(100) = 89.6$; for colectomy: $(.03)(0) + (.97)[(.20)(78) + (.80)(100)] = 92.7$. Therefore, colonoscopy is the arm with the highest utility, at 94.6.

5. a. The incidence of pertussis and the incidence of severe adverse effects believed to be attributable to the vaccine, such as encephalopathy, seizures, and HHE, are the most important assumptions.

b. The current schedule is better, because the proposed schedule would result in a 42% increase in incidence of pertussis.

c. There is no difference as far as attributable events are concerned, but there is a higher rate of chance events under the proposed schedule. Although a larger number of deaths occur under the present schedule, approximately 50% more seizures would occur under the proposed schedule. Therefore, the proposed schedule appears to be preferable if the overall desire is to decrease deaths from pertussis.

CHAPTER 15

1. A
2. D
3. B
4. E
5. D
6. C
7. C
8. C
9. A
10. E
11. C
12. A
13. D
14. C
15. A
16. C
17. A
18. E
19. D
20. A
21. D
22. B
23. C
24. B
25. B
26. A
27. B
28. C
29. C
30. B
31. D
32. E
33. A
34. E
35. C
36. A
37. B
38. B
39. B
40. D
41. C
42. B
43. B
44. E
45. D
46. D
47. A
48. B
49. A
50. E

Appendix C:
Flowcharts for Relating
Research Questions to
Statistical Methods

Flowchart to Use   **Research Question**

C-1   Is there a difference (ordinal or numerical measures)?
C-2   Is there a difference (ordinal or numerical measures, 3 or more groups)?
C-3   Is there a difference (nominal measures)?
C-4   Is there an association?
C-5   Are there 2 or more independent variables?

C-1: Is there a difference? Ordinal or numerical measures.

C-2: Is there a difference? Ordinal or numerical measures, 3 or more groups.

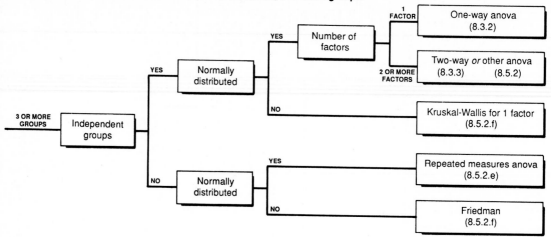

C-3: Is there a difference? Nominal measures.

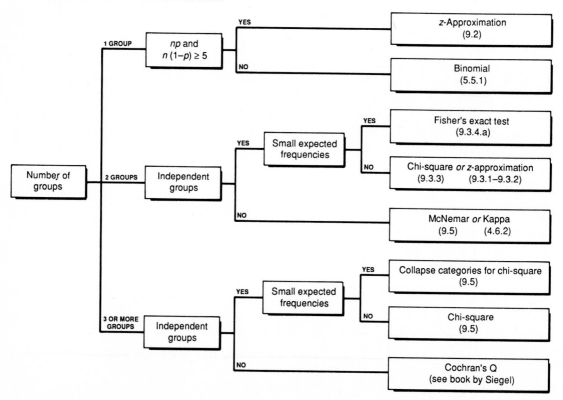

C-4: Is there an association?

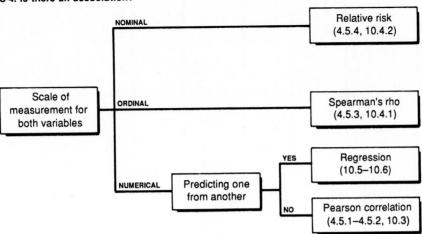

C-5: Are there two or more independent variables?

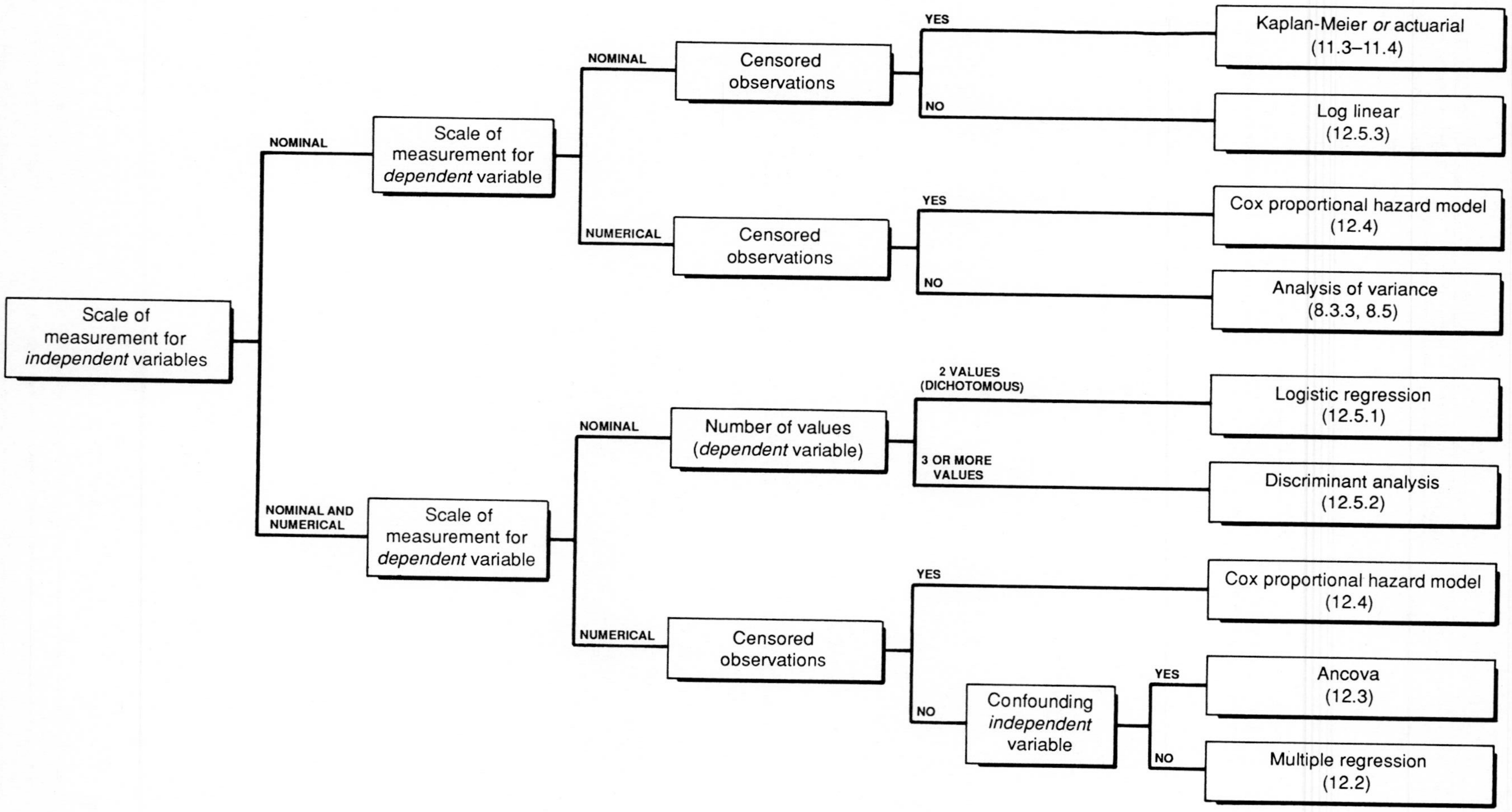

Index

absolute value, 47
actuarial analysis, 187, 190–192
addition rule, 67–68
adjusted rates, 52–54
agreement, measures of, 58–59
AIDS, 64, 239
alpha error, 95–96
alpha level for multiple tests of hypothesis, 125, 132–133
alpha value, 93, 96
Alzheimer's disease, 124
ANA, 229
analysis of residuals, 177–178
analysis of variance (ANOVA), 124–140, 225–226, 272
  assumptions for use, 130
  block design, 137–138
  fixed-effects model, 128
  formulas in, 129–130
  Freidman two-way, 139
  interactions in, 131–132
  Kruskal-Wallis one-way, 138–139
  mean square, 126
  nested, 138
  nonparametric, 138–139
  one-way, 125–130
  random-effects model, 128
  repeated measures, 128
  residual, 128
  sum of squares, 126–128, 130
  table, 127
  two-way, 130–132, 209
ANCOVA, 215–218
aneurism, 245
angina pectoris, 43
APGAR scores, 21
aplastic anemia, 14, 64
arthritis, 21, 38
artificial intelligence, 258

bar chart, 21, 23
Bayes' theorem, 69, 236–237
  (Also see revising probabilities)
bell-shaped curve, 48, 77
Berkson's fallacy, 268
Bernoulli, Jacob, 74
Bernoulli process, 74
Bernoulli trial, 74
beta error, 93, 96
beta-adrenergic drug, 43

bias, 11, 13, 15–17, 265–272
  admission rate (Berkson's fallacy), 268
  compliance, 270
  detection, 270
  "fishing expedition", 272
  insensitive measure, 270
  membership, 268
  migration, 272
  nonresponse or volunteer, 268
  prevalence/incidence (Neyman), 267
  procedure, 269–270
  recall, 269–270
  in study design, 265–272
  in subject selection, 267–269
  Will Rogers phenomenon, 270
bile, supersaturation, 9, 22, 28–30, 38
bimodal, 45
binomial distribution, 74–75, 80
binomial formula, 75
biostatistics
  meaning of term, 1
bivariate plots, (See scatterplots)
blind, 12, 58, 270–271
block design in ANOVA, 137
blood gases, 82
blood pressure, 17, 43
BMDP, 202–204
bone marrow transplantation, 14, 64
Bonferroni t procedure, 133
box and whisker plots, 27–29, 38
Breslow test, 196
burn healing, 18

calculations, accuracy of, 4
cancer
  breast, 1, 15, 18
  cervical, 21
  colorectal, 20–21, 24–28, 33, 37–39, 245
  liver, 229
  lung, 11, 124
  pancreatic, 16–17
canonical correlation analysis, 226
carcinoid syndrome, 8, 16
  tumors, 99
cardiac arrest, 20, 22–23, 33, 37
case-control studies, 6–8, 10–12, 16–18, 20, 265, 276
case-series, 6–8, 10, 17–18, 265

categorical data, (See nominal data)
cause-specific mortality rate, 51
central limit theorem, 84–86
central tendency, 43–46
  (Also see specific measures, such as mean, median)
cerebral infarction, 39
chance agreement, 59
checklist for reading the medical literature, 275–276
chi-square
  distribution, 149
  formula, 149, 151
  test, 148–154, 209, 221
Chlamydia trachomatis, 245
cholesterol, 9, 43–45, 54–55, 83
  gallstones, 20, 22
chronic lymphocytic leukemia, 10
  (Also see leukemia, 186)
cirrhotic patients, 13
clinical decision making, 2–3, 229–230, 245–250
  (Also see medical decision making)
clinical trials, 12–13, 15, 66, 264, 275–276
cluster analysis, 225
cluster random sample, 72
coefficient of determination, 55, 164
coefficient, regression, 170–172, 174, 210–211
coefficient of variation, 48
cohort, 9
cohort life table, 187
cohort studies, 6–7, 9–12, 16–18, 264, 276
  comparison with case-control, 10–12
combinations, formula, 75
comparison, 132
  a priori, 132–133
  post hoc, 133–135
complementary events, 67
compliance bias, 270
computer programs, 59–61, 119–121, 139–140, 157–159
  for correlation, 59, 61, 180–182
  for descriptive statistics, 59–60
  for means, 119–121
  for proportions, 157–159
  for regression, 180–182

computer programs (*cont.*)
  for survival analysis, 202–204
  for tables and graphs, 33–36
  (Also see specific packages)
conditional probability, 68
confidence bands, 175–176
confidence intervals, 91–92, 96–97, 103–104, 108–109, 112–114, 143–148, 168–169, 174–175, 272
  compared to hypothesis tests, 96–97
  for mean difference, 108–109
  for odds ratio, 168–169
  for proportions, 143–148
  for relative risk, 168–169
confidence limits, (See confidence intervals)
confounding, 52, 136–137, 215–220
contingency table, 21, 23, 29, 38, 153–154
continuity correction, 151–152
continuous probability distribution, 74
continuous scale, 22
contrasts, 132
control, 12, 15
control subjects, 7
  case-control studies, 16–17
  concurrent, 12, 15
  external, 13
  historical, 13, 15
  randomized, 12–13, 15, 264
  self, 13
controlled trials, 12
coronary artery disease (Nurses' Health Study), 18, 42, 269
coronary artery surgery study (CASS), 13, 65, 76
coronary atherosclerosis, 18, 43, 269
correlation, 54–56, 162–168, 179
  assumptions, 166
  coefficient, 54–56, 60, 163–164
  confidence interval, 166
  statistical test for, 164–167
Cox-mantel logrank statistic, 199
Cox regression model, 218–219
critical ratio, 87
cross-validation, 214–215
cross-over studies, 13
cross-product ratio, 60
cross-sectional studies, 6, 8–9, 17–18, 20, 38, 265
crude rate, 51
Cutler-Ederer life table method, 190
cystitis, 142

data, progressively censored, 187
decision making, 245–259, 276
  components of a decision, 246–247
  diagnostic strategies, 247
  health policy, 250–251

protocol evaluation, 253–257
sensitivity analysis, 249–250
utilities, 247–248
decision theory topics
  artificial intelligence, 258
  Markov models, 258
  multiple testing strategies, 257–258
decision trees, 234–236, 238, 247–255
  designing trees, 246–247, 250–251
  determining probabilities, 247, 251–255
definitions, 4
degrees of freedom, 47, 102, 149, 152
Demoivre, Abraham, 77
dependent variable, 126, 162, 209
depression, 124
descriptive statistics, 43, 100, 272
designs used in studies, (See study designs as well as specific types of studies)
detection bias, 270
determination of sample size, 118–119, 156–157
diagnostic tests, evaluation of, 231–240
dichotomous observations, 21
direct method of rate adjustment, 52–53
discrete probability distribution, 74
discrete scale, 22
discriminant analysis, 220
dispersion, 43, 46–50
  (Also see specific measures, such as standard deviation, range)
distribution, 46
dot plot, 29, 38
double-blind, 12, 271
dummy coding, 211–212
Duncan's new multiple range test, 135
Dunnett's procedure, 135

effect size, 222–223
eneuresis, 8, 207
epidemiologic studies, 3, 18, 38
epidemiology
  meaning of term, 1
errors in measurement, 58
estimation, 90–92
events, 66–69
experimental studies, 6, 12–13, 18
"eyeball" or graphic test for means, 104, 109, 113

factor analysis, 225
factorial designs in ANOVA, 137
false negative test, 95, 231–232, 240
false positive test, 95, 231–232, 240
F distribution, 87, 116, 126–127
Fisher, Ronald, 137

Fisher's exact test, 150–151
Fisher's z transformation for the correlation, 165–166
formula
  binomial, 75
  chi-square, 149, 151
  coefficient of variation, 48
  combinations, 75
  confidence limits, 91–92
  correlation coefficient, 54
  mean, 44–45
  mean squares, 130
  normal distribution, 77
  standard deviation, 47
  standard error, 85
  standard error of the difference between two means, 112
  sum of squares, 130
  t test, 104, 109, 111, 114
  variance, 47
Framingham study, 10, 16–17
F ratio
  (See F test)
frequency distribution, 25–29
frequency polygon, 28–29, 38
frequency table, 25–26, 38
  estimating mean in, 44–45
Friedman two-way ANOVA, 139
F test, 115–116, 127, 129–130

gallstones, 9, 20, 22
Galton, Francis, 170
gastrointestinal tract bleeding, 19, 38
Gauss, Carl Frederick, 77
gaussian distribution, 77–78
  (Also see normal distribution)
Gehan statistics, 196–199
generalized Wilcoxon statistic, 196–199
geometric mean, 45
goodness of fit, 155–156
Gosset, William, 102
graphs, (See specific types of graphs)
graphs, misleading, 30–32
Greenwood's formula for standard error, 191
groups
  independent, 111–115
  paired, 107–111
guidelines
  comparing means, Flowcharts C-1, C-2
  comparing proportions, Flowchart C-3
  evaluating associations, Flowchart C-4
  multiple independent variables, Flowchart C-5
  predicting, Flowcharts C-4, C-5

hazard function and survival analysis, 193–194
health maintenance organizations (HMO), 207

hemoccult test, 245
hemoglobin, 38
hepatitis B virus (HBV), 16, 161
herpes simplex virus, 7
herpes zoster, 7
histograms, 26–28, 38
historical cohort studies, 10
historical controls, 13, 15
human immunodeficiency virus
    (HIV), 10, 64, 239
hypertension, 43, 65, 100, 267–268
hypothesis testing, 90, 92–96
    correlation, 164–166
    difference between independent
        means, 114–118
    mean difference, 109–111
    proportion, 145–150

immune complexes, 161
immune function, 124
incidence, 51
independent events, 67
independent groups t test, 114–115
independent means, 111–115
independent variable, 126, 162, 209
indirect method of adjusting rates,
    53–54
inference, 82–83, 100
    (Also see estimation and hypoth-
        esis testing)
inflammatory lung disease, 42
interactions, 131–132
intercept of the regression line,
    170–172, 174
interquartile range, 49
interval estimation, (See confidence
    intervals)
interval scales, 22
intraobserver reliability, 58, 269
intrarater reliability, 58, 269
ischemic heart disease, (See coro-
    nary artery surgery stu-
    dy - CASS)

joint probability, 69

Kaplan-Meier product limit, 192–
    193
kappa, 59
kidney transplants, 186, 200
Kruskal-Wallis test, 196
kyphoscoliosis, 99, 265

LaPlace, Pierre-Simon, 77
latin square design, 138
leaf plots, (See stem and leaf plots)
least significance difference (LSD)
    test, 135, 137
least squares regression, 170–171
leukemia, chronic lymphocytic, 10,
    186
levels of a factor, 128
life table, 187

life table analysis, 187, 190–192
likelihood, 69
likelihood ratio, 237–238
linear combination, 211
linear regression, 162, 170–179,
    210–211
    assumptions, 173
    linear (straight line) assumption,
        173, 178
    multiple, 180, 211–215
    residuals, 177–180
    simple, 170–177
logistic regression, 219–220
log linear analysis, 219–221
log rank statistic, 199–201
longitudinal studies, 7
    (Also see cohort and case-control
        studies)
lung cancer, 11, 124

main effects, 131
Mann-Whitney U test, 116–118
Mann-Whitney-Wilcoxon rank
    sum test, 116–118
Mantel logrank statistic, 199–200
Mantel-Haenszel statistic, 200, 216
marginal probabilities, 67, 221
Markov models, 258
mastectomy, 188
matched (paired) studies, 107
matched t test, (See paired t test)
matching, 8
McNemar test, 154–155
mean, 22, 44–45
    calculation of, 45
    comparison of means, (See t test,
        Wilcoxon tests and
        ANOVA)
    use of, 45–46
mean square among groups, 126,
    129–130
mean square within groups, 126,
    129–130, 133
measurement error, (See bias)
median, 21, 45
medical decision making, 229–230
    (Also see clinical decision
        making)
meta-analysis, 222–224, 276
    potential bias in, 224
midpoint of distribution, 27
MINITAB, 33, 59–60, 119–120,
    157
missing data, 214
mode, 45
models, 128, 170–171
modified addition rule, 68
morbidity rate, 51
mortality rate, 51
multiple R, 213
multiple regression, 180, 211–215
    backward elimination, 214
    forward selection, 213
    regression coefficients, 210–211
    stepwise, 213–214
multiple variables, statistical

method for analysis,
    210–228, 273
multiplication rule, 67–69
multivariate methods, 210–227
multivariate analysis of variance,
    225–226
mutually exclusive events, 67
myocardial infarction, 43

negatively skewed, 46
nephrostolithotomy, 13
nephrotic syndrome, 161
nested designs, 138
Newman-Keuls procedure, 134–136
nominal data, 21–23, 38, 50, 56
nonlinear observations, 178
non-mutually exclusive events, 68
nonparametric ANOVA, 138–139
nonparametric procedures, 106,
    110–111, 116–118, 138–
    139
nonprobability sampling, 72
nonrandomized controls, 13
nonrandomized studies, 13, 266
normal distribution, 77–80
    use of, 80
null hypothesis, 93, 126
numerical data, 22–30, 38, 44, 46,
    50, 54

observation, censored, 189
observational studies, 6
odds, 69
odds ratio, 57, 268
    confidence limits, 168–169
one-tailed test, 93
one-way ANOVA, 125–130
ordinal data, 21–23, 26, 29, 38

paired t test, 109–110
parameters, 73
parenteral nutrition, 208
Pearson, Karl, 137, 170
Pearson product moment correla-
    tion coefficient, (See cor-
    relation coefficient)
pelvic inflammatory disease, 39
peptic ulcer disease, 15
percentages, 21–22, 26, 28, 33, 50
percentiles, 48–49
pictographs, 23
pie charts, 23
plots
    frequency polygons, 28–29, 38
    histograms, 26–28, 38
    scatterplots, 29–30, 162–163
    stem and leaf, 24–25, 28, 38
point estimates, 90–91
Poisson distribution, 76–77
Poisson, Simeon D., 76
polynomial regression, 178, 214
population, 69
    sampled versus target, 73, 266
portal hypertension, 13

positively skewed, 46
post hoc comparison procedures, 133–135
posterior probability, 233
power, 95–96, 118–119, 156–157, 215, 271–272
predictive value, 232, 240
prevalence, 3, 51, 230
prevalence studies, 8
  (Also see cross-sectional studies)
prior probability, 230
probability, 66–69
  addition rule, 67–69
  in decision trees, 247, 249–250
  distributions, 73–78
  independent, 67
  joint, 67
  marginal, 67, 221
  multiplication rule, 67–69
  mutually exclusive, 67
  revision, 232–240, 248, 249
probability of disease, 230–240
probability samples, 70
proportional hazards model, 218–219
proportion, 21, 50, 143–155
  dependent, 154–155
  independent, 143–154
prospective studies, 10
P-value, 93–94, 272–273

qualitative data, (See nominal data)
quantitative data, (See numerical data)
quartile, 27

RAND Corporation study, 207
random assignment, 72, 266
random numbers, 70–71
random sampling, 70–73, 266
random variable, 73–74
randomized block design, 137
randomized clinical trials, 12–13, 264
randomized factorial designs, 137
range, 46
rank correlation, (See Spearman rank correlation)
rank order scale, 21
  (Also see ordinal scale)
rank statistics, (See nonparametric procedures)
rank-sum test, 116–118
rank transformation, 106
rate, 50–51
  adjusted, 52–54
ratio, 50
ratio scale, 22
reading the medical literature, 1–2, 30–32, 97, 100, 125, 143, 162, 187, 230, 265–276
  checklist for, 275–276
regression, 162, 170–180, 210–211
  assumptions, 173
  coefficient, 170–172, 174, 210–211

common errors, 179
  multiple, 180, 210–215
  relationship to correlation coefficient, 171–180
  standardized coefficient, 212–213
regression toward the mean, 178–179
relative risk, 56–57
  compared with chi-square, 152
  confidence interval, 168–169
  odds ratio as estimate of, 57
reliability of measurements, 58–59, 269–271
repeated measures design, 138
research report, 265–276
  abstract, 265
  checklist, 275–276
  discussion, 274
  introduction, 265–266
  methods, 266–272
  results, 272–274
respiratory failure, 99
response rates, 17
representative sample, 17
retrospective, 8
retrospective cohort, 10
revising probabilities, 232–238, 241–242
  Bayes' theorem, 236–237
  decision tree method, 234–236, 238
  likelihood ratio, 237–238
  2 × 2 table method, 232, 234, 238
rheumatoid arthritis, 18, 38, 82
rho, Spearman's, (See Spearman rank correlation)
risk factors, 3, 7–9
risk ratio, (See relative risk)
robustness, 112, 130
ROC curves, 240–241

Salk polio vaccine trials, 268
sample, 69–73
  reasons for using, 70
sample size determination
  importance of, 271–272
  in studies with multiple measures, 215
  one mean, 118–119
  one proportion, 156
  two means, 119
  two proportions, 156–157
  (Also see power)
sampling distribution, 83, 87, 143
  of the mean, 83–84, 87
sampling methods
  cluster, 72
  nonprobability, 72
  random, 70–73
  simple, 70–71
  systematic, 71
  stratified, 71–72
SAS, 33, 59
scatterplot, 29–30, 54–55, 162–163

scales of measurement, 20–22, 272
  nominal, 21–23, 38
  numerical, 22–29, 38
  ordinal, 21–23, 26, 29, 38
Scheffe's procedure, 134
sensitivity, 3, 95, 231, 240
  determination of, 231–232
sensitivity analysis, 249–250
shock, 142
sigma (population standard deviation), 77
simple random sampling, 70–71
signed-ranks test, 110–111
skewed distribution, 46, 50
  negative, 46
  positive, 46
Spearman rank correlation (rho), 56, 166–168
specificity, 3, 95, 231, 240
  determination of, 231–232
sports injuries, 208
SPSS, 33–34, 36, 59, 119–120, 139
standard deviation, 46–48, 86–87
  calculation of, 47
  distinction with standard error, 85–86
  use of, 48, 86–87
standard error
  of regression, 173
  of the mean, 85–87
standard mortality ratio, 53–54
standard normal distribution (z), 78–79, 87–90, 92, 143–148
standardized rates, 52–54
statistical hypothesis, 93
statistical methods, articles
  common problems encountered, 272
  "fishing expedition", 272
  migration bias, 272
  multiple significance tests, 272
statistics, 73
  meaning of word, 1
STATISTIX, 59, 120–121, 157–159
stem and leaf plots, 24–25, 28, 38, 45
stratified random sample, 71–72
stroke, 207–208
Student's t, 102–103
  (Also see t-test)
study designs, 6
  advantages and disadvantages, 15–18
  bias in, 267–271
  case-control, 6–8, 10–12, 16–18, 20, 265
  clinical trial, 12–13, 15, 264
  cohort, 6–7, 9–12, 16–18, 264
  controlled, 12–13, 15–17
  cross-over, 13
  cross-sectional, 6, 8–9, 17–18, 20, 38
  experiment, 6, 12–13, 18
  latin square, 138
  nested, 138
  prevalence, 9

randomized block, 137
randomized factorial, 137
repeated measures, 138
self-controlled, 13
survey, 9, 17
uncontrolled, 12, 14
sum of squares, 126–128, 130
suppression of zero, 31
survey, 9, 17
survival analysis, 190–202
    comparing two distributions,
        194–200
    curves, 194–200
    hazard function, 193–194
symbols used in this book, 5
    (Also see inside back cover)
symmetric distribution, 46, 50
SYSTAT, 34, 121, 139, 158–159
systematic sampling, 71
systemic lupus erythematosus
        (SLE), 229

tables for displaying data, 22–26,
        33–36, 38
t distribution, 87, 101–103
test statistic, 93
tests of hypothesis, 92–97
    correlation, 164–166
    one mean, 104–105
    one proportion, 145–146
    two means, dependent, 109–111
    two means, independent, 114–
        118
    two proportions, dependent,
        154–155
    two proportions, independent,
        147–152
    two variances, 115–116

testing threshold, 230–231
thermography, 9, 142
threshold, 230–231
thrombosis, deep vein, 9, 142
tobacco, 11
transformations, 105–106
treatment threshold, 230–231
trial
    in probability, 66
    (Also see clinical trials)
t test for correlation, 164–165
t test for means, 104–105, 109–110,
        114–115, 209
    assumptions in, 111–112
    independent groups, 114–115
    one sample, 104–105
    paired, 109–110
    two samples, 114–115
t test for regression, 174–175, 212
Tukey's honestly significant differ-
        ence (HSD) test, 134
two-tailed test, 93
type I error, 95–96, 271
type II error, 95–96, 271

ulcers, gastric, 142, 266–267
unbiasedness of statistics, 91
uncontrolled studies, 12, 14
uniform distribution, 84
utility analysis, 247–248

variable, 73–74
    confounding, 215–219
    dependent, 126, 162, 209
    explanatory, (See variable, inde-
        pendent)
    independent, 126, 162, 209

multiple, 210
    outcome, (See variable, de-
        pendent)
    random, 74
    response, (See variable, de-
        pendent)
variance, 47
    homogeneous, 130
    test of, 115–116
    (Also see F test)
variation, 57–58
varices, esophygeal, 13
venography, 9, 142
ventricular wall motion, 207
vital statistics, 2–3, 51–52
volunteer effect, 268

weighted average, 44
Wilcoxon rank-sum test, 116–118,
        139
Wilcoxon signed-ranks test, 110–
        111, 139
Wilcoxon test for survival curves,
        (See Gehan statistic)
Wilks' lambda, 220

X-axis, 23, 26, 29

Y-axis, 23, 26, 29

z-distribution, 78–79, 87–90, 92,
        143–148
z-score, 77
z-test for proportions, 145–148